LANDSCAPE PLANNING
Environmental Applications

LANDSCAPE PLANNING
Environmental Applications
Fifth Edition

William M. Marsh
University of British Columbia
University of Michigan (Emeritus)

WILEY

John Wiley & Sons, Inc.

ACQUISITIONS EDITOR   Ryan Flahive
ASSOCIATE EDITOR   Veronica Armour
EDITORIAL ASSISTANT   Meredith Leo
MARKETING MANAGER   Margaret Barrett
SENIOR PRODUCTION EDITOR   Richard DeLorenzo
SENIOR DESIGNER   Kevin Murphy
PHOTO EDITOR   Sarah Wilkin and Hilary Newman
COVER   A.M. Mewett
CHAPTER HEADER ILLUSTRATION   Brian Larson

This book was set in 10/12 Garamond Light by MPS Limited, A Macmillan Company and printed and bound by Courier-Westford.
The cover was printed by Phoenix Color Corporation.

This book is printed on acid-free paper.

ISBN: 978-0-470-57081-4

Printed and bound in the United States of America

10 9 8 7 6 5 4 3 2

Dedicated to Three Men Who Made a Difference:

William R. Marsh (1918-2009)

James W. Mewett (1928-2009)

M. Gordon Wolman (1924-2010)

FOREWORD

Landscape Planning: Environmental Applications by William M. Marsh is a thoroughly splendid book, an exhaustive selection and amalgamation of the salient principles and processes derived from geology, hydrology, soil science and ecology, expressly oriented to the needs of landscape architecture.

It contains insightful descriptions of physiographic regions, each with characteristic geomorphology, hydrology, soils and ecologies. It presents fluvial and soil processes, watersheds, flood plains, erosion, sedimentation, each with enlightening text and diagrams. It could well be described as "How to Design with Nature." I give thanks to the author as should you.

I Commend *Landscape Planning* to your attention.

Ian L. McHarg
6th October 1997

University of Pennsylvania
The Graduate School of Fine Arts
Department of Landscape Architecture
and Regional Planning

PREFACE

In the 1970s, as faculty and students in my college pressed a dean for funding to support environmental studies, he suggested that "all this environmental stuff is probably just another academic fad." Apparently, either fads are slow to die in academia, or the dean was wrong. And he surely was wrong. The environmental fields, both on and off campus, have exploded since the 1970s, and the evolution of this book is one small measure of that explosion. Every edition has had to be larger and more thoughtful than its predecessor. Indeed, the principal problem I have faced over the years is trying to keep pace with landscape planning in all its various forms. Whether defined as environmental planning, landscape planning, environmental design, land use planning, or landscape ecology, or whether it appears in landscape architecture, geography, planning, applied ecology, forestry, or any of a dozen other sister programs, the problem is the same: understanding and dealing reasonably with the land in the face of expanding land use and changing technology. More than ever, the land begs us for solutions.

Most who will read this book are in the early stages of their landscape careers, and you are the ones who must advance thought and applications over the next half century. The task is daunting, for real progress will require reaching beyond the universities, consulting firms, and government offices. Like it or not, responsibility will require participating in the power centers of the global economy because all economic activity is based on land use and because virtually all land uses, no matter how distant and abstract they seem, are played out in the land. And if we are to survive, the land must be treated less like an economic soccer field and more like our home and habitat. Landscape is our habitat and we must help land users, large and small, to learn and relearn the rules of living with it as a partner.

Happily, more people than ever are taking responsibility for the landscape. Colleges and universities are blessed with growing numbers of students committed to building a better world. Almost everywhere citizens of every stripe are stepping up to the plate to address problems with watersheds, urban streams, community development, public transportation, safe food, and many more aspects of our habitat. Although we may be losing ground on certain fronts, by and large progress is being made toward creating more livable landscapes for a great many of us.

ACKNOWLEDGMENTS

Credit goes to many people for bringing this edition of *Landscape Planning* to fruition. At the top of the list are two women, Veronica Armour and Alison Mewett. Veronica took hold of this project early on at Wiley headquarters and guided it through its setup, organization, and production. Alison, my wife, worked with me at every stage, taking on tasks large and small whenever they arose, including the cover design. Close behind these two are Sara Wilkin at Wiley, Stephanie Brown at the University of

British Columbia, and Ed Dionne and Fred Dahl at MPS who researched photographs and websites, edited text and graphics, and politely pulled it all together. My sincere thanks to these folks for doing their jobs very well.

I also extend my thanks to the many people who have contributed to my thinking about landscapes and how we live in them. My brothers, James G. Marsh and Bruce D. Marsh, devoted endless hours to conversations about life, land, and thought, as did Doug Paterson, Paul deGreeff, Tom Dishlevoy, Jack Goodnoe, and many other friends and colleagues in British Columbia, Michigan, and other places. Along with my wife, they patiently listened to my rants and then added thoughtful observations, with a touch of leavening, that allowed me to make something more civilized of it all.

Heartfelt thanks also go to the landscape architects, geographers, planners, and many others inside and outside the classroom who have stood by this book through the years. Their support has been an endless source of motivation to keep nudging the book ahead with new chapters and revisions to old ones. If this edition is better than earlier ones, much of the credit goes to its readers.

Regrettably, this decade ends with the loss of two giants in the landscape field: Ian L. McHarg in 2002 and M. Gordon Wolman in 2010. We all owe a great deal to these men not only for their wisdom and their intellectual generosity, but also for their abiding conviction that it is all worth it. As McHarg said, "The world is a glorious bounty." It certainly is. Fight for it.

WMM
Comox, British Columbia

CONTENTS

CHAPTER 7
Groundwater Systems, Land Use Planning, and Aquifer Protection

CHAPTER 8
Runoff and Stormwater Management in a Changing Landscape

CHAPTER 9
Watersheds, Drainage Nets, and Land Use Planning

CHAPTER 14
The Riparian Landscape: Streams, Channel Forms, and Valley Floors

CHAPTER 15
The Coastal Landscape: Shoreline Systems, Landforms, and Management Considerations

CHAPTER 16
Solar Climate Near the Ground: Landscape and the Built Environment

CHAPTER 21
Wetlands, Habitat, and Land Use Planning 447

CHAPTER 22
Framing the Land Use Plan: A Systems Approach 467

AN INTRODUCTION TO THE BOOK, THE FIELD, AND SOME PERSPECTIVES

0.1 OPENING STATEMENT

A predictable sameness is creeping over the face of the North American landscape. Since the early 1970s, highways, shopping centers, residential subdivisions, and most other forms of development have taken on a remarkable similarity from coast to coast. Not only do they look alike, but modern developments also tend to function alike, including the way they relate to the environment, that is, in the way land is cleared and graded, stormwater is drained, buildings are situated and presented, and landscaping is arranged.

Landscape diversity

We know that this facade of development masks an inherently diverse landscape in North America. If we look a little deeper, it is apparent that *landscape diversity* is rooted in the varied physiographic and ecological character of the continent, and this in turn, reflects differences in the way the terrestrial environment functions. Does it not seem reasonable then, that development and land use should also reflect these differences if they are to be responsive to the environment? Herein lies one of the most important missions for landscape planning: to guide development toward environmentally responsive planning and toward design schemes that avoid mismatches between land uses and environment and that yield sustainable landscapes in the long run.

Landscape as Human Habitat. Twenty-five years ago, when I first scratched out the early chapters of this book, I figured we were well on our way toward achieving a better balance between land use and environment. Though progress has admittedly been made on certain fronts, as a whole the relationship between landscape and people's lives has not improved much and, in many respects, has actually weakened. We should ask why. Standing out among the reasons is the loss of our traditional knowledge about landscape. Increasingly, we are a dislocated population, often urban or suburban refugees with little or no firsthand experience in reading land, choosing a place to live, and relating it to our lives.

But we are also a society of individuals with the freedom, motivation, and pride to act independently. This includes deciding on a place to live. Often as not, however, our choices are guided not by a sense of place, but by automobile access, images of the good life, displays of affluence, and other things unrelated to land. Or, the choice of place is made by somebody else, such as a developer and a real estate agent, for whom place is merely a location on a map.

Land, life, and values

The point is that too many of us have no real opportunity to discover and kindle the core values that are so essential to building meaningful and lasting relationships between life and landscape. Among those values are membership in a *community of people*, a sense of *morality* about the land and its community of organisms, and a *spiritual attachment* to place as part of our personal sense of meaning and well-being. As a result, we risk losing something fundamental to life itself, something necessary not only to understanding our place in the land, indeed on the planet itself, but to understanding our responsibility as stewards of the landscape (Fig. 0.1).

0.2 BACKGROUND CONCEPT

Virtually every modern field of science makes contributions toward resolving societal problems. In some cases the contributions are not very apparent, even to the practitioners in their respective fields, because they are made via second and third parties. These parties are usually members of the applied professions, such as urban planning, landscape architecture, architecture, and engineering, who synthesize, reformat, refine, and adapt knowledge generated by scientific and technical investigations.

Applied professions and sciences

The Landscape Fields. Geography, geology, hydrology, soil science, ecology, remote sensing, and many other fields have such a relationship with the various fields of landscape planning and design, especially landscape architecture and urban and

Fig. 0.1 Land use detached from the landscape. A manufactured community that shows little or no sensitivity to local landscape, thereby denying residents the opportunity to build life relationships with the land.

regional planning. For more than a century, earth and environmental science professionals have studied the world's biophysical features, learning about their makeup, how to measure them, the forces that change them, and how we humans interact with them. Landscape planning is concerned with the use of resources, especially those of the landscape, and how to allocate them in a manner consistent with people's goals. Thus landscape planning and the environmentally-oriented sciences are linked together because of a mutual interest in resources, land use, and the nature and dynamics of the landscape.

0.3 CONTENT AND ORGANIZATION

This book addresses topics and problems of concern to planners, designers, scientists, and environmentalists. The focus is on the environmental problems associated with land planning, landscape design, and land use. The coverage is broad, though not intended to represent the full range of existing applications; such a book would be too large and cumbersome for most users.

Scope of study   The choice of topics was guided by three considerations:

- First, the main components of the landscape should be represented including topography, soils, hydrology, climate, vegetation, and habitat.
- Second, the topics should be pertinent to modern landscape planning as articulated by landscape architects, urban planners, and related professionals: for example, stormwater management, slope classification, and wetland interpretation.
- Third, the topics should not demand advanced training in analytical techniques, data collection, mapping, and field techniques.

Changing Content of the Field. Each edition of this book has covered more topics than its predecessor because the field of landscape planning and design is continually growing. This reflects both the development of the field professionally and the expanding legitimacy of environmental considerations in planning and design. For example, only 30 years ago it was barely legitimate for professionals outside civil engineering to address stormwater management. In the past decade or two, environmental planners and landscape designers have critically addressed not only stormwater as a runoff problem, but stormwater as a major source of water pollution, as well as stormwater infrastructure as a constraint in community design. In addition, the field has taken up wetland planning, urban climate, and groundwater management, and a variety of ecological topics, including biodiversity and landscape ecology.

With this edition

This Edition. With this edition of *Landscape Planning* several topics have moved up the agenda. Center stage is site-adaptive planning and design. Although this concept was introduced in earlier editions, here it is featured in many chapters as an approach necessary to building sustainable landscapes. It follows that much greater emphasis is placed on systems in this edition. This is based on the argument that the only way to create truly sustainable landscapes is through sustainable systems because earth's landscapes, including the artifacts we place in them, are driven by systems. And in this context planners and designers need to weigh the effects of climate change. Though we have long talked about climate change in the urban context, the problem has taken on regional and global dimensions with serious implications for land use. Although it is not clear how to address this problem in our local planning and design schemes, we very much need to continue the search for solutions.

0.4 LANDSCAPE PLANNING, ENVIRONMENTALISM, AND ENVIRONMENTAL PLANNING

The term *landscape planning* is used in this book to cover the macro environment of land use and planning activity dealing with landscape features, processes, and systems. Only four decades ago, the term *land use planning* was generally used for this sort of activity, but today, given new knowledge, the recognition of new problems, the changing needs of society, and the modern proliferation of specialty fields, several new and alternative titles have emerged.

Rise of environmentalism

Environmentalism. The *Environmental Crisis* of the 1960s and 1970s was brought on by a flurry of concern over the quality of the environment. Much of this concern took the form of a political movement to protect the "environment" from the onslaught of industry, government, urban sprawl, and war. Loosely translated, the term *environment* was taken by the movement to mean things of natural origin in the landscape, that is, air, water, forests, animals, river valleys, mountains, canyons, and the like. From this emerged the *environmentalist*, a person who believes in or works for the protection and preservation of the environment. *Environmentalism*, it follows, is a philosophy—a political or social ideology—that implies nothing in particular about a person's training, knowledge, or professional credentials in matters related to the environment. Organizations such as the Sierra Club, Greenpeace, and Friends of the Earth practice environmentalism.

Emergence of new fields

The environmental crisis also paved the way for stronger and broader environmental legislation at all levels of government. New types of professional skills were needed to provide services in connection with environmental assessments, waste disposal planning, air and water quality management, and other areas. In response, several new so-called environmental fields emerged, and many established fields, such as civil engineering and chemistry, developed environmental subfields. Taken as a whole, the resultant environmental fields fall roughly under three main headings: environmental science, environmental engineering and technology, and environmental planning. Our concern falls within the latter.

Environmental Planning. This is a catchall title applied to planning and management activities in which environmental rather than social, cultural, or political factors, for example, are central considerations. The term is often confused with environmentalism and with the preparation of environmental assessments and impact studies. In reality, environmental planning covers an enormous variety of topics associated with land development, land use, and environmental quality: not only late twentieth century topics such as landscape ecology and the management of wetlands,

Landscape planning but also more traditional problems, such as watershed management and planning municipal water supply systems. To some extent, *landscape planning* is also a term of convenience used to distinguish what we might call the landscape fields (such as geography, landscape architecture, geomorphology, and urban planning) from other areas of environmental planning, many of which have come to be more closely tied to environmental engineering and public health.

0.5 THE SPATIAL CONTEXT: SITES AND REGIONS

Because we need to address topics and problems of the landscape that are pertinent to modern planning as articulated by the practicing professional, most of the material in this book is presented at the site or community scale. *Sites* are local parcels whose sizes usually range from less than an acre to hundreds of acres, with a simple ownership or stewardship arrangement (individuals, families, or organizations). They are the spatial units of land use planning, the building blocks of communities.

Whose region? **Regions.** In the vocabulary of land use planning and landscape design, *regions* are variously defined as the geographic settings of communities, either a single community and its rural hinterland, several communities and the systems connecting them (roads or streams, for example), or a metropolitan area with its inner city, industrial, and suburban sectors. This concept differs from the geographer's notion of a region, which encompasses a much larger area—for example, the Midwest, the Great Plains, or the Hudson Bay region.

Many environmental problems have a regional scope (geographer's version)—for instance, acid rain in the eastern midsection of the continent and groundwater supplies for irrigation in the Great Plains. For a variety of reasons, however, planning programs at this scale have generally not been very effective in North America. Most examples of effective or promising landscape and environmental planning are of regional (planner's version), community, or site scales. But you will also find these are often terms of convenience that may not accurately address the true scale of the problems and environment under consideration.

0.6 FINDING THE APPROPRIATE SCALE

Geographic, or *spatial, scale* is an essential part of all planning problems, but it is one of the least effectively used dimensions in modern land use systems. Illustrations of the misuse of scale pervade landscape and society throughout the developed world. Most stem from mismatches among the functions or processes that a program is supposed to address, the magnitude of the facilities and systems actually designed to serve it, and the environment in which it is placed and upon which it is dependent.

Glaring mismatches **Missing the Mark.** In the United States and Canada, "bigger is better" has become a rule of thumb for many decision makers in this and the last century. Public education, for example, has shifted toward larger and larger schools that have lost touch with students as individuals and with the communities they serve (Fig. 0.2*a*). Cities in both developed and developing nations have expanded into regional entities with such large populations (Mexico City is expected to reach 50 million and Shanghai 100 million in this century) that they may now exceed the limits of manageability as physical and social systems. In the United States, Canada, and parts of Europe, expressway systems originally designed for national defense and interregional travel have become daily escape conduits for an urban-based workforce seeking to scale life down in smaller, more traditional communities beyond the urban region.

In the environment, the U.S. federal government has approached flood management with the construction of massive dams and the manipulation of river channels while largely ignoring local and regional watershed management and flood-sensitive

(a)

(b)

Fig. 0.2 Oversized facilities permeate the North American landscape: (*a*) a huge suburban school dependent solely on buses and automobiles, separated from the community it serves, where parking lots cover more area than the school itself; (*b*) huge dams and related flood control facilities often alienate communities, both human and natural, from the river as a functioning part of their landscape.

land use planning. The result is the extensive loss of riparian habitat, woodland, and farmland while "protected" floodplain settlements have become increasingly susceptible to impacts from large floods because of a naive reliance on engineered structures (Fig 0.2*b*). The rise of local watershed management groups and stream restoration associations is in many ways an attempt to address these problems at a grassroots level and to counterbalance the top-down approach of federal, state, and provincial programs.

Misapplications   At the same time, communities have borrowed environmental policies from other, often distant, communities and applied them locally without regard for the differences in the scale, fabric, and operation of the landscape systems for which they were originally designed. Instead of problems such as stormwater drainage, flooding, water quality, and habitat conservation being mitigated or resolved, many remain unchanged or worsened. Engineered stormwater systems designed for eastern North America, for example, are applied to topographic and hydroclimatic environments in the West, where they are not only inappropriate and damaging to property, stream channels, and habitat, but also an economic burden to communities, especially small ones.

Contributing Factors. Why these problems of scale? There are many reasons, and the following examples are a few of the significant ones. Consider national policies that lead to sweeping applications of standard infrastructure systems, such as dams and expressways, irrespective of the character of land use and environment at local and regional scales. Or consider economics based on benefit-cost rationale that show savings with large large-scale development schemes. And then there is the rising threat of professional liability facing planners and engineers for not building bridges, dams, and other facilities large enough to withstand all environmental contingencies. Consider national policies that lead to sweeping applications of standard infrastructure systems, such as dams and expressways, irrespective of the character of land use and environment at local and regional scales. Or consider economics based on benefit-cost rationale that show savings with large large-scale development schemes. And then there is the rising threat of professional liability facing planners and engineers for not building bridges, dams, and other facilities large enough to withstand all environmental contingencies.

Underlying causes

Fig. 0.3 Interstate highway I-93 where it passes through Franconia Notch in the White Mountains of New Hampshire. It is a scaled-down section of expressway designed to fit a narrow valley.

Not the least among the reasons for scale problems is a lack of understanding by decision makers and their technical advisors about the workings of the environment. Often they have little or no understanding of the processes and systems that shape and sustain the landscape, including the scales at which they operate. From place to place, the environment is different in fundamental ways. Unless these differences are made part of the information base for decision making and design, we will continue to build mis-sized and unsustainable infrastructures and land use systems.

Franconia Notch. By way of example, the people of New Hampshire sensed the limitations of scale when they stopped the federal and state transportation departments from building a standard four-lane interstate highway through Franconia Notch, a narrow pass in the environmentally prized White Mountains (Fig. 0.3). The project was held up for more than a decade until the highway planners agreed (at the request of the courts) to modify their standard approach based on federal rules and guidelines and to adopt an alternative plan better suited to the topography, drainage, and recreational features of the pass.

Scaling down The result was a scaled-down stretch of interstate highway. It was not a perfect solution, but it was decidedly better than a conventional, large, limited-access expressway with huge exit and entry ramps squeezed into the center of the valley (Fig. 0.4). Franconia Notch serves to remind us that scale can be applied effectively if the makeup, fabric, and operation of local and regional landscapes and the magnitude and potential impacts of the proposed development are understood, and if a compelling case is presented to decision makers based on documented evidence.

0.7 THE SUSTAINABLE SYSTEMS DILEMMA

Great expectations Modern society prides itself in technological innovation and in the development of energy, communication, financial, transportation, water, and many other systems. Modern systems are unbelievably efficient in moving energy, matter, words, and

25

Fig. 0.4 To better fit the scale of the valley and help preserve recreational and scenic features, access routes and ramps were scaled down along I-93 in New Hampshire.

pictures, and we have grown so dependent on them that today's land uses, no matter where they are located, cannot survive without them. We expect these systems to be fast, dependable, and tireless, and when they wobble (because of power outages or fuel shortages, for example), tremors of panic rush through society, and corporate and government heads are threatened.

The Challenge. One of the jobs of land use planning—whether in building design, facility siting, infrastructure engineering, or landscape planning and design—is to provide access to a full complement of up-to-date technological systems. At the same time, however, we are expected to guide land use toward an equitable and sustainable balance with natural systems, systems that by their very nature are typically

Finding balance not as fast, dependable, and tireless as we have come to expect with technological systems. And herein lies a rub, for we tend to apply the performance standards of the manufactured to the natural.

The costs of this misapplication are evident all around us. As a whole, the misapplications are expensive, environmentally damaging, and not sustainable. One example that stands out is runoff systems. To improve the "efficiency" of stream systems, engineers have for centuries deepened, straightened, and extended natural channels, resulting in huge capital expenditures, serious management dilemmas, extensive habitat damage, and the false impression by society that the threat of flooding has been reduced or eliminated. In the Florida Everglades, for example, after decades of building a massive system of drainage canals, we are now trying to stall the effects of the degradation of this international treasure by decommissioning the canals. Unfortunately, other examples of applying the performance standards of manufactured systems to natural systems are almost endless.

Sustainable Landscapes. If we are to move toward sustainable landscapes, we must address systems, both natural and manufactured. In landscape planning and

Addressing systems design we must bring the two together on a parcel of land in such a way that they can provide appropriate services and perform according to their inherent capabilities.

The sustainable (green) buildings that we hear so much about are overwhelmingly dependent on manufactured systems.

Sustainable landscapes, on the other hand, are dependent mainly on natural systems, which are usually larger, more complex, and more challenging as planning and design problems. Thus, the union of architecture and landscape architecture around the sustainability theme is from the outset fraught with difficulties. The success of sustainable exercises with green buildings tempts us to apply the same philosophy to landscape planning and design. In reality the opposite may be called for, that is, less technological intervention and more attention to nature's solutions in creating sustainable systems.

0.8 CASE STUDY

Building Sustainable Communities on Sustainable Landscape Systems

W. M. Marsh

There is a widespread belief, especially among political leaders, that growth is necessary in order for a town or city to prosper. In other words, to reach and maintain a prosperous state, a community must be continually expanding, drawing in more resources, and using more land. But there is good reason to doubt this belief. Studies reveal that the cost of growth in North American communities commonly exceeds the income from new tax revenues. Growth requires building new infrastructure, such as roads and sewers, and expanding services such as education and policing. The costs of added infrastructure and services are very high and often are not offset by the tax revenue generated by the new development, especially residential development. Often the community, despite outward signs of prosperity from new construction activity, declines in quality and economic well-being.

There is another route to prosperity: building on the existing resource base. This can be achieved by upgrading existing infrastructure, improving the efficiency of services, and improving the efficiency, quality, and profitability of land use. A simple way of improving land use efficiency and profitability, for example, is to infill vacant lands, which most communities have in abundance and which the established infrastructure already services. In other words, communities can move forward to a more sustainable state by creating prosperity from within through making better use of what they already have.

Building sustainable communities is fundamental to building sustainable landscapes because only when a community approaches internal sustainability is it in a position to seriously address sustainability in the larger landscape of which it is a part. How does a community go about this? What factors does it address in planning for a sustainable landscape?

A community can begin by identifying and measuring the systems on which it depends, such as its watersheds, ecosystems, and groundwater aquifers. The most meaningful measure of landscape sustainability is the performance of these systems—not just one, but several or more interrelated systems, both natural and manufactured. For a community dependent on the forest industry, for example, this would mean finding an enduring balance in a forest ecosystem subject to lumbering activity. One approach that is actively used by foresters is to establish a balance between wood cut and new wood grown. By using this measure, if the wood harvested is equal to the wood grown, then the system is in balance and considered sustainable.

But this is a crude and largely inaccurate measure of sustainability for it does not take into account the many other systems of the landscape that are part of the forest ecosystem and hence part of the community, including streams, groundwater, soil, animal habitat, climate, as well as various land use activities. For instance, if we maintain a balance between wood harvested and wood grown but do it in such a way that it increases runoff rates and soil erosion, then our forest management practices are not sustainable (Fig. 0.A). Increased runoff and soil erosion can reduce the

Fig. 0.A Forest clear-cutting, a standard forestry management practice in the Pacific Northwest, commonly leads to higher runoff rates, increased slope failures and soil erosion, degraded landscape scenic quality, and reduced recreational opportunities.

capacity of the land to support forests, leading to reduced growth rates and lower-quality forest. In the long run, the whole system winds down as yields fall and weedy tree species replace marketable ones, a trend that in turn can threaten the sustainability of the community. Similarly, forest management practices that reduce or eliminate other economic activities can drive a community and its landscape toward an unsustainable state. For example, cutting practices that degrade a landscape's scenic quality, damage streams, and drive away game can hurt economic activity dependent on tourism, fishing, and hunting.

Because of the endless variations in the conditions and operations of communities and their landscapes, there is no universal formula for measuring and evaluating sustainability. Settlements and environments are different from place to place, and these differences demand that each community tailor its own version of a sustainability plan. No matter where we are, however, we have to address key systems and their performance as the starting point on the road to sustainability.

0.9 SELECTED REFERENCES FOR FURTHER READING

Barbato, Joseph, and Weinerman, Lisa (eds). *Heart of the Land: Essays on Last Great Places*. New York: Vintage Books, 1994.

Berry, Wendell. *A Continuous Harmony: Essays Cultural and Agricultural*. New York: Harcourt Brace Jovanovich, 1972.

de Steiguer, J. E. *The Age of Environmentalism*. Boston: WCB/McGraw-Hill, 1997.

Fodor, Eben. *Better, Not Bigger*. Stony Creek, CT: New Publishers, 1999.

Fowler, Edmund P. *Cities, Culture and Granite*. Toronto: Guernica, 2004.

Jacobs, Jane. *Dark Age Ahead.* Toronto: Vintage, 2004.

Spirn, A. W. *The Language of Landscape.* New Haven, CT: Yale University Press, 1998.

Thompson, G. F. (ed). *Landscape in America.* Austin: University of Texas Press, 1995.

Turner, Frederick. *Spirit of Place: The Making of an American Literary Landscape.* Washington, DC: Island Press, 1989.

Related Websites

American Society of Landscape Architects et al. Sustainable Sites Initiative. 2008. http://www.sustainablesites.org/report/
An interdisciplinary initiative setting benchmarks for sustainable landscapes. The guidelines focus on site-specific design, ecosystem services, and measuring outcomes.

Forest Stewardship Council of Canada. 1996. http://www.fsccanada.org/
An organization that certifies wood as sustainable. The site gives regional standards for certification including ecosystem-based management, climate-change-adapted practices, and social/economic sustainability.

Portland Development Commission (PDC). http://www.pdc.us/default.asp
PDC is responsible for urban revitalization in Portland, Oregon, which is a model of sustainable development. The site covers infill and process details and several examples of successful projects.

LANDSCAPE PLANNING: ROOTS, PROBLEMS, AND CONTENT

1.1 INTRODUCTION: ROOTS

Quest for order

People have probably engaged in some form of environmental planning for as long as organized society has been around. There is ample evidence, for example, that the ancient Mesopotamians devised elaborate planning schemes for distributing irrigation water in the desert and that the Romans of Caesar's time programmatically drained wetlands to gain additional farmland and reconfigured harbors to improve navigation. It is clear, however, that most of the ancients' interests in environmental planning were purely practical, having to do with things like trade, food supplies, water, and defense with little or no regard for what we would call environmental impact. In many respects, planning and engineering programs were seen as attempts to bring order to an otherwise disordered natural environment, even to correct perceived defects in nature. Curiously, a great many of the measures applied to these "defects," such as irrigation of desert soils by the Mesopotamians and reconfiguring harbors by the Romans, failed for reasons the builders never anticipated, often from environmental feedback within the system affected.

The Rise of Romantic Values. Prior to the fifteenth century and with few exceptions, nature itself was given little regard as part of the environment. Arguably the lowest point in Western civilization's attitude toward nature occurred in Europe in the Middle Ages, when nature was commonly viewed with suspicion, fear, and ignorance. Forests, for example, were seen as dangerous places haunted by beasts and thieves, and a common person's total familiarity with local geography would

Early attitudes

not extend much beyond the village of his or her birthplace. However, with the Renaissance (beginning in the fifteenth century) and the Enlightenment (in the seventeenth and eighteenth centuries), humans and nature came to be on friendlier terms. The enlightened mind saw nature as having logic and order, a line of thought facilitated by Newton's model of the natural universe as a perfectly structured and perfectly functioning machine.

In the eighteenth and nineteenth centuries, the concept of nature was extended to include the pleasure and the enjoyment of natural things. This marked the beginning of a love affair with the environment, called the **Romantic Movement**. With the Romantic Movement we find nature is given consideration for its own sake and

Celebrating nature

for its beauty, spiritual meaning, and influence on the quality of life. In landscape design a new school of thought, called Landscape Gardening, emerged in England in which the landscapes of rural estates were made to look "natural" by using curved lines in gardens, field edges, and water features (Fig. 1.1). The arts of the Romantic Movement also reflect the rise in environmental consciousness; nineteenth-century painting, music, and literature illustrate this awareness especially well with, among other things, bucolic scenes and pastoral moods.

In the United States the first village improvement associations, which began in the 1850s, applied Romantic concepts to communities. They beautified streets, cemeteries, and town squares, and they promoted laws for the protection of songbirds and the creation of parks. In cities, public parks emerged in an attempt to bring nature into urban centers such as New York City as part of nineteenth-century social reform efforts. Urban parks followed the British Landscape Gardening concepts, with curving landscape forms and features reminiscent of idealistic rural scenery.

Overall, the Romantic Movement can be credited with elevating the concept of nature and the natural environment to the status of an important human value, thereby forming an important underpinning for modern landscape planning. Indeed, the environmental crisis of the 1960s and 1970s was founded largely on environmental quality and decline as moral issues.

Controlling disease

Public Health and Environment. Another development of the nineteenth century also provided an important underpinning for environmental planning: namely, the scientific understanding of the environment's role in **public health**.

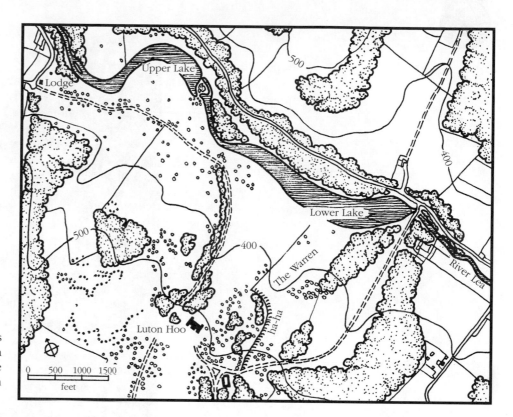

Fig. 1.1 The curved edges of this landscape design by Capability Brown at Luton Hoo, England, illustrate one of the central themes of English Landscape Gardening in the 1700s.

This understanding came about through the documentation of environmentally sensitive diseases, such as malaria, dysentery, and typhoid fever (Fig. 1.2). It resulted in an improved public and institutional understanding of the relationship between human impacts on the environment (such as the decline in water quality from sewage discharges) and the health and well-being of society. One manifestation of this perception was the planning and development of municipal sanitary sewers. Chicago's system, built in 1855, was one of the first in North America. Another outcome was the development of public water supply systems that could deliver safe drinking water to cities. Water purification by filtration and chlorination was introduced in the 1930s.

Wise use concept

The Conservation Movement. A third underpinning for environmental planning was the **Conservation Movement**. It, too, began in the 1800s, growing out of a concern for the damage and loss of land and its resources as a result of development and misuse. Tied to both Romantic and scientific thought, the Conservation Movement, led by environmental stalwarts such as John Muir and J. J. Audubon, initiated the national park movement. Yellowstone, the first U.S. national park, was established in 1872, and Banff, the first Canadian national park, was established in 1885. The movement subsequently led to many other major conservation programs, including the U.S. Forest Service, the U.S. Natural Resources Conservation Service, the U.S. Bureau of Land Management, and the Canadian Department of Oceans and Fisheries, as well as many state and local programs.

Land use applications

The Conservation concept also influenced community land use planning in the United States, Canada, the United Kingdom, and elsewhere. Conservation-sensitive land use planning in the 1970s adopted an ecological perspective in which uses were assigned to the land according to its carrying capacities, environmental sensitivity, and suitability as a human habitat. Although the Conservation concept has been practiced

Fig. 1.2 An illustration from the mid-1800s in New York City depicting the sort of living conditions that led to the public health movement.

in one form or another for many decades, its application to community development was advanced significantly by Ian L. McHarg, a landscape planner and designer, who in the 1960s, 1970s, and 1980s championed the concept of integrated landscape planning as a means of striking a balance between land use and the environment.

The Environmental Movement. In the face of rapidly expanding cities, highway development, and a burgeoning industrial sector after 1945, these movements crystallized in the **Environmental Crisis** of the 1960s and 1970s (Fig. 1.3). Although the environmental crisis is often remembered for protest movements and social upheaval, one of its most lasting effects in the United States was the creation of a massive body of environmental law, the National Environmental Policy Act. This act addresses air quality, water quality, energy, the work environment, and a host of other areas. Similar bodies of environmental policy were subsequently enacted by many states as well as by other developed countries and some developing countries.

New policy

From its explosive beginning the **Environmental Movement** also led to broad social and cultural changes that are still in motion. In the early 1970s environmental

Fig. 1.3 The social protest of the Environmental Crisis of the late 1960s and early 1970s marked the beginning of environmentalism as a major political force in North America.

New agendas

courses and degree programs emerged in colleges and universities. For the first time, large-scale land use development projects, which until this time had been almost exclusively the domain of engineers and architects, began to draw on the services of other professionals, especially landscape architects, geographers, planners, geologists, and ecologists. At the same time, "environmental thinking" began to creep into the public school system, nurturing a whole generation of environmentally sensitive citizens. In the midst of it all, however, came a refrain that still echoes, now worldwide: "Can we afford it?" Astonishingly, the cost argument persists despite the warning signs.

1.2 THE PROBLEM: CHANGE AND IMPACT

Waves of change

The rate at which North Americans have developed this continent is unprecedented in the history of the world. In scarcely 100 years, from about 1800 to 1900, settlers spread across the continent, probing and transforming virtually every sort of landscape. In the vast woodlands and grasslands of the continent's midsection, scarcely a whit of the original landscape remained by the opening of the twentieth century. In its place came crop farms, ranches, towns, cities, and connecting railways and highways. Moreover, since 1950 or so, a second wave of landscape change has swept across North America, especially in the region east of the Mississippi and along the Pacific Coast, one that has transformed the rural landscape of farms and small towns into the satellite communities, dormitory settlements, and playgrounds of huge urban complexes. And today, a third wave of landscape change is underway as population shifts from the middle sections of the continent toward the coasts and urban centers. The wholesale transformation of the great North American landscape represents only part of the story, however. The introduction of synthetic materials and forms represents the rest.

Techno-infusion

Manufactured Landscape. The age of materialism and economic expansionism has produced a colossal system of resource extraction that, for the benefit of the United States and Canada, reaches over most of the world. At the output end of the system is the manufacture of products and residues of various compositions, many decidedly harmful to humans and other organisms. Virtually everything ends up in the landscape:

- Steel, glass, concrete, and plastics in the form of buildings and cities.
- Waste residues in the form of chemical contaminants in the air, water, soil, and biota.
- Solid and hazardous wastes in landfills, waterbodies, and wetlands.

The waste stream is enormous. In the United States, for example, the annual production of synthetic organic compounds, which includes scores of hazardous substances such as pesticides, exceeds 100 million tons. The output of solid waste from urban areas alone approaches a billion tons annually.

Landscape change

The landscapes that are ultimately created are essentially new to the earth. Cities, for example, are often built of materials that are thermally and hydrologically extreme to the land and whose structural forms are geomorphically atypical in most landscapes. It is a landscape distinctly different from the one it displaced and, in many respects, decidedly inferior as a human habitat. The modern metropolitan environment that results tends to be less healthy, less safe, and less emotionally secure than most people desire. Moreover, the very existence of such environments poses a serious uncertainty for future generations, owing to the high cost of maintaining both the environment and the quality of human life within them. In addition, their relationship with the natural environment of water, air, soil, and ecological systems is a lopsided one, hardly a fit with our notion of a sustainable balance between an organism and its habitat. Herein lies much of the basis for environmental and land use planning, landscape design, and urban and regional planning.

Mismatches with Landscape. The planning problems we are facing today are many and complex, and not all, of course, are tied directly to the landscape. For those that are, most seem to result from mismatches between land use and environment. The mismatches stem from five main origins:

1. *Initially poor land use decisions* because of ignorance or misconceptions about the environment, as exemplified by the person who unwittingly builds a house on an active fault or ignores warnings about hurricane-prone shore property.

2. *Environmental change* after a land use has been established, as illustrated by the property owner who comes to be plagued by flooding or polluted water because of new upstream development or by the coastal properties flooded and eroded because of a sea level rise related to global warming.

3. *Social change*, including technological change, after a land use has been established and represented, for example, by the resident who lives along a street initially designed for horse-drawn wagons but now used by automobiles and trucks and plagued by noise, air pollution, and safety problems.

Poor decisions and change

4. *Poor planning and design*, as in the case of the road redesigned for greater efficiency and safety which instead induces more accidents among cars, bicycles, and pedestrians than the roadway it replaced.

5. *Violations of human values* concerning the mistreatment of the environment, such as the eradication of species, the degradation of streams, and the alteration of historically valued landscapes (Fig. 1.4).

Fig. 1.4 Value conflict. Increasingly people are asserting their environmental values in both the public and private arenas at all scales, from the local to the global.

1.3 THE PURPOSE OF PLANNING

Why plan?

In general, the primary objective of planning is to make decisions about the use of resources. Over the past 25 years, the need for land use and environmental planning has increased dramatically, as demonstrated by the rising competition for scarce land, water, biological, and energy resources; by the need to protect threatened environments; and by the desire to maintain or improve the quality of human life. The problems and issues are diverse in both type and scope—ranging from worldwide issues, such as permafrost decay with global warming and desertification with overgrazing and deforestation, to problems of draining and filling a 2-acre patch of wetland on the edge of a city or piping and diverting a stream to simplify infrastructure design in a residential subdivision. In North America, despite the political undertones historically associated with public planning, environmental planning has gained real legitimacy in the past several decades, though more as a reactive than as a proactive process—that is, more as a system of restrictive (should-not) policy than as one of constructive (how-to) policy.

Officials and agents

Who Does Planning? Actually, professional planners probably do not do the majority of planning. Most of it is done by corporation officers, government officials and their agents, the leaders of educational and religious institutions, the military, and various other organizations, including citizens' groups. Professional planners (those with formal credentials in planning or related areas) usually function in a technical and advisory capacity to the decision makers, providing data, defining alternative courses of action, forecasting impacts, and structuring strategies for the implementation of formal plans. The overall direction of a plan, however, always represents some sort of policy decision, one based on a formal concept of what a company, university, city, or neighborhood intends for itself and thus will strive to become. These concepts about the future are called *planning goals*, and they are the driving force behind the planning process.

1.4 PLANNING REALMS: DECISION MAKING, TECHNICAL, AND DESIGN

Three broad classes of activity make up modern planning: decision making, technical planning, and landscape design.

Decision Making. This first activity—the decision-making process itself—is usually carried out in conjunction with or directly by formal bodies such as planning commissions, town councils, and corporate boards. It involves first defining the issues and problems of concern, and second providing the directions and means necessary for making decisions. Among the tasks commonly undertaken in **decision-making planning** are the:

Planning activities

- Formulation of policies.
- Articulation of goals.
- Definition of alternative courses of action.
- Selection of preferred plans.
- Sponsoring of technical studies

Technical Planning. The second class of planning activity, **technical planning**, involves various processes and services in support of both decision making and design:

Mapping, testing, and analysis

- Environmental inventories, such as habitat, soil, and land use mapping.
- Forecasting change, such as stream flooding, sea level rise, and population growth.
- Engineering analysis, such as soil testing for construction and facility cost and life cycle estimating.
- Assessment of the environmental impacts related to proposed land use development.

Technical planning is usually carried out by a variety of specialists, including cultural geographers, physical geographers, geologists, ecologists, hydrologists, wildlife biologists, archaeologists, economists, and sociologists, as well as by professional planners from landscape architecture, urban planning, and architecture and civil engineering. The line separating decision-making from technical-planning activities may be distinct or indistinct, depending on organizational arrangements and the nature of the problem.

Landscape Design. Following the decision-making process and the first wave of technical support studies, we move into the arena of design. **Landscape design**—which is mainly the domain of landscape architects but is also practiced by architects and planners—entails the laying out on paper or on the computer screen the configuration of the uses, features, and facilities that are to be built, changed, or preserved based on

Giving form

the decision maker's directives. Design may call for additional technical studies, such as stormwater calculations, soil testing, map refinements, and economic analysis.

Therefore, the planning processes and the relationships among the three areas of professional activities should be regarded not as a linear sequence—though they are often executed that way—but rather as an interrelated circuit, as depicted in Figure 1.5. Resolving a planning problem usually involves several iterations of the circuit in which a check-and-balance relationship often emerges among decision makers, technical planners, and designers.

1.5 ENVIRONMENTAL IMPACT ASSESSMENT

The use of technical planners or specialists in planning has increased sharply in the past three decades as a result of environmental impact legislation. Indeed, the enactment of the **National Environmental Policy Act of 1969 (NEPA)** is directly tied to

Master policy

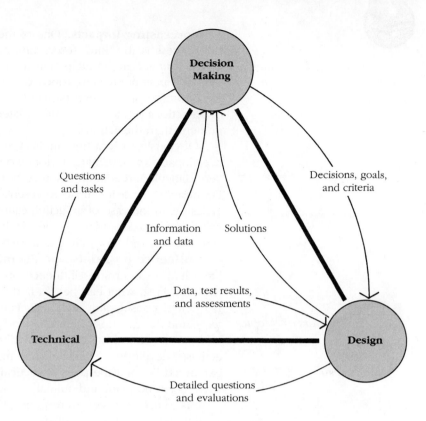

Fig. 1.5 The relationship among the technical, decision-making, and design realms of modern planning. Much of the activity of environmental planning falls within the technical sphere.

the emergence of environmental planning as a formal area of professional practice. This act calls for planners to forecast and evaluate the potential impacts of a proposed action or project on the natural and human environments. Although the act covers only projects involving federal funds (such as sewage treatment systems, highways, and domestic military facilities), NEPA initiated the enactment of more broadly based environmental impact legislation at state and local levels in many parts of the United States. As a whole, the various bodies of impact legislation brought on a flurry of activity in environmental planning in the 1970s and served to establish environmental factors as legitimate considerations in urban, regional, transportation, and other types of planning.

The Methodology. The basic **environmental impact methodology** can be summarized in five steps, or tasks, that are normally performed sequentially:

Basic EIS steps

1. Select variables or factors that are pertinent to the problem, record them in an inventory, and identify their interrelationships.
2. Formulate alternative courses of action.
3. Forecast the effects (or impacts) of the alternatives.
4. Define the differences between the alternatives: that is, specify what is to be gained and lost by choosing one alternative over another.
5. Evaluate and rank the alternatives and select the preferred one.

The report prepared from this assessment is called an **environmental impact statement (EIS)**. It must identify the unavoidable adverse impacts of the proposed action, any irreversible and irretrievable commitments of resources as a result of the proposed action, and the relationship between short-term uses of the environment and long-term productivity. In addition, the EIS must include among its alternatives one calling for no action, and this, too, must be subjected to analysis and evaluation.

Forecasting Impacts. One of the most important and challenging tasks in the EIS process is the third: forecasting the impacts of alternative actions. An *environmental impact* can be defined as the difference between (1) the condition or state of the environment given a proposed action and (2) the condition expected if no action were to take place. Impacts may be direct (resulting as an immediate consequence of an action) or indirect (resulting later, in a different place, and/or in different phenomena than the action).

Major challenge

Obviously, forecasting indirect impacts, along with their correlative, cumulative impacts (where many factors work in combination to produce change), can be very difficult and is often a source of much uncertainty in environmental assessment. Furthermore, since an impact represents an environmental change, the problem that also arises (as in the case of an action calling for eradicating vegetation that contains both valued and noxious plant species) is that of deciding which are desirable and undesirable impacts and how different impacts should be weighted for relative significance.

Influence on Landscape Planning. The EIS methodology and requirements have had an enormous influence on land planning thought and practice in North America. First, what had been a fairly closed process in the decades after World War II opened considerably with EIS. Participation expanded beyond engineers, architects, and the captains of business and government to include many other professions and parties, not only scientists of various stripes, but environmentalists, community activists, developers, educators, and many others. The range of topical considerations expanded beyond the purely practical to include phenomena of sociocultural value such as prized plant and animal species, archeological resources, ethnic factors, and local land use practices. In response, the land use planning methodology in both the public and private sectors changed so that environmental inventories, impact assessments, and some form of public participation became more or less standard in most large planning projects.

New players and topics

1.6 AREAS OF ACTIVITY IN LANDSCAPE PLANNING

The scope of modern landscape planning is surprisingly broad and encompasses a great many areas of professional activity. Some activities with environmental inventories and impact assessments, as suggested, are common to most land use projects in both rural and urban settings. Others, such as risk management planning and restoration planning, deal with particular problems and settings. Yet others are broadly based and oriented more toward community and regional land use planning. The following paragraphs describe a number of these activities.

Environmental Inventory. One of the best-known activities in landscape planning and design is the so-called **environmental inventory**, an activity designed to provide a catalog and description of the features and resources of a study area. The basic idea behind the inventory is that we must know what exists in an area before we can formulate planning alternatives and design schemes for it. Among the features consistently called for in environmental inventories are water features, slopes, microclimates, floodplains, soil types, vegetation associations, and land use, as well as archaeological sites, wetlands, valued habitats, and rare and endangered species. Environmental inventories are a major part of environmental impact statements, in which the inventory must also include an evaluation of the recorded phenomena based on criteria such as relative abundance, environmental function, and local significance. This is supposed to indicate the comparative importance or value of a feature or resource (Fig. 1.6).

Surveying the stage

Site Analysis and Evaluation. A second type of planning activity is aimed at the discovery of **opportunities and constraints**. This activity is often undertaken

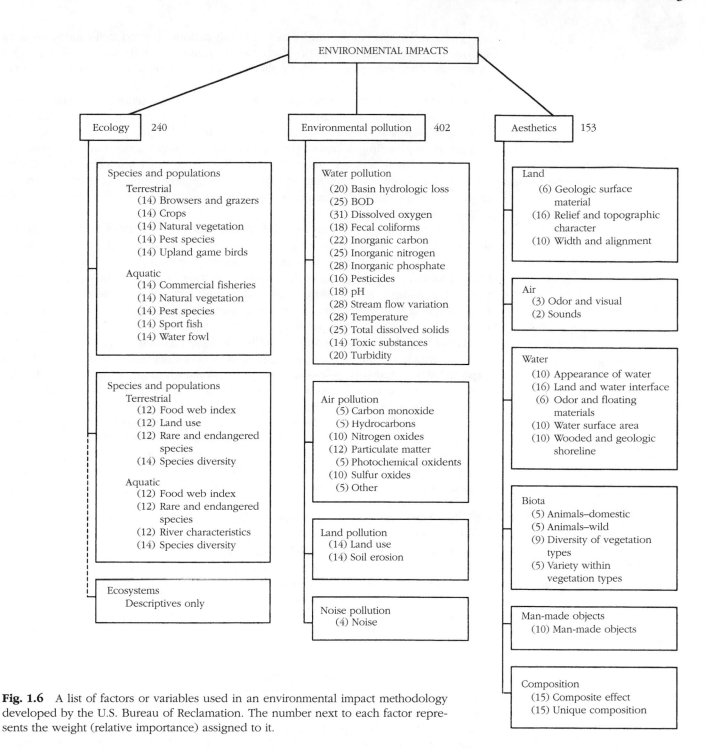

Fig. 1.6 A list of factors or variables used in an environmental impact methodology developed by the U.S. Bureau of Reclamation. The number next to each factor represents the weight (relative importance) assigned to it.

Pros and cons after a land use program has been proposed for an area but the density, layout, and appropriate design of the land use program are still undetermined. The study involves searching the environment for features and situations that would (1) facilitate a proposed land use and those that would (2) deter or threaten a proposed land use. Basically, the objective is to find the potential matches and mismatches between land use and environment and to recommend the most appropriate relationship between the two. This may involve a wide range of considerations including off-site ones where the site is affected by systems and actions beyond its borders, such as stormwater runoff and air pollution from development upstream and upwind, as well as phenomena covered by public policies, such as protected species, floodplains, and wetlands.

Site assessments

Documenting opportunities and constraints is often included in **site assessments**. These are prepurchase or preplanning environmental profiles of sites highlighting whatever conditions might be germane to land value, purchase agreements, and program planning. A central part of modern site assessments, particularly on the urban/suburban fringe, are inspections for hazardous wastes such as buried storage tanks and possible safety problems such as shallow mine shafts, contaminated groundwater, potentially contentious border conditions, risky floodplains, and unsafe soils and slopes.

All of the above are commonly part of **site analysis**, an activity that investigates the makeup and operation of a site vis-à-vis a proposed use program. The program serves as an initial problem statement or a testable proposition. The analysis is tailored to the particular requirements of the program; to the character of the site itself; to the site's relationship to surrounding lands, waters, and facilities; to community concerns; to policy issues; and to other factors. Site analysis may also include several of the remaining areas of activity as well.

Land Capability Assessment. Studies designed to determine the types of use and how much use the land can accommodate without degradation are referred to as **capability** or **suitability assessments**. For sites or study areas composed of different land types, the objective is to differentiate buildable from nonbuildable land and to define the development capacity, or **carrying capacity**, of different land units or sub-

Land capacity

areas (Fig. 1.7). Capability studies may also be performed to determine best use, such as open space, agricultural, or residential, for different types of land over broad areas. Capability and capacity studies are fundamental to **sustainability planning**, in which the overriding objective is to build land use systems that are environmentally enduring and balanced for both humans and other organisms over the long term. The concept of sustainability in development planning has become more or less interchangeable with the "green" facilities theme, most notably advanced by architects in their building standards called LEEDS (Leadership in Energy and Environmental Design).

Hazard Assessment and Risk Management. The objective in **hazard assessments**, which are a specialized type of constraint study, is to identify dangerous zones in the environment where a use is or would be put in jeopardy of damage or destruction. Hazard research has been concerned both with the nature of threatening environmental phenomena (floods, earthquakes, and storms) and with the nature of human responses

Sizing up threats

to these phenomena. Zoning and disaster relief planning for hurricane and flood-prone areas, for example, have benefited from hazard assessment at the national, state, and local levels. Another benefit is the emergence of **risk management planning** as a part of development programs, which involves building strategies and contingency plans for coping with hazards and providing emergency relief services. Both hazard assessment and risk management planning are gaining serious national attention in the United States in response to recent disasters such as the 1993 Mississippi flood, the 1994 Northridge earthquake near Los Angeles, the North Dakota flood of 1997, the 2004 rash of hurricanes in Florida, and the Katrina and Ike disasters in Louisiana and Texas.

Impact Forecasting. Hazard assessment, environmental impact assessment, opportunity/constraint studies, and site assessments all depend on another activity: **forecasting impacts**. This activity involves the identification of the changes that a proposed action calls for or implies, followed by an evaluation of the type and magnitude of the environmental impact. The process is tough because of the difficulty in

Best guess?

deriving accurate forecasts by analytical means. As a result, forecasts of impacts are usually best estimates, and the significance assigned to them seems to be as much a matter of perspective (such as an engineer's versus an environmentalist's) as anything

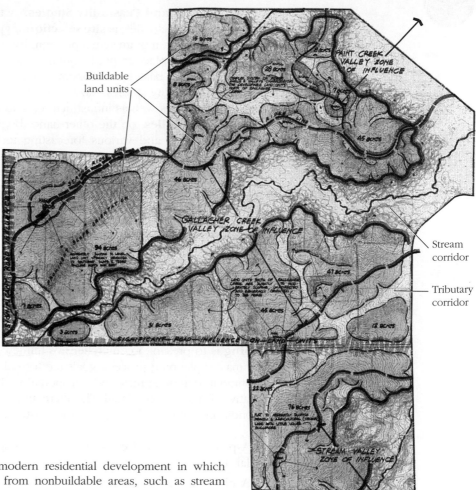

Buildable
land units

Stream
corridor

Tributary
corridor

Fig. 1.7 Land capability map for a modern residential development in which buildable land units are differentiated from nonbuildable areas, such as stream corridors.

else. Nevertheless, the *process* is an important one because it often leads to clarification of complex issues and their environmental implications, to modification of a proposed action to lessen its impact, or to the abandonment of a proposed project.

Special Settings. Delineation, analysis, and evaluation of **special environments,** such as wetlands, unique habitats, and archaeological sites, is a rapidly rising area of planning activity. Though logically a part of impact assessment, capability studies, and most other planning activities, special environments have gained increased attention with the enforcement of wetland protection laws, rare and endangered species laws, and similar ordinances relating to prized environmental resources. The focus of activity to date is overwhelmingly empirical, dealing mainly with the field identification and mapping of the feature or organism in question. The results usually center on the question of the presence or absence of, for example, a threatened species or a valued habitat, as the basis for deciding whether a proposed land use can or cannot take place in or near the area under consideration.

Places and values

Other kinds of special environments are those in need of some kind of restorative action. **Restoration planning** addresses environments such as wetlands, wildflower habitats, stream channels, and shorelines that have been degraded by land use activities. Perhaps the most common restoration planning activity is associated with damaged wetlands, degraded streams, and waste disposal sites. Wetland mitigation includes both the restoration of damaged wetlands and the construction of essentially new wetlands.

Where to go

Site Selection and Feasibility Studies. A task common to development planning, among other activities, is **site selection**. Typically, we would begin with an idea for a land use program or an actual program for a facility or enterprise, and attempt to find an appropriate place to put it. Often the search entails no more than an exercise in location analysis based on economic factors, but, properly done, it should also include land capability studies, hazard assessments, and various other types of environmental studies, as well as infrastructure, land use, and policy evaluations.

Feasibility studies, on the other hand, begin with a known site, and, with the aid of field studies and various forecasting techniques, attempt to determine the most appropriate use for it. Increasingly, planners and developers are interested in learning about a site's limitations based on environmentally protected areas and features, such as wetlands and habitats, as a part of feasibility studies.

Finding techno-fits

Facility Planning. This activity involves siting, planning, and designing installations that depend on structural and mechanical systems. Sewage treatment plants, industrial installations, airfields, and health complexes are typical examples. Not many years ago, little consideration was given to environmental matters in facility planning beyond water supply, sewer services, and, in some cases, an environmental impact statement. Today, however, environmental analysis and site planning, both onsite and offsite, are given serious consideration as they influence environmental quality, public image and community relations, landscape management, and especially public and worker liability.

Grand schemes

Master Planning and Framework Planning. The overriding aim of **master planning** is to present a comprehensive framework to guide land use changes. Early in the master planning process, goals are formulated relating to land use, economics, environment, demographics, and transportation. Existing conditions are analyzed, and alternative plans are formulated. The alternatives are then tested against goals and existing conditions, and one is adopted. The master plan usually comprises three parts:

1. A program proposal consisting of recommendations, guidelines, and proposed land uses.

2. A physical plan or design scheme, showing the recommended locations, configurations, and interrelationships of the proposed land uses.

3. A scheme for implementing the master plan that identifies funding sources, enabling legislation, and guidelines for how the changes are to be phased over time.

The master planning process may be preceded by an activity called **framework planning**. Indeed, the framework plan often sets the stage for building the master plan by providing a conceptual foundation for it. It usually involves experimenting with various schemes aimed at fitting program to site, a process that necessitates a critical examination of both the proposed program and the site. The principal product of this activity is a recommendation about a conceptual layout that includes circulation, open space, and envelopes of space earmarked for various facilities as well as the basic logic behind the concept.

Beyond planning and design

Management Planning. Finally, we must consider **management planning**. Although environmental planning is normally associated with the early phases of the planning process, it is becoming apparent that it must also be part of the design, construction, and operational phases of projects. In the construction phase, management plans must be formulated to minimize environmental damage from heavy equipment, material spills, soil erosion, and flooding. Similarly, once construction is over and the land use program is operational, environmental management programs are often required to achieve lasting stability among the landscape, built facilities, and environmental systems such as drainage, airflow, and ecosystems. As with master plans, management plans must be comprehensive to be effective, and they must meet the criteria of landscape sustainability through compatibility with larger land use–environmental systems.

1.7 METHODS AND TECHNIQUES

The methods used in environmental and landscape planning are basically no different from those in other areas of planning and design. The fact that environmental phenomena are closely associated with the natural sciences does not mean that this area of planning is necessarily more rigorous than, for example, open space or recreation planning. The differences lie rather in the perspectives, particularly in what components of the plan are given the greatest emphasis and in the analytic techniques used to generate data and to test the planning and design schemes.

Groundwork

Inventory and Analysis. The questions and topics that are the focus of analysis in environmental planning originate in all phases of projects and problem solving, and they are of varying complexity and sophistication. In the early phases of a project, the emphasis is generally on gathering and synthesizing data and information. Planners often refer to this listing as an **environmental inventory**, following the language of EIS methodology. The idea behind the inventory is to learn all we can about the character of a project site and its setting. Although this normally includes field inspection and field measurement, the generation of quantitative data usually is not the primary objective. Instead, the sources of most data are secondary (published) sources: topographic maps, soil maps, aerial imagery, climatic data, and streamflow records.

On analysis

When detailed field measurement is undertaken early in a project, it is usually in connection either with a known or suspected engineering, safety, or health problem, such as soil stability or buried waste, or with policy problems, such as wetlands or threatened species. For the most part, environmental analysis in the early phases of a project is typically not analytic in the scientific sense; that is, it is more concerned with defining distributions, densities, and relations among the various components of the environment than with rigorous testing of cause-effect relationships.

Later in the project, the process becomes more **analytic** as problems and questions arise relating to formulating and testing planning and design schemes. The techniques employed vary widely. For some problems, quantitative models are used, such as hydrologic models to forecast changes in streamflow and flood magnitudes in connection with land development in a watershed. For others, hardware models are called for, such as windtunnel analysis of building shapes and floodflow simulations in stream tanks (Fig. 1.8). For still others, a statistical analysis to test the relationships between two or more variables (such as runoff and water quality) is appropriate.

Preparing the story

Synthesis and Display. Generating results from the various data-gathering, descriptive, and analytic efforts does not mark the end of the environmental planner's or designer's responsibility. Ahead lies the difficult task of **integrating** the various and sundry results in a meaningful way for decision making. No calculus has been invented that satisfactorily facilitates such a difficult integration—a dilemma faced in all planning problems. Integrating almost always requires some sort of a screening and evaluation to determine the relative importance and meaning of the results.

The actual integration is usually a qualitative rather than a quantitative process and typically centers on a visual (graphic) **display** of some sort. This may be a matrix, a flow diagram, a set of map overlays, or a gaming simulation board. Above all, it is important to understand that the final outcomes are found not in the results of specific procedures or techniques (as we might be led to believe from our experience in a college science laboratory class). Rather, they emerge from a less exact and more eclectic process that invariably rests on a decision maker's or decision-making body's perspectives and values concerning the problem, as well as on related political and financial agendas.

Fig. 1.8 A hardware model of a portion of the Mississippi River used by the U.S. Army Corps of Engineers to simulate the behavior of floodflows in a partially forested floodplain. The rows of cards produce an effect on flow similar to that of trees.

1.8 ENVIRONMENTAL ORDINANCES (BYLAWS)

Policy development

Environmental regulations in the United States and Canada abound at all levels of government. As noted earlier, the mother ordinance in the United States is the National Environmental Policy Act of 1969 (NEPA), which provides for a massive regulatory program under the U.S. Environmental Protection Agency. In the 1970s, broadly similar environmental protection programs were developed by many states and provinces. In the 1980s and 1990s, as the need for environmental and land use regulation continued to rise, additional policies were legislated, reflecting new and changing state and local values for groundwater protection, waste disposal management, soil erosion control, watershed management, hillslope development, and many other issues.

The search for order

Playing by the Rules. Today, a great multilayered umbrella of environmental rules, regulations, bylaws, and guidelines covers land use activities and their interplay with the environment. Taken together, the mass of environmental regulations borders on the incomprehensible. As with most bodies of law, the enforcement process has become highly selective, and the ordinances actually enforced tend to change over time with shifts in national, state/provincial, and community interests; with politics; with the latest needs; and with economic conditions. Responding to environmental regulations when implementing land use, engineering, and related projects can be a bit of a shell game, and the appropriate response often requires professional assistance in identifying, interpreting, and negotiating environmental policy. In most parts of Canada and the United States, local specialists, such as environmental attorneys, urban planners, and policy specialists have emerged to provide such services.

Interpreting policy

Meeting the requirements of environmental regulations is a very important part of land use and environmental planning. Countless projects have failed because of inattention to ordinances/bylaws or because of judgmental errors in interpreting and negotiating regulations. Projects have also failed in some communities because of excessive and cumbersome regulatory demands placed on applicants seeking approval to make a land use change. The difficulties commonly stem from philosophical differences in what the ordinances are intended to do. For example, some communities and their officials expect strict adherence to specified measures, such as a slope requirement or setback from a stream, with little or no latitude in interpretation. *Variances*—that is, approved exceptions—are expected to be minor and few. Other communities apply the law in a more liberal way, intending that its spirit or intent must be satisfied and that reasonable interpretation may be applied.

Performance concept

Performance-Based Approach. Most environmental ordinances are either *restrictive*, meaning that they are aimed at what you should not do, or *prescriptive*, meaning that they tell you what you should do. Some, however, allow you to propose, within certain limits, your own way of meeting environmental requirements. One such approach is called **performance-based planning**. Unlike prestrictive ordinances that typically specify the measures to be used (such as detention ponds for stormwater management), the performance-based approach allows planners and designers a certain amount of freedom to devise alternative means of coming up with the desired end.

To justify an alternative approach, however, the applicant must provide reliable evidence that the measures will actually work. Evidence from parallel projects in other locations, preferably nearby ones, is usually helpful, as are relevant reports from the professional, scientific, and engineering literature. However, in the absence of such evidence, demonstration projects are often necessary before the full-scale implementation can take place. Demonstration projects are often expensive and may require several years to carry out. Thus, communities and planners are limited in using this approach and therefore usually overwhelmingly rely on conventional restrictive and prescriptive regulations.

Beyond regulations

Problem Solving. In the end—no matter how detailed, applicable, and seriously enforced they may be—regulations do not themselves solve planning and design problems. At best they are a safety net for society, providing some insurance that the affected areas, systems, features, and artifacts of public value and concern are addressed as a part of land use change. The real process of landscape planning should take place at a much higher level with far more elegance than is called for by public rules and regulations. Among other things, it should seek **integrated solutions** that, for example, do not end with wetland and floodplain mapping, but address wetlands, floodplains, streams, groundwater, and stormwater in a comprehensive way, focusing on water systems and their management, their relations to land use systems, and the creation of sustainable landscapes.

1.9 THE PLANNING PROFESSIONS AND PARTICIPATING FIELDS

Traditionally, only three planning and design fields—urban planning, landscape architecture, and architecture—are recognized for training the professionals who guide the formal planning processes. These fields focus mainly on the decision-making and design aspects of planning.

Urban planning

Planning and Design Fields. Urban planning has the broadest scope, with concern for entire metropolitan areas. Most professional activity in urban planning revolves around policy development and regulatory practices in the public sector related to economic development, social programs, land use, and transportation planning. Most urban planners work for planning and related agencies in cities, townships, and counties, although the number working in the private sector for consulting firms, banks, realtors, and developers has increased in the past two decades.

Landscape architecture

Landscape architecture tends to be more site oriented than urban planning, with a stronger emphasis on design. Landscape architects work with both the natural and built elements of the landscape, seeking to blend the two into workable and pleasing environments. Professional activity covers the full range of settings from urban to wilderness landscapes and includes projects as small as residential site planning and others as large as national park planning and design. Landscape architects work in both the private and public sectors, and their activities typically include aspects of urban planning, civil engineering, architecture, horticulture, and scientific fields such as geography, botany, and geomorphology.

Architecture

Architecture has the narrowest focus in landscape planning, dealing mainly with buildings and their internal environments. Architecture is concerned with the landscape mainly as a setting for buildings and related facilities but also as a source of environmental threats to building stability and safety (such as floods and earthquakes). Because of today's broad environmental regulations and the high costs of energy, water, and other building resources, architecture is generally forced to pay more attention to environmental matters than has been the tradition in the field, and this is reflected in the rise of sustainability practices leading to green concepts in building design.

Supporting fields

Technical Fields. With each decade a growing number of scientific disciplines, or what planners and designers often call **technical fields**, participate in planning in North America. The 1970s and 1980s saw increased participation from geography, geology, biology, chemistry, anthropology, and political science in the formal arenas of planning, mainly in connection with environmental assessment and impact activities. The 1980s nurtured the development of *technical subfields* in response to the increased complexity of planning problems and the need for specialists in areas such as hazardous waste management, groundwater protection, and wetland evaluation and restoration.

Technical subfields have emerged in both traditional planning fields and the participating sciences. Many of the subfields sponsored by the scientific disciplines have counterparts, more or less, in the traditional planning fields. Linked by common research interests, these subfields form an important source of data and information for the decision-making and design processes. In landscape architecture, for example, there are ties with geography, botany, and ecology over issues such as watershed management, habitat planning for urban wildlife, and wetland restoration.

Geography

Both architecture and geography are interested in the microclimates of building masses and urban environments. **Geography** also shares an abiding interest with landscape architecture and planning in remote sensing, environmental assessment, wetland mitigation, stream restoration, land use mapping and monitoring, and many other problems and activities. And high on the list in most areas of activity in these fields is the application of Geographic Information Systems (GIS), which in the last several decades has emerged as a major technical subfield.

Engineering

Civil Engineering. Larger and more influential than urban planning, landscape architecture, or any of the subfields, **civil engineering** has played a role in both environmental and land use planning. Civil engineering is involved in a major way in most private and public development projects including site assessment, environmental impact analysis, site design, risk management, transportation planning and design, stormwater management, and wetland mitigation. The field has had a major influence on environmental policy, particularly in the areas of stormwater management, flood control, waste management, channel and harbor modification, shoreline protection, and water supply planning. It has also made major research contributions in the fields of hydrology, geomorphology, and pollution control. Because civil engineers control infrastructure planning and design in development projects, they have by default come to exercise a major control on community design, but often to the point of neglecting landscape architects, urban planners, architects, and environmental interests.

1.10 CASE STUDY

Adapting the EIS to Issues of Natural Gas Drilling in New York State

Stephan Schmidt and Shannon Stone

The adaptation and implementation of an environmental impact statement (EIS) is not without its problems. For example, because an EIS is often not completed for more general programs or policies, it is often criticized as not being able to assess cumulative impacts over time. In response to some of these criticisms, a generic (or programmatic) EIS has been introduced in order to provide opportunities to mitigate or abandon environmentally unsound concepts before they became specific site projects. Recent developments in New York State highlight how the EIS process is evolving to meet changing demands.

In 2008, as energy prices spiked, natural gas companies mobilized to tap into the Marcellus Shale, a natural gas formation as deep as 8000 feet below ground stretching from West Virginia throughout Pennsylvania and Ohio to southern New York (Fig. 1.A). The Marcellus Shale is estimated to contain 50 trillion gallons of recoverable natural gas, enough to supply the entire United States for two years.

To extract the gas, energy companies proposed a hyperintensive type of hydraulic fracturing process, known as *slickwater fracturing*, that sends vast quantities of water mixed with sand and chemicals at high pressure to fracture the rock more than a mile underground. New York's Department of Environmental Conservation (DEC) requires that the complete chemical composition of additives be disclosed before a permit is issued. According to Argonne National Laboratories, water produced from natural gas wells contains higher concentrations of aromatic hydrocarbons such as benzene, toluene, ethylbenzene, and xylene than oil drilling byproducts and are 10 times as toxic.

A Generic Environmental Impact Statement (GEIS), most recently updated in 1992, regulates drilling for natural resources in New York. It requires most drilling

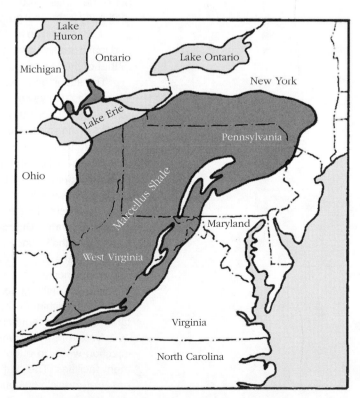

Fig. 1.A The geographic extent of Marcellus Shale in eastern United States and southern Ontario.

applications to submit only a short environmental assessment form to the DEC. A GEIS is more general or conceptual in nature than a site-specific EIS and typically is used to consider either multiple separate actions within a geographic area, a sequence of related actions, or programs or plans that have wide application or restrict the range of future alternative policies.

However, in this case the GEIS contained a number of shortcomings. It did not address the specific issue of horizontal drilling that is necessary to access the Marcellus Shale. Nor did it include a cumulative impact analysis of hydraulic fracturing on the environment. In addition, the oil and gas industry was exempt from all local ordinances and community oversight, regardless of how incompatible the drilling was with established zoning. A common complaint from local governments was, and still is, that the state has not passed appropriate enabling legislation allowing them to regulate or mitigate the drilling process.

In response to public concerns, New York's DEC proposed updating the GEIS in 2008 with a Supplemental GEIS to further review the environmental impacts of the hydraulic fracturing process and the impact of drilling in important watersheds. The first step involved outlining (or "scoping") the factors to be included in the Supplemental GEIS:

■ The potential impacts of withdrawing large volumes of water from both surface and groundwater sources needed for the drilling process.
■ The transport of water to and removal of fracturing fluid from well sites (particularly the use of heavy machinery in the transport process, which may damage rural roads).
■ The use of additives in the hydraulic fracturing fluid.
■ The potential cumulative effects on wetlands, air quality, aesthetics, noise, traffic and community character, and local governments.

To assist in the development of the scoping document, the DEC held a series of public hearings throughout the affected area in the fall of 2009. The hearings heard testimony from local governments, environmental organizations, and concerned citizens from Colorado and Wyoming with experience and knowledge of chemical spills and leaks caused by gas drilling in the Barnett Shale, a natural gas formation similar to that of the Marcellus.

The scoping process did not imply a formal moratorium on the issuance of drilling permits. However, it required the completion of an individual, site-specific environmental impact study to receive a horizontal drilling permit until the Supplemental GEIS was finished, a requirement covered under of the State Environmental Quality Review Act. After the Supplemental GEIS is completed, the permit applicant will not need additional environmental review, and the Supplemental GEIS will likely be the policy template for Marcellus Shale mining in New York. The draft Supplemental GEIS addresses certain specific aspects of the natural gas drilling process, including two particularly contentious issues: wastewater disposal and water supply.

Wastewater Disposal. The hydraulic fracturing process requires that fluids returned to the surface are pumped into tanks or lined open pits (Fig. 1.B). As open pits are susceptible to overflowing from strong rainstorms and local runoff, inspectors must monitor open pits to ensure they are not leaking or overflowing. Upon cessation of drilling operations at a well site, the open pits must be cleaned up and remediated within 45 days. The draft Supplemental GEIS will examine whether steel tanks should be required to contain returned fluids to prevent overflows.

A state-licensed waste transporter must truck the water from the drilling site to injection wells, to out-of-state industrial treatment plants, or to local sewage treatment facilities. The draft Supplemental GEIS will review the suitability of fluid disposal at municipal wastewater treatment facilities, potentially require the reuse or

Fig. 1.B An aerial view of wastewater fluid from a natural gas drill site being emptied into an open, lined pit near Dimock, Pennsylvania. A single Marcellus Shale horizontal gas well treatment can require 2.5 million gallons of water, of which 1 percent (25,000 gallons) is hazardous chemicals, including acid and biocide. (Photograph by J. Henry Fair)

recycling of the flowback fluid, or necessitate a waste disposal plan to check that the proposed disposal site has available enough capacity.

Water Supply. The water required for fracturing a single well ranges between 0.8 and 2.5 million gallons. This water is either pumped from wells or delivered on-site via tanker trucks from permitted surface water intakes or municipal supplies. Freshwater is required but possible alternatives are reused flowback water, wastewater treatment plant effluent, cooling water, and saline aquifers. The sheer volume of water needed in the slickwater fracturing technique is also a concern if well water is used, as using this source could lead to a drop in the local water table.

The draft Supplemental GEIS will assess the potential consequences of large surface water intakes that slickwater fracturing would require, including the combined impact of water withdrawals upstream and downstream and an evaluation of the effects on fish and wildlife. Possible mitigation measures include a reduction or stoppage of water withdrawals during low-flow periods, water intake design to minimize effects on aquatic wildlife, and controls or treatment to prevent the spread of invasive or nuisance species.

Local Policy Implications. A separate (yet related) issue, made apparent through the introduction of the Marcellus Shale drilling, is the lack of tools that local governments have at their disposal to regulate or mitigate this and similar processes. The New York State Legislature has considered new legislation to rectify this by broadening powers afforded to municipalities with regard to gas drilling. For example, one bill includes language to enable towns and villages to require that gas developers provide bonds against damages to local roads. Though the bill does not authorize the municipalities to prohibit gas development, it allows them to impose a six-month moratorium on gas development activities.

Stephan Schmidt is an assistant professor in the Department of City and Regional Planning at Cornell University and teaches a course in environmental planning. Shannon Stone is a graduate student in the same department and has previously worked with environmental nongovernmental organizations in New York State.

1.11 SELECTED REFERENCES FOR FURTHER READING

Argonne National Laboratory. *A White Paper Describing Produced Water from Production of Crude Oil, Natural Gas, and Coal Bed Methane.* Washington, DC: U.S. Department of Energy, 2004.

Glacken, C. J. *Traces on the Rhodian Shore.* Berkeley: University of California Press, 1990.

Godschalk, David R. *Planning in America: Learning from Turbulence.* Chicago: APA Planners Press, 1974.

Hargrove, Eugene C. *Foundations of Environmental Ethics.* Upper Saddle River, NJ: Prentice-Hall, 1989.

Holling, C. S. (ed.). *Adaptive Environmental Assessment and Management.* Hoboken, NJ: Wiley, 1978.

Leopold, Aldo. *A Sand County Almanac.* Oxford: Oxford University Press, 1949. (Reissued by Ballantine Books, 1970.)

Lynch, Kevin, and Hack, Gary. *Site Planning.* Boston: MIT Press, 1984.

Marsh, William M., and Grossa, John M. *Environmental Geography: Science, Land Use, and Earth Systems,* 3rd ed. Hoboken, NJ: Wiley, 2005.

McHarg, Ian L. *Design with Nature.* Hoboken, NJ: Wiley, 1995.

Newton, N. T. *Design on the Land: The Development of Landscape Architecture.* Cambridge, MA: Belknap Press, 1971.

Ortolano, Leonard. *Environmental Regulation and Impact Assessment.* Hoboken, NJ: Wiley, 1997.

Platt, Rutherford H. *Land Use and Society: Geography, Law, and Public Policy.* Washington, DC: Island Press, 1996.

West Publishing. *Selected Environmental Law Statutes.* St.Paul, MN: West Publishing, 1995.

Related Websites

Baker, Douglas C., Sipe, Neil G., and Gleeson, Brendan J. "Performance-Based Planning." *Journal of Planning Education and Research. 25,* 4, 2006, pp. 396–409. http://jpe.sagepub.com/cgi/content/abstract/25/4/396
The pros and cons of applied performance-based planning. The article also examines the implications for future use of this alternative to traditional zoning, mainly a more consistent process.

Canada Mortgage and Housing Corporation. International Experiences with Performance-Based Planning. 2000. https://www03.cmhc-schl.gc.ca/b2c/b2c/init.do?language=en&shop=Z01EN&areaID=0000000035&productID=00000000350000000147
An examination of performance-based planning internationally in a downloadable document. The site also provides implications for how these examples could foster innovation for development in a changing Canada.

MetropolisMag.Com, "Q & A: William Saunders." 2009. http://www.metropolismag.com/pov/20090427/q-a-william-saunders
An interview with William S. Saunders, author of the book Urban Design, in which he talks about the evolving roles of landscape architects and architects in urban design. It goes on to discuss the roles of government in urban greening measures.

University of New Brunswick. The Industrial City in Transition: A Cultural and Environmental Inventory of Greater Saint John. http://www.unbsj.ca/cura/index.php
A look at how organizations work together to compile a complete social, environmental, and economic analysis. This comprehensive research will contribute to the future planning of this and other areas.

THE PHYSIOGRAPHIC FRAMEWORK OF THE UNITED STATES AND CANADA

2.1 INTRODUCTION

There is an astounding geographic variety to the North American landscape. We celebrate this as part of our national heritage in both Canada and the United States. For two centuries, geographic diversity and dreams of economic opportunity have driven westward movement and settlement. Geographic diversity is still at the root of many of our life decisions. We carry out our lives in many different locations, and the notion of interesting, picturesque, and even intriguing geographic settings is never far from our thoughts when we consider making a move

Ignoring local landscape

The Diversity Dilemma. Yet, despite our preoccupation with the land, development in the past 50 years has largely ignored much of the inherent geographic diversity in the North American landscape. We build land use systems with little regard for place-to-place differences in the scale and types of landforms, drainage systems, and ecosystems. Since 1960 or so our highways, shopping centers, and residential developments from coast to coast look and function as though they came from one set of cookie-cutter models. The huge bank of knowledge we have acquired from tens of thousands of studies on the North American landscape simply has not become part of the professional culture of planning and engineering.

Many community and environmental groups sense this when examining development proposals. They voice an uneasiness about the layouts they see despite the fact that planners have usually met all the local and regional environmental rules and planning regulations. What they sense is a lack of fit with the local landscape, an unexplainable awkwardness to modern land use plans.

Missing the mark

Why this state of affairs? Part of the answer is due to the precedent tradition in law and public policy that encourages communities to build environmental and land use regulations according to what has worked for other communities. In other words, most ordinances and bylaws, such as those covering stormwater management, road design, and land platting, are borrowed from other jurisdictions and not tailored to fit local landscapes. Part of the answer is also due to the belief among public officials that planning and design problems are solved by rules and regulations; therefore, developers are expected to go no further in their understanding of landscape than the regulations require. And, of course, part is also due to a genuine lack of knowledge about local and regional environment, its essential processes, systems, and features, and how it all fits together.

In this chapter we briefly survey the physiography of the United States and Canada. Our purpose is to gain some understanding of the different types of terrain, resources, and environmental conditions that make up the landscapes beyond our own theaters of operation. Although we are barely able to scratch the surface of this huge body of knowledge, we want at least to understand the macro framework and maybe learn enough to ask tough questions of ourselves and others. It is important in this context to realize how mobile we are as modern professionals and that much of our project work in landscape planning and design is carried out in other places, often far away from our home, office, or university.

2.2 THE PHYSIOGRAPHIC FRAMEWORK

Physiographic regions are large geographic entities defined by the composite patterns of the *features* that give a landscape its individual character: landforms, drainage, soils, climate, vegetation, and land use.

Landscape as systems

Beyond Forms and Features. We must also remember that the physiography of any region represents more than just its landscape features; it must also include a host of processes that operate at or near the earth's surface. These processes are arranged in various systems characterized by flows of matter driven by energy: drainage systems, climatic systems, mountain-building (geologic) systems, geomorphic systems, ecosystems, and land use systems that overlap and interact in different ways and at different rates.

The systems that govern the landscape and drive the processes have developed over various periods of time in North America, ranging from millions of years (in the case of the geologic systems that built the Appalachians and the Rockies) to only a few decades or centuries (in the case of the land use systems that built our settlements, farms, and highways). The physiographic patterns and features we see today represent an evolving picture—at this moment a mere slice of the terrestrial environment at the intersection of many time lines representing the effects of different forces and systems.

The Stage. Generally, regional geology provides a useful framework for describing the gross physiography of North America. These regions are well known: the Canadian Shield, the Appalachian Mountains, the Interior Highlands, the Coastal Plain, the Interior Plains, the Rocky Mountains, among others (Fig. 2.1*a*). The geologic structure of each region sets the drainage trends and patterns and in turn the general character of landforms. Added to this is the role of climate as it influences vegetation, soils, runoff, permafrost, and water resources (Fig. 2.1*b*).

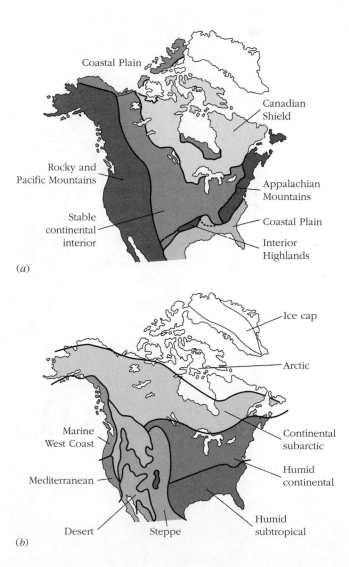

(*a*)

(*b*)

Fig. 2.1 (*a*) The major structural (geological) divisions and (*b*) the major climatic zones of the United States and Canada.

Regions and provinces

When we speak of a physiographic region, we are referring to a vast geographic entity defined by a particular combination of landforms, soils, water features, vegetation, and related resources. Ten major physiographic regions are traditionally defined for the United States and Canada (Fig. 2.2*a*), and they are broken down into smaller regions called **physiographic provinces** (Fig. 2.2*b* and Table 2.1). At the level of provinces, physiography begins to take on real meaning for many of the issues and problems in landscape planning. We begin with the Canadian Shield.

Fig. 2.2 (*a*) The physiographic regions of the United States and Canada and (*b*) their provinces classified as mountains, plateaus, or lowlands.

Arctic Coastal Plain
Alaska-Yukon Region
Pacific Mountain System
Rocky Mountain Region
Interior Plains
Canadian Shield
Hudson Bay
Intermontaine Region
Appalachian Region
Interior Highlands
Atlantic Coastal Plain

(*a*)

Brooks Range
Yukon Lowland
Arctic Coastal Plain
Arctic Lowlands
Baffin Upland
Arctic Lowland
Aleutian Islands
Alaska Range
Yukon Basin
Coast Mountains
Bear-Slave Churchill Uplands
Hudson Bay
Labrador Highlands
St. Lawrence Lowland
Canadian Rockies
Frazier Plateau
Great Plains
Hudson Platform
Puget Sound-Williamette Valley
Laurentian Highlands
Cascade Mountains
Superior Uplands
Coast Ranges
Northern Rockies
Northern Appalachians
Columbia Plateau
Middle Rockies
Appalachian Plateau
North Appalachians
Adirondacks
Sierra Nevada
Central Lowlands
Ridge and Valley
Basin and Range
Southern Rockies
Ozark Plateaus
Blue Ridge
Central Valley
Colorado Plateau
Ouachita Mts.
Piedmont
Outer Coastal Plain
Inner Coastal Plain
Baja California
Coastal Plain
Mississippi Embayment
Inner Coastal Plain

Mountains
Plateaus
Plains

(*b*)

Table 2.1 Physiographic Regions and Provinces of the United States and Canada

Region	Provinces
Canadian Shield	Superior Uplands
	Laurentian Highlands
	Labrador Highlands
	Hudson Platform
Appalachian Mountains	Blue Ridge
	Piedmont
	Ridge and Valley
	Appalachian Plateaus
	Northern Appalachians
Interior Highlands	Ozark Plateaus
	Ouachita Mountains
Atlantic and Gulf Coastal Plain	Outer Coastal Plain
	Inner Coastal Plain
	Mississippi Embayment
Interior Plains	Central Lowlands
	Great Plains
	St. Lawrence Lowlands
Rocky Mountain Region	Canadian Rockies
	Northern Rockies
	Middle Rockies
	Southern Rockies
Intermontane Region	Colorado Plateau
	Columbia Plateau
	Basin and Range
Pacific Mountain System	Alaska Range
	Coast Mountains
	Frazier Plateau
	Cascade Mountains
	Coast Ranges
	Sierra Nevada
	Central Lowlands
	Puget Sound–Willamette Lowlands
Alaska–Yukon Region	Brooks Range
	Yukon Basin
Arctic Coastal Plain	North Slope
	Mackenzie Delta
	Arctic Lowlands

2.3 THE CANADIAN SHIELD

The **Canadian Shield** is a large physiographic region in the northcentral part of the continent (Fig. 2.1). It is composed of the oldest rocks in North America (older than a billion years) and is the *geologic core* of the continent. Geologically, the Canadian Shield is extraordinarily complex, with intersecting belts of highly deformed crystalline (igneous and metamorphic) rocks throughout. These rocks have been subjected not to one or two, but to many ancient episodes of tectonic deformation and metamorphism. Thus most of the rocks are hard, tightly consolidated, and diverse in mineral

Geology

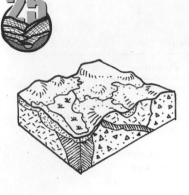

composition, including iron ore, nickel, silver, and gold. These minerals are the object of major mining operations, which is one of the principal economic activities of the Shield.

Bedrock and Glaciation. Most of the Canadian Shield has been *geologically stable* (relatively free of earthquakes and volcanic activity) for the past 500 million years or so. During that time, erosional forces have worn the rocks down to rough plateau surfaces, such as the Superior Uplands, the Algonquin Uplands, and the Labrador Highlands, which have the ruggedness of low mountains. In addition, large sections of the Shield lie at lower elevations and are covered with sedimentary rocks. The largest of these areas—though it does not belong to the Canadian Shield region per se—is the broad interior of the continent stretching from the U.S. Midwest through the Canadian Interior Plains northward to the Arctic Lowlands of northern Canada.

Surface forms   Over this vast area the Shield rocks are buried under a deep cover of sedimentary rocks, thousands of feet thick in most places, and hence have essentially no influence on the surface environment.

A significant recent chapter in the long and complex physiographic development of the Canadian Shield was the *glaciation* of North America. Great masses of glacial ice formed in the central and eastern parts of the Shield and, in at least four different episodes in the past 1 to 2 million years, spread over all or most of the region. From the interior, the ice sheets advanced in a radial pattern, scouring the Shield's surface, *Glaciation*   removing soil cover, and rasping basins into the less resistant rocks. The last ice sheet, the Wisconsin Glaciation, which covered the whole region 18,000 years ago, melted from the Shield's interior only 6000 to 8000 years ago when the continent was inhabited by the first North Americans.

On the fringe of the Shield, the ice scoured large freshwater basins, called *shield-margin lakes*. On the south are the Great Lakes, and along the western margin are lakes Manitoba, Athabasca, Great Slave, and Great Bear. In the center of the Shield, *Lakes*   marked by the Hudson Bay region, the great mass of ice depressed the earth's crust below sea level. As the ice melted back, Hudson Bay took shape. But for the past several thousand years the Bay has been shrinking as the underlying crust rebounds from glacial unloading and pushes the water back. This process accounts for the vast, flat coastal plain that is slowly growing on the Bay's southern margin.

Over much of the Shield glaciation left the land with an irregular and generally light soil cover, interspersed with low areas occupied by lakes and wetlands. When we combine this characteristic with the already diverse surface geology, we can easily see why the Canadian Shield is one of the most *complex landscapes* in North America. From the standpoint of environmental planning and engineering, the Shield *Landscape diversity*   presents few easily definable patterns and trends in landforms, drainage, and soils. Considerable variation can be expected even at a local scale of observation; therefore, careful field work is called for in virtually all planning and engineering problems.

Climate and Vegetation. The Canadian Shield lies principally in the *continental subarctic climate* zone. It is marked by six to seven months of freezing temperatures, *Permafrost*   fiercely cold midwinters, and short cool summers. A great proportion of the Shield north of James Bay (at the southern end of Hudson Bay) is occupied by *permafrost*. Most permafrost is *discontinuous*, meaning that its distribution is irregular, usually occurring in the form of patches of different sizes and depths. The largest patches are generally found in the northern zones of the Shield and/or in areas favored by cool microclimates and topography and soils conducive to cold ground (see Chapter 18 for details on permafrost).

The vegetative cover is predominantly *boreal forest* except in the extreme north, where it is tundra. Boreal forest is the least biologically diversified of the earth's major forest ecosystems, containing only a small fraction of the number of species in the *Vegetation and soils*   subtropical forests 2000 miles to the south on the Atlantic Coastal Plain. In addition, it is of the smallest stature and the least productive of the world's major forest systems.

The *annual organic productivity* of the boreal forest averages 800 grams per square meter of ground, and much of that is provided by the ground cover such as mosses, sedges, and various shrubs.

The boreal forest is dominated by white spruce, birch, aspen, and poplar in upland areas and by white spruce, black spruce, and tamarack (larch) in the lowlands (Fig. 2.3). Trees are very slow growing and often appear in dwarf forms in wet areas and on the tundra fringe. *Bogs* are found throughout lowland areas and on lake and stream margins in upland areas. They are characterized by zonal vegetation patterns grading from small trees on the edges to sedges and mosses toward the center. Bogs and other wetlands are dominated by organic soils (*Histosols*), whereas uplands are dominated by *Podsols, Brunisols,* and *Cryosols* (Canadian classification) or *Spodosols* (American classification).

Drainage and Settlement. Throughout much of the Shield are extensive outcrops of resistant rock, which exert a strong influence on drainage. These outcrops *Deranged drainage* vary in shape and size and in places impart to the terrain an almost chaotic fabric. Not surprisingly, drainage patterns are very irregular—so much so that the term *deranged drainage* is used to describe them (Fig. 2.4). Stream systems are interrupted by topographic depressions and bedrock barriers that slow the flow in some areas and hasten it in others. Lakes, streams, and wetlands are abundant, and freshwater is clearly one of the Shield's greatest resources, a fact that has not gone unnoticed by the Canadian government, which has promoted the development of hydroelectric dams over much of the southern part of the region.

Settlement on the Canadian Shield is light, and land use is sparse throughout most of the region, including the two sections that lie within the United States: the Adirondack Mountains and the Superior Uplands. Where settlements are found, they are usually related to an extractive economic activity, typically mining, forestry, *Land use* or fishing. It is, however, an alluring recreational landscape, and it has attracted the development of parks, resorts, and summer homes along the Shield's southern

Fig. 2.3 Boreal forest of the Canadian Shield, the least disturbed of the world's great forests.

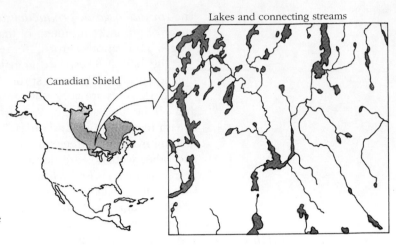

Fig. 2.4 An example of the complex character of the deranged drainage of the Canadian Shield.

margin in Ontario and Quebec. Herein lies a steadily growing challenge for planners and designers. Among other things, it requires understanding the Shield's unique physiography, the need for careful site analysis, and the use of terrain adaptive approaches in designing sanitary, water supply, road, and energy systems.

2.4 THE APPALACHIAN REGION

The **Appalachian Region** is an old mountain terrain that formed on the southeastern side of the Canadian Shield several hundred million years ago (Fig. 2.2). Although the Appalachians date from around the same time as the Rocky Mountains, they developed appreciably different terrains. The Appalachians are lower, less angular, and less active geologically. Whether these differences have always existed or are due to a long period of inactivity in the Appalachians during which erosional forces have worn them down,

Landforms   we cannot say. In any case, the Appalachians are characterized by *rounded landforms* that are generally forest-covered in their natural state and rarely exceed 6000 feet in elevation. Unlike the Canadian Shield, however, distinct topographic trends and patterns are discernible in the Appalachian Region, making it somewhat easier to deduce general physiographic conditions at the regional planning scale (Fig. 2.5).

    The Blue Ridge Province. The Appalachian Region stretches from northern Alabama in the American South to Newfoundland in the Canadian Maritime provinces. It is subdivided into five provinces on the basis of landforms (Table 2.1). The highest

Main axis   and smallest province (in area) is the **Blue Ridge**, which stretches in a narrow band from New York to Georgia. It is composed of folded metamorphic rocks, that is, rocks hardened from the heat and pressure of mountain building, and strongly linear mountain forms. As the backbone of the Appalachians, the Blue Ridge forms the *principal drainage divide* of the Appalachian region, marking the eastern rim of the Mississippi Basin.

    Because of its high elevation (4000 to 6000 feet), climate over much of the Blue Ridge province is distinctly wetter and cooler than lands to the east and west. This is especially pronounced in its highest reaches, the Great Smoky Mountains of North Carolina, Tennessee, and Georgia, where there is a marked *orographic effect* on the western slopes that yields 80 inches or more annual precipitation. In addition, at higher elevations the cooler climate favors a more northern-like forest association made up of hardwoods (birch, maple, beech, elm, and oak) mixed with conifers (hemlock, white pine, spruce, and fir). In some places the mountain summits are barren (and are called *balds*), not because the climate is too cold for trees, but more because they are covered with rock rubble and the soil is sparse. The province can claim only one city, Asheville, North Carolina, at an elevation around 2000 feet

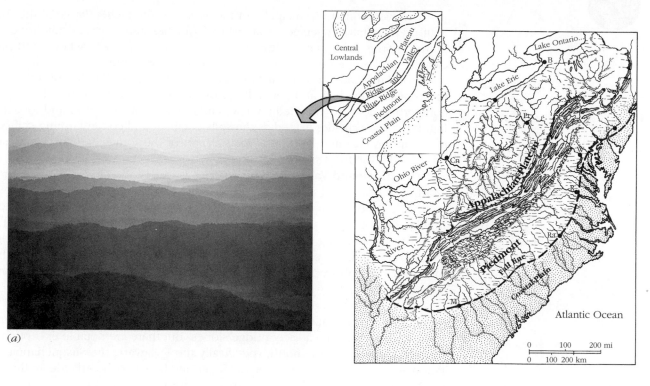

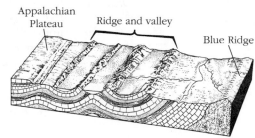

Fig. 2.5 (*a*) The Great Smoky Mountains of the Blue Ridge, looking east from Grandfather Mountain in North Carolina. (*b*) The Appalachian Region of eastern United States including the Blue Ridge, the Piedmont, the Ridge and Valley and the Appalachian Plateau. In the southeast lies the Atlantic Coastal Plain. The lower diagram shows the relationship among the Blue Ridge, Ridge and Valley, and Appalachian Plateau provinces.

Physiography

(600 meters). State and federal parks are common to the Blue Ridge, including the popular Great Smoky Mountain National Park in North Carolina and Tennessee.

The Piedmont Province. East of the Blue Ridge is the **Piedmont Province** of the Appalachians (Fig. 2.5). This province is composed mainly of metamorphic rocks covered by a soil mantle of variable thickness. Although large parts of the Piedmont are fairly level, the terrain is best characterized as a hilly plateau, especially in the south. From the Blue Ridge, the Piedmont slopes gradually eastward until it disappears under the sedimentary rocks of the Coastal Plain, along its eastern and southeastern border. *Drainage* follows this incline toward the Atlantic.

The valleys of large streams, such as the Savannah and the Peedee, are marked by major forest corridors that run all the way to the sea. Stream and *river gradients* are relatively steep on the Piedmont, but where they cross onto the Coastal Plain about midway between the Blue Ridge and the coast, they decline somewhat and the flow becomes less irregular. For early settlers and traders moving up rivers from the Atlantic, the first fast water (rapids) they would encounter started with the eastern edge of the Piedmont. The eastern border of the Piedmont became known as the *Fall Line*, and on some rivers it became a place of settlement. Richmond, Virginia; Raleigh, North Carolina; and Macon, Georgia are Fall Line cities (Fig. 2.5).

Climate and soils

The Piedmont lies principally in the *humid subtropical climate* of the American South. Precipitation averages from 50 to 60 inches annually, and intensive summer thunderstorms are common. Runoff rates are high, especially where land has been cleared, and *soil erosion*, which was once very widespread, can be locally severe today on farmland and construction sites. Soils belong to the *Ultisol* order and tend to be heavily leached and generally poor in nutrients. Early in the development of Southern agriculture, the Piedmont was a favorite area for cotton and tobacco farming, but these activities waned as the soils declined and cotton farming shifted to the Mississippi Valley and later into Texas. Much of the former farmland now supports mixed oak-pine forests.

The Ridge and Valley Province. This is one of the most distinctive terrains in North America. It lies west of the Blue Ridge, stretching from middle Pennsylvania to middle Alabama in a belt generally 50 to 75 miles wide. The Ridge and Valley Province topography is controlled by folded and faulted sedimentary rocks that have

Landforms

been eroded into long ridges separated by equally long valleys (Fig. 2.5). Ridges and valleys run parallel to each other and are more or less continuous for several hundred miles, broken only occasionally by stream valleys, called *water gaps*, or dry notches, called *wind gaps.*

Drainage lines generally follow the trend of the landforms, with trunk streams flowing along the valley floors, forming *parallel* or *subparallel patterns*, and their tributaries draining the adjacent ridge slopes. But there are notable exceptions. Some of the large rivers in the north, specifically the Delaware, the Susquehanna, and the Potomac, drain across the grain of the ridges and flow into the Atlantic. In the southern

Drainage

part of the Ridge and Valley, most streams drain into the Cumberland and Tennessee rivers, which are part of the Mississippi System. Settlements and farms in the Ridge and Valley are concentrated in the valleys, where soil covers are heavy and water is abundant. The upper ridges have traditionally been left mostly in forest, being too steep for much else. However, modern residential and recreation development, which is attracted to the forests, rugged terrain, and excellent vistas, has pushed its way onto some Appalachian Ridges with varying degrees of environmental disturbance related to road building, forest clearing, and slope excavation.

Appalachian Plateaus. West of the Ridge and Valley Province is a section of elevated sedimentary rocks into which rivers have cut deep valleys. This province is referred to as the **Appalachian Plateaus** (the *Allegheny Plateau* in the north and the

Landforms

Cumberland Plateau in the south), and it extends from western New York to northern Alabama. The plateau surface dips westward from a high elevation of 3000 feet or more along its eastern border. This border is marked by a prominent escarpment called the *Allegheny Front*, which rises 500 to 1000 feet above the adjacent valley and extends the full length of the plateaus, about 700 miles.

The sedimentary rock formations of the Appalachian Plateaus are flat-lying for the most part. Stream valleys are incised several hundred feet into these rocks, leaving flat-topped uplands between valleys. Where the density of stream valleys is high, however, the uplands are narrow, more ridgelike, and the plateaus' topography is steeply hilly or almost mountainous, especially in the higher sections. Sandwiched

Coal measures

among the limestone, shale, and sandstone strata within the upland areas are extensive coal deposits, generally considered to be the largest single field of bituminous reserves in the world.

Where the coal deposits lie close to the surface, the favored extraction method is strip mining (Fig. 2.6). Where the coal is too deep for strip mining, shaft mining is used, in which a horizontal tunnel is dug into the exposed end of the formation along a valley wall. Strip mining has produced serious environmental problems in the United States. Not only are topsoil, vegetation, and habitat destroyed, but also the spoils (coal debris and subsoil) are subject to erosion and weathering, resulting in

Fig. 2.6 The effects of strip mining in the Appalachian Plateaus, a land that was originally forested. Today mining companies are required to restore damaged landscape, but extensive areas of badly damaged land from former operations remain.

sedimentation and chemical pollution of streams as well as unsightly and unproductive landscapes.

The **Northern Appalachians**, north of New York, are made up of several separate mountain ranges composed principally of crystalline rocks. The most prominent of these ranges are the *Green Mountains* and *White Mountains* of Vermont, New Hampshire, and Maine; the *Notre Dame Mountains* of the Gaspe Peninsula of Quebec; and the *Long Range Mountains* of northern Newfoundland (Fig. 2.7). The Northern Appalachians are *geologically complex* and made up principally of metamorphic rocks similar to those of the Blue Ridge and Piedmont provinces. (The Adirondack Mountains of northern New York, despite their geographic location within the Northern Appalachian Region, are a southern extension of the Canadian Shield.)

Landforms

Unlike the Blue Ridge and Piedmont provinces, the Northern Appalachians were glaciated, leaving abundant rock exposures and an irregular cover of glacial deposits. These characteristics, combined with extensive northern conifer forests and abundant lakes and wetlands, give much of the Northern Appalachian landscape a character similar to the uplands of the Canadian Shield. The lakes have been the victim of the acid rain phenomenon attributed mainly to the sulfur dioxide emissions from power plants and industrial sources fed by coal from the Appalachian Plateaus. In addition, this section of the Appalachians borders on the Atlantic Ocean, producing an especially rugged coastline of rocky headlands and deep embayments.

Drainage

The comparison to the Canadian Shield extends to more than the general character of the landscape. In addition, the climate is noted for its long, cold winters, especially in the interior from Maine to Newfoundland (where there are fewer than 120 frost-free days a year), and settlement is sparse with a strong orientation toward lumbering, fishing, and tourism. Moreover, the fauna of the province, highlighted by moose, beavers, Canada geese, lynx, and black bears, are more like those of the Shield than the Appalachians to the south.

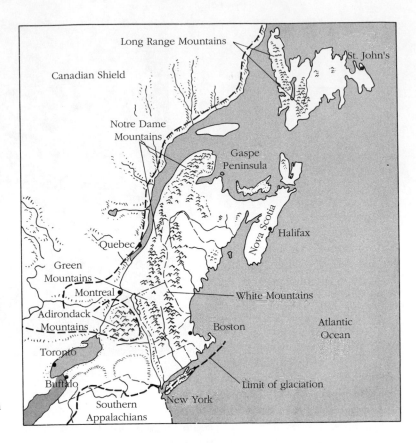

Fig. 2.7 The Northern Appalachians stretching from Massachusetts to Newfoundland.

2.5 THE INTERIOR HIGHLANDS

West of the Cumberland Plateau, in southern Missouri and northern Arkansas, lies a small region of low mountainous or plateau terrain that closely resembles the Appalachians. This region is called the **Interior Highlands**, and it is made up of two main provinces: the *Ozark Plateaus*, whose landforms are similar to the Appalachian *Landforms* Plateaus and the *Ouachita Mountains*, whose landforms are very similar to those of the Ridge and Valley section except that the grain of the terrain is east–west trending (Fig. 2.2). The highest elevations in the **Ozarks**, as this area of plateaus is commonly called, lie between 2000 and 3000 feet; those in the Ouachita Mountains lie between 3000 and 4000 feet. Most ridges in the Ouachita Mountains are forested and too steep for settlement; in the Ozarks, however, ridges are often flat-topped and cleared for farming. Ozark stream valleys, on the other hand, are relatively deep and steep sided and, like those in the Appalachian Plateaus, are usually forested. Unlike the Appalachian Plateaus, however, the Ozarks are not underlain by extensive coal deposits.

Drainage in the Ozarks follows a radial pattern with streams on the north draining to the Missouri, those on the west draining to the Arkansas River, and those on the south draining to the Black River. Limestone is the predominant bedrock throughout the Ozarks, and it has given way in many areas to *karst* (dissolved rock) topography *Drainage* dominated by cavern systems. These caverns are tied to stream valleys, many of which have become attractive places for large reservoirs. This opportunity has not escaped the federal government, and today both the Ozarks and the Appalachian Plateaus are laced with reservoirs where forested stream valleys once existed.

2.6 THE ATLANTIC AND GULF COASTAL PLAIN

Geology

The **Atlantic and Gulf Coastal Plain** forms a broad belt along the U.S. Eastern Seaboard and the Gulf of Mexico from Cape Cod on the north to the mouth of the Rio Grande in the south (Fig. 2.2). This region is composed entirely of sedimentary rock formations that dip gently seaward, disappearing at depth under the shallow waters of the continental shelf. Where resistant sedimentary rock formations such as sandstone outcrops occur within the Coastal Plain, they form belts of hilly topography with soils that are compositionally similar to the bedrock (Fig. 2.8). Conversely, weak formations, such as chalky limestones, form lowlands and deeper, richer soils. Overall, however, the *topographic relief* of the Coastal Plain is very modest, and the highest elevations [mainly along the inner (landward) edge] reach only 300 feet or so above sea level. The Coastal Plain is generally divided into three provinces: the *Inner Coastal Plain*, the *Outer Coastal Plain*, and the *Mississippi Embayment* (Fig. 2.9).

Landforms

The Outer Coastal Plain. Bordering the sea, in the province called the **Outer Coastal Plain**, the land is generally low, wet, and heavy with stream, wetland, and shoreline sediments. Some of the largest *wetlands* in North America, such as the Everglades in Florida, the Great Dismal Swamp of Virginia and North Carolina, and the delta swamps of Louisiana, are found here. In addition to extensive swamps, lagoons, estuaries, and islands are abundant. These are subject to frequent incursions by floodwaters from streams, storm waves, and hurricane surges. Offshore islands, called *barrier islands*, which are mostly accumulations of sand from shallow water deposition by waves, are especially prone to damage from hurricanes and shore erosion. However, this has not deterred development. Each year residential and recreational land uses push farther onto the barrier islands throughout the Atlantic and Gulf coasts.

Principal sections

The Outer Coastal Plain can be subdivided into at least five sections (Fig. 2.9). In the *northern section*, from Cape Hatteras north to Cape Cod, extensive lowland areas are drowned, the result of a rising sea level since the last continental glaciation. The principal features of this section are the large, ecologically rich *estuaries*, such as Chesapeake Bay, which formed as the ocean flooded the lower reaches of river valleys. The next section south, the *cape-island section*, stretches from Cape Hatteras to northern Florida and is characterized by long sandy beaches (e.g., Cape Fear) and by

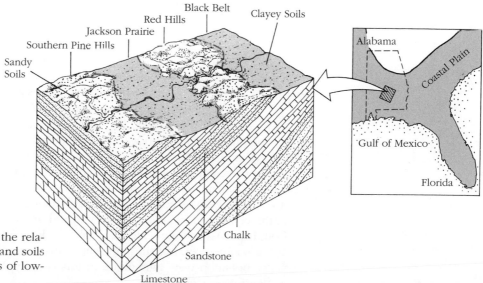

Fig. 2.8 A block diagram showing the relationship among geology, landforms, and soils in the Coastal Plain resulting in belts of lowlands and intervening hills.

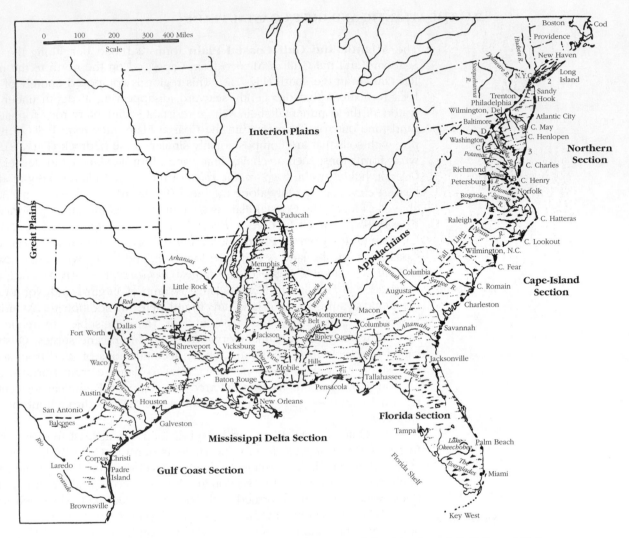

Fig. 2.9 The principal features and sections of the Atlantic and Gulf Coastal Plain.

offshore islands such as those described earlier. The most impressive of these islands are the Sea Islands, which apparently formed as the land in this area slowly subsided and sea level rose over the past several thousand years.

The *Florida Peninsula section* owes its origin to a great arch in the sedimentary rocks of the Coastal Plain. The arch is composed of limestone, and its axis, which reaches about 300 feet above sea level, extends roughly down the center of the penin-sula. The combination of Florida's wet climate and limestone bedrock has resulted in extensive karst topography. Much of the drainage is subterranean (groundwater is *Florida section* the state's main water supply) via *cavern systems* that discharge into springs and sink-holes. Sinkholes are common throughout the peninsula, but they are the most prominent in the center of Florida where they form hundreds of large inland lakes. The outflow from Lake Okeechobee, Florida's largest lake (750 square miles), feeds the *Everglades* at the southern end of the peninsula. Diversion of this flow by canals, coupled with water pollution from agriculture in this century, has degraded this nationally valued wetland ecosystem.

Westward from the Florida Peninsula is the *Gulf Coast section*. Fed with massive amounts of river sediments, principally from the Mississippi, the Gulf Coast is conducive to barrier island formation. These low, narrow *longshore islands* parallel most of the coast and behind them are equally long lagoons with connecting wetlands, bays, and related habitats. These areas abound with waterfowl and other birds, including those that migrate between the Gulf Coast (and regions farther south) and the Arctic coastal plains in Canada.

The Mississippi Section. The *Mississippi River Delta* lies roughly in the middle of the Gulf Coast section. One of the most diverse environments on the continent, the *outer delta* is actually a composite of several smaller deltas built mainly in the past 2000 years. Each finger of the delta contains many distributary channels (or bayous)

The Great Delta

bordered by natural levees, behind which lie swamps, lagoons, and various types of lakes. The deep swamps support trees such as cypress and tupelo, whereas the shallow ones support oaks, hickory, other hardwoods, as well as various reeds, sedges, and palmetto. Not surprisingly, *habitat diversity* is extraordinary in this waterland of fresh, brackish, and saltwater in wetland, lake, and stream habitats. Because of changes in the Mississippi's water quality, reductions in its sediment load, and wetland eradication for land use activity, however, the delta is now showing serious signs of habitat degradation. Over the past half century or more, it has lost massive areas of wetland and is now losing 20 square miles of wetland a year. This shrinkage, coupled with losses to shore erosion, has reduced the delta's resistance to hurricane overwash such as that produced by Hurricane Katrina in 2005.

The Mississippi River brings more than a million tons of sediment to its delta every day. This process has been going on for more than 50 million years and has created a long, narrow alluvial lowland called the **Mississippi Embayment**, which begins

Illinois southward

at the southern tip of Illinois and extends 500 miles south to the modern delta (Fig. 2.10). Virtually all streams entering the Embayment as tributaries of the Mississippi are diverted southward by the great river so that they parallel the Mississippi in their lower courses. Several, including the Black, Yazoo, and Ouachita rivers, follow the outer edge of the Embayment for more than 100 miles before joining the Mississippi. The result is a valley unique among those in North America in that it is made up of a system of several large river channels occupying a single valley. During large floods, the Mississippi often turns floodwater back into the tributaries, which are actually slightly lower than the Mississippi itself where they parallel the main channel.

The Inner Coastal Plain. On the **Inner Coastal Plain**, the land is generally higher and better drained than along the coast and in the Mississippi Embayment. There are two opposing trends to the topography here. *First*, systems of hills and valleys parallel the coast that have formed along the outcropping edges of different rock

Landforms and drainages

formations. The lowlands are often marked by belts of deep, dark soils, such as the Black Belt of Alabama, and the uplands, such as the Southern Pine Hills, are characterized by sandy soils (Fig. 2.8). *Second*, intersecting these bedrock-controlled landforms are ribbons of lowlands formed by the valleys of streams draining to the coast. Large lowlands are found along all the major river valleys, and most contain large areas of wetland and forests where they have not been drained and cleared for agriculture. West of the Mississippi River, the inner border of the Coastal Plain is clearly marked by the southern fringe of the Interior Highlands and more abruptly by the Balcones Escarpment in Texas.

Coastal Plain Summary. Flooding is frequent, widespread, and severe in all large river lowlands of the Coastal Plain. The damage it wreaks on property and life is

Flooding

enormous and has increased steadily throughout the twentieth century. The causes of this problem are manifold and include: (1) *high runoff* rates under the rainy climate of the South; (2) frequent tropical *storms and hurricanes* in coastal areas, which force heavy runoff and streams to back up in the Outer Coastal Plain; (3) extensive and growing *residential development* in high-risk zones along river lowlands and coastal

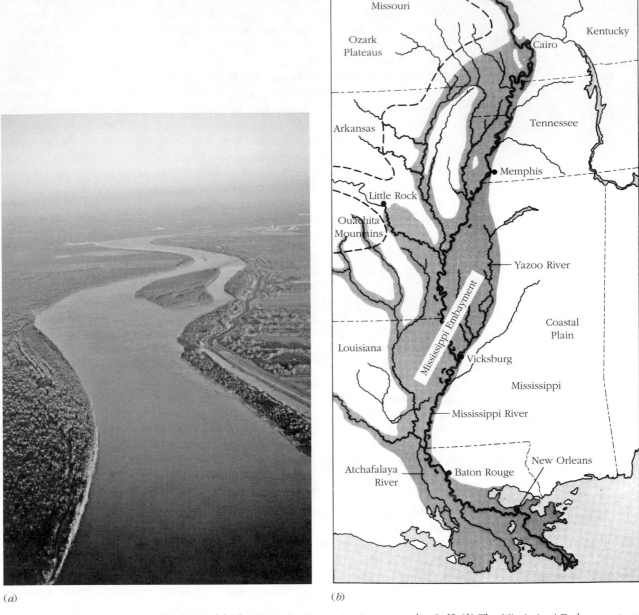

(a)   (b)

Fig. 2.10 (*a*) The Mississippi coursing its way to the Gulf. (*b*) The Mississippi Embayment and connecting lowlands. Notice the paralleling tributaries such as the Yazoo River.

areas; and (4) *engineering (structural) changes* such as the construction of levees in river valleys, which can constrict flows and increase the levels of floodwaters in certain locations (Fig. 2.11).

The Coastal Plain lies almost entirely within the *humid subtropical climatic zone. Precipitation* generally ranges from 40 to 60 inches a year, with most occurring in the summer from thunderstorms. All the Coastal Plain is subject to hurricanes, and each section of the Outer Coastal Plain can expect at least one highly destructive event per decade. *Soils* tend to vary from sandy to clayey to organic depending on the underlying bedrock, the nearness to the coasts, and river valley location. Most soils tend to be

Fig. 2.11 A constructed levee designed to help control flooding along the Mississippi River in southern Louisiana. The floodplain on the right is frequently below river level.

heavily leached, especially in nonwetland locations, and are reddish in color from iron oxide residues below the topsoil. *Vegetation* patterns tend to follow the patterns of topography and drainage, with pines and oaks generally on the higher ground and tupelo, gum, and bald cypress in the wet lowlands. Only in Texas is the vegetation significantly different, with forest grading into parkland and then into desert shrub from east to west.

Climate and vegetation

Planning Realities. In dealing with the problems in the Coastal Plain, it is critical that planners first differentiate between upland and lowland terrain. The differences are often subtle, especially on the edges, but vegetation and soil patterns can be helpful as indicators. Next, planners need to delineate units of upland ground and define their relationship to drainage systems, including channels, wetlands, floodplains, and habitats. This should include an assessment of the relative stability of different types of land related to soils, karst formations, and flooding from various sources.

Drainage first

Finally, it is important to define the essential processes and systems that operate in various environments, particularly in the river lowlands and coastal areas, and to understand the risks posed by each (mainly floods, storm surges, and hurricanes). Inattention to these risks—bordering on blatant disregard by real estate interests, energy developers, and communities bent on building tax bases—has set the stage not only for massive disasters like Katrina in 2005 and Ike in 2008, but also for widespread property damage of an incremental nature by many smaller, yet highly powerful, storm and flood events that happen year in and year out. The record of seemingly ongoing setbacks calls into question the effectiveness of land use planning efforts in this region. It also points up the need, in the face of rising sea levels and increasing storm sizes and numbers, to advance regional planning with renewed vigor and, in light of the environmental unknowns facing this region, to make resilience a mainstay of land use planning.

Disaster risk

2.7 THE INTERIOR PLAINS

The heart of the North American landscape is a broad region of rolling terrain called the **Interior Plains** (Fig. 2.2). This region is made up of two large provinces, the *Central Lowlands* and the *Great Plains*, and one small province, the *St. Lawrence Lowlands*.

Provinces

Roughly defined, the region has the shape of a great triangle with its base along the Rocky Mountains and its apex to the east pointing up the St. Lawrence Lowland. The western and northeastern borders are sharply defined by the Rocky Mountain Front and Canadian Shield, respectively. The southern border varies from a neat edge, such as along the Balcones Escarpment in Texas, to a more transitional boundary along the Appalachian plateaus.

Bedrock and Surface Deposits. All three provinces are underlain with *sedimentary rocks* composed principally of limestone, sandstone, and shale. These formations are flat-lying or subtly bent into broad arches and basins. Michigan, for example, is at the center of a basin, whereas central Tennessee and Kentucky are situated across an arch. Bedrock exposures appear only in escarpments along the edges of basins and arches, in the walls of large stream valleys, and in areas of karst topography. But these exposures are small compared to the area covered by surface deposits.

Surface deposits

The surface deposits of the region are diverse in origin, composition, and thickness. In the Central Lowlands, north of the valleys of the Ohio and Missouri rivers, the deposits are mainly *glacial* and vary from sandy to clayey to mixed materials. In Illinois, Iowa, northern Missouri, Kansas, and Nebraska, deposits of rich windblown silt, called *loess*, cover the glacial deposits and are the dominant surface material (Fig. 2.12). Much of the agricultural wealth of the Central Lowlands and Great Plains can be attributed to the fertile loess-based soils. Other deposits of the Interior Plains include sand dunes in northwestern Nebraska, southwestern Saskatchewan, and around the Great Lakes, clayey lake beds near the Great Lakes and Lake Winnipeg, and river deposits (alluvium) along most stream valleys throughout the region. These deposits, mainly glacial drift, loess, and alluvium, provide most of the parent materials for the fertile agricultural soils of the Interior Plains.

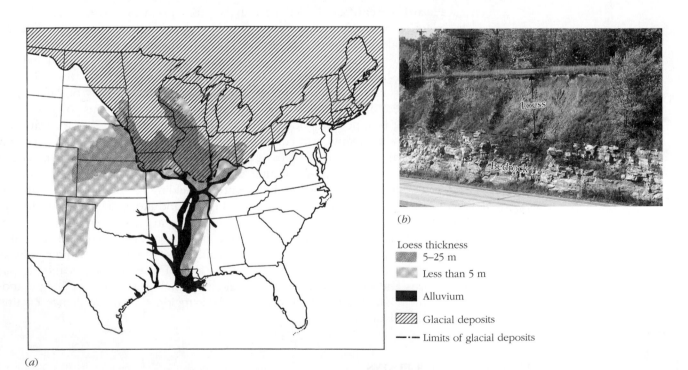

(*a*)

(*b*)

Loess thickness

5–25 m

Less than 5 m

Alluvium

Glacial deposits

Limits of glacial deposits

Fig. 2.12 (*a*) The surface deposits of the Interior Plains. Glacial drift is found north of the Missouri and Ohio rivers. Loess covers the drift in the central Midwest and extends over the adjacent section of the Great Plains. (*b*) Loess deposits (on top of the bedrock) exposed in a road cut in Missouri.

Water Resources. The Interior Plains are drained by three major watersheds: (1) the *Mississippi* and its tributaries upstream from the head of the Mississippi Embayment; (2) the *St. Lawrence*, which drains the Great Lakes Basin; and (3) the *Nelson, Churchill,* and *Mackenzie* drainage basins, which drain the Canadian Plains and small portions of North Dakota and Minnesota. Water is generally abundant in the Central Lowlands, especially the Great Lakes area, but it declines westward into the Great Plains as mean annual precipitation falls and evaporation rates rise. From the Ohio River to the Rocky Mountain Front, the mean annual precipitation drops from more than 40 inches to about 20 inches. At the 100th meridian (which is roughly in the middle of the Great Plains) the area takes on a *semiarid* climate, which dominates the Great Plains from central Texas to southern Alberta and Saskatchewan.

Major watersheds

Surface water is locally plentiful in the many stream valleys that cross the Great Plains. Streams such as the Saskatchewan, Yellowstone, Platte, Arkansas, and many similar ones that follow the eastward slope of the plains, rise and fall with seasonal runoff from the Rockies. In some valleys such as the Missouri and Platte, systems of dams have been constructed for flood control and agricultural water supply. In addition, groundwater is abundant throughout much of the Great Plains; one aquifer, the *Ogallala,* which is one of the largest in the world, stretches from South Dakota to northern Texas. Not surprisingly, agriculture in the Great Plains is overwhelmingly dependent on irrigation and is growing more so each decade.

Water supply

Climate and Soils. There is also a marked *north–south climatic gradient* in the Great Plains related to temperature and evapotranspiration rates. At the southern end of the Great Plains, in the area of the Llano Estacado and Edwards Plateau, lake and reservoir evaporation rates exceed 5 feet a year and the climate approaches the dryness of a true desert. Northward, evaporation declines with temperature and about 300 miles north of the U.S.–Canada border, scattered trees (mainly aspens) appear on upland surfaces. A little farther north, the plains are covered by boreal forest. Here the climate is subarctic, and although annual precipitation is no greater than farther south, evapotranspiration rates are much lower and the moisture balance is favorable. At the very northern end of the Great Plains Province, more than 2500 miles from the Edwards Plateau of Texas, the boreal forest gives way to tundra and we enter the Arctic Coastal Plain (Fig. 2.2).

Soils in the Central Lowlands are extremely diverse, especially in the upper Mississippi Valley, the Great Lakes region, and the St. Lawrence Lowland, owing to the extraordinary mix of glacial deposits left during the last glaciation 10,000 to 20,000 years ago. Environmental surveys for planning must recognize this diversity even at the local scale and also that, in some deposits, soil materials change substantially with depth. Beyond the local variations related to different deposits (parent materials), soils also show a broad, regional trend related to the east–west bioclimatic gradient. Figure 2.13 illustrates these regional trends with the 100th meridian marking the change from humid region soils (mainly *Alfisols* and *Spodosols*) to dry region soils represented by *Mollisols* (or *Chernozems* in Canada). *Mollisols* are dark, organic rich soils that form under grass covers in the semiarid climatic zone.

Spatial variation

Landscape Character. In the northern half of the Central Lowlands, lakes and wetlands are abundant and soils tend to be sandy in the newer glacial terrain. This province is second only to the Canadian Shield for its huge population of inland lakes. The landscape is generally diverse with mixed conifer–hardwood forests among the lakes and wetlands, but westward this diversity gives way to a more uniform landscape, the *Prairies*, and farther west, to the *Great Plains* (Fig. 2.13).

In the prairies and plains, the terrain tends to fall into two physiographic classes: river lowlands (for example, the valleys and floodplains of the Illinois, Mississippi, Missouri, and Iowa rivers), and broad, level, or gently rolling uplands between the river valleys, which make up the bulk of the landscape. The valleys support the principal tree covers, typically great corridors of willows, cottonwoods, sycamore, and other

Physiography

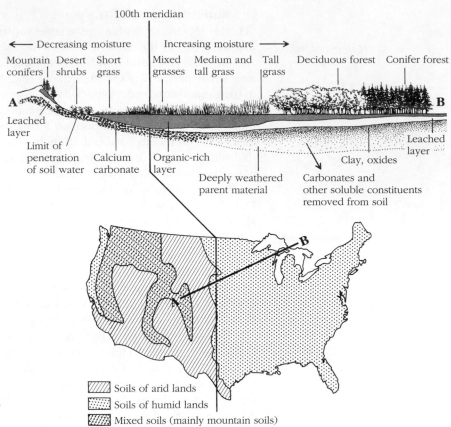

Fig. 2.13 Landscape change across the Interior Plains showing the influence of climate and vegetation on soil.

riparian species. The uplands originally supported *parkland* (mixed trees and tall grasses) that reached as far east as western Indiana, and *prairies* (tall, medium, and short grasses) that extended to the Rocky Mountains.

Little if any of the original prairie remains. Virtually all of it has been plowed and/or grazed and either destroyed or degraded. Today the region supports a massive agricultural economy dependent mainly on corn and wheat. The rural landscape that once contained great habitat corridors of grassland between linear corridors of woodland, now tends to be highly fragmented by farms, roads, and settlements. Both the Great Plains and the Central Lowlands are among the richest agricultural regions in the world, but in the past 50 years the Central Lowlands has also become a region of massive urbanization.

Planning Issues. Superimposed over a large part of this North American heartland are major urban/industrial corridors, such as the one linking Toronto, Buffalo, Cleveland, Toledo, and Detroit. The growth of these corridors in the past several decades has resulted in many serious environmental problems, including (1) conflicts *Corridors of conflict* between urban and agricultural land uses; (2) misuses and the eradication of floodplains, wetlands, and shorelands; and (3) widespread air and water pollution, including threatened groundwater resources from buried wastes and agricultural residues. The Central Lowlands and the St. Lawrence Lowlands are major sources of acid rain, and water pollution from both point (concentrated outfalls) and nonpoint (geographically diffused) sources is critical in both urban and agricultural areas.

Of major concern to environmental planning and management in the Interior Plains are the problems arising at the interface between the expanding urban corridors *Damage and restoration* and the rural landscape. As in the Eastern Seaboard, few planning problems in the Central and St. Lawrence lowlands involve primeval landscapes (virgin or near virgin lands). Rather, most involve landscapes that have been subjected to several previous land uses, usually some combination of lumbering, early farming, modern farming, residential, and/or industrial development. Almost invariably, these used landscapes have suffered environmental damage: wetland drainage, stream channelization, habitat loss and fragmentation, and waste burials. Therefore, modern environmental planning in these provinces increasingly involves landscape repair and restoration as a part of community and economic development projects.

2.8 THE ROCKY MOUNTAIN REGION

Provinces

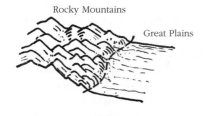

The western border of the Great Plains is formed by the **Rocky Mountain Front**, one of the most distinct physiographic borders in North America. From the western Great Plains (at an elevation of 3500 to 5500 feet) the terrain abruptly rises 3000 to 6000 feet, and we enter the rugged Rocky Mountain Region. This region stretches about 2000 miles from central New Mexico to near the northern border of British Columbia. It is made up of four provinces: the *Southern, Middle,* and *Northern provinces* in the United States and the *Canadian Province* in Canada (Fig. 2.2). The Canadian Rockies form a long, narrow belt 50 to 100 miles wide that extends 500 miles northward from the Canadian–U.S. border. The American provinces, by contrast, are wider and far more irregular geographically.

Landforms and Drainage. The Rocky Mountains are considered to be relatively young mountains because they are still active, though many of the rocks are of ages comparable to those in the Appalachians. Geologically, they are the most diverse in the American provinces, where volcanic, metamorphic, and sedimentary rocks are all very prominent. The highest terrain in the Rockies, which is around 13,000 to 14,000 feet (4000 to 4300 meters) elevation, is found in Colorado, Wyoming, and along the British Columbia–Alberta border. The largest areas of low terrain are the *Wyoming Basin* (5000–7500 feet elevation), which is underlain by sedimentary rocks and lies between the Northern and Southern Rocky Mountain provinces, and the *Rocky Mountain Trench*, a remarkably long, narrow valley stretching along the western edge of the Canadian Province in southeastern British Columbia (Fig. 2.14).

Landforms

Fig. 2.14 A view of the Rocky Mountain Trench looking southeastward. This great valley contains the upper reaches of both the Fraser and the Columbia rivers.

Big rivers

Americans have traditionally recognized the Rocky Mountains as the so-called continental divide—the high ground that separates drainage between the Pacific and Atlantic watersheds. To the Gulf–Atlantic via the Mississippi go the Missouri, Platte, Arkansas, and many other large rivers; to the south goes the Rio Grande; to the Pacific via the Colorado and Columbia rivers go the Okanagan, Snake, Green, Gunnison, and of course the heads of the Colorado and Columbia themselves. By contrast, in the northern part of the Canadian Province, the Rockies do not form the continental divide. The Peace and Mackenzie Rivers, which drain into the Arctic Ocean, both have headwaters in the Rocky Mountain Trench west of the Rockies in British Columbia.

Bioclimate and Soils. Owing to their diverse geology and rugged, high topography, the Rocky Mountains are highly varied in soils, climate, and biogeography. The American Rockies lie generally in an arid/semiarid climatic zone and most of the basins between ranges are relatively dry and dominated by prairie grasses and/or desert shrubs. But moisture conditions improve with elevation, not only because precipitation tends to increase higher up, but also because evapotranspiration rates decline at cooler temperatures. Around 7500 feet elevation in the Southern Rockies and 4000 to 5000 feet elevation in the Northern and Canadian Rockies, trees appear; forests of pine, fir, and aspens extend upslope several thousand feet until they are limited by cold temperatures.

Vertical zonation

The elevation of the *tree line* varies from 11,000 feet or more in New Mexico to only 5000 to 6000 feet in the northern Canadian Rockies where the mountain forests merge into the boreal forest system. At the timberline, trees give way to *alpine meadow* consisting of grasses, wildflowers, and dwarf (shrub-sized) trees. For most mountains, alpine meadow marks the uppermost bioclimatic zone, but for mountains above 12,000 feet elevation, there is also a zone of permanent snowfields and/or glaciers (Fig. 2.15*a*).

Besides the vertical zonation of climate, bioclimatic conditions are often quite different from range to range depending on location and orientation to prevailing winds and the sun. West slopes usually receive greater precipitation in their upper reaches than east-facing slopes because of the prevailing westerly airflow; and south-facing slopes are measurably drier owing to greater solar heating from the southerly exposures. Thus, the vegetation zones on individual mountains often occur at different elevations from west to east as well as from south to north (Fig. 2.15*b*).

Water Resources. The Rocky Mountains are the primary source of water for most of the American West and the Great Plains. The mountains induce relatively high precipitation rates (30–50 inches annually in many areas), much of which accumulates as snow that is released to streams in the spring and summer. West of the Rockies, competition for a limited water supply is especially keen and growing more so with the expansion of agriculture and urban development. The water of the Colorado River System, for example, is partitioned by law between the Rocky Mountain states of Colorado, Wyoming, Utah, and New Mexico and three states that border on the lower Colorado River, California, Nevada, and Arizona. Eight major dams have been constructed on the Colorado system, and their reservoirs provide huge amounts of water not only to agriculture and cities but for evaporation into the dry desert air (Fig. 2.16). By the time the Colorado reaches the Mexican border, little or no water remains.

Land Use. Settlement in the Rocky Mountains is light; there are no large cities in the region. (Denver, Salt Lake City, and Calgary lie on the borders of the region.) Most land is publicly owned. In both Canada and the United States, extensive tracts have been set aside as national parks, national monuments, and national forests (Fig. 2.17). Increasingly, landscape planning and management activity here and farther west are concerned with competition among mining interests, ranchers, preservationists, and others with interests in the use of government lands. National forests and other federal lands are open to grazing and mining, both of which are sources of contention in many areas of the West because of their impacts on rangeland ecology and stream water quality.

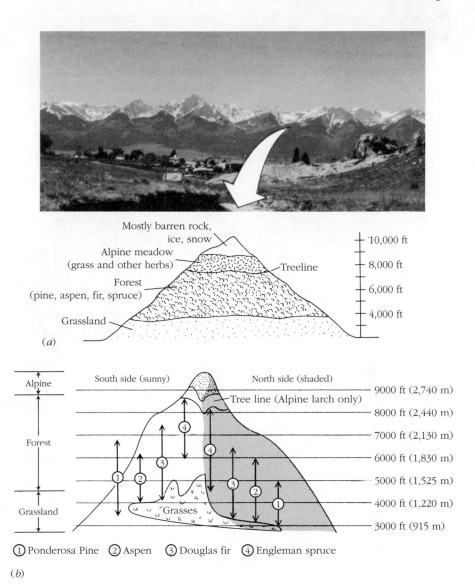

Fig. 2.15 (a) A simple model of the vertical zonation of vegetation in the Northern Rocky Mountains, a response to climate change with elevation. (b) The influence of sunny and shaded slope on vegetation zones.

2.9 THE INTERMONTANE REGION

Plateaus

Between the southern province of the Rockies and Pacific Mountain Region to the west lies an elevated region of plateaus and widely spaced mountain ranges. There are two major plateaus in the **Intermontane Region**: The *Colorado* lies at elevations around 6500 feet and is composed of sedimentary rocks; the *Columbia* lies at elevations around 5000 feet and is composed of basaltic rock. The remainder of the region, about half its total area, is the *Basin and Range Province*, which stretches from the Columbia Plateau on the north to the Rio Grande Valley on the southeast (Fig. 2.18).

The Basin and Range Province. This province, which is also known as the **Great Basin**, is characterized by disconnected mountain ranges that trend north–south and that were formed by the faulting and tilting of large blocks of diverse rock. The ranges are 50 to 75 miles long, 10 to 25 miles wide, and generally 7000 to 10,000 feet in elevation. The basins between the ranges are as broad as or broader than the mountain ranges and filled with thick deposits of sediment (thousands of feet deep) eroded from the adjacent mountains. The basin floors lie at elevations

Landforms

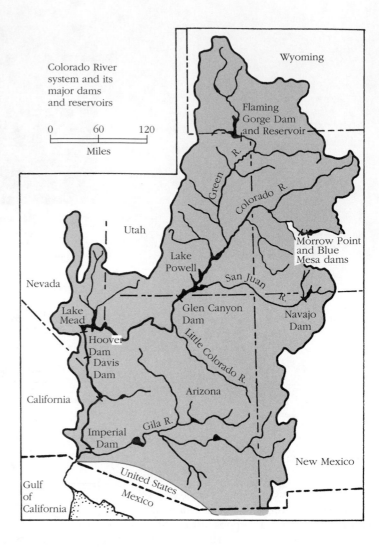

Fig. 2.16 The Colorado watershed and stream system, including major dams and reservoirs.

of 3000 to 5000 feet, except in the southern part of the province where they are generally much lower (Fig. 2.18).

The Basin and Range Province is severely arid (with less than 10 inches of annual precipitation) except for the upper reaches of the higher ranges, which receive enough rain and snowfall to produce runoff. Some of the runoff delivered to the *Climate and soil* mountain slopes in winter and spring reaches the basin floors where it is trapped and evaporates, forming dry lake beds called *playas*. The runoff also brings sediment to the lower mountain slopes where, as flows decline, it is deposited, forming *alluvial fans*, one of the distinctive landforms of the region. Soils on the valley floors belong to the *Aridisol* order; in and around the playas they are heavily saline. On the alluvial fans, the soils are far less salty, compositionally varied, and generally belong to the *Entisol* order (see Fig. 5.5).

The northern half of the Basin and Range is the only area of *closed drainage* in the United States and Canada, that is, with no outlet to the sea. The largest stream *Drainage* in the area, the Humboldt River, rises in northeastern Nevada and ends in playas on the western side of the state. Most streams, however, are short, ending in local basins only miles from their headwaters. In the southern half of the Basin and Range, by contrast, drainage is *open* and dominated by the Colorado River and its lower tributaries (Fig. 2.15). The lower Colorado Valley is one of the driest areas of North America, where reservoir evaporation (such as from Lake Mead) exceeds 100 inches a year.

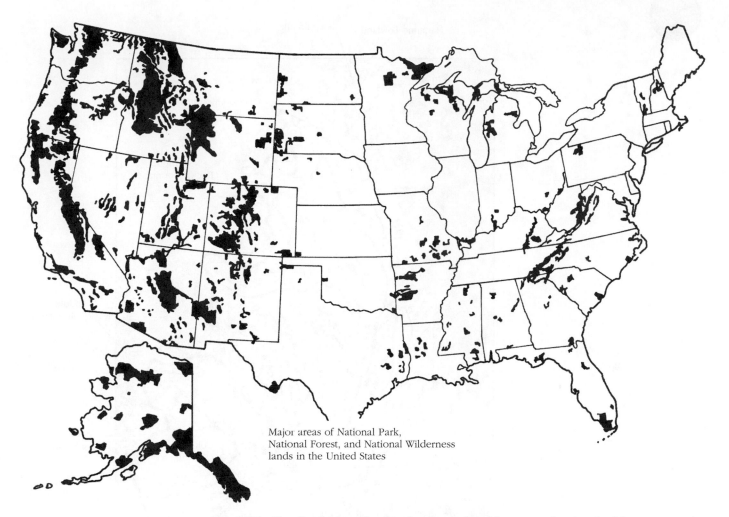

Major areas of National Park,
National Forest, and National Wilderness
lands in the United States

Fig. 2.17 The distribution of national parks, national forests, and national wilderness areas in the United States. In Canada, most of the country is held in public lands.

In the past several decades, population has grown tremendously in the southern part of the Intermontane Region, principally in Arizona and southern Nevada. This has resulted in both urban and agricultural development, which has placed great demands on the limited water supplies in this arid region. Groundwater aquifers are being rapidly depleted in some areas, and Arizona is being forced to turn to aqueduct systems fed by the Colorado to transport water to cities and farmlands. Los Angeles also uses the Colorado for part of its water supply.

The Colorado Plateau. Between the Rocky Mountain and Basin and Range provinces lies the **Colorado Plateau**, one of the truly celebrated physiographic provinces in the United States. It is a land of canyons, colorful landscapes, and a rich

Sacred landscapes Native American history, and it is the home of the Grand Canyon, America's premier landform. The Colorado Plateau is a platform of sedimentary rocks with large areas of volcanic rocks and extensive surface deposits. The general elevation of the plateau is 5000 to 6000 feet, but some mountain ranges within it, such as the Henry Mountains, reach 11,000 feet.

The Colorado River cuts through the heart of the plateau where it and its tributaries have carved a great system of canyons (Fig. 2.18). At its deepest cut—the Grand Canyon—4000 to 5000 feet of rock are exposed in the canyon walls, revealing a spectacular geologic cross section. The picture of geology and landforms in the province

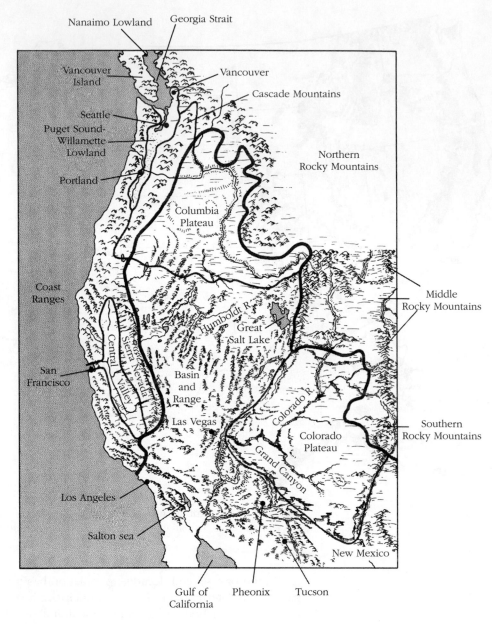

Fig. 2.18 The Intermontane Region, its provinces, and the southern part of the Pacific Mountain System.

is enhanced by the arid climate (average precipitation is less than 15 inches a year), which limits vegetation over most of the plateau to desert shrubs (mainly creosote bush and mesquite) and bunch grasses. Trees grow at higher elevations with pine forest beginning around 7500 feet elevation.

2.10 THE PACIFIC MOUNTAIN REGION

Along the Pacific margin of North America is a complex region composed of many, large mountain ranges (Fig. 2.2). This region, known as the **Pacific Mountain System**, contains just about every geologic structure and mountain type imaginable.

Also, by virtue of its location on or near major fault zones on the western edge of the North American tectonic plate, it is the most geologically active part of the continent. Earthquakes are commonplace throughout the region, and volcanism is active in the Cascade Range (which produced the Mount St. Helens eruptions of 1980 and 2004), in the Alaskan Peninsula, and in the Aleutian Islands, the most volcanically active area of North America.

Mountain diversity

The region is made up of eight provinces that form a belt 200 to 400 miles wide extending from the Mexican border almost to the Russian border at the end of the Aleutian Islands. Some of the more prominent mountain ranges include the *Coast Mountains* of British Columbia and Alaska, the *Sierra Nevada* of California, the *Cascades* of Washington and Oregon, the *Coast Ranges* of California and Oregon, and the *Alaska Range* of southern Alaska. The Coast Mountains and the Sierra Nevada Mountains are mostly great masses of resistant granitic rock, whereas the Cascades are basically huge piles of weak (erodible) volcanic material. The Coast Ranges of California and Oregon are mainly deformed sedimentary rocks aligned into a series of ridges and valleys paralleling the coast. The Alaska Range, in the northernmost part of the region, is a complex body of mountains built by folding, faulting, and volcanic activity and is the highest range in North America with many mountains exceeding 15,000 feet in elevation. Southwest of the Alaska Range are the Alaska Peninsula and the Aleutian Islands, which comprise a chain of volcanic mountains.

The Pacific Coast. Unlike the Atlantic Coast, the **Pacific Coast** lacks a coastal plain. In most places the mountains or their foothills run to the sea where wave erosion has carved a rugged and picturesque shoreline. Owing to ancient sea level changes and geologic uplift of the land, much of the coastline is terraced, that is, sculpted into steplike formations. In addition, the coastline is frequently intersected by streams that have cut narrow canyons down to the shore. Sandy beaches with sand dunes are found in the bays and near stream mouths. Accessibility to the coastline is difficult, and development is risky because of slope instability, the limited area suitable for building, and the often fragile character of the ecological environment.

Rugged coastline

Seismic Risk. Earthquakes are a threat throughout the entire Pacific Mountain Region. The region is laced with fault lines, and many have an eventful history of activity. The most active earthquake zones are found in California and Alaska. California is clearly the more hazardous of the two, especially south of the San Francisco Bay area, not only because of the prominent fault systems there (such as the San Andreas), but also because of the massive urban development lying on or near active fault zones. California's urban/suburban population in this area approaches 20 million or more people. Earthquakes of relatively modest magnitudes in and near Los Angeles and San Francisco in the past few decades portend the destructiveness of truly large earthquakes in these areas.

Cities and faults

Major Lowlands. Two major lowlands are found in the Pacific Mountain Region. In the south is the *Central Valley of California*, which lies between the Sierra Nevada and the Coast Ranges (Fig. 2.18). To the north is the *Willamette Valley–Puget Sound–Georgia Basin lowland*, which lies between the Coast Ranges and the Cascades in Oregon, Washington, and British Columbia. These lowlands are floored with deep deposits of sediment washed down from the surrounding mountain ranges over millions of years. Only the Willamette Valley houses a major stream system that follows the north–south axis of the lowland. The floors of the other two valleys are largely filled with ocean water. The three largest cities of the Pacific northwest—Portland, Seattle, and Vancouver—are located in the Willamette–Puget Sound–Georgia Basin lowland.

Climate and Vegetation. The climate, vegetation, and soils of the Pacific Mountain Region are as diverse as its geology and landforms. In Alaska and Canada, heavy precipitation in the Coast Mountains—the annual average exceeds 100 inches—nourishes large glaciers at elevations above 6500 feet and great conifer forests of hemlock, spruce, and cedar at lower elevations. The conifer forests extend down the

Diverse bioclimate

coast into Washington and Oregon, where the dominant tree is Douglas fir. Farther down the coast, in northern California, stands of huge redwoods become the dominant coastal forests. Precipitation declines sharply farther down the California coast, and between San Francisco and Los Angeles, where annual precipitation falls below 20 inches, the redwoods give way to a scrubby forest, called *chaparral*. Still farther south, in extreme southern California and the Baja California peninsula of Mexico, the chaparral gives way to bunch grass and shrub desert (Fig. 2.19).

In California, the combination of subtropical climatic conditions and water supplies from mountain streams has produced one of the most diverse and productive

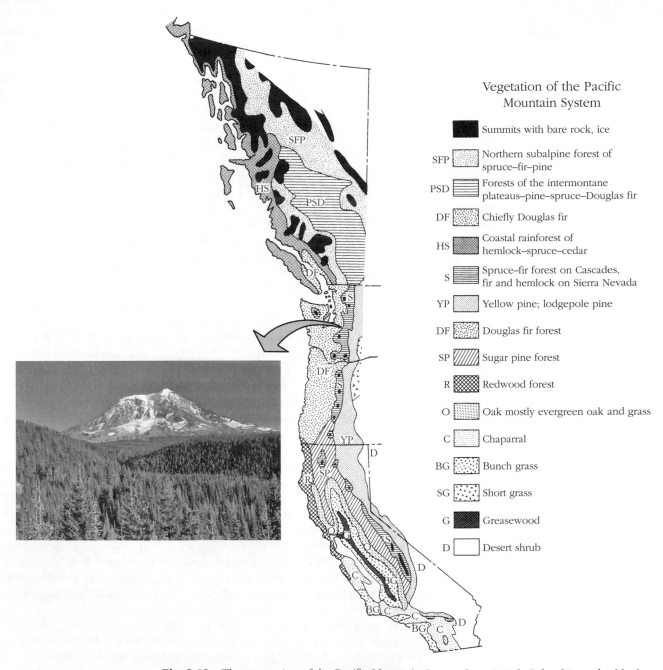

Vegetation of the Pacific Mountain System

	Summits with bare rock, ice
SFP	Northern subalpine forest of spruce–fir–pine
PSD	Forests of the intermontane plateaus–pine–spruce–Douglas fir
DF	Chiefly Douglas fir
HS	Coastal rainforest of hemlock–spruce–cedar
S	Spruce–fir forest on Cascades, fir and hemlock on Sierra Nevada
YP	Yellow pine; lodgepole pine
DF	Douglas fir forest
SP	Sugar pine forest
R	Redwood forest
O	Oak mostly evergreen oak and grass
C	Chaparral
BG	Bunch grass
SG	Short grass
G	Greasewood
D	Desert shrub

Fig. 2.19   The vegetation of the Pacific Mountain System from British Columbia to the Mexico border. Photograph shows a typical scene in the Cascade Mountains.

agricultural regions in the world. Grains, vegetables, grapes, and fruits are grown extensively in the southern two-thirds of the state, mostly with the aid of irrigation. The forest industry traditionally flourishes in the area between San Francisco and the Alaska Panhandle. In this area, as in the Rocky Mountains, publically owned forests

Farms and forest and parks occupy large tracts of land in both Canada and the United States. In the United States, only about 10 percent of the original (old growth) forest remains, and a contest is being waged among loggers, environmentalists, and the federal government for control of these forests. Elsewhere in the Pacific Region, south of Canada, most lumbering is carried out in areas of second growth forest.

Environmental Challenges. The Pacific Mountain Region abounds with environmental planning and management problems. The scenic landscapes and pleasant climates are attractive to settlement, and the population is growing in most areas. California's population now exceeds 38 million, more than 5 million more than all of Canada. Growing population and expanding land use, including some of the most notorious urban sprawl yet devised and coupled with a diverse and active environment, produce virtually every type of environmental problem known today and at

Land use dilemma a wide variety of geographic scales. California heads the list with serious air pollution, water supply, land stability, and habitat loss problems, and the state contains more species on the federal list of rare and endangered species than any other in the union. But the Pacific Northwest just to the north, including the Vancouver area in British Columbia, is rapidly developing its own list that includes forest management, urban sprawl, coastal development, water supply, and a host of pollution problems.

Among the region's most alarming problems is the dramatic decline of salmon. Over the past 20 years, most watersheds have lost a large share of returning populations of coho, and some species have declined to the point where they no longer have viable breeding stock. The cause of the decline is most certainly tied to multiple factors. At the top of the list are watershed changes—including degraded water quality and barriers to fish migration—and overfishing, but climatic change and increased predation may also be involved (Fig. 2.20).

2.11 THE YUKON AND COASTAL ARCTIC REGION

South of Alaska's North Slope lies the Brooks Range (and its eastern limb in Canada, the British Mountains), a low, east–west trending mountain range. South of the Brooks Range and occupying the large interior of Alaska and the adjacent portion of the Yukon Territory is the **Yukon Basin** (Fig. 2.2). The Yukon River drains this large basin, flowing westward into the Bering Sea. Fairbanks, the principal city of cen-

Yukon Basin tral Alaska, is located near the center of the Yukon Basin. *Permafrost* is found over most of the Basin, but its coverage is discontinuous and its thickness highly variable. The landscape of the Yukon Basin is dominated by boreal forests in the south, but northward the tree cover grows patchy and gives way to tundra. The treeless tundra stretches in a broad belt along the entire Arctic Coast.

The Arctic Ocean is fringed by a coastal plain similar in topography to the Coastal Plain of southern United States. The **Arctic Coastal Plain**, however, is narrower and more desolate, being extremely cold and locked in by sea ice most of the year. The North Slope of Alaska, now famous for its oil reserves, is part of the Arctic Coastal

Arctic Coastal Plain Plain, as are the Mackenzie River Delta (just east of the Canadian–Alaskan border) and the plain that fringes the northern islands of Canada. Virtually the entire Arctic Coastal Plain is underlain by permafrost, which in some areas extends offshore under the shallow waters of the Arctic Ocean.

Climate, permafrost, and ecology are the principal planning considerations of this region. The growing season, which is less than 60 days, prohibits agriculture,

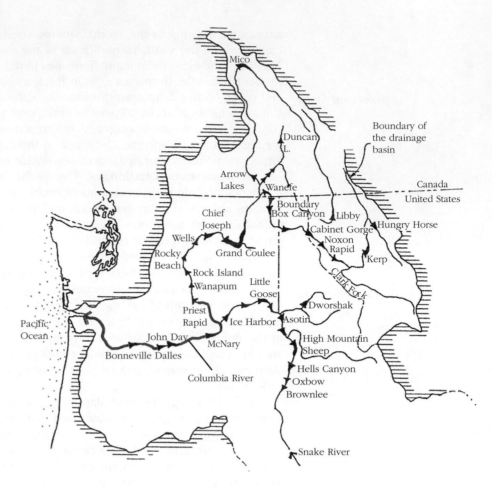

Fig. 2.20 Dams on the Columbia River and its principal tributaries. Once a Mecca for Pacific salmon, the dams and other factors have reduced annual migrations to a dribble.

and permafrost limits development over much of the region because it leads to infrastructural damage. As one of the last great wilderness reserves on the continent, the tundra is given highest priority by environmentalists for long-term protection of its ecosystems. Conflict over economic development proposals and programs is destined to continue for decades.

Climate change Climate change is another serious problem of this region. The best evidence to date indicates that the vast Arctic and subarctic regions of North America are undergoing some of the strongest effects of global warming with mean annual temperatures as much as 2–3°C higher than the 30-year average. Sea ice is declining, leaving shores exposed to wave erosion from a rising sea, and permafrost is receding, leading to changes in drainage and landscape ecology.

2.12 SELECTED REFERENCES FOR FURTHER READING

Atwood, W. W. *The Physiographic Provinces of North America*. Boston: Ginn, 1940.

Berry, Wendell. *The Unsettling of America: Culture and Agriculture*. New York: Avon, 1978.

Bird, J. B. *The Natural Landscape of Canada*. Hoboken, NJ: Wiley, 1972.

Birdsall, S. S., and Flovin, J. W. *Regional Landscapes of the United States and Canada*. Hoboken, NJ: Wiley, 1985.

Bowman, Isaiah. *Forest Physiography.* Hoboken, NJ: Wiley, 1909.

Chapman, L. J., and Putnam, D. F. *The Physiography of Southern Ontario,* 3rd ed. Ontario: Ministry of Natural Resources, 1984.

Conzen, M. P. (ed). *The Making of the American Landscape.* Boston: Unwin Hyman, 1990.

Fenneman, N. M., and Johnson, D. W. *Physical Divisions of the United States* (U.S. Geological Survey Map). Washington, DC: U.S. Government Printing Office. 1946.

Hunt, C. B. *Natural Regions of the United States and Canada.* San Francisco: Freeman, 1974.

King, P. B. *The Evolution of North America.* Princeton, NJ: Princeton University Press, 1959.

Leighly, John (ed.). *Land and Life: A Selection from the Writings of Carl Ortwin Sauer.* Berkeley: University of California Press, 1963.

Paterson, J. N. *North America: A Geography of the United States and Canada.* New York: Oxford University Press, 1989.

Pirkle, E. C., and Yoho, W. H. *Natural Landscapes of the United States,* 4th ed. Dubuque, IA: Kendall, 1985.

Schroeder, Walter A. *Opening the Ozarks.* Columbia, MO: University of Missouri Press, 2002.

Thornbury, W. D. *Regional Geomorphology of the United States.* Hoboken, NJ: Wiley, 1965.

Related Websites

Colorado Partners in Flight. "Colorado Land Bird Conservation Plan." 2000. http://www.rmbo.org/pif/bcp/intro/exsum.htm

Website of collaborative work looking at physiographic regions and associated bird species. The goal of this project is to protect the native bird populations for each region.

Geological Survey Canada. "Permafrost: Communities and Climate Change." 2007. http://gsc.nrcan.gc.ca/permafrost/communities_e.php

A wealth of information about permafrost, climate change, and their implications for northern development. The downloadable brochure from this link gives a picture of what must be considered for the future of infrastructure in these areas.

Gilliam, J. W., Osmond, D. L., and Evans, R. O. "Selected Agricultural Best Management Practices to Control Nitrogen in the Neuse River Basin." North Carolina Agricultural Research Service Technical Bulletin 311. Raleigh, NC: North Carolina State University, 1997. http://www.soil.ncsu.edu/publications/BMPs/physio.html

Physiographic-specific design to reduce nonpoint source river pollution in North Carolina. Regional and provincial knowledge is critical in effective, sustainable planning.

Grand River Conservation Authority. "Forest Types and Physiographic Regions," from the *State of the Watershed Report.* 1997. http://www.grandriver.ca/forestry/ForestPlan_Ch12_Types.pdf.

A look at the current state of watersheds in different physiographic regions. The site uses indicator species (mainly songbirds) to evaluate the health of watersheds and correlates the physiographic region with plant species.

Worsley School. "The Canadian Shield." 1996 – 2010. www.worsleyschool.net/socialarts/shield/canadianshield.html

A look at the inherent challenges of developing the Canadian Shield, including photos of sample bedrock after excavation.

3

LANDSCAPE FORM AND FUNCTION IN PLANNING AND DESIGN

3.1 INTRODUCTION

Losing touch

It is no news that we are in serious trouble with the landscape. There are few places in North America where we have been able to achieve a truly lasting balance among land use activities, facilities, and environment. Part of the explanation for this state of affairs has to do with the direction of modern urban life. In the past generation or two, people have generally lost touch with the land and in turn with most of the traditional knowledge about how the landscape works. Planning for the use and care of the local landscape is no longer part of personal and family tradition and responsibility. To a large extent it has been relegated to second and third parties and has been transformed into bureaucratic and business processes made up of inventories, checklists, permits, and contracts that require little understanding of the landscape's true character.

Systems-driven change

The route to finding the true character of the landscape lies in understanding how the land functions, changes, and interacts with the life it supports. One of the most fundamental points of understanding is that the landscape is more dynamic than it is static, with forms and features in a continuous state of change. Change is driven by systems of processes that include rainstorms, streamflow, fires, soil formation, plant growth, and, yes, the human activities that we call land use. These processes shape the landscape now as they did in the past. Increasingly apparent is the need for successful landscape planners and designers to respond to more than mere shapes and features, but also to the processes themselves and to the systems that drive them. Indeed, land use sustainability itself is rooted in this perspective.

3.2 ESSENTIAL SYSYEMS AND PROCESSES OF THE LANDSCAPE

One of the prevailing misconceptions about the landscape is that it is made up of features, especially natural features, that date from times and events millions of years ago. This view is especially common for landforms and soils that are typically regarded as products of the geologic past. Therefore, any attempt to understand their origins and development requires special knowledge of ancient chronologies and events. For most landscapes, however, this view is probably not valid.

Landscape of the here and now

Landscape Diagnosis. By and large, the forms and features we see in the local landscape are the products of the processes and systems that presently operate there. This perspective is important because it implies that we can, more or less, understand the landscape according to the workings of the environment in *the present era*. Using an analogy from medical science, we know that the body is a product of long-term biological evolution but we also know that individual organs have sizes, shapes, and compositions related to their present function. As with the organs in the human body, form and function in the landscape go hand in hand. It follows that, as the clever physician is able to read basic functional (physiological) problems from changes in organ shape and composition, so the insightful student of landscape can read changes in the functional character of the land from observable changes in landforms, soil, drainage features, vegetation, and related parts of the landscape.

Formative systems

What are the essential or *formative systems* of the landscape? They include longshore or drift systems in the coastal zone, prevailing and seasonal wind systems on the seashore and in deserts, glacial and permafrost systems in the mountains and polar lands, and runoff-watershed systems almost everywhere. Runoff actually includes a family of water systems all fed directly or indirectly by precipitation. These include overland flow, streamflow, soil water, and groundwater, and together the work they accomplish, measured by the total amount of material eroded from the land, exceeds that of all other formative systems manifold, even in dry environments.

Therefore, we can safely conclude that the landforms we see in most landscapes are mainly water carved, water deposited, or influenced by water in some significant way (Fig. 3.1). Even the landscapes shaped by other systems such as glaciers and wind are also worked on by running water.

Building landscape

Landscape Differentiation. The *differentiation of terrain* into various physiographic zones and habitats begins with weathering and the sculpting of landforms by runoff. As the landforms take shape, different moisture environments emerge, such as wet valley floors, mesic (intermediate) hillslopes, and dry ridge tops. These in turn give rise to different plant habitats, and it is the combination of moisture conditions, vegetation, and surface sediment that yields soil. Thus, in our search for fundamental order in the landscape, we must get cause and effect in the right order by first examining landforms and drainage. In most instances the remainder of the landscape—that is, vegetation, soils, habitats, and biota—will fall into place once the essential system of landforms and runoff has been worked out (Fig. 3.2). Moreover, this rule applies at essentially all spatial scales from that of microtopography to the macro forms of the great landform systems.

Deducing function

Form and Function Relations. If we agree that the landscape represents a basic *form–function* relationship, then we should be able to deduce a great deal about the processes that operate in it from the forms we can observe. This concept is a very important part of terrain analysis and environmental assessment because in land planning projects we rarely have the time and resources to undertake our own scientific investigations. We are usually unable to launch in-depth studies to generate firsthand data leading to the analysis of processes and systems and their effects on the landscape. Therefore, we must learn to turn the approach around and read the formative process of the land from the forms and features that can be observed and measured in the field and on maps and imagery.

Retaining balance

The *form–function concept* also reveals that any actions taken as a part of land use planning and engineering that result in changed landscape forms, such as cut-and-fill grading, stream channel reconfiguration, and vegetation alteration, must produce changes in the way the landscape processes work. If certain balances exist on a slope among, for example, runoff, slope inclination, soils, and vegetation, then an alteration of one component without appropriate counter alterations in the other three will result in an imbalance. The slope may begin to erode, sediment may accumulate on it, or vegetation may decline. In any case, maintaining balance in the

Fig. 3.1 Water-carved landscapes. Runoff is the most effective natural agent in shaping the landscape over most of the earth, especially in terms of landform, soil, and habitat formation.

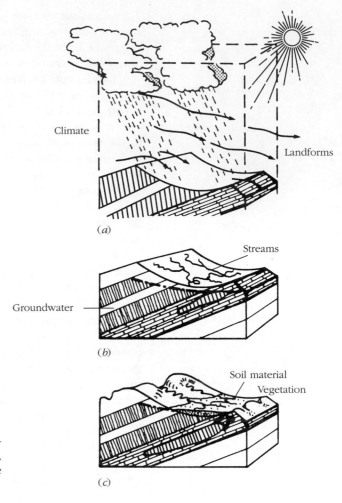

Fig. 3.2 A schematic diagram illustrating the development of landscape beginning with (*a*) basic landforms and climatic influences, followed by (*b*) the development of water systems, and (*c*) the formation of soil, vegetation, and habitat.

landscape in the face of physical changes brought on by land use demands that we view the land as a dynamic game board rather than a static stage setting. As a rule, our goal in landscape planning should be to guide change in such a way as to maintain the long-term performance of the critical processes and systems of the landscape—in other words, to guide change toward a sustainable landscape in which our actions are largely invisible to nature.

3.3 THE NATURE OF LANDSCAPE CHANGE

There is an old debate in science about the character of change in nature. Does nature change gradually over long spans of time, as traditional evolutionists thought, or does nature change in short bursts, as many earth scientists think? Both views are correct to some extent, but the second one appears to be more appropriate for the larger features of the landscape such as stream channels, hillslopes, and land use.

Magnitude and force

Magnitude and Frequency Concept. The essential processes and systems of the landscape work at highly uneven rates, rising and falling dramatically over time. Each rise or fall in streamflow or wind, for example, can be described as an *event*. Each event can be measured in terms of its magnitude (its size), such as the discharge of a stream or the velocity of wind in response to a storm. A huge streamflow that produces massive flooding is a high-magnitude event, and it exerts much greater force on the environment than a small or modest event.

Very significant to our understanding of landscape change is the fact that the *force* exerted on the environment by an event increases *geometrically* with its magnitude. Therefore, to accurately interpret the potential for change (or work) by a force such as wind or running water, we must understand, for example, that a threefold-magnitude increase from, say, level 2 to level 6 represents an increase in force as great as 25-fold. This is termed an exponential relationship in which force increases as some power function of event magnitude.

Who does the work?

If we examine the relationship between the *magnitude and frequency* of events in the landscape, we find that almost regardless of the agent (process) involved (e.g., streamflow, wind storms, rainfalls, fires, earthquakes, snowfalls, oil spills, car accidents, or disease epidemics), the pattern is basically the same. There are large numbers of small events, much smaller numbers of medium-sized events, and very few large events. The truly giant events, which can render huge amounts of change, are very scarce indeed. As it turns out, the events that do the most work *in the long run* are not the giants (they are too infrequent) and not the high-frequency events (they are too small even when counted together), but the fairly large events of moderately low frequency. In streams, for example, these events may be bankfull discharges and modest floodflows that occur once a year, every other year, or even less frequently.

Finding evidence

Reading the Clues. An inherent pitfall of landscape analysis for planning purposes is that we usually have no chance to observe or measure the events that really shape the landscape. For hydrologic and atmospheric processes, these events, such as a very intensive thunderstorm, a floodflow, or a massive snowfall, have a duration or life ranging from an hour or two to several days and a frequency of occurrence of once every few years or less. The events we are apt to see on the typical field visit (usually a fairly nice day) are probably meaningless in terms of total effectiveness in shaping the landscape. Therefore, unless we are careful we run the risk of grossly misinterpreting how the site actually functions, how its features are shaped, and how things relate to one another. We may be led to infer that the processes we happen to observe are really the essential ones or that the landforms and related features were shaped so long ago that the processes operating there in this era cannot possibly have anything to do with the site as we see it. Both conclusions would be mistaken.

From form to function

How do we avoid the pitfalls of misinterpreting the landscape? The answer lies in understanding form–function relations. When we describe the forms and features of the landscape, we are actually observing the artifacts and fingerprints of the formative processes and systems (Fig. 3.3). Through insightful field investigation we can begin to deduce which processes at which levels created, shaped, or affected different landscape features, as well as which processes and events are meaningful to both existing and future land uses and facilities.

An example

Stream valleys and channels provide some of the best illustrations of this approach. Studies show that the stream channel is shaped principally by relatively large flows, mainly those that occur once or twice a year or once every two or three years. These may be floodflows or flows that fill the channel with modest spillover into adjacent low areas. Such flows are capable of scouring the channel bed, moving heavy sediment loads, eroding banks, and causing the channel to shift laterally, as illustrated in Figure 3.3. Bank vegetation may be undercut, and, where flows overtop the banks, waterborne debris such as dead leaves is often stranded on shrubs and trees, marking the floodwater level. Taken together, these features serve as markers of flow depths and extent as well as indicators of the distribution of energy and the work accomplished by running water. They reveal that stream channels and valleys are highly dynamic environments and that the formative events are both destructive and constructive—important information for landscape planning and management.

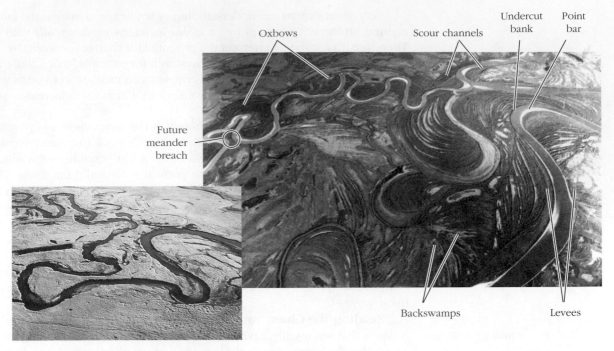

Oxbows · Scour channels · Undercut bank · Point bar · Future meander breach · Backswamps · Levees

Fig. 3.3 A section of the Mississippi River floodplain showing the artifacts (forms and features) left in the landscape as a record of the river's formative processes. Inset shows channel formation in action where a stream is cutting into the bank along a series of bends.

3.4 THE CONCEPT OF CONDITIONAL STABILITY

Whether a landscape is stable under the stress of the various forces applied to it depends not only on the strength of the forces, but also on the resisting strength of the landscape. Resisting strength is provided by forces that hold the landscape together, that is, forces that keep soil from washing away, slopes from falling down, and trees from toppling over. Among the resisting forces—which include gravity, chemical cementing agents, and vegetation—living plants are the most effective in holding the soil in place.

Critical balance In most natural landscapes, a state of balance exists between the driving forces, represented by water, wind, human, and other processes, and the resisting forces. Only when this balance is broken—usually because a powerful event exceeds the strength of the resisting force—is the landscape prone to massive change such as wholesale soil erosion and slope failure. Most landscapes, however, are resistant to breakdown from all but the strongest events. In some places, however, stability is maintained by an extremely delicate balance that is *conditional* on a special ingredient in the environment. That ingredient, such as a mat of soil-binding roots on an oversteepened slope, functions as the stabilizing linchpin in the landscape. If the linchpin is weakened or released, the landscape can literally fall apart under the stress of an event of even modest magnitudes.

Conditional stability **Tipping the Balance.** Recognizing such conditional situations is critical to landscape planning and design because it is necessary to guide the change brought on by land use without pulling a linchpin and triggering a chain of damaging events. Thus, in evaluating a site for a planning project, it is important to identify features that may be pivotal to the overall stability of the landscape. These may not be the most apparent features in terms of size or coverage. Among the conditional, or *metastable*, features commonly noted are steep, tree-covered slopes; vegetated sand dunes; stream

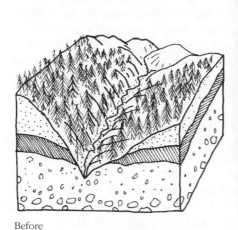

Before   After

Fig. 3.4 Slope failure following clear-cutting of forested terrain in the American Northwest. These are not sand dune slopes, but they can exhibit the same unstable tendency when they are deforested and logging roads are cut into weak materials.

banks made up of root-bound soil; grass-covered prairie soils; swales armored with boulders; perched wetlands; and groundwater seepage zones.

An example   An example of the concept of conditional stability is provided by a forest-covered sand slope in a coastal setting. In an unvegetated state, sand slopes (such as a wave-eroded bank or a dune face) can be inclined at an angle no greater than 33 degrees. When a cover of woody plants is added to the slope, especially trees, the angle can be raised to as much as 45 to 50 degrees because the roots and trunks lock the sand into a stable slope form. Such oversteepening of wooded slopes is a common occurrence in coastal environments where sand is added to an existing slope by wind or runoff. The alteration or removal of vegetation for roads, trails, structures, or lumbering can initiate slope failure, erosion, and a host of related problems on conditionally stable slopes (Fig. 3.4).

3.5 PERSPECTIVES ON SITE AS LANDSCAPE

Site as real estate   Invariably, projects and problems in land use planning involve a parcel of space in the landscape called a *site* or a *project area*. It may range in size from less than 1 acre to thousands of acres. Its shape is usually some sort of rectilinear form, a product of mapping systems and the surveyor's coordinate lines. As a physical entity, the site has meaning mainly as a piece of real estate whose value is governed principally by its size and location. In land planning it is the envelope of space for which a use is sought or to which some land use has been assigned and within which a plan will be designed.

The Functional Site. From the standpoint of the environment and its functions,
Transparent borders   however, the site as conventionally defined has limited meaning and generally *cannot* be used to define the scope of environmental analysis for planning and design purposes. The reason is that as a parcel of space the site usually has little to do with the workings of the site's environment, that is, with the processes and systems that shape and characterize the landscape of which the site is a part. Air, water, and organisms, for example, move in spaces and patterns that typically show little or no relationship to the space defined by a site. Therefore, as we pursue planning problems involving

sites, we must deal with many different envelopes of environmental space as they ultimately relate to a prescribed piece of real estate.

Site in three dimensions Although we commonly view the site as a two-dimensional plane defined by the surface of the ground and contiguous water features, in reality it is distinctly three dimensional. The third dimension, height and depth, extends the site upward into the atmosphere and downward into the ground. The relevance of atmospheric and subterranean phenomena to planning and design problems is generally secondary compared to surface phenomena. However, concern for groundwater contamination, air quality, and climate change, for example, increasingly calls for serious consideration of these phenomena.

In addition, we should appreciate that the atmospheric and the subsurface realms are the source areas of forces that drive many of the surface systems and processes. For example, groundwater is the principal source of stream discharge, and solar radiation is the primary source of surface heat. Alteration of these driving forces directly or indirectly affects terrestrial processes such as runoff, erosion, evaporation, and photosynthesis, which may in turn change the fundamental balance of the surface environment including slope stability, ground-level climate, wetland water balance, streamflow regimes, as well as the conditions of facilities such as building foundations, roadbeds, and utility lines.

3.6 DIMENSIONS OF THE SITE AS DYNAMIC SPACE

Sites are traditionally described according to their static attributes, that is, according to their forms and features and the spatial relations among them. However, sites can also be described according to their dynamics, that is, according to systems and processes that shape their forms and features. In addition to running water, rainfall, wind, animal movements, and other natural systems, there are also human systems to consider, namely, the various land use activities and their by-products such as stormwater systems, vehicle traffic, noise, and air pollution.

The Systems Context. Each system documented at the site scale is part of a larger landscape, such as a watershed, which ties the site to a larger envelope of space. Think of this larger system as a *network* or flow scheme that can be defined in terms of its spatial dimensions (including its places of origin and destination), its pattern of movement, the work it performs, and its linkages with other landscape systems and features.

Intersecting systems Standing on a site, no matter where it is located, you can imagine that you are at the intersection of several systems that occupy different levels or strata of space at, above, and below the surface. Each system or flow network usually originates somewhere beyond the site, and it goes somewhere else after it has crossed the site. As it moves through the site, the flow is subject to change as it interacts with the various forms and features of the site, both natural and human made (Fig. 3.5). Think of the site as a link in the system's overall balance, and, for the system to maintain *continuity of flow*, any changes to the site as a result of development or any sort of intervention must maintain that continuity and balance if we are to achieve sustainability in the landscapes we occupy.

The boundary layer **The Upper Tier.** Air occupies the upper level, or tier, of the site. Airflow usually originates far beyond the site because most winds are part of regional scale systems driven by pressure differences among air masses. The movement of air over the surface can be described as a fluid layer called the **atmospheric boundary layer**. This layer, which measures about 1000 feet deep, is dragged over the earth's surface by the general motion of the larger atmosphere. As it slides over the surface, airflow in the lower levels is slowed substantially because of the frictional resistance imposed by the landscape. As a result, wind velocities at ground level are usually a small fraction of those just 10 or 20 meters (33 to 66 feet) above the surface (Fig. 3.6).

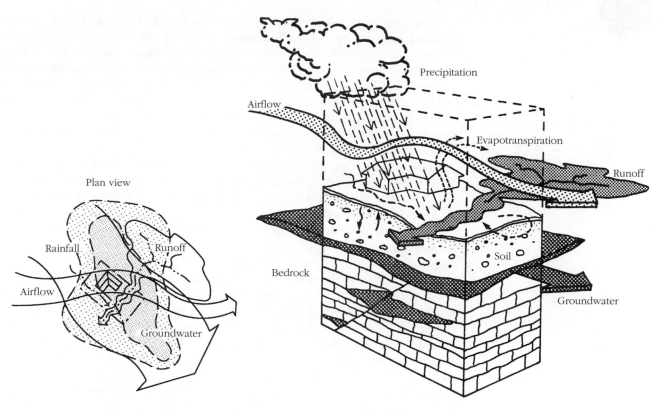

Fig. 3.5 Several important systems that intersect at a planning site: surface runoff, precipitation, evapotranspiration, groundwater, and airflow.

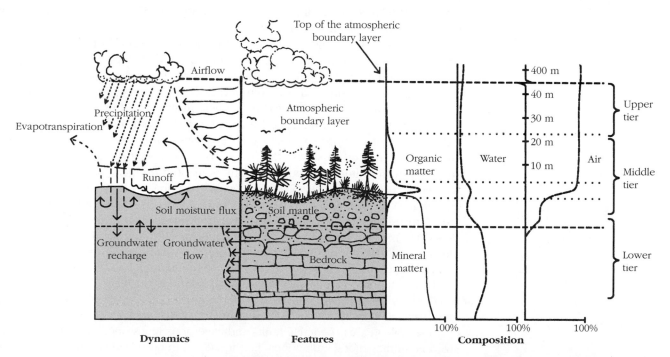

Fig. 3.6 A schematic illustration of the dynamics, features, and composition of the landscape in the upper, middle, and lower tiers of the landscape.

The change in wind velocity with height above the ground is described by a standard curve called the *normal wind velocity profile*. This curve reveals that the greatest change in velocity can be expected in the lowest 100 feet of the boundary layer where most planning and design activities are focused. The exchange of heat, moisture, and pollutants with the overlying atmosphere takes place across this zone, and the faster the airflow is, the greater the rate of exchange will be. This explains the requirement that factory and power plant exhaust stacks be heightened to reach faster streamlines as a means of increasing the rate of pollutant diffusion and reducing local pollutant concentrations near the ground.

The active landscape

The Middle Tier. The middle tier of this model is the landscape itself, which extends from the upper limits of the vegetation canopies and tops of built structures to the lower limits of root systems, buildings, and utility systems (Fig. 3.6). This is decidedly the most active layer of the landscape, with a host of flows taking on complex motions and exchanges. Although most movement is horizontal, some is decidedly vertical. Water, the dominant landscape agent, is delivered vertically as precipitation, then moves laterally as runoff. *Runoff* assumes a variety of flow forms as it moves over the surface, within the soil, or deeper down, depending on slope, soil, and vegetation conditions. On the surface water may flow in thin sheets at velocities as low as 0.1 foot per second or in streams where it can flow 50 to 60 times faster.

The dominant agent

In site planning, runoff is one of the most serious considerations in terms of both water quantity and quality. Generally, land development of any sort results in increased surface runoff (stormwater) and decreased water quality. Since every site is part of a flow system, local loading of runoff is always passed on, and the problem tends to become cumulative downhill and downstream. Engineered drainage systems facilitate this trend. But where drainage is impeded in some way—say, by beaver dams or soil excavation—water may collect locally and initiate wetland formation. These sites encourage the development of special plant communities with distinctive floristic compositions and strong organic productivities. The organic matter added to the moist soil from plant remains further improves the water-holding capacity of the wetland and advances its development and stability. This is one way in which the landscape becomes diversified as it is differentiated into zones of particular hydrologic, soil, and ecological character.

Meaning at depth

The Lower Tier. The lower tier of the site is comprised of the soil mantle and underlying bedrock (Fig. 3.6). Once considered relevant to planning only in terms of building foundations, utility lines, and water supply, today the subsurface environment must be given more detailed attention because of our heavy reliance on contaminant-free groundwater. *Groundwater* is the largest reservoir of fresh, liquid water on the planet, but it is widely threatened by contamination from surface pollutants and buried waste. Unlike surface water, which is flushed from the land relatively quickly, groundwater moves so slowly that once contaminated it remains polluted for decades and even centuries. Groundwater movement in *aquifers* (underground materials with large, usable water supplies) is limited to a fraction of a foot per day; therefore, exchange (flushing) times are typically decades or centuries. In addition, because of their large sizes, aquifers almost always extend well beyond the scale of individual sites (Fig. 3.5). In site planning, therefore, it is useful to know what part of an aquifer the site lies over. For example, does the site lie over the input (water recharge) zone or over the output (water discharge) zone of the aquifer?

Below the soil mantle lies the zone of *bedrock*. Compared to soil material, we usually view the presence of bedrock as a sign of ground stability in land use planning, especially in infrastructure development. The bedrock can, however, be decidedly unstable in areas of active faults and cavernous limestone. Faults usually occur in swarms running along fault zones. Earthquakes can occur anywhere in the fault zone, and their destructiveness is governed by the magnitude of the energy release as well as by the nearness of the earthquake to the surface and its proximity to urban type development. In the case of cavernous limestone, the concern is with surface collapses and associated groundwater flows. Not all limestone is cavernous, but where it is, such as in central Florida, collapse features and caverns should be located and taken into account in land use planning.

3.7 SOURCES OF ENVIRONMENTAL DATA FOR SITE PLANNING

Most landscape planning and design problems require a host of information. Rarely, however, is the pertinent information readily available and what is available is usually not in a form suitable for application. Therefore, we are constantly faced with the problem of finding and/or producing data and transforming it into information, usually in the face of serious time and resource constraints. Clearly, making the right choices about how to go about this is essential early on in the project.

Firsthand observation

The Field Context. Despite the technological advances in the acquisition and processing of landscape data—and here we refer mainly to Geographic Information Systems (GIS)—the analysis and evaluation of the landscape still rely heavily on firsthand field investigations. This is especially so for problems involving small to medium-sized sites (several acres to several hundred acres), for which the resolution of secondary data sources, such as satellite imagery, soil maps, and standard topographic maps, may be marginal. Although these sources are helpful in understanding the regional context, the internal character of such sites may fall between the cracks, and it is important in mapping not to make things look more precise than the data allow.

For large problem areas we must, of course, rely mainly on secondary sources to gain a sense first of physiography and the scale, types, and diversity of landscape features, and second of the prevailing landscape systems, such as watersheds. Here GIS is especially well suited. Nevertheless, field observation is also necessary, but it must usually follow the examination of secondary sources and serve as a ground-truth exercise to test the validity of our initial ideas about the makeup and operation of the environment. Among the secondary sources, topographic contour maps are probably the most valuable.

Scale and coverages

Topographic Contour Maps. *Topographic contour maps* are published by the U.S. Geological Survey (Department of Interior) as part of the U.S. comprehensive national mapping program. In Canada similar topographic maps are issued by the Department of Energy, Mines, and Resources and by various provincial mapping programs. However, Canada has no national mapping program comparable to that of the United States and therefore the coverage is piecemeal. In the United States, coverage ranges from the entire coterminous United States on one sheet to coverage of a local area of about 55 square miles. The latter sheets, called *7.5-minute quadrangles* (because they cover about 7.5 minutes of latitude) are the most useful for planning purposes. The comparable sheets in Canada are the 1:10,000 (1-centimeter:100-meter) topographic base maps.

The 7.5-minute quadrangles and their correlatives in Canada are excellent sources of information on drainage systems, topographic relief, and slopes, and they are helpful in locating land use features, water features, and wooded areas. The 7.5-minute maps are printed at a scale of 1:24,000 (about 0.4 mile to the inch) with a contour interval of 10 feet (some newer maps may include a smaller contour interval). Today most planning projects involving major facilities call for highly detailed topographic maps at much larger scales (often as large as 1 inch to 100 feet or 1 inch to 50 feet). For these maps, aerial mapping companies must be specially contracted.

Information and resolution

Soil Maps. *Soil maps* in the United States are prepared by the U.S. Natural Resources Conservation Service (formerly the U.S. Soil Conservation Service) and published county by county in booklets called *county soil reports*. These map reports give the classification and description of soils to a depth of 4 to 5 feet as well as an indication of the representative slope of the ground over the area covered by each soil type. The soil type and slope are printed on the map in a letter code. Most boundary lines between soil types are highly generalized and should be scrutinized in the field for site planning problems. Unfortunately, the scale of the soil maps (usually at 1:20,000) is larger than that of the 7.5-minute U.S. Geological Survey topographic maps, prohibiting easy compilation of soil and topographic data on a single map. Chapter 5 discusses soil maps in more detail.

Joint photo program

Aerial Photographs. *Aerial photographs* are available for virtually all areas of the United States and Canada. In the United States they were regularly produced by various governmental agencies, including the U.S. Natural Resources Conservation Service, the U.S. Forest Service, and the U.S. Bureau of Land Management. Today a joint program called the National Aerial Photography Program (NAPP) works on behalf of these and other federal agencies, and the photographs (both old and new) are available from the EROS Data Center in Sioux Falls, South Dakota. Standard aerial photographs (or orthophotographs) are available in 9-inch-by-9-inch formats in black-and-white prints that are suitable for stereoscopic (three-dimensional) viewing. Individual photographs can be enlarged for planning purposes to any desired scale, and, although they cannot be used as a source of precise locational information (because of inherent photographic distortions), aerial photographs are an excellent source of information on vegetation, land use, and water features, especially when used in combination with other sources of geographic information.

Scanners and radars

Satellite Imagery. In addition to aerial photographs, a wide variety of *nonphotographic imagery* is available today. This imagery is provided mainly by scanners and radars mounted in satellites that relay data to earth receiving stations. Much of this imagery is produced for meteorological purposes and lacks the spatial resolution needed for most planning problems. However, several satellite systems show promise for landscape planning, including the *Landsat* satellites, which have been in operation since 1972, and the *SPOT* satellites, which have been in operation since 1986. The newest among the Landsat satellites is a scanner system called *Thematic Mapper*, which provides resolution as fine as 30 by 30 meters in six bands. The high-resolution SPOT satellite provides even finer resolution (10 by 10 meters) in one visible band. Landsat, SPOT, and other satellite imagery can be acquired from the EROS Data Center.

Special Sources. Increasingly, *special sources* of data and information are available, especially for populous regions. Many states and counties contract their own aerial photographic surveys on a regular basis with imagery in black and white, color, and infrared formats. In addition, information on water resources, wetlands, and other landscape features and resources is available for selected areas as a result of research projects, environmental impact reports, planning projects, and the efforts of local environmental organizations such as watershed and streamkeeper groups.

Governmental sources

Among the special sources are various offices and programs within *governmental organizations*. The U.S. Geological Survey, for example, is active in research in every state. Besides topographic maps, the Survey also offers a wide range of other maps, as well as data and reports for various local and regional problems and resources. These include earthquake hazard maps, stream discharge records, groundwater surveys, maps of geological formations, and regional- and state-based reports on the nation's water resources. The same generally holds for the U.S. Environmental Protection Agency (EPA) and the National Oceanic and Atmospheric Administration (NOAA), each of which produces a variety of maps, reports, and data on different topics for different regions and communities. The body of special information sources is huge and growing rapidly and space prohibits us from describing them here. Indeed, it is probably fair to say that any individual would find it impossible to keep track of this mass, especially for areas beyond his or her own theater of operation. Therefore, in the face of a planning problem in any location, it is advisable to move quickly to local clearinghouses such as planning commissions, colleges and universities, and environmental organizations (both governmental and nongovernmental) to find the special sources.

Nongovernmental sources

Nongovernmental organizations (NGOs) have become valuable sources of environmental data and information. NGOs have been around a long time, but until the past decade or two they concentrated almost exclusively on government lobbying activity. Now, however, many are also sponsoring projects that generate useful environmental data. Those that operate at a global scale are of limited value in site planning, but those that operate locally are often surprisingly helpful in site planning projects. Hundreds of organizations in a state or province might be involved, for example, in stream and watershed monitoring and restoration, wetland protection, and parkland acquisition and management.

3.8 THEORETICAL PERSPECTIVES ON LANDSCAPE

We should not conclude our overview on landscape without touching on some of the major conceptual statements that have emerged over the past century or so. Broadly speaking, these concepts constitute theoretical statements on landscape change and development, and, although no one of these concepts is avidly followed today, each has contributed to our understanding of landscape and its formation. To some extent, the concepts represent the framing paradigms of landscape studies and as such lend perspective to landscape planning and design.

Evolution-based thinking

If we ignore early theological and geological debate on the origins of the earth, the first major theoretical statements on landscape emerged in the late 1800s at a time when scientific thought was strongly influenced by evolution theory. In its simplest form, the central idea in evolution theory was change over time. The manner of this change was often conceived as a developmental process in which organisms, or whatever phenomena were being examined, tended to change sequentially or stagewise through time. In the late 1800s, for example, Russian soil scientists proposed that soils, like organisms, evolve toward a mature state in response to their environment, particularly the bioclimatic environment. A little later, the American geographer W. M. Davis articulated a unified model of landscape evolution based on stream erosion, which was to become a major theme in twentieth-century educational circles.

Landform evolution

The Geographic Cycle. Davis called his concept the **geographic cycle,** and in it he proposed that landscapes evolve through a series of developmental stages as streams deepen and widen their valleys. He envisioned three main stages of development, which he named *youth, maturity,* and *old age.* In the youthful stage the landscape is rugged and valleys are V-shaped as streams cut into the land. In maturity, valleys have begun to widen and take on U-shapes, and in old age valleys are very wide with no intervening uplands (Fig. 3.7).

The fluvial landscape

Davis's interests were in landforms, but it is apparent that the geographic cycle also applies to the other components of landscape. From youth to old age, the landscape becomes increasingly fluvial, or stream controlled, evolving into a low-relief landscape dominated by broad floodplains, deep soil mantles, and great corridors of riparian vegetation. Indeed, we find these landscapes in abundance across the earth, but they tend to be associated more with geographic position in watersheds than with a stage of evolutionary development of a land mass. Likewise, mature and youthful landscapes are also plentiful, but they too appear less stage controlled and more related to position in drainage systems. Generally, terrain is more youthlike in the upper reaches of watersheds and more old agelike in the lower reaches. The Mississippi system generally follows this pattern. Davis gave the term *peneplain* to the broad lowland of old age, and he proposed that over geologic time peneplains were eventually uplifted and the cycle of erosion and landscape evolution renewed.

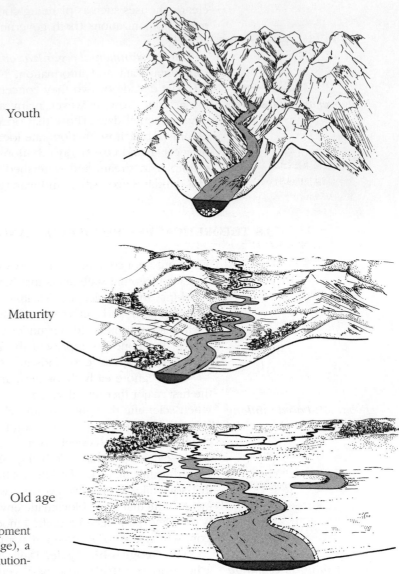

Youth

Maturity

Old age

Fig. 3.7 The three main stages of landscape development in the geographic cycle (youth, maturity, and old age), a concept of landscape change based roughly on evolutionary theory.

Some contributions

Although the geographic cycle is not widely supported in modern scientific circles, several of the concepts on which it is based are still embraced. One of these is *base level*, the lowest elevation to which a stream can downcut into the land. For rivers terminating in the sea, base level is set by sea level. Inland, the base level may be set by a lake, wetland, another stream, or a reservoir. Reservoirs set artificial base levels that disrupt the normal relationship of a stream to the landscape and interrupt the flow of sediment. A related concept is *grade* or *graded profile*, which is the upsloping longitudinal profile of a river. From mouth to headwaters, streams gradually increase in elevation, thus forming a long concave profile (Fig. 3.8). The graded profile reflects an equilibrium condition in a stream system representing a balance among channel slope (gradient), rate of flow, and the ability to move sediment. In other words, as the stream develops a graded profile, it moves toward an equilibrium or near equilibrium condition.

Dynamic Equilibrium Concept. The observation that stream systems trend toward equilibrium prompted another theoretical concept about river-sculptured

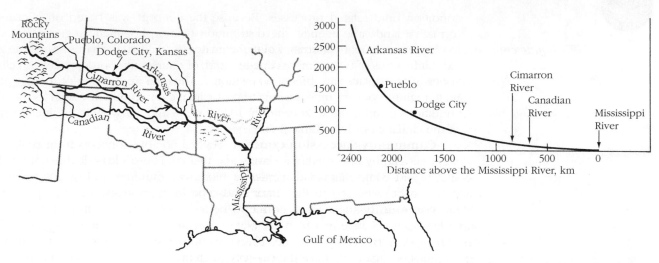

Fig. 3.8 The elevation profile of the Arkansas River illustrating the concept of the graded profile. The Arkansas' base level is the Mississippi River, whereas the Mississippi's is the ocean (Gulf of Mexico).

Landscapes as energy systems

landscapes based on the energy systems theory. This concept, called **dynamic equilibrium**, was advanced around 1960. In sharp contrast to the evolutionary perspective of the geographic cycle, the dynamic equilibrium concept argued that stream systems (and their landscapes) function as energy systems and as such are always trending toward a steady state. Instead of winding down to lower and lower energy levels (toward entropy) with the advancing stages of the geographic cycle, according to this concept the stream's energy system may increase or decrease with changes in the factors controlling its energy: climate (more precipitation means greater potential energy), runoff (more runoff means greater kinetic energy), and land elevation (higher elevation means greater potential energy).

Anticipating outcomes

Landscape change within watersheds need not be progressive and toward flatter, lower ground (with deep soil mantles), but can move in any direction (e.g., valleys may be downcut, filled, or maintained) depending on changes in the driving forces in the system. We see such trends in today's landscape, such as increased channel erosion and valley deepening at the local scale in response to greater stormwater runoff with land clearing and urban development. With global warming in the twenty-first century, sea level rises are projected to raise the base levels of streams draining to the ocean and in turn reduce potential energy in the lower parts of streams. The implications for landscape change include increased flooding and sedimentation in coastal lowlands.

Climate-based landscapes

Morphogenetic Regions Concept. Neither the geographic cycle nor the dynamic equilibrium concept addressed the influence of earth's varied climatic conditions on landscape development. W. M. Davis himself recognized this apparent shortcoming, and he and others wrote about an "arid land cycle" that addressed landform development peculiar to deserts. Later an attempt was made to build a broadly based concept incorporating more or less the full range of climate types, as well as the geomorphic processes and landscape types associated with each. This concept, called **morphogenetic regions**, examined landscape as the product of different climatic regimes rather than only stream erosional systems.

With this concept, much of the emphasis was on surface processes such as weathering, and it underscored the differences, for example, between the wet tropics with strong chemical weathering and periglacial environments dominated by permafrost

Climate change

conditions and related processes. Because the concept was based on climate as a formative landscape agency, the distribution of morphogenetic regions should follow the broad zonal patterns of global climates. But climates change. So some zones exhibit features of two or more climates, and, of course, the borders between climatic zones in most places are broadly transitional. Thus the application of the morphogenetic regions concept must recognize past climatic environments, such as the glacial conditions of midlatitude zones only 10,000 years ago, as well as the inexact nature of the climatic borders (Table 3.1).

More evolution-based thinking

Community-Succession Concept. Several notable concepts from ecology and plant geography also qualify as landscape theory. These ideas deal mainly with the spatial and compositional dynamics of biotic communities and/or ecosystems, but they can be extended to the larger landscape in many respects. The first of these, the **community-succession concept**, belongs with the evolution-based concepts of landscape change in which the plant cover and related biota on a parcel of land change stage by stage over time. This concept is based on the observation that communities of organisms have the capacity to change the surface environment in terms of soil conditions, moisture, ground-level climate, and so on, making it suitable for other communities. When these other communities become established, they in turn render further change making things suitable for yet other communities.

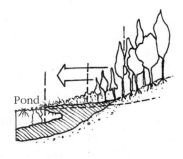

Pond

The succession model usually begins with a pond or a wetland or a denuded (barren) surface, such as ground newly exposed by a retreating glacier, a harsh fire, or intensive wind erosion, onto which a community of resilient organisms, called *pioneers*, becomes established. The plants are usually small and hardy (e.g., algae, mosses, and lichens), and they prepare the site, as it were, for another community, which is larger and more complex than the first. Each community represents a *successional stage*, and each changes the landscape, making it more favorable for other biological communities, until eventually a relatively enduring community, called a *climax community*, is established. This community represents a state of bioequilibrium within the larger climatic environment.

Successional stages

The community-succession concept has been applied to all sorts of landscapes and at scales ranging from the local to the broadly regional. Like the geographic cycle,

Table 3.1 Morphogenetic Regions and Their Characteristics

Region	Present Climate	Past Climates	Active Processes[a]	Landforms
Zone of glaciers	Glacial (cold; wet)	Glacial	Glaciation	Glacial
Zone of pronounced valley formation	Polar, tundra (cool; wet, dry)	Glacial, polar, tundra	Cryogenic processes; stream erosion; mechanical weathering; (glaciation)	Box valleys; patterned ground; glacial forms
Extratropical zone of valley formation	Continental (cool, temperate; wet, dry)	Polar, tundra, continental	Stream erosion; (ground frost processes) (glaciation)	Valleys
Subtropical zone of pediment and valley formation	Subtropical (warm; wet, dry)	Continental, subtropical	Pediment formation; (stream action)	Planation surfaces and valleys
Tropical zone of planation[b] surface formation	Tropical (hot; wet, wet–dry)	Subtropical, tropical	Planation; chemical weathering	Planation surfaces and laterite

[a] And former processes
[b] Processes leading to a low plainlike surface.

Source: Adapted from Büdel, 1963.

Validity questions

it is intellectually seductive because of its simplicity and apparent validity based on casual field observations in old farm fields, brownfields, and the like. Plant distributions do indeed change over time and space, but whether this concept accurately describes or even approximates the nature of that change is questionable based on rigorous field research. Rather, the sequence can apparently begin with any one of various assortments of biota and trend toward simpler or more complex communities depending on the types and magnitudes of forces shaping the habitat.

Environment pushes back

Disturbance Theory. This concept is in most respects a counterstatement to the community-succession concept. Rather than plants and related organisms acting as the principal change agents in a relatively passive or nurturing environment, disturbance theory holds that external (mainly abiotic) forces exert a shaping or controlling influence on the makeup and distribution of plant and animal communities in the landscape. The forces, including fire, floods, drought, diseases, volcanic eruptions, hurricanes, land use, and pollution, operate at different magnitudes and frequencies around and in ecosystems. The organisms that make up ecosystems survive or perish depending on whether the impact of these forces, acting individually or collectively, exceeds their tolerance limits. When these forces are strong, such as with a hurricane, volcanic explosion, or construction activity, ecosystems in the affected area may be damaged, reduced, or destroyed.

Adjusting to change

Disturbance theory argues that the earth's surface is characterized by a more or less continuous string of such events operating at different magnitudes and frequencies and that ecosystems, including soils, drainage, and other elements of the landscape, are continuously adjusting to these forces. In other words, the distributional patterns of ecosystems we see on the earth's surface are in a constant state of flux, and progressive change ending in a climax state, as the succession concept argues, is too simple an explanation for this process. Thus, the disturbance concept views the environment's behavior as more irregular or chaotic than the succession concept does (Fig. 3.9).

View on time

Much of what we see as evidence of succession or disturbance depends on our perspective of time. If we look at the segments of time in the intervals between powerful events, succession appears to be a reasonable model. These intervals are relatively quiescent periods when growth and infilling can take place. However, if

(a)

(b)

Fig. 3.9 (*a*) The Pisgah Forest in New England, a 300-year-old climax forest, before the 1938 hurricane. (*b*) The forest was destroyed in one day by the hurricane, placing the notion of stability in climax communities in doubt.

we take a longer view of time, then ecosystem changes look more like fluctuations in response to environmental perturbations, and a particular spatial trend may or may not be apparent.

Humans as a disturbance

No review of landscape concepts could ignore the human factor. When we consider the actions of humans, the time scale is especially critical. Set into geologic time (millions of years), the human epic on earth is a single event or disturbance of global proportions, the end of which we have yet to see. It is one of the major events of earth's recent geologic history. Set into historical time (several thousand years), however, the human epic can be broken down into thousands of events in different places with different effects. Examples are:

- The land degradation in the Mediterranean Basin in ancient Greek and Roman times.
- Industrialization and urban development in Europe in the eighteenth and nineteenth centuries.
- Plowing of the North American prairies in the nineteenth and twentieth centuries.
- The destruction of large areas of tropical rainforest in the twentieth and twenty-first centuries.

Limited recovery

In many instances, some ecological recovery, which might be regarded as succession, takes place after major events; however, it is rarely complete. The Mediterranean lands can no longer support the forests and grass covers that existed before Christ. The lands in the Great Plains, ravaged by erosion during the North American Dust Bowl of the 1930s, once again support grasses, but the cover is weaker with fewer species (Fig. 3.10). Overall, the trend in historical times has been toward increased magnitude and frequency of disturbance by humans as our numbers, our consumption of resources, and our occupancy of the planet's surface have expanded. The periods or windows of time available for recovery, in turn, are shortened, leading to a lower order of quality in landscapes with fewer species, lower productivity, and often less resilience to future disturbances.

Systems disorder?

Chaos Theory. Some observers argue that the global environment has become more chaotic in the past several decades, and indeed **chaos theory** is receiving serious attention today. Among other assertions, chaos theorists cite human-induced

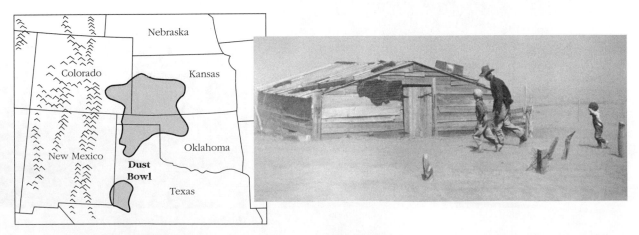

Fig. 3.10 A severe landscape disturbance. The North American Dust Bowl of the 1930s where a combination of dry land plowing and drought resulted in degraded prairie ecosystems, extensive soil erosion, and reduced agricultural potential.

Fig. 3.11 Devastation from the 2010 Haiti Earthquake and Hurricane Ike (2008). Such disasters may suggest a trend toward a riskier and more chaotic world.

environmental change such as increased runoff and flooding related to urban development, increased ocean storminess related to global warming, and advancing land degradation related to desertification. Whether environmental systems such as those of the atmosphere, watersheds, periglacial lands, coastlines, and biota are trending toward less orderly states is uncertain, but plenty of indicators are cause for serious concern. At the top of the list is atmospheric warming, which appears to be giving rise, among other things, to heightened storm magnitude and frequency, including hurricanes, the breakdown of permafrost landscapes, and the increased desiccation of arid and semiarid lands.

Vulnerable locations   Added to this is a world population growing at nearly 80 million persons a year, with more people occupying disturbance-prone environments than in the past. Marginal environments that were traditionally beyond the bounds of significant agriculture and settlement, such as steep mountain slopes, deserts, stormy coastlines, and flood-prone river valleys, are clearly subject to more erratic changes and disturbances than nonmarginal environments. As population growth and economic development drive land uses farther into marginal environments, our susceptibility to disturbance rises. Natural disasters not only seem more common and more destructive, but perhaps they are (Fig. 3.11). Not surprisingly, as these experiences mount, nature appears to be less knowable, less predictable, and more chaotic. Chaos theory may emerge as the leading landscape concept of the twenty-first century.

3.9 CASE STUDY

Dams, Sustaining Streamflow, and Landscape Equilibria: An Illustration from Fossil Creek, Arizona

Charlie Schlinger and Lorrie Yazzie

We have long recognized the relationship among a watershed, the size of the streams draining it, and the throughput of sediment. This is one of nature's most basic landscape systems. Its work involves the transfer of water and sediment downslope, ultimately moving both to the sea as parts of much larger earth cycles. To do this work, the watershed/stream system must undergo constant adjustment as it seeks to balance itself in response to a wide array of changes in its environment. In nature, these changes take the form, for example, of variations in discharge with different rainfall events or changes in sediment supply with changes in vegetation related to forest fires, or changes in channel roughness when trees or boulders fall into the stream.

A dam throws such a system seriously out of balance. Among other things, the continuity of flow is interrupted, and, because flowing water is needed to move sediment, the transfer of sediment is also interrupted. The sediment, of course, is deposited in the slack water behind the dam, interrupting or even breaking the train of sediment transport through the system. Below the dam the channel is starved of sediment.

How does the system respond to this sudden change? It seeks a new equilibrium to correct the imbalance. For Fossil Creek in northern Arizona, this took the better part of a century, beginning in 1916 when the Fossil Springs Diversion Dam was built across the stream to supply water for a hydroelectric power facility. The dam was sited at a location where there already was a considerable drop over a bedrock ledge in the channel (Fig. 3.A). The dam created a 25-foot-tall barrier to streamflow, with a reservoir extending upstream approximately 700 feet and narrowing from 100 feet at the dam to about 20 feet at the upper end. The valley consists of steep bedrock on one side, and on the other is a cliff face formed by a large travertine mound.

With the dam in place, almost the entire spring flow was diverted, and the 25,000 cubic yards of reservoir storage was available to trap sediment brought in by storm flows. Within several decades, the reservoir was filled with sediment. (This was of little concern to the power company because it did not impact their water diversion.) The sediment mass created a new landform with a surface elevation even with the crest of the dam. On this surface a stable stream channel and riparian ecosystem developed, and once again sediment moved through this reach of channel. A new equilibrium had been established (Fig. 3.B).

Fig. 3.A The Fossil Creek dam as it appeared in 2002, nearly 85 years after its construction.

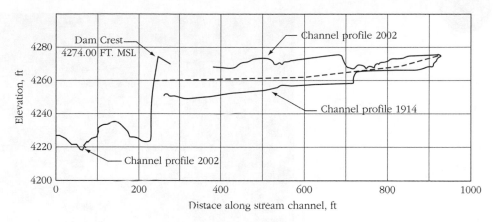

---- Approximate channel profile after removal of upper dam

Fig. 3.B Channel profiles (gradients) in 1914, 2002, and after partial dam removal.

With the new channel came a new valley floor with all the attributes, more or less, of a natural stream valley: a meandering channel with fluctuations in discharge and cutting and filling; an alluvial plain (floodplain); riparian vegetation; native fish and other fauna, including a transient community of hikers, campers, naturalists, and sightseers.

Toward the end of the twentieth century, environmental groups—including American Rivers, Center for Biological Diversity, Northern Arizona Audubon Society, The Nature Conservancy, and the Sierra Club–Grand Canyon Chapter, along with the Yavapai-Apache Nation—began campaigning for the restoration of flow to Fossil Creek and hydroelectric facility decommissioning, including removal of the Fossil Springs Diversion Dam. By 1999, an agreement for flow restoration and facility decommissioning was reached with the owner, Arizona Public Service Co.

Three dam removal alternatives were considered and evaluated: no action, complete dam removal, and partial dam removal. Much of the decision hinged on the fate of the sediment mass and how it would respond, first, to the stream's modest 46-cubic-feet-per-second (cfs) baseflow, and second, to much larger storm flows. Forecasts of sediment loss were developed, and it appeared entirely possible that large flows of 500 cfs or more would have the power to sweep large volumes of sediment downstream if the dam were lower or removed. This analysis, while only approximate, provided regulators and the concerned agencies with some idea of the risks, such as downstream sedimentation, imposed by the dam removal options. However, more vexing questions arose:

- What would be the impacts to the equilibrium environment the stream had created over 90 years?
- How much time would be needed to establish a new set of equilibria?
- What would be the quality of the next generation of environments?

These questions were not so easily answered and did not drive the decision making.

Based on its long-term natural resource management objectives for this remote watershed, the U.S. Forest Service chose to remove the top 14 feet of the dam. Lowering was completed in September 2008, and the stream, sustained primarily by its base flow, began downcutting, creating a fairly narrow and steep-walled channel through the sediment wedge as it worked its way toward a new base level. But large amounts of sediment remained upstream of the dam, and a truly large

Fig. 3.C The aftermath of the December 2008 stormflow channel lowering, alluvial plain erosion, and sediment removal. Notice that the riparian area is now stranded above the valley floor.

flood event could wash much of it away overnight. In late December of 2008, a moderate rainstorm struck the region, driving Fossil Creek to a peak flow estimated at 1200 cfs and removing much of the sediment and riparian vegetation above the dam. This left the alluvial plain and its riparian ecosystem stranded above the channel and thus without water and sediment inputs from the stream (Fig. 3.C).

What is the lesson? The problems we face in environmental remediation and restoration are complex. At one level is the argument that we should pursue every opportunity to remove prior human interventions, such as dam construction, and rehabilitate the environment. And there are many sound reasons behind this argument, not the least of which is the risk of dam failure or the potential to reestablish fish migration routes. But there are also opposing arguments, especially when nature has incorporated past actions into its working whole and has established new equilibria not just for the stream, but for many other systems tied to the stream environment.

In this case, the Forest Service was unwilling to manage a dam that had outlived its function as an engineered facility, citing public safety as one of the key motivations, and environmental organizations supported the dam's removal based on a long-held principle that we as a society should make every effort to free stream systems of such heavy-handed interventions. The partial removal of the Fossil Creek Dam represents a reasonable compromise that met the majority of stakeholders' objectives but that leaves to future stakeholders the question of what to do about the remainder of the dam and the systems that will become dependent on it.

Charlie Schlinger is a faculty member in the Department of Civil and Environmental Engineering at Northern Arizona University and Lorrie Yazzie is a Field Inspector for the Arizona Department of Environmental Quality.

3.10 SELECTED REFERENCES FOR FURTHER READING

Brunsden, D., and Thornes, J. B. "Landscape Sensitivity to Change." *Transactions of the Institute of British Geographers. 4*, 1979, pp. 463–484.

Clements, F. E. *Plant Succession: An Analysis of the Development of Vegetation.* Washington, DC: Carnegie Institution No 242, 1916.

Davis, W. M. *Geographical Essays* (D. W. Johnson, ed.). New York: Dover Publications, 1954.

Hack, J. T. "Interpretation of Erosional Topography in Humid Temperate Regions." *American Journal of Science.* Bradley 258-A, 1960, pp. 80–97.

Marsh, W. M., and Dozier, J. "Magnitude and Frequency Applied to the Landscape." In *Landscape: An Introduction to Physical Geography.* Hoboken, NJ: Wiley, 1981.

Mitchell, C. W. *Terrain Evaluation,* 2nd ed. New York: Longman and Wiley, 1991.

Nikioroff, C. C. "Reappraisal of Soil." *Science.* 3383, 1959, pp. 186–196.

Peltier, L. C. "The Geographic Cycle in Periglacial Regions as It Is Related to Climatic Geomorphology." *Annals of the Association of American Geographers. 40,* 1950, pp. 214–236.

Raup, H. M. "Vegetational Adjustment to the Instability of the Site." *Proceedings and Papers of the Sixth Technical Meeting.* Edinburgh: International Union for the Conservation of Nature and Natural Resources, 1959.

Selby, M. J. *Earth's Changing Surface: An Introduction to Geomorphology.* Oxford: Clarendon Press, 1985.

Wilcock, P. R., and Iverson, R. M. (eds.). *Prediction in Geomorphology.* Washington, DC: American Geophysical Union, 2003.

Wolman, M. G., and Gerson, R. "Relative Time Scales and Effectiveness of Climate in Watershed Geomorphology." *Earth Surface Processes. 3,* 1978, pp. 189–208.

Wolman, M. G., and Miller, J. P. "Magnitude and Frequency of Forces in Geomorphic Processes." *Journal of Geology. 58,* 1960, pp. 54–74.

Related Websites

Environmental Protection Agency. "Wetlands, Oceans, and Watershed." 2009. http://www.epa.gov/owow/nps/
Information about healthy watersheds and developing watershed plans to achieve healthy systems. The site offers data sources and looks at planning process.

Natural Resources Canada. "Earth Sciences." 2009. http://ess.nrcan.gc.ca/index_e.php
Canadian government website providing geographic maps for the country. Follow links to view hydrogeological and watershed maps. The mapping link provides aerial photos, satellite imagery, and other maps. Also view landscape photos with descriptions of the causal processes.

Natural Resources Canada. "Geological Survey of Canada." 2009. http://gsc.nrcan.gc.ca/index_e.php
Governmental agency researching Canada's geology. A "surficial materials" map can be downloaded showing the geology and hydrography of Canada.

United States Geological Survey. Home page. 2008. http://www.usgs.gov/
Source on geology and hydrology information. For helpful description and diagrams about groundwater, follow the "Water" link under "Science Areas."

United States Geological Survey. "Earth Resources Observation and Science (EROS) Center." 2009. http://eros.usgs.gov/
A governmental site providing maps and imagery for the United States. Follow the data link to view aerial photos, satellite imagery, land cover maps, and elevation maps. Following the "Science" link will lead you to landscape dynamics information.

TOPOGRAPHY, SLOPES, AND LAND USE PLANNING

4.1 INTRODUCTION

Slope attraction

Given the choice of a place to live, most of us will choose hilly over flat ground. This is not surprising, for to most people, hilly terrain is more attractive because it has greater variations in vegetation, ground conditions, and water features, to say nothing of the opportunities it affords for vistas and privacy in siting houses. Our success with establishing and nurturing land uses on hillslopes, however, is not equal to our love for them. In fact, for many land uses, slopes are decidedly inferior places as building sites because construction costs, risks of facility damage, and environmental damage are typically greater there. Many communities recognize these limitations and require information on slopes to help guide development decisions. However, a great many do not and should.

Slope and landscape

Conventional residential development is best suited to slopes of less than 15 percent (8 degrees) or so, though modern engineering technology encourages building on much steeper ground. Industrial and large-scale commercial developers prefer sites with slopes of less than 5 percent. The influence of slopes on modern roads depends on the class of the road: the higher the class, the lower the maximum grades allowable. Interstate class expressways (divided, limited access, four or six lanes) are designed for high-speed, uninterrupted movement and are limited to grades of 4 percent, that is, 4 feet of rise per 100 feet of distance. On city streets, where speed limits are 20 to 30 mph, grades may be as steep as 10 percent, whereas driveways may be as steep as 15 percent.

Slopes also have a pronounced influence on the biophysical character of the landscape. High on the list is their influence on runoff and ground stability. Runoff typically flows faster and with greater erosive power on large, steep slopes. Likewise, steeply sloping ground is usually the first to fail in slides and mudflows when the land is disturbed by earthquakes, road building, deforestation, and/or saturation from heavy irrigation, leaking water lines, or heavy rainfall. And the influence of slopes on microclimate and landscape ecology is often meaningful. In the midlatitudes, slopes with southern exposures are warmer and drier with fewer and smaller trees than their north-facing counterparts. The resultant differences in soil formation, runoff, plant species, and animal habitat are often striking.

Slopes and planning

On balance it is not surprising that the slope map is one of the most useful tools available for land use planning and site design in communities concerned with finding the right fit between land uses and landscape, with all that implies in terms of environmental impact, infrastructure costs, and safety. But building a meaningful slope map and related planning and design guidelines require more than reading a topographic contour map and coming up with a convenient set of slope classes. It also requires targeting the appropriate problems, understanding the nature of slopes as parts of landscape systems, and translating it all into meaningful information for planners and designers.

4.2 THE NATURE OF SLOPE PROBLEMS

Misuse of slopes

The need to consider topography in planning is an outgrowth of the widespread realization not only that land uses have slope limitations but also that slopes have been seriously misused in modern land development. The misuse arises from two types of practices:

1. The placement of structures and facilities on slopes that are already unstable or potentially unstable.

2. The disturbance of stable slopes leading to failure, accelerated erosion, and/or ecological deterioration of the slope environment.

The first practice can result from inadequate survey and analysis of slopes in terrain with a history of slope instability. More frequently, however, it probably results from public planning agendas that overlook slopes or inadequate planning controls (for example, zoning and environmental ordinances/bylaws) on development. In some instances, admittedly, surveys reveal no evidence of instability, and the failure of a slope catches inhabitants completely unawares.

Slope Disturbance. The disturbance of slope environments is probably the most common source of slope problems in North America. Three types of disturbances stand out:

Causes of disturbances

- *Mechanical cut and fill*, in which slopes are reshaped by heavy equipment. This often involves steepening and straightening, resulting in a loss of the equilibrium associated with natural conditions. In Canada and the United States this is best exhibited along highways and logging roads and in mining areas.

- *Deforestation* in hilly and mountainous terrain by lumbering operations, agriculture, fire, and urbanization. This results not only in a mechanically weakened slope because of the reduced stabilizing effect of vegetation, but also in changes in the soil moisture balance and groundwater conditions leading to increases in erosion and slope failure (see Fig. 3.4).

- *Drainage alteration* related to improper site grading and facility siting and construction, leading to changes in runoff patterns, flow types, discharge rates, and water table levels, all commonly resulting in landscape degradation, soil erosion, and/or slope failure (Fig. 4.1).

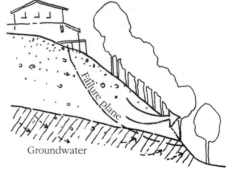

Fig. 4.1 Slope failure and erosion resulting from alteration of drainage, vegetation, and soil associated with residential development.

4.3 BUILDING SLOPE MAPS FOR LAND USE PLANNING

Years ago the topographic configuration of the land could be measured only by field surveying. In its simplest form, this involved projecting a level line into the terrain from a point of known elevation and then measuring the distances above and below the line to various points along the ground. Once elevation points were known and registered geographically to establish their locations, a contour map could be constructed.

The Data Base. Topographic contour maps are comprised of lines, called *contours*, connecting points of equal elevation. In modern mapping programs, such as the one practiced by the U.S. Geological Survey, the contours are drawn from specially prepared sets of aerial photographs. These photographs and the optical apparatus used to view them enable the mapper to see an enlarged, three-dimensional image of the terrain. Based on this image, the mapper is able to trace a line—the contour—onto the terrain at a prescribed elevation. The contour elevation is calibrated on the basis of survey markers, called *bench marks*, placed on the land by field survey crews before the aerial photographs were flown.

Topographic contours

To determine the inclination of a slope from a topographic contour map, we must know the scale of the map and the elevation change from one contour to the next, called the *contour interval*. With these, the change in elevation over distance can be measured, and in turn a ratio or percentage of the two numbers can be calculated:

Percent slope

$$percent\ slope = \frac{change\ in\ elevation}{change\ in\ distance} \times 100$$

Converting and Translating Slope. Percentage is one of the two quantitative measures for slope inclination, the other being degrees. Conversions can be made for degrees or percentages with the aid of a trigonometric table or with a nomogram, such as the one in Figure 4.2. Notice that the values on the two scales differ significantly; for example,

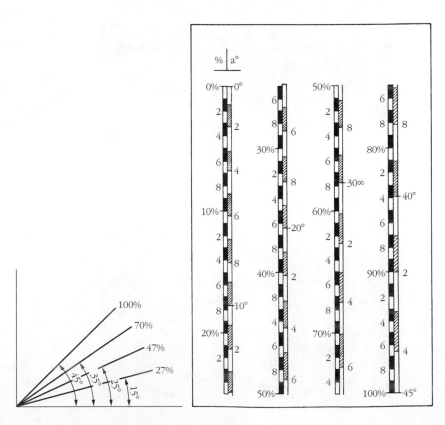

Fig. 4.2 Degree equivalence of percent slope up to 100 percent.

100 percent is equal to 45 degrees. Degrees are commonly used in engineering calculations involving slope stability problems, whereas percentages are the standard units used in landscape planning and design for slope classification, mapping, and grading.

Setting slope limits   To avoid damage to the environment and costly site manipulation related to cut-and-fill and drainage alterations, it is necessary in land use planning to make the proper match between land uses and slopes. In most instances this is simply a matter of assigning to the terrain uses that would (1) not require modification of slopes to achieve satisfactory performance and (2) not themselves be endangered by the slope environment and its processes. Generally speaking, topographic contour maps alone do not provide information in a form suitable for most planning and design problems. The contour map must instead be translated into a map made up of slope classes tailored to a particular planning problem such as the upper limits of residential development. The utility of such slope maps is a function of (1) the criteria used to establish the slope classes and (2) the scale at which the mapping is undertaken.

The scale of mapping and the level of detail that are obtainable are strictly limited by the scale and contour interval of the base map. When maps of two or three different scales are available, the scale chosen should be the one that best suits the scale of the problem for which it is intended, such as community master planning, highway planning, or site design.

Defining mapping classes   **Defining Slope Classes.** The criteria used to set the slope classes depend foremost on the problems and questions for which the map will be employed. The area mapped in Figure 4.3 is hilly and subject to heavy off-road vehicle traffic. The problem is how to manage this traffic with the least environmental damage such as soil

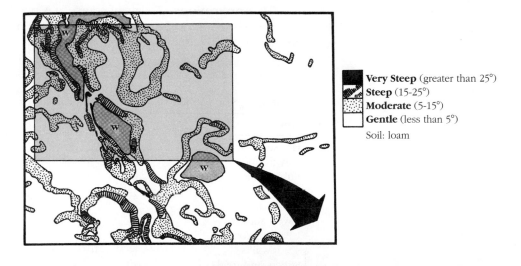

■ **Very Steep** (greater than 25°)
▨ **Steep** (15-25°)
▥ **Moderate** (5-15°)
□ **Gentle** (less than 5°)
Soil: loam

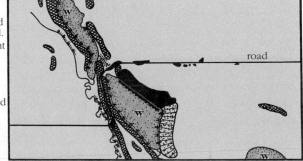

■ **Presently active** with erosion and/or failure underway.

▨ **Highly susceptible** to failure and erosion should forest be removed.

▨ **Erosion imminent** under present use.

□ **Least susceptible** to failure and erosion under agricultural, residential, and related uses.

⟋ Natural waters presently influenced by sedimentation from eroded slopes.

⟋ Natural waters highly susceptible to sedimentation should nearby slopes be activated.

Fig. 4.3 Slope classification for the purpose of identifying slopes prone to erosion and failure. The area around the water features is abandoned farmland.

erosion and lake sedimentation. Information on slope, relating to both steepness and condition, is essential to development of a management plan.

For areas under the pressure of suburban development, the question is which land uses and facilities are appropriate and inappropriate for which slopes? One set of criteria in addressing this question is the maximum and minimum slope limits of the various activities and facilities. Another is the natural limitations and conditions of the slopes themselves, which are taken up in the next section. With respect to land use activities, we want to know the optimum slopes for parking lots, house sites, residential streets, playgrounds and lawns, and so on (Table 4.1).

In addition to the selection of appropriate slope classes and a proper base map, the preparation of a slope map also involves:

Building slope maps

1. *Definition of the minimum size mapping unit.* This is the smallest area of land that will be mapped, and it is usually fixed according to the base map scale, the contour interval, and the scale of the land uses involved. For 7.5-minute U.S. Geological Survey quadrangles (1:24,000), units should not be set much smaller than 10 acres, or 660 feet square.

2. *Construction of a graduated scale* on the edge of a sheet of paper, representing the spacing of the contours for each slope class. For example, on the 7.5-minute quadrangle, where 1 inch represents 2000 feet and the contour interval is 10 feet, a 10 percent slope would be marked by a contour every 1/20 inch.

3. *Placement of the scale on the map* in a position perpendicular to the contours to delineate the areas in the various slope classes (Fig. 4.4).

4. *Coding or symbolizing.* Each of the areas is delineated according to some cartographic scheme, preferably one that is tailored to the problem and audience.

GIS Applications. Today, of course, it is possible to use Geographic Information Systems (GIS) to plot slope maps and GIS does offer advantages, especially for large areas. But it also has some disadvantages. One is the loss of firsthand experience and

Pros and cons

Table 4.1 Slope Requirements for Various Land Uses

Land Use	Maximum (%)	Minimum (%)	Optimum (%)
House sites	20–25	0	2
Playgrounds	2–3	0.05	1
Public stairs	50	—	25
Lawns (mowed)	25	—	2–3
Septic drainfields	10–12[a]	0	0.05
Paved surfaces			
Parking lots	3	0.05	1
Sidewalks	10	0	1
Streets and roads		—	1
20 mph (32 km/h)	12		
30	10		
40 mph (64 km/h)	8		
50	7		
60 mph (96 km/h)	5		
70	4		
Industrial sites			
Factory sites	3–4	0	2
Lay-down storage	3	0.05	1
Parking	3	0.05	1

[a] Special drainfield designs are generally required at slopes above 12–15%.

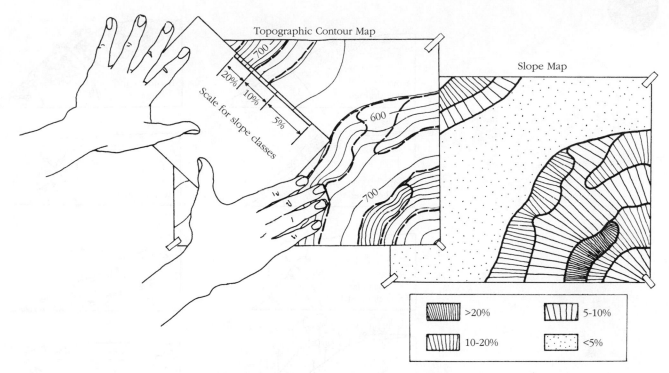

Fig. 4.4 Schematic diagram showing the use of a graduated scale for mapping slopes from a contour map. The lower map shows the results.

the attendant knowledge that comes with poring over a contour map and discovering associations among slopes, landforms, drainage features, land use facilities, and other site attributes. Another disadvantage is the odd, angular polygons that appear on maps generated from most computer mapping programs, polygon forms that tend to lack credibility as expressions of natural phenomena. On balance, it is probably advisable to use manual techniques for mapping small sites and to set up a manual test set for large ones before enlisting the computer.

4.4 INTERPRETING SLOPE STEEPNESS AND FORM

In addition to inclination, we must often know a number of other things about slopes in order to make meaningful interpretations for land use planning and design. A first consideration is slope composition or lithology, that is, the soil and rock material that comprise the slope.

Angle of repose

Composition and Steepness. For any earth material, there is a maximum angle, called the **angle of repose**, at which it can be safely inclined and beyond which it will fail. The angle of repose varies widely for different materials, from 90 degrees in strong bedrock to less than 10 degrees in some loose, unconsolidated materials. Moreover, in unconsolidated material it may vary substantially with changes in water content, vegetative cover, and the internal structure of the particle mass. This is especially so with silty and clayey materials. A poorly compacted mass of saturated clay

Soil considerations

and silt may give way at angles as low as 5 degrees (8 percent), whereas the same mass with high compaction and much lower water content can sustain angles greater than 45 degrees (100 percent). As a rule, in compositionally complex materials, angles of repose are difficult, if not impossible, to define with much accuracy because influencing factors such as moisture content and compaction are so variable.

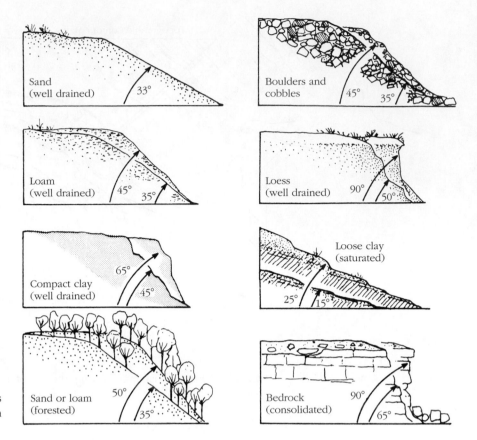

Fig. 4.5 Angles of repose for various types of slope materials. Angles are given in degrees.

In surface deposits of more homogeneous composition, however, the angle of repose can be defined with more confidence. This is especially so for coarse materials, such as sand, pebbles, cobbles, boulders, and bedrock itself, which are less apt to vary with changes in water content and compaction. Representative angles of repose for some of these materials and others are given in Figure 4.5. Beyond the angles shown, these materials are susceptible to failure in which the slope ruptures and slides, slumps, falls, or topples, or in the case of saturated materials of clay, silt, and/or loamy composition, flows downslope.

Influencing factors

Vegetation Controls. The influence of vegetation on slopes is highly variable depending on the type of vegetation, the cover density, and the type of soil. Vegetation with extensive root systems undoubtedly imparts added stability to slopes composed of clay, silt, sand, and gravel, but the influence is limited mainly to the surface layer where the bulk of the roots are concentrated. For very coarse materials such as cobbles, boulders, and bedrock blocks, the influence of vegetation is typically not significant unless large trees buttress loose rock. On sandy slopes such as sand dunes or coastal bluffs, the presence of a forest cover can increase the inclination by 10 to 15 degrees above the 33 degree repose angle, producing a *metastable* slope condition. *Metastable* refers to a state of conditional stability in which an oversteepened slope is held in place by an ingredient such as plant roots. Such slopes are sometimes marked by a distinctive convex profile.

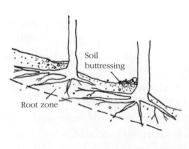

Dewatering factor

In addition to the mechanical influences of plant roots, vegetation also helps reduce soil water, which in large amounts can add significantly to slope instability.

Through transpiration, a forest can dewater soil by as much as 40 percent of the annual precipitation in some regions. Removal of vegetation can result in substantial rises in soil water and shallow groundwater, saturating and weakening surface materials. In California the loss of mechanical stabilizing and dewatering effects have frequently been cited as the causes of slope failure after the destruction of foothill vegetation by fire.

The Meaning of Form. Slope form or shape can also be an important factor in assessing slope stability and land use potential. Form is expressed graphically in terms of a *slope profile*, which is basically a silhouette of a slope drawn to known proportions with distance on the horizontal axis and elevation on the vertical axis. The vertical axis is often exaggerated to accentuate topographic details (Fig. 4.6).

Common slope forms

Five basic slope forms are detectable on contour maps: straight, S-shape, concave, convex, and irregular (Fig. 4.7). These forms often provide clues about a slope's makeup, past behavior, and potential stability. High on the list of considerations is an early map and aerial survey to identify typical and atypical slope shapes and conditions. For example, in areas where slopes are comprised of unconsolidated materials (soil materials and various types of loose deposits) and bedrock is not a controlling factor, slope form is usually smooth, the product of the interplay and balance among vegetation, soil composition, and runoff processes. Where this model is thrown out of equilibrium, that is, destabilized in some way, slope form and surface conditions are changed, often radically. Destabilization can be caused by natural and/or human-made influences, including undercutting by rivers, excavations for roads and buildings, wave erosion, gullying, earthquakes, and complications from past slope failures.

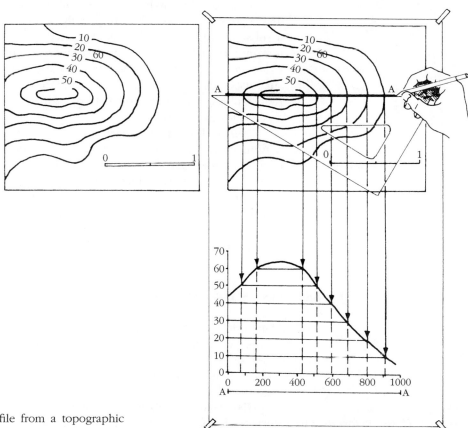

Fig. 4.6 Construction of a slope profile from a topographic contour map.

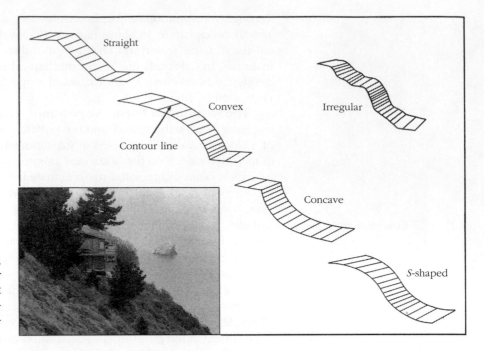

Fig. 4.7   Five basic slope forms: straight, convex, concave, *S*-shaped, and irregular with corresponding contour spacing as it appears on a topographic map. Inset photograph shows a dangerously steep convex slope leading down to the ocean.

S-shaped slopes   Smooth *S-shapes* usually indicate long-term slope stability and a state of equilibrium among stabilizing and destabilizing slope forces. Such slopes rarely exceed 45 degrees inclination and are usually secured against heavy erosion and failure by a substantial plant cover. Steep *S*-shaped slopes are commonly associated with hilly terrain, mountain foothills, and coastal areas where there is a permanent forest cover.

Concave slopes   *Concavities* in otherwise straight or *S*-shaped slopes are often signs of former failures, such as slides or slumps, and may indicate past disturbance from logging operations, road building, pipeline construction, and/or destabilizing subsurface conditions such as groundwater seepage.

Convex slopes may indicate the presence of resistant bedrock in the midslope, loading of the upper and middle slope by deposits (natural or human), and/or retreat

Convex slopes   of the footslope due, for example, to a road cut or wave erosion of a bluff (Fig. 4.8*a*). *Irregular* slopes may reflect a variety of conditions. Where bedrock is a factor, variations in the resistance of adjacent strata to weathering and erosion may produce alternating gentle and steep slope segments. In other instances rock units or deposits within a slope are unstable and behave erratically with changing moisture contents, giving rise to a chaotic slope profile.

4.5 ASSESSING SLOPES FOR STABILITY

Many criteria should be taken into account in assessing the susceptibility of slopes to failure. At the top of the list are slope angle, composition, and history of instability. Steep slopes composed of rock formations, surface deposits, and/or soil types with a discernible *record of instability* stand a greater chance of failure when subjected to high magnitude natural forces and disturbance from land use activity (Fig. 4.1).

History of Instability. As planners in California have shown, simply mapping

Unstable formations   unstable formations and their positions in steep slopes can provide highly valuable information for land use planning. These formations are often marked by visible failure scars imprinted into vegetation and soil covers (Fig. 4.8*b*). In time they are

Failure scars

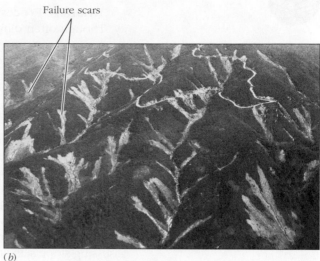

(a)

(b)

Fig. 4.8 (*a*) Long, convex slopes on the Nova Scotia coast. The toe is cut back by wave erosion and the middle is held in place by bedrock. (*b*) Scars left by failures and erosion on hillslopes stand out from the surrounding landscape.

modified by runoff and erosion, but even ancient scars are often detectable on the basis of drainage and topographic irregularities and subtle variations in the types and patterns of vegetation.

Deforestation　　**Geobotanical Indicators.** Plant cover is another important criterion inasmuch as devegetated slopes show a much greater tendency to fail than fully vegetated ones. Studies of slopes in the American and Canadian West indicate that deforested slopes and slopes cut by logging roads in some areas fail more frequently under the stress of heavy precipitation than do fully forested, undisturbed ones. From the air, scars caused by slope failure or severe erosion can often be identified on the basis of breaks in the plant cover. Fresh scars are marked by light tones on aerial photographs *Slope scars*　and older scars by different plant species and forms (Fig. 4.8*b*).

Although the stabilizing effects of vegetation are often not appreciated, there is no doubt about the overall effectiveness of a vigorous, dense plant cover, especially in reducing shallow failures and surface erosion. In addition to the strength added by roots, vegetation has many indirect effects on slope stability, including, as noted earlier, a remarkable capacity to dewater wet soils through transpiration. Accordingly, irregularities in an otherwise continuous and healthy plant cover are often clues to slope problems or potential problems.

Road cuts and shaking　　**Undercutting and Earthquakes.** Slope undercutting and earthquake activity are also significant. Active erosion at the foot of a slope by waves, rivers, or human excavation produces steeper inclinations and less confining pressure on the lower slope, thereby increasing the failure potential (Fig. 4.8*a*). Road cuts are particularly important on steep hillsides not only because the slope is physically weakened by the cut, but the cut may also intersect groundwater, activating seepage and runoff that destabilize surface materials. When earthquakes jar rock and soil material, interparticle bonds may be weakened and the material's resistance to failure reduced. Some of the worst disasters from slope failure have been triggered by earthquakes, especially where material such as snow, rock debris, and sediments is perched on high slopes that are already near the failure threshold. The 1994 Northridge, California, earthquake (magnitude 6.7) triggered more than 11,000 slope failures over an area of 6,200 square miles.

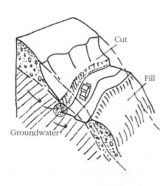

Saturation and runoff

Drainage Processes. Drainage is a major consideration in assessing slope stability. Though often difficult to evaluate, runoff processes, soil water, and groundwater can have a pronounced influence on slope conditions and stability. A number of factors contributed to the failure shown in Figure 4.9, but typically with a great many West Coast failures, water was the triggering factor. Although most rainfalls and runoff events are incapable of weakening slopes, especially vegetated ones, heavy rainfalls can lead to soil saturation, transforming surface material into a muddy slurry prone to shallow mudflows. (Researchers in California have found that once winter rainfall exceeds 10 inches, an intensive rainfall delivering more than 4 inches in 6 hours to mountain slopes can trigger mudflows.) With surface saturation, the soil loses its infiltration capacity, and rainfall is converted directly into surface runoff. The runoff easily erodes the muddy surface, and where the flows concentrate and quicken, they may cut gullies into the slope face (Fig. 4.10). At depths of only 3 to 6 feet, gullies can intercept shallow groundwater, allowing it to seep out, further weakening soil material.

Seepage and pressure

Groundwater Processes. Groundwater seepage is a critical consideration in assessing slope stability. It can undermine slopes by processes known as *sapping and piping* (see inset in Fig. 4.10), and because groundwater flow is driven by hydrostatic pressure (in the spaces between soil particles), called *pore-water pressure*, this force may weaken the skeletal strength of materials within a slope, inducing failure. Pore-water pressure can be a significant cause of slope failure around reservoirs where the water table and hydrostatic pressure have been raised under reservoir side slopes. Pore-water pressure can also weaken ground where irrigation water or water from leaking sewer and water pipes has increased seepage along a slope. Both pore-water pressure and deep soil saturation leading to muddy soil masses were at the heart of the devastating slope failures in the Pacific Northwest in the winter of 1996–1997.

Drainage alterations

Land Use Alterations. Finally, there is the land use factor to consider. Where slopes are cut by roads, mining operations, utility lines, or building foundations (see the examples in Figure 4.9) resulting in the alteration of slope form and the disruption of drainage and vegetative cover, the overall stability may be compromised. Alterations in drainage may be especially significant. As already noted, the raising of reservoirs is a well-established cause of slope failures. Earthen dams are known to fail by the same mechanism, namely, seepage and pore-water pressure. Drainage changes around roads, pipelines, and bridges can also cause increased seepage, and redirected streamflows can weaken slopes and cause serious failures.

Difficulties

Synthesis. Unfortunately, the means to integrate analytically all these variables have not been developed, and for large areas field and laboratory testing are not economically feasible. Engineering models for calculating stability are generally not reliable either because most natural slopes (as opposed to slopes constructed with fill material) are too complex in terms of composition and drainage to fit the equations. In addition, the engineering formulas are site and slope specific and do not lend themselves to areawide mapping programs. Therefore, evaluations of slope stability for purposes of land use planning and environmental management must be based on some sort of systematic review using the criteria cited above—steepness, composition, history of instability, and the like. A basic checklist of these criteria is provided in Table 4.2. Whether all these criteria can be used in mapping depends on the availability and reliability of data and one's abilities to synthesize and interpret them.

Data sources

Generally, topographic contour maps, aerial photographs, geologic maps, seismic maps, and soil maps are used as data sources, but they require interpretation and adaptation because of differences in scale, resolution, and units of measurement. When the desired criteria can be employed, the map overlay technique can be used to delineate areas of different levels of slope stability, and GIS techniques offer the best means of accomplishing such tasks.

Finally, slope stability is always the product of a combination of conditions, some transient, some permanent, some visible, some not. Not surprisingly, failure events

Fig. 4.9 The La Conchita landslide near Santa Barbara, California, March 1995. This slope had all the markings of a high-risk setting, including steep inclination (>60%), evidence of former failure, susceptibility to instability under heavy rainfall, concentrated stormwater runoff from the upper slope, and land use disturbance (note the road cut near midslope). In 2005 after a heavy rainfall, it failed again, killing 10 people and destroying more than 15 homes.

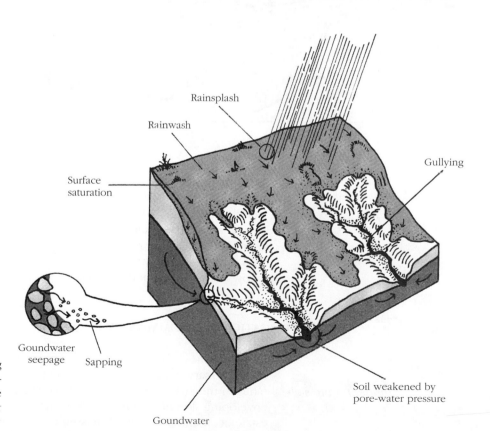

Fig. 4.10 Schematic diagram showing the influence of drainage on slope stability and erosion. The slope is least stable where gullies cut deeply and groundwater seeps from the ground.

Table 4.2   Slope Stability Checklist

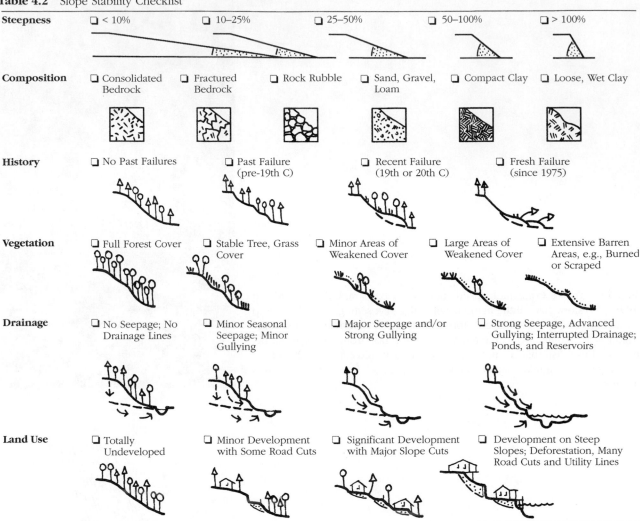

Playing the odds   are difficult to forecast. Some California slopes with no record of instability fall apart, as it were, after vegetation loss to fires is coupled with ground saturation from prolonged, heavy winter rains; other slopes that are steeper, wetter, and more denuded hold their own. But this is a probability game and the surest bet in the long run is to follow the results of one's analysis.

4.6 APPLICATIONS TO COMMUNITY PLANNING

Inattention to slope   In the absence of effective planning controls on development and the base information necessary to help guide decision making, many American and Canadian communities in the past several decades have seen development creep over the landscape, enveloping sloping and flat ground alike. The main force driving development is real estate economics. Usually dismissed is attention to topography, drainage, and related factors—beyond extreme cases where they add significant value to real estate or major costs to construction. Unlike development in an earlier era when the decision maker, developer, and resident were one and the same, in modern development these roles are often played by different and separate parties.

Added costs

The Lesson. As a result, in the modern, large-scale development project, the developer often makes poor decisions on siting, and, upon sale of the property, the risk is passed on to the buyer. When a problem eventually arises, such as ground movement and foundation failure, it becomes the responsibility of the owner/resident. Complicating matters further, the property may have a second or third owner by the time the problem actually surfaces. The social and financial costs for remedial action are typically high and must be borne by the current property owner, by the community, by some higher level of government, or by all three parties. And then there are additional costs to the community related to, among other things, shortened life cycles of public infrastructure. These harsh lessons have taught communities that the urge to develop beyond the terrain's capacity must be curbed by land use regulations, stringent design guidelines, and the education of developers and buyers.

Planning, design, and environmental ordinances (or bylaws) based on slope have been enacted in many North American communities. The rationale behind such policies varies. In some communities, they are aimed at growth control, in others at environmental protection, and in still others at risk management. In any case, the implementation of slope-based regulations almost always necessitates (1) building a slope map showing suitable and unsuitable terrain and (2) establishing a procedure to review and evaluate development proposals and site plans for areas at risk.

Slope averaging

Regulatory Approaches. Communities use several different approaches in designing slope-related regulations. One approach is based on development density and average slope inclination within specified land areas or units. Terrain units (such as valley walls or foothills) comprising many individual but similar slopes are assigned a maximum allowable density of development according to the average slope inclination (Fig. 4.11). The percentage of ground to be left undisturbed, as required by three California communities, is shown in Table 4.3. In Chula Vista, for example, terrain with slopes averaging above 30 percent requires that for each acre of development 9 acres be left in an undisturbed state.

Slope specific

A second approach is based strictly on the inclination of individual slopes. Each slope over a specified area, usually a proposed development site, is mapped according to the procedure described earlier in the chapter. Four or five classes are typically used, and for each there is a maximum allowable density set by ordinance. In Austin, Texas, for example, residential density is limited to one unit per acre for slopes in the 0 to 15 percent class. For slopes in steeper classes, 15 to 25 percent and 25 to 35 percent, development is allowable only upon approval of a formal request.

Slope condition

The third approach employs several slope characteristics in addition to inclination. For example, slopes comprised of unstable soils, conditionally stable forested slopes, and scenically valued slopes are mapped in combination with slope inclination. Development plans are compared directly to the distribution of each factor. This approach is commonly employed in areas of complex terrain where regulations based on inclination alone would prove inadequate.

Slopes in watersheds

Systems-Based Approach. None of the methods used by communities treats slopes as parts of landscape systems. Virtually all slopes belong to watersheds, and slopes are critical to their operations as both hydrologic and geomorphic systems inasmuch as they govern the movement and delivery of water and sediment. The geographic patterns of slopes are not randomly distributed in the watershed. Rather, steep ground (rugged topography) and its associated soil, drainage, and ecological conditions occur mostly in the upper parts of watersheds, in valley heads, foothills, and on mountain slopes.

Ordinances need to protect and manage these areas (zones) not only to reduce land use and environmental damage, but also to hold down servicing costs, protect

Fig. 4.11 An example of terrain units in an area of mountainous topography. The highest densities would be allowed in class I where slopes average 10 percent or less.

Table 4.3   Undisturbed Area (Open Space) Requirements for Sloping Ground in Three California Communities

Percent Slope (avg.)	Chula Vista (%)	Pacifica (%)	Thousand Oaks (%)
10	14	32	32.5
15	31	36	40
20	44	45	55
25	62.5	57	70
30	90	72	85
35	90	90	100
40	90	100	100

Costs and wise use   water supplies, manage flooding, and conserve open space (Fig. 4.12). All communities are situated in a watershed of some size and shape, and most are located at the receiving (downstream) end of the basin. Suburban and rural residential development, on the other hand, are pushing into the upper, more scenic and environmentally risky parts of watersheds. Unless communities recognize this trend and the need to formulate watershed-based approaches to land use planning that address landforms, slope stability, drainage, flooding, soils, forests, and so on, in an integrated framework, individual slope ordinances themselves will be of little value in the long run. We examine watersheds and their drainage systems later in Chapters 8 and 9 and once again in Chapter 22, which presents a system-based approach to land use planning.

4.7 WHY SLOPE MAPS MAY NOT WORK IN SOME AREAS

One of the chief rationales used by communities for limiting development on steep slopes is that facilities pose threats to the slope environment and related areas downhill.

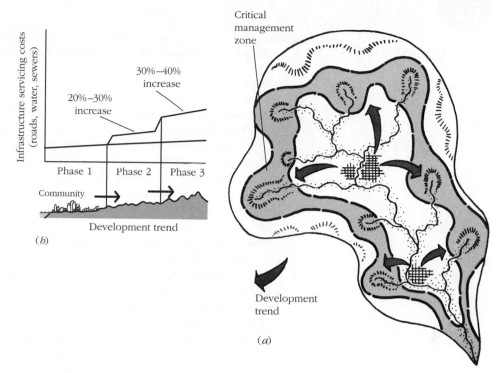

Fig. 4.12 (*a*) The distribution of unstable and potentially unstable ground in a watershed (defined by the critical management zone). Development trends, shown by the arrows, are often toward this zone. (*b*) As residential building reaches higher and more rugged terrain, servicing costs are substantially greater.

Drainage concerns

Among the commonly cited concerns is that development on slopes drives up runoff rates, resulting in increased soil erosion, greater flooding and decreased water quality downslope. This assumes, of course, that the rate of hillslope runoff and related environmental damage rise with slope steepness.

Slope and Runoff. For a number of reasons, the relationship of slope inclination to runoff is often misunderstood. First, the total volume of rainfall on a slope actually decreases at steeper slope angles because the slope surface (catchment) area exposed to the sky is smaller for steeper slopes. For a vertical cliff, there is no catchment area. Second, the area above a slope is often much more significant than the slope itself as a source of runoff. Some gentle slopes may be more hazardous than nearby steep slopes because of large upslope contributions of runoff. Third, standard slope maps do not take slope runoff patterns into account, that is, whether runoff concentrates or diffuses as it spills over the slope face. The reason is that slope maps are only two-dimensional expressions. Three-dimensional expressions of slopes give us quite another view.

Noses and hollows

Slopes in Three Dimensions. As three-dimensional forms, most slopes are either concave or convex (they form either a nose or a hollow), and these forms control the patterns of runoff. On a nose, runoff diffuses (spreads) downslope, whereas in a hollow, runoff concentrates downslope (see Fig. 9.10). Therefore noses, which are often steeper than hollows, tend to be dry, whereas hollows tend to be wet. Thus, the gentler slope form, the hollow, is in fact a more risky place for development from the standpoints of both facility damage and environmental degradation. The steeper slope, on the other hand, is drier with less upslope–downslope continuity of drainage. It follows that strict adherence to conventional slope maps (based only on steepness) could result in decisions opposite the ordinance's intent where the objective is to limit the impact of land use on slope drainage, water quality, and related factors such as habitat.

4.8 CASE STUDY

Slope as a Growth-Control Tool in Community Planning

W. M. Marsh

Years ago, when rural communities in the United States and Canada were setting up their systems of government, large areas of undeveloped land were often declared part of the community by virtue of the exaggerated alignment chosen for the town or village limits at the time of incorporation. These areas were often drawn into maps of the community with little or no attention to the nature of the terrain and, in most cases, with no realistic expectation on the part of the town fathers that the community itself would ever expand into them. Some communities even went so far as to plat these areas—that is, subdivide them on paper into residential lots, city blocks, streets, and neighborhoods—but most left the areas undeveloped.

In recent decades, many of these empty areas have begun to take on development. Because they were already platted, the communities often find that they have little control over where development takes place. Their only recourse is to formulate planning policy to help manage new development and keep the community in balance with its resource base.

Planning ordinances and bylaws aimed at regulating new development can take various forms. One form is a growth-control regulation that simply sets limits on the amount, type, and rate of development. This can be accomplished, for example, by setting limits on land use density or by setting quotas on the approval of building permits. Another form is the environmental ordinance that limits hillslope development based on habitat, open space, landscape scenic quality, or a related consideration. Servicing is also a regulatory tool. Some communities use water supply and wastewater servicing as limiting factors; that is, if servicing does not already exist, building applications cannot be approved.

In California, hillslope development is a major concern not only in terms of mudslide risk, but in terms of landscape aesthetic as well. Some communities have given this issue serious attention and have enacted design guidelines to minimize the visual impact of hillslope development. Santa Clarita, for example, recommends keeping building rooflines below the slope profile or horizon, breaking up roofs to create a more textured surface, and using natural building materials and colors (Fig. 4.A).

For the northern Michigan community shown in Figure 4.B, platting was carried out in the late 1800s or early 1900s when steep slopes and other terrain obstacles were often ignored despite the fact that they were hard to get up, difficult to build on, and even more difficult to farm. Today, however, building on these slopes is technically feasible and because of the vistas they provide over Lake Superior, they are highly attractive to prospective home builders. To protect this valuable part of the community environment, a slope ordinance (bylaw) is clearly needed.

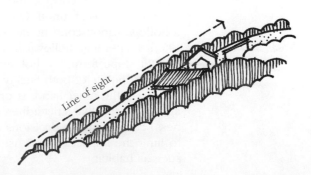

Fig. 4.A Slope and landscape aesthetics. Rooflines are limited to elevations below the slope horizon.

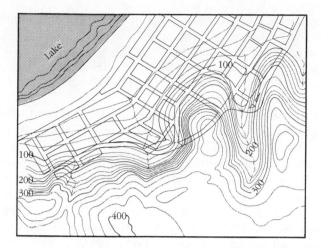

Fig. 4.B Early platting in some communities placed streets on exceedingly steep slopes, clearly beyond the capability of most nineteenth- and early twentieth-century small-town engineering.

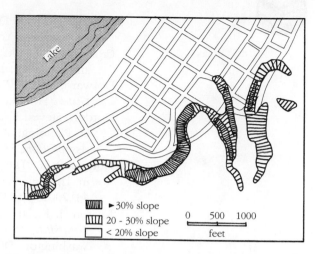

Fig. 4.C Documenting steep slopes is the first step in a terrain-based approach to community planning and development.

To be most effective, the ordinance should be based on rational criteria that justify the controls imposed on land users. In other words, the reasons for not building on slopes should appeal to the citizen's logic from both a personal and community standpoint. In this case, the slopes are composed of sandy soil held in place by hardwood forests, and changes in the balance among the forces acting on the slope face could easily lead to instability and extensive damage downslope. This can be demonstrated by documentation from scientific and planning literature reporting on problems in similar settings. In addition, residential development on these slopes would not be served by the city sewer systems and therefore would require on-site sewage disposal.

Evidence is abundant and compelling on the severe limitation that steep slopes pose to on-site (septic) sewage systems. Also, as in California, these slopes form scenic backdrops to the community below, and development that clears away patches of forest often degrades this amenity. Based on these and other reasons, there is ample justification in this community for invoking slope as a growth-limiting factor for residential and other types of development and for drafting a map (such as the one in Fig. 4.C) along with the appropriate rationale to demarcate slopes that are off limits.

4.9 SELECTED REFERENCES FOR FURTHER READING

Briggs, R. P., et al. "Landsliding in Allegheny County, Pennsylvania." *Geological Survey Circular.* 728, 1975.

Campbell, R. H. "Soil Slips, Debris Flows, and Rainstorms in the Santa Monica Mountains, Southern California." U.S. Geological Survey Professional Paper 851. Washington, DC: U.S. Government Printing Office, 1975.

Carson, M. A., and Kirkby, M. J. *Hillslope Form and Process.* Cambridge, UK: Cambridge University Press, 1972.

Cooke, R. U., and Doornkamp, J. C. *Geomorphology in Environmental Management,* 2nd ed. New York: Oxford University Press, 1990.

Crozier, M. J. "Field Assessment of Slope Instability." In *Slope Instability* (D. Brunsden and D. B. Prior, eds.). Hoboken, NJ: Wiley, 1984, pp. 103–142.

Dorward, Sherry. *Design for Mountain Communities: A Landscape and Architectural Guide.* Hoboken, NJ: Wiley, 1990.

Gordon, Steven I., and Klousner, Robert D., Jr. "Using Landslide Hazard Information in Planning: An Evaluation of Methods." *Journal of the American Planning Association.* 52, 4, 1986, pp. 431–442.

Guthrie, R. H. "The Effects of Logging on the Frequency and Distribution of Landslides in Three Watersheds on Vancouver Island, British Columbia." *Geomorphology.* 43, 2002, pp. 273–292.

Keaton, J. R., and DeGraff, J. A. "Surface Observation and Geologic Mapping." In *Landslides: Investigation and Mitigation* (K. A. Turner and R. L. Schuster, eds.). Washington, DC: National Academy of Science Press, 1996, pp. 178–230.

Milne, R. J., and Moss, M. R. "Forest Dynamics and Geomorphic Processes on the Niagara Escarpment, Collingwood, Ontario." In *Landscape Ecology and Management.* Montreal: Polyscience Publications, 1988.

Olshansky, R. B. *Planning for Hillside Development.* Chicago: American Planning Association, 1996.

Schuster, Robert L., and Krizek, Raymond J. *Landslides: Analysis and Control.* Washington, DC: National Academy of Science Press, 1978.

Terzaghi, K. "The Mechanism of Landslides." In *Application of Geology to Engineering Practice* (S. Paige, ed.). Geological Society of America, Berkey Volume, 1950.

Turner, K. A., and Schuster, R. L. *Landslides: Investigation and Mitigation,* Special Report 247. Washington, D C: National Academy of Science Press, 1996.

Related Websites

Natural Lands Trust. "Growing Greener." http://www.natlands.org/categories/category.asp?fldCategoryId=1

Model ordinances for conservation design. These specify slope and watershed standards for development. The site also has more information on conservation planning and land management.

State of Washington Department of Ecology. "Puget Sound Landslides." http://www.ecy.wa.gov/programs/sea/landslides/index.html

Sample slope stability maps for the coastal lands in Puget Sound. The site also gives general information about landslides, including excellent diagrams of signs of unstable land and types of landslides.

U.S. Geological Survey. 'Landslide Hazards Program.' 2009. http://landslides.usgs.gov/

Information on causes of landslides and mitigation suggestions. Research about hazard assessments determines the likelihood of landslides. The site also provides links for further information on this topic.

5

ASSESSING SOIL FOR LAND USE PLANNING AND WASTE DISPOSAL

5.1 INTRODUCTION

Complacency about soil

The relationship between soil composition and agriculture is apparent at practically any scale of observation, especially in areas of more traditional agriculture. The relationship between soil composition and other types of land uses, however, is often not so apparent. In North America, at least part of the reason for this is that modern developers and community planners have given precious little attention to soils. Among the contributing factors is a sense of complacency created by a popular impression that modern engineering technology can overcome problems with the soil.

Granted, the technology exists to build practically any sort of structure in any environment. However, the added costs of construction and environmental mitigation in areas of weak, unstable, and/or poorly drained soil are often prohibitive. In addition, soils play a pivotal role in landscape systems, especially drainage and ecosystems, and when their role is ignored or misunderstood, the systems can be compromised. No matter how we look at it, the result is a less sustainable landscape.

Hidden waste

An additional consideration is the heightened concern over solid and hazardous waste disposal in the soil. Finding new disposal sites is a perennial problem requiring, among other things, the careful evaluation of soil, topography, and drainage. It has also come to light that most urban areas have waste hidden in scores of disposal sites and dumps that pose not only an environmental threat but also a serious liability to development. If buried waste is encountered in constructing facilities, such as roads, buildings, and utilities, the site may have to be abandoned and/or the waste excavated and properly disposed somewhere else. For many landowners, the costs of mitigation are proving to be insurmountable; hence the demand for predevelopment soil surveys in many areas.

Hitting the target

Barring special needs such as dealing with buried waste or unstable slopes, planners nevertheless tend to treat soils less seriously than other major environmental factors. Too often data and information from agricultural soil surveys, such as those conducted by the U.S. Natural Resources Conservation Service, are taken at face value and transposed to site maps without much regard for inherent limitations related to mapping scale and resolution and to the nature of the problem under consideration. In other instances, soils are purposely left to the engineers to be investigated as a part of facility design and construction, but the investigations typically miss a lot because they are usually restricted to building sites and subsoil materials. Neither approach is acceptable in landscape planning, although both agricultural soil surveys and engineering investigations can be important sources of soil data for planning purposes.

5.2 SOIL COMPOSITION

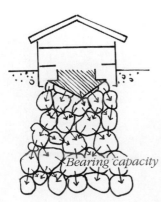

Bearing capacity

Several soil features, or *properties*, are used to describe soil conditions for problems involving land development. Of these, texture and composition are generally the most meaningful; from them we can make inferences about bearing capacity, internal drainage, erodibility, and slope stability. (We will discuss texture in the next section.) **Composition** refers to the materials that make up a soil. Basically, there are just four compositional constituents: mineral particles, organic matter, water, and air.

Mineral Constituents. *Mineral particles* comprise 50 to 80 percent of the volume of most soils and form the all-important skeletal structure of the soil. This structure, built of particles lodged against each other, enables the soil to support its own weight as well as that of internal matter such as water and the overlying landscape, including buildings. Sand and gravel particles generally provide the greatest stability and, if packed solidly against one another, usually yield a relatively high bearing capacity. *Bearing capacity* refers to a soil's resistance to penetration from a weighted object such as a building foundation. As a whole, clays tend to have lower bearing capacities than sands and gravel. Loosely packed, wet masses of clay tend to compress and slip laterally under the stress of loading (Table 5.1).

Organic Matter. The quantity of *organic matter* varies radically in soils, but it is extremely important for both negative and positive reasons. Organic particles usually provide weak skeletal structures with very poor bearing capacities. Organic matter tends to compress and settle differentially under roadbeds and foundations, and, when dewatered, it may suffer substantial volume losses as well as decomposition, wind erosion, and other effects. Understandably, deep organic soils, which may reach *Mixed review* thicknesses of 20 feet or more, pose the most serious limitations to facility development and to land use in general.

On the positive side, organic matter is vital to the fertility and hydrology of soil. Indeed, the loss of topsoil from agricultural lands by runoff and wind erosion is viewed as one of the most serious environmental issues of our time. The role of organic matter in the terrestrial water balance is underscored by the water storage function of topsoil and organic deposits in wetlands. Topsoil absorbs significant amounts of precipitation and therefore helps reduce runoff rates. Organic deposits often serve as moisture reservoirs for wetland vegetation as well as points of entry for groundwater recharge.

5.3 SOIL TEXTURE

Definition The size of mineral particles found in soil ranges enormously from microscopic clay particles to large boulders. The most abundant particles, however, are sand, silt, and clay, and they are the focus of examination in studies of soil texture (Fig. 5.1). **Texture** is the term used to describe the composite sizes of particles in a soil sample, usually several representative handfuls. To measure soil texture, the sand, silt, and clay particles are sorted out and weighed. The weight of each size class is then expressed as a percentage of the sample weight.

Textural Classes. Since not all the particles in a soil are likely to be clay or sand or silt, additional terms are needed for describing various mixtures. Soil scientists use *Loam and...* 12 basic terms for texture, at the center of which is the class *loam*, an intermediate mixture of sand, silt, and clay. According to the U.S. Natural Resource Conservation Service, the loam class is comprised of 40 percent sand, 40 percent silt, and 20 percent clay. Given a slightly heavier concentration of sand, for instance 50 percent, with

Table 5.1 Bearing Capacity Values for Rock and Soil Materials

Class		Material	Allowable Bearing Value (tons per square foot)
1		Massive crystalline bedrock, e.g., granite, gneiss	100
2	Rock	Metamorphosed rock, e.g., schist, slate	45
3		Sedimentary rocks, e.g., shale, sandstone	15
4		Well-compacted gravels and sands	10
5		Compact gravel, sand/gravel mixtures	6
6		Loose gravel, compact coarse sand	4
7		Loose coarse sand; loose sand/gravel mixtures, compact fine sand, wet coarse sand	3
8	Soil materials	Loose fine sand, wet fine sand	2
9		Stiff clay (dry)	4
10		Medium-stiff clay	2
11		Soft clay	1
12		Fill, organic material, or silt	(Fixed by field tests)

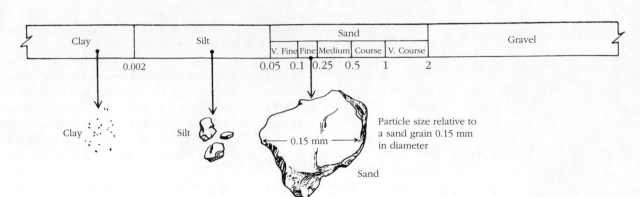

Fig. 5.1   A standard scale of soil particle sizes. The illustrations give a visual comparison of clay and silt with a sand particle 0.15 millimeter in diameter.

10 percent clay and 40 percent silt, the soil is called *sandy loam*. In agronomy, the textural names and related percentages are given in the form of a triangular graph (Fig. 5.2). If we know the percentage by weight of the particle sizes in a sample, we can determine the appropriate soil name from the graph.

Separating particles   **Sieve Analysis.** To produce particle size data requires a laboratory procedure called **sieving,** in which the dried soil sample is filtered through a set of sieves, each with a different mesh size. The sample is thus sorted into particle size classes, which can then be weighed and expressed as a percentage of the total sample weight. This procedure works well for silt, sand, and larger particles, but it is not appropriate for clay. Clay particles are too small to sieve accurately. Therefore, in clayey soils the fine particles are measured according to their settling velocity when suspended in water. Since these microscopic particles settle out very slowly, they are easily segregated from the larger sand and silt particles. The clay-rich water is then drawn off and evaporated, leaving a residue of clay, which can be weighed.

Sample test forms   **Field Hand Test.** For site planning and design purposes, however, this procedure is usually too elaborate, and we must turn to other ways of estimating texture. One such method is the **field hand test.** This procedure involves extracting a handful of soil and squeezing it into three basic shapes: (1) a *cast,* a lump formed by squeezing a sample in a clenched fist; (2) a *thread,* a pencil shape formed by rolling soil between the palms; and (3) a *ribbon,* a flattish shape formed by squeezing a small sample between the thumb and index finger (Fig. 5.3). The behavioral characteristics of the soil when molded into each of these shapes, if they can be formed at all, provide the basis for a general textural classification. The sample should be damp to perform the test properly.

Behavior of the soil in the hand test is determined by the amount of clay in the sample. Clay particles are highly cohesive, and, when dampened, they behave as a plastic. Therefore, the higher the clay content is in a sample, the more refined and durable the shapes into which it can be molded. Table 5.2 gives the behavioral traits of five common soil textures.

5.4 SOIL MOISTURE AND DRAINAGE

The water content of soil varies with composition, texture, local drainage and topography, vegetation, and climate. At the microscale, most water in the soil occupies the spaces between the particles; only in organic soils and certain clays do the particles themselves actually absorb measurable amounts of water.

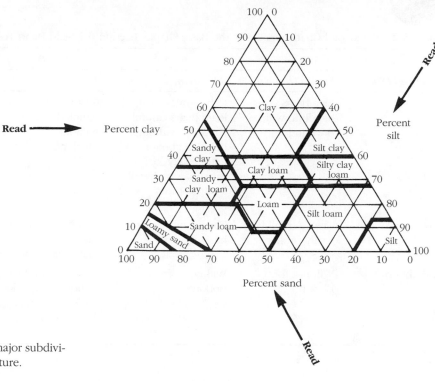

Fig. 5.2 Soil textural triangle, including the major subdivisions, used by the U.S. Department of Agriculture.

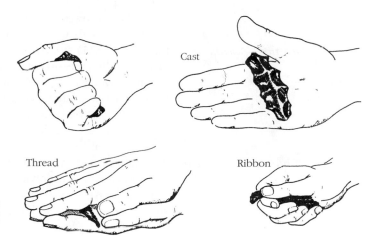

Fig. 5.3 The basic shapes used in the field hand test. See Table 5.2 for evaluation criteria in determining soil textural type.

Forms of Soil Water. Two principal forms of water occur in both mineral and organic soils: capillary and gravity. **Capillary water** is a form of molecular water, so-called because it is held in the soil (around and between particles) by the force of cohesion among water molecules. This force, however, is not great enough to prohibit plants from drawing on this water; indeed, it is the principal source of moisture for most plants. Neither is it great enough to prohibit water molecules from moving along moisture gradients in the soil, that is, from moist spots to dry spots. In the summer, most capillary water transfer is upward toward the soil surface as water is lost to the atmosphere in evaporation and transpiration.

Gravity water is liquid water that moves in the interparticle spaces not occupied by capillary water. It moves in direct response to the gravitational force, percolating downward through the soil, until at some depth in the subsoil and/or underlying

Capillary water

Gravity water

Table 5.2   Behavioral Characteristics of the Basic Shapes Used in the Field Hand Test and the Soil Type Represented by Each

Field Test (Shape)	Sandy Loam	Silty Loam	Loam	Clay Loam	Clay
			Soil Type		
Soil cast	Cast bears careful handling without breaking	Cohesionless silty loam bears careful handling without breaking; better-graded silty loam casts may be handled freely without breaking	Cast may be handled freely without breaking	Cast bears much handling without breaking	Cast can be molded to various shapes without breaking
Soil thread	Thick, crumbly, easily broken	Thick, soft, easily broken	Can be pointed as fine as pencil lead that is easily broken	Strong thread can be rolled to a pinpoint	Strong, plastic thread that can be rolled to a pinpoint
Soil ribbon	Will not form ribbon	Will not form ribbon	Forms short, thick ribbon that breaks under its own weight	Forms thin ribbon that breaks under its own weight	Long, thin flexible ribbon that does not break under its own weight

bedrock, it accumulates to form groundwater. *Groundwater* completely fills the interparticle spaces, saturating soil and rock materials and forcing out most air.

Soil Drainage. References to drainage in soil reports are usually to gravity water and to a soil's ability to transfer this water downward. Three terms are used to describe this process: (1) *infiltration capacity*, the rate at which water penetrates the soil surface (usually measured in centimeters or inches per hour); (2) *permeability*, the rate at which water in the soil moves through a given volume of material (also measured in centimeters or inches per hour); and (3) *percolation*, the rate at which soil takes up water in a soil pit or in a pipe in the soil (used mainly in wastewater absorption tests and measured in inches per hour).

Drainage parameters

Poor soil drainage means that the soil is frequently or permanently saturated and may often have water standing on it. Terms such as "good drainage" and "well drained" mean that gravity water is readily transmitted by the soil and that the soil is not conducive to prolonged periods of saturation. Soil saturation may be caused by several conditions: the local accumulation of surface water (because of river flooding or runoff into a low spot, for example); a rise in the level of groundwater in the soil column (because of the raising of a reservoir or excessive application of irrigation water, for example); or the particles in the soil are too small to transmit infiltration water downward (because of impervious layers within the soil or clayey soil composition).

What's well drained?

5.5 SOIL, LANDFORMS, AND TOPOGRAPHY

Nearly all soils are formed in deposits laid down by geomorphic processes of some sort, such as wind, glacial meltwater, ocean waves, river floods, and landslides. In Canada and the United States, three types of surface deposits dominate the landscape: *glacial drift* (mainly from the continental glaciers), *alluvium* (from streams), and *aeolian* (from wind, mainly loess and dune sand) (see Fig. 2.12).

Surface deposits

Parent Material. These deposits form the **parent material** for most soils, and it is important to know not only which material is likely to occur in a particular location

Spatial variability

but what sort of spatial variability it is likely to have. Broadly uniform terrain, such as the central Great Plains, exhibits little variability in soil materials over large areas; loess blankets virtually all surfaces except for stream valleys. In diverse terrain such as the Great Lakes Region or the mountainous West, however, it is not uncommon to find deposits of markedly different compositions over areas as small as 10 acres or so. For example, in river valleys, deposits of sandy gravels and organic matter are often complexly intermingled across floodplains; and in glaciated terrain, deposits may range from loose sands to dense clays over a distance of only several hundred meters or less. Each deposit, however, tends to be associated with certain indicators; principally, landforms, water features, and/or vegetation types (Fig. 5.4).

Soil and Subsoil. Once in place, the surface layer of all deposits is subject to alteration by soil-forming processes associated with climate, vegetation, surface drainage, and land use. In time, these processes produce a complex medium characterized by a sequence of layerlike zones, or *horizons*, of different physical, chemical, and biological compositions. This medium is usually 3 to 8 feet deep and is known as the **solum,** or so-called true soil. This is what soil scientists measure and map in national, regional, and county soil surveys. [Appendix A (see the book's website) provides information on the principal soil classification systems used in mapping the solum in the United States and Canada.]

The solum

Below the solum, in what is generally termed the *subsoil*, the parent material is largely undifferentiated, that is, without horizons. This lack of differentiation is especially significant in land development problems because landfills, basements, roadbeds, slope cuts, piping systems, and the like are built in these materials. The Unified Soil Classification System (USCS), which is widely used in civil engineering, applies to these materials (Table 5.3).

Solum

Subsoil

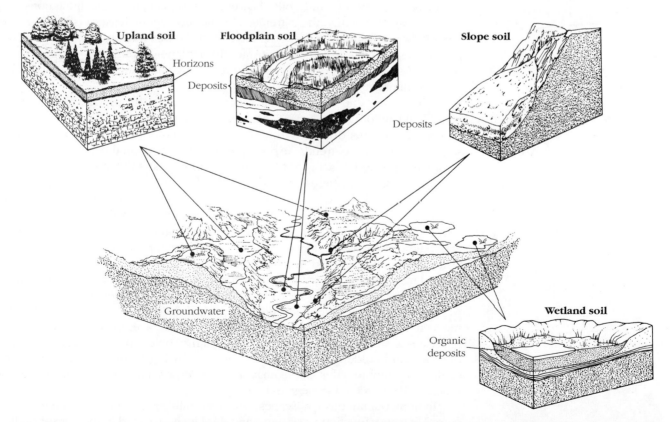

Fig. 5.4 Common sources and settings of deposits in which soils form: wetland, slope, floodplain, and upland deposits.

Table 5.3   Unified Soil Classification System

Letter	Description	Criterion	Other Criteria	
G	Gravel and gravelly soils (basically pebble size, larger than 2 mm in diameter)	Texture	Based on uniformity of grain size and the presence of smaller materials such as clay and silt	
S	Sand and sandy soils	Texture	W:	Well graded (uniformly sized grains) and clean (absence of clays, silts, and organic debris)
			C:	Well graded with clay fraction, which binds soil together
			P:	Poorly graded, fairly clean
M	Very fine sand and silt (inorganic)	Texture, Composition	Based on performance criteria of compressibility and plasticity:	
C	Clays (inorganic)	Texture, Composition	L:	Low to medium compressibility and low plasticity
O	Organic silts and clays	Texture, Composition	H:	High compressibility and high plasticity
P	Peat	Composition		

Landform relations

Soil-slope trends

Mountain slope soils

Landforms and Soil. Geomorphologists have developed a basic taxonomy of the landforms associated with various types of deposits, making it possible to infer certain things about soil materials based on a knowledge of landforms. Any attempt to correlate landforms and soils, however, must recognize the limitations of geographic scale. Generally, the extremes of scale provide the least satisfactory results. At the scale of individual landform features, such as wetlands, sand dunes, or outwash plains, the correlation can be quite good for planning purposes. Appendix B (see website) lists a number of landform features, along with their associated soil composition and drainage.

Also, predictable trends in soil makeup are related to topographic gradients on individual landforms (Fig. 5.5). These trends are attributable to the fact that all terrains function as systems driven by runoff moving over slopes of different gradients. Soil scientists term these trends **toposequences** because of the apparent correlation between topography (actually position on a slope) and soil type. Toposequences can be found at several different scales in the landscape.

In Figure 5.5 the toposequence is the product of a drainage system that moves water and sediment from a mountain ridge to an adjacent valley. The principal landform here is an *alluvial fan*, a common depositional feature along dry mountain valleys. The texture of the deposits changes from the top of the fan to the valley floor as a function of slope and runoff. Coarse materials (sands and gravels) dominate the upper fan, where runoff volume and velocity are high, whereas fines (clays and silts) dominate the lake on the valley floor, where runoff is relatively modest.

Another type of mountain side deposits, *talus slopes*, are comprised of rock fragments that have fallen into place from an upper rock face. In contrast to alluvial fans, particle size tends to increase downslope, with pebble- and cobble-size fragments at the top and boulder-size rubble near the toe. Owing to the coarse, angular nature of the rock rubble, talus slopes may achieve surprisingly steep angles of repose, as much as 100 percent (45 degrees).

In nonmountainous landscapes, vegetated hillslopes are usually the focus of concern. They, too, function as systems, and this fact is reflected in the corresponding soil changes across the slope, as illustrated in Figure 5.6. The toposequences, of course, are usually more subtle than on mountain slopes and limited largely to the surface layer.

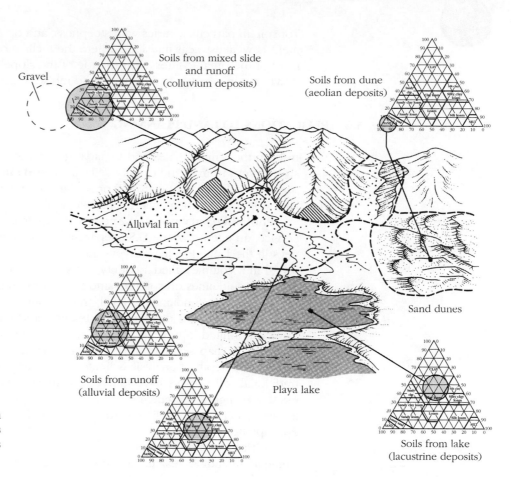

Gravel

Soils from mixed slide
and runoff
(colluvium deposits)

Soils from dune
(aeolian deposits)

Alluvial fan

Sand dunes

Soils from runoff
(alluvial deposits)

Playa lake

Soils from lake
(lacustrine deposits)

Fig. 5.5 The relationship between soils and landforms in mountainous terrain. Note the texture changes with distance into the valley.

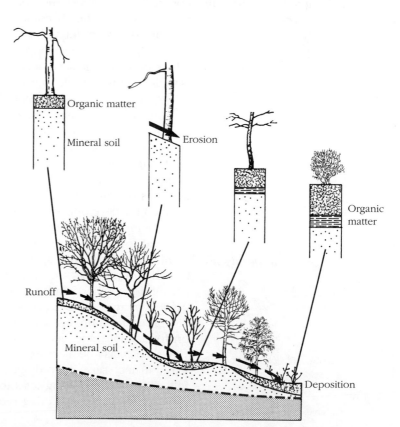

Organic matter

Mineral soil

Erosion

Organic
matter

Runoff

Mineral soil

Deposition

Fig. 5.6 A typical example of soil variation associated with a toposequence across a hillslope. Topsoil is thin on steep sections and thickens on the gentler parts of the slope where runoff is slower.

Topsoil, in particular, varies with steepness, and on *S*-shaped slopes, topsoil development is weakest near midslope where the inclination is greatest and runoff removes most of the organic litter. Near the toe of the slope, topsoil thickens as runoff slows down, organic matter is deposited, and moisture accumulates.

5.6 APPLICATIONS TO LAND PLANNING

In both the United States and Canada, government agencies have conducted extensive surveys to map and classify soils. Such **soil surveys** were originally designed to serve agriculture, but since 1960 or so they have been expanded to serve community land use and environmental interests as well.

Published Surveys. As we noted in Chapter 3, the Natural Resources Conservation Service, an agency of the U.S. Department of Agriculture, is responsible for soil surveys in the United States. Surveys are organized by county, and the results are published in a volume called a county soil survey that contains soil descriptions, various data and guidelines on soil uses and limitations, and maps of soil distributions.

Soil survey data   At the most fundamental level, the soil surveys provide data and information in three key areas: soil drainage, soil texture, and soil composition. The surveys are, however, limited in three significant respects: (1) major metropolitan areas are omitted; (2) the map scale and accuracy are marginal for sites smaller than 100 acres; and (3) the depth of analysis is limited to the upper 4 to 5 feet of soil. Furthermore, traditional soil surveys have inherent accuracy problems owing to a mapping process that is highly inferential in most areas. Therefore, caution should be exercised in applying soil polygons directly to Geographic Information System (GIS) files without appropriate field checks.

Initial Considerations. In planning problems dealing with community development, including residential, industrial, commercial, and related land uses, soil *Dealing with organic soil* composition is one of the first considerations. Organic soils are at the top of the list because they are highly compressible under the weight of structures and they tend to decompose when drained. Decomposition and compression lead to subsidence, and studies show that subsidence rates of 1–2 meters over several decades are common where organic soils have been drained, farmed, or built over. Where the organic mass is shallow, it is possible to excavate and replace the soil with a better material, but this is often expensive and ecologically harmful, and it may lead to drainage problems.

Warning   Always bear in mind that organic soils invariably form in areas where surface and subsurface water naturally collects. Therefore, excavation does not eliminate these conditions; in fact, it may exacerbate them by releasing the confining pressure on inflowing water. From an ecological standpoint, we must recognize that organic soils are often indications of the presence of wetlands that may support valued communities of plants and animals. On balance, then, sites with organic soils should be avoided for development purposes.

Mineral soils   For mineral soils, texture and drainage are the primary considerations. Coarse-textured soils, such as sand and sandy loam, are preferred for most types of development because bearing capacity and drainage are usually excellent (see Table 5.1). As a result, foundations are structurally stable and usually free of nuisance water. Clayey soils, on the other hand, often provide poor foundation drainage, whereas bearing capacity may or may not be suitable for buildings. In addition, certain types of clays are prone to shrinking and swelling with changes in soil moisture, creating stress on foundations and underground utility lines (Fig. 5.7). Therefore, development in clayey soils may be more expensive because special footings and foundation drainage may be necessary. Accordingly, site analysis often requires detailed field mapping, followed by engineering tests to determine whether clayey soils may pose bearing capacity, drainage, and other problems.

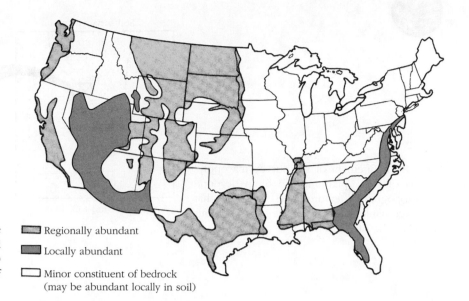

Fig. 5.7 The distribution of near-surface rock containing clay prone to swelling and shrinking. In soil, this clay (montmorillonite) is responsible for stress and the cracking of foundations and utility lines.

Regionally abundant

Locally abundant

Minor constituent of bedrock
(may be abundant locally in soil)

Mapping Scale and Resolution. In preparing soil maps, the scale of analysis is often as important as the type of analysis. For regional-scale problems, Natural Resources Conservation Service (NRCS) maps are generally acceptable especially if they are used in conjunction with topographic maps, aerial photographs, and surface geology or landform maps. At the site scale, however, NRCS reports and maps are usually too general for most projects. In these cases, a finer grain of spatial resolution is called for, which can be obtained in part by refining the NRCS soil boundaries with the aid of a large, site-scale topographic map.

Site and scale

This is based on the fact that many borders of soil types on NRCS maps follow topography, and at a site scale of detail, topographic trends and details are much more apparent than the soil scientist could detect in building the original map. Therefore, where adjacent soils are clearly terrain specific, the configuration of the border can be articulated much more accurately on large-scale site maps, as illustrated in Figure 5.8.

Geomorphic Factors. Beyond NRCS considerations, or in areas not covered by NRCS maps, soil analysis for site planning should begin with two basic considerations. The first is the recent geomorphic history of the area to gain some idea of the types of materials to expect. This information can be obtained in a telephone call to a local university (geography, geology, or soil science department), the state geologist's office, or the county NRCS office. In any case, we need to know at the outset whether the area is one of marine clay, sand dune deposits, glacial outwash, or other type of material.

Information sources

The second consideration is local topography and drainage. Using topographic maps and field inspection, drainageways (stream valleys, swales, and floodplains) and areas of impeded drainage (wetlands, ponds, and related features) should be mapped. These are the zones where poorly drained and organic soils are most likely found. The remaining area can be defined as slopes and upland surfaces, and, although soil texture may vary in these zones, drainage should be better and organic deposits less likely.

Topography and drainage

Field Testing. Within this framework, a limited testing program can be carried out by digging holes at selected points along swales, floodplains, hillslopes, and ridgetops. This is best accomplished with a backhoe, which can quickly excavate to a depth of 2 meters or more, but a shovel or soil auger will suffice for shallow depths. Once a test pit is excavated, the following questions can be answered:

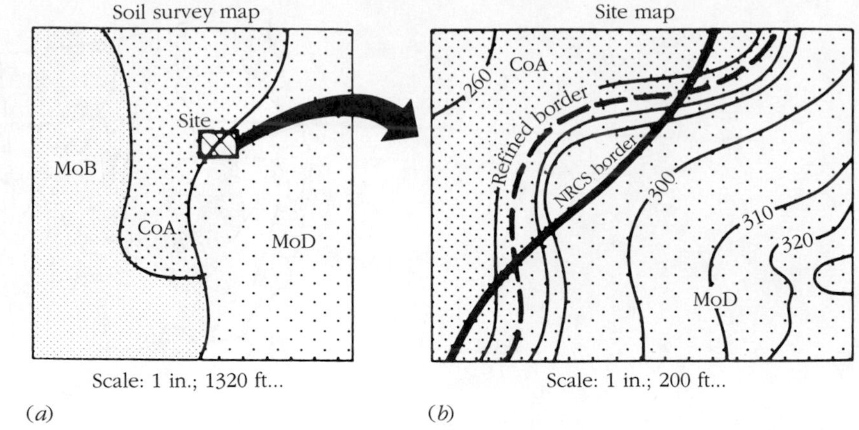

Soil survey map

Site map

Fig. 5.8 Refinement of soil survey map polygons (*a*) for site planning can often be done using topographic features, such as slopes that are discernable only on large-scale contour maps like the site map (*b*).

Scale: 1 in.; 1320 ft...

Scale: 1 in.; 200 ft...

(*a*)

(*b*)

Key questions
- Is the material below the surface mainly organic?
- Does water rapidly seep into and eventually fill most of the pit?
- Do the sides of the pit slump away rapidly with excavation and seepage?
- Is buried waste (such as agricultural debris, industrial rubble, municipal garbage, or hazardous waste) present?

If the answer to any of these questions is yes, as it is for the excavation in Fig. 5.9, the soil should generally be classed as unsuitable for development. Further analysis may be required for engineering purposes, but for site planning, this level of investigation is adequate for laying out use zones, circulation systems, and defining building sites.

Seasonal conditions   A note of caution: What this test does not tell you is anything about seasonal conditions. For example, a soil that is dry and stable in summer may become saturated and may soften and liquify in winter and spring. Therefore, a single sampling in one season may be misleading in locations subject to seasonally extreme ground conditions.

When to bore   Soil borings are test holes used to measure the strata and composition of soil materials at depths greater than 10 feet. In most instances, borings are necessary only where facilities are to be built or where there is a suspected problem with groundwater, bedrock, or buried waste. The depth to which a boring is made depends on the building program requirements (for example, heavy buildings or light buildings), the types of materials encountered underground, and the depth to bedrock. If the soil mantle is relatively thin—say, less than 30 feet—and the bedrock is stable, the footings for large buildings are usually placed directly on the bedrock. In areas of unstable soils, such as poorly consolidated wet clays, or weak bedrock, such as cavernous limestone, special footings are usually required to provide the necessary stability.

5.7 PLANNING CONSIDERATIONS IN SOLID WASTE DISPOSAL

One of the most pressing land use problems in urbanized and industrial areas today is the disposal of various types of waste: municipal garbage, chemical residues from industry, rubble from mining and urban development, various forms of industrial and agricultural debris, and, most recently, nuclear residue from power plants and military manufacturing and development installations.

Fig. 5.9 Organic matter and groundwater make this soil unsuitable for development. The organic layer is hidden by a layer of mineral soil, quite possibly fill material.

Solid waste

Waste Types and Disposal. Waste material is classed as either *solid waste* or *hazardous waste*, depending on whether it contains toxic materials. Solid waste is assorted materials variously described as trash, garbage, refuse, and litter. It is considered nonhazardous, that is, not harmful to humans and other organisms, and most (about 80 percent) is produced in rural areas by mining and agriculture. The remainder is *municipal solid waste* and *industrial solid waste*, which is disposed of in landfills in and around urban areas and is a major concern in developed countries.

Landfills

Landfills (or sanitary landfills) are managed disposal sites. Waste is buried in the ground, and this is the preferred method of solid waste disposal in the United States and Canada (Fig. 5.10). Most other disposal methods are either environmentally unacceptable or too expensive for most waste. Examples include burning, open pit dumping, and ocean dumping.

Urban and industrial wastes, though low in total output compared to agriculture and mining, present the most serious solid waste disposal problem for the planner, for several reasons. At the top of the list are groundwater contamination and public health hazards, surreptitious hazardous waste within the solid waste, the conflict over aesthetic and real estate values, the limited availability of land around cities for disposal sites, and the high cost of garbage collection and hauling in urban areas. In addition, the production of urban and industrial waste has been rising in most parts of the United States and Canada.

Siting criteria

Site Selection and Containment. *Siting criteria* The selection of a disposal site for urban and industrial solid waste is one of the most critical planning processes faced in suburban and rural areas today. Properly approached, it should be guided by three considerations: (1) *cost*, which is closely tied to land values and hauling distances; (2) *land use and environment* in the vicinity of the site and along hauling routes; and (3) *site conditions*, which are largely a function of soil and drainage. In both urban and industrial landfills, the chief site problem is containment of the liquids that emanate from the decomposing mass of waste.

(a)

Fig. 5.10 A sanitary landfill operation. (a) The mass of garbage is mounded up and interlaid with clayey soil. (b) The final mass is covered with a soil sheet several feet thick and then planted to stabilize it and improve its visual quality.

(b)

Leachate

These fluids, collectively called **leachate**, are composed of heavy concentrations of dissolved compounds that can contaminate local water supplies. Leachates are often chemically complex and vary in makeup according to the composition of the refuse. Moreover, the behavior of leachates in the hydrologic system, especially in groundwater, is poorly understood. Therefore, the general rule in landfill planning and management is to restrict leachate from contact with either surface or subsurface water. Groundwater contamination by leachates typically results in the loss of potable water for many decades (Fig. 5.11).

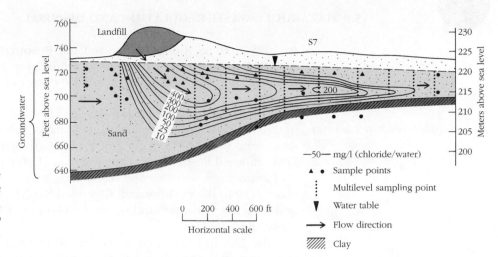

Fig. 5.11 A leachate plume emanating from a landfill. The plume traveled 2200 feet over 35 years, contaminating local groundwater to a depth of 40–80 feet. (From U.S. Geological Survey.)

Leachate containment

Leachate containment may be accomplished in some cases by merely selecting the right site. An ideal landfill site should be excavated in soil that is effectively impervious, that neither receives nor releases groundwater, and that is not subject to contact with surface waters such as streams or wetlands. Dense (compact) clay soils and relatively high ground are the preferred site characteristics. In addition, the clay should not be interlayered with sand or gravel, not subject to cracking upon drying, and stable against mass movements such as landslides.

Site preparation

Where these conditions cannot be met, the site must be modified to meet performance requirements. If the soils are sandy and permeable, as was the case illustrated in Figure 5.11, a liner of clay or a synthetic substance such as vinyl must be installed to gain the necessary imperviousness. The same measure must be taken in stratified soils, wet soils, sloping sites, and sites near water features and wells. In many areas of the United States and Canada, however, the decision to modify a site is not left up to the landfill developer or operator because strict regulations often govern which sites may even be considered for landfills. Thus, inherently poor sites are often eliminated from serious consideration by virtue of planning policy.

Management Considerations. In addition to a formal site selection and preparation plan, a growing number of local and state/provincial governments require the preparation of a management plan for the design and operation of landfills. In the case of sanitary landfills for municipal garbage, the planner is asked to address the following:

Design criteria

1. Compartmentalization of the landfill into cells or self-contained units of some sort.

2. Phasing of the operation, such as excavation and filling of a limited area at any one time.

3. Landscaping, pest control, and protection of the site during operations.

4. Limiting the total thickness of refuse by interspersing layers of soil within the garbage.

5. Backfilling over the completed fill using a layer of soil with a provision for vent pipes to release gases, if necessary (Fig. 5.10*b*).

6. Preparation of a plan for grading, landscaping, and future use of the site.

5.8 HAZARDOUS WASTE REGULATION AND DISPOSAL

In 1976, the U.S. Congress enacted the **Resource Conservation and Recovery Act (RCRA)**, which provided a legal definition of hazardous wastes and established guidelines for managing, storing, and disposing of hazardous materials in an environmentally sound manner.

Hazardous waste cleanup

The Superfund Program. The most widely known regulatory program for hazardous waste management in the United States is the **Superfund** program. It was created in 1980 by an act of Congress (under the formal title Comprehensive Environmental Response Compensation and Liability Act) for the purpose of cleaning up hazardous waste sites and responding to the uncontrolled release of hazardous materials into the environment. Central to Superfund is that it empowers the federal government to provide funding for the cleanup of severely contaminated hazardous waste sites.

The EPA has identified more than 47,000 hazardous waste sites in the United States that may need cleanup. As of 2007, 1,569 sites were placed on the National Priorities List. More than 10,000 waste sites are awaiting evaluation for possible priority status, and, although most will not gain Superfund status, many will. So the list of Superfund sites is bound to grow.

Liability concerns

Polluter-Pay Rules. In response to the growing concern of prospective property owners, banks, and developers about the liability of hidden wastes on a site, several states have enacted **polluter-pay laws**. These laws place the responsibility for site cleanup on past property owners under whose ownership the waste was buried. Such laws are proving to be effective because the cost of remediation is added to real estate, thereby returning pollution costs to the market system rather than passing them along to the state or federal government. Polluter-pay provisions are, not surprisingly, playing a major role in reusing (recycling) urban land, especially industrial *brownfields*, that is, land given up by old industrial operations.

Storage and processing

Disposal and Treatment Options. Several strategies are available for managing hazardous waste, though none is without drawbacks. In general they fall into two classes: disposal and treatment. *Disposal* involves collecting, transporting, and storing waste with no processing or treatment. *Treatment* involves submitting the waste to some sort of processing to make it less harmful. Because of increasingly stringent health and safety requirements, as well as endless local land use and public opinion problems, land disposal has become a very costly undertaking. For this and other reasons, the trend in hazardous waste management is toward on-site treatment, but off-site disposal is still the most common management method.

Secure landfill designs

The most widely used land disposal method is the **secure landfill**. The objective of this method is to confine the waste and prevent it from escaping the disposal site. The first step in planning a secure landfill is the selection of an appropriate site, one that minimizes the risk of groundwater and surface water contamination. Normally this is a very lengthy and complex process involving a host of field tests, analyses, reviews, public hearings, and permit applications. Site selection is followed by landfill design, which must include (1) a clay liner backed up (underlain) by a plastic liner; (2) a clay dike and clay cap; (3) a leachate collection and drainage system; and (4) monitoring wells around the site to check that leachate is not leaking into the soil and groundwater. Despite these measures, it is argued that secure landfills can still leak. Therefore, it is extremely important that only the very best sites are used—namely, those with deep, dense-clay soils, little topographic relief (elevation variation), large setbacks from water features and residences, and water tables at great depth (Fig. 5.12).

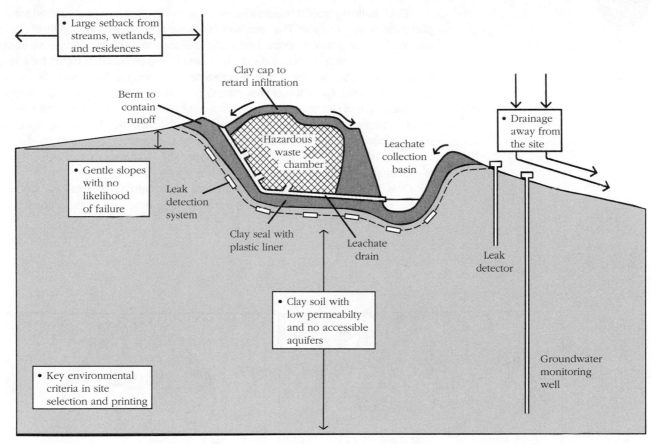

Fig. 5.12 Environmental criteria and design features of a secure landfill site according to current standards in the United States.

5.9 CASE STUDY

Mapping Soil at the Site Scale for Private Development

W. M. Marsh

The aim in mapping soils for site planning and architectural projects is clearly different than it is for agricultural purposes. Where the main question is the suitability of soil for facilities such as roads and buildings, rather than suitability for crops, the scope of the study and parameters investigated must be defined differently. This requires a clear understanding of the development program: What is to be built? How much floor space is called for? What ancillary facilities will be needed (parking and the like)? What utility systems are necessary? What open space, habitat, and landscaping areas are scheduled? And so on. These elements of the program must be known to determine (1) the scales at which mapping must be conducted; (2) the soil features (parameters) that should be recorded; and (3) the depths to which soil must be examined.

Each building project requires its own soil survey, that is, one designed for that particular project and site. The results of this survey, along with the results of related studies such as drainage, slope, and traffic, are used to formulate alternative design schemes for development of the site. An example is a project that called for a large research and office facility in a campuslike setting along the Mohawk River in New York State. In this case, it was important to determine not only certain standard soil parameters, such as texture and drainage, but also the nature of the soil environment in terms of the processes and factors that shape soil formation, especially runoff processes. The resultant soil map (Fig. 5.A) was instrumental in defining buildable land units (Fig. 5.B), which in turn served as a framework for the formulation of alternative development plans.

The fact that soils are closely related to physiography, especially topography, drainage, and vegetation, is critical to the logic of defining land units. Land units are

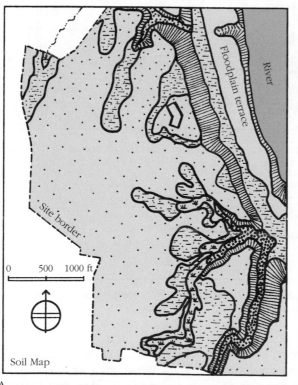

A.

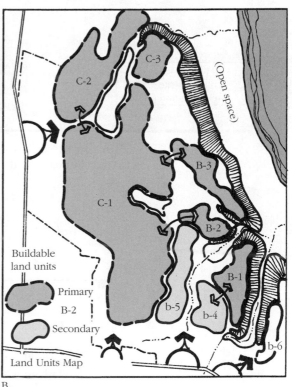

B.

Upland soil: Silty loam; well drained in most months but may be moist at depths of 1-2 ft; bearing capacity generally good, but foundation drainage may be needed.

Valley soil: Diverse with heavy concentrations of large boulders; channel and slope deposits; prone to flooding and high water table.

Slope soil: Dry surface; rapid runoff; prone to failure and erosion if disturbed and/or deforested.

Wetland soil: Silty clay loam with appreciable organic fraction; serious drainage problems in most months; unsuitable for building.

Transition soil: Silty loam; seasonally very wet; runoff collection zones; good potential for storm-water management facilities; poorly suited for building.

Fig. 5.A & B Site-scale soil map in which soil types are based mainly on physiographic criteria. A land units map based largely on soil and physiography, designed to provide a framework for site planning

spatial entities that possess a set of unifying physical traits such as upland topography, well-drained soil, and old field plant cover. Although the focus is on buildable land units, other types of land units are also significant in development planning. In particular, units featuring wetlands, slopes, and valley floors are important to other parts of development programs such as habitat conservation, landscaping, and stormwater management.

Unlike private development of the past, when buildings were all that mattered, much private development today strives to create comprehensive programs that relate to many types of land. Creating alternative land use layouts leading to a master plan should aim to integrate buildable and nonbuildable land units into a workable whole. Following the selection of a preferred site plan, more detailed and spatially focused soil analyses should be conducted. This involves soil borings taken at specific locations designated for buildings and other facilities in the plan, as well as tesing soil fertility and moisture conditions for landscape and habitat restoration.

5.10 SELECTED REFERENCES FOR FURTHER READING

Briggs, David. *Soils.* Boston: Butterworths, 1977.

Colonna, Robert A., and McLaren, Cynthia. *Decision-Makers' Guide to Solid Waste Management.* Washington, DC: Government Printing Office, U.S. Environmental Protection Agency, 1974.

Davidson, Donald A. *Soils and Land Use Planning.* New York: Longman, 1980.

Gordon, S. I., and Gordon, G. E. "The Accuracy of Soil Survey Information for Urban Land Use Planning." *Journal of the American Planning Association.* 47, 3, 1981, pp. 301–312.

Hillel, Daniel. *Out of the Earth: Civilization and the Life of the Soil.* Berkeley: University of California Press, 1992.

Hills, Angus G., et al. *Developing a Better Environment: Ecological Land Use Planning in Ontario, A Study Methodology in the Development of Regional Plans.* Toronto: Ontario Economic Council, 1970.

Hole, F. D. "An Approach to Landscape Analysis with Emphasis on Soils." *Geoderma.* 21, 1978. pp. 1–13.

Hopkins, Lewis D. "Methods of Generating Land Suitability Maps: A Comparative Evaluation." *Journal of the American Institute of Planners.* October 1977, pp. 388–400.

Montgomery, D. R. *Dirt: The Erosion of Civilizations.* Berkeley: University of California Press, 2007.

Pettry, D. E., and Coleman, C. S. "Two Decades of Urban Soil Interpretations in Fairfax County, Virginia." *Geoderma.* 10, 1973, pp. 27–34.

Schaetzl, Randall J., and Anderson Sharon. *Soils: Genesis and Geomorphology.* New York: Cambridge University Press, 2007.

Soil Society of America. *Soil Surveys and Land Use Planning.* Madison, WI: Soil Society of America, 1966.

Related Websites

Environmental Protection Agency. Superfund. 2009. http://www.epa.gov/superfund/ *The basics about hazardous waste sites. The brownfields, environmental indicators, and redevelopment links are especially useful. The site also provides a map with the national superfund sites*

Gregg Drilling and Testing, Inc. United Soil Classification System. http://www.geology.wmich.edu/fhydro/HFC%20Docs/Gregg%20-%20Unified%20Soil%20Classification.pdf

A simple chart showing the United Soil Classification System. The site also provides field identification tips for soil.

NASA. Goddard Space Flight Center. Soil Science Basics. 2002. http://soil.gsfc.nasa.gov/basics.htm

Basics about soil structure, its formation, and characteristics. Geared toward youth, this site gives information about different depositional environments and landforms.

State of Connecticut. Hazardous Waste Sites and Soil Contamination. 2009. http://www.ct.gov/dph/cwp/view.asp?a=3140&q=387462

A public-health perspective on hazardous waste and soil contamination issues. Has links to its publications such as "Public Health and Brownfields," and their local hazardous site assessment program.

USDA. Natural Resources Conservation Services. http://soils.usda.gov/

An exhaustive site about soil, including soil surveys, soil taxonomy, and an overview of soils.

SITE, SOILS, AND WASTEWATER DISPOSAL SYSTEMS

6.1 INTRODUCTION

Until the twentieth century, soil served as the primary medium for the disposal of most human waste. Raw organic waste was deposited in the soil via pits or spread on the surface by farmers. In either case, natural biochemical processes broke the material down, and water dispersed the remains into the soil, removing most harmful ingredients in the process. But this practice often proved ineffective where large numbers of people were involved because the soil became saturated with waste, exceeding the capacity of the biochemical processes to reduce the concentration of harmful ingredients to safe levels.

Disease and human waste

The effects were occasionally disastrous to a city or town. Water supplies became contaminated and rats and flies flourished, leading to epidemics of dysentery, cholera, and typhoid fever. In response to these problems, efforts were made in the latter half of the 1800s to dispose of human waste safely. In cities, where the problems were the most serious, sewer systems were introduced, allowing waste to be transported in water through underground pipes to a body of water, such as a river or lake, beyond the city. Chicago was one of the first to build a sewer system in 1855.

Water pollution

This method removed waste from cities and thus helped solve the immediate public health problem. But as cities grew and sewer discharges increased, pollution loading in streams, lakes, and harbors rose dramatically, and some sort of end-of-pipe mitigation became necessary. This gave rise to mechanical treatment systems designed to reduce organic solids. But by midcentury the populations of North American cities had outgrown these primary level systems, and in the 1970s the U.S. federal government spent huge amounts of money to expand and upgrade treatment facilities to secondary and tertiary levels.

Site-scale treatment

Outside cities, pit-style privies survived well into the twentieth century. With the growth of suburban neighborhoods after 1930, however, the outdoor toilet proved unacceptable. In its place came small-scale water-based systems for individual homes, which used the soil to absorb and treat effluent on site. These systems are designed to function entirely underground and depend on the soil's capacity to take in and filter wastewater. Thus, in areas not served by municipal sewer systems, it is critical to distinguish suitable from unsuitable soil conditions as part of site surveys and assessments, not only for planning purposes but also as a site design consideration.

6.2 THE SOIL-ABSORPTION SYSTEM

Soil as a medium

The septic drainfield method of waste disposal is one version of a **soil-absorption system (SAS)**, so-called because it relies on the soil to absorb, filter, and disperse wastewater. The system is designed to keep contaminated water out of contact with the surface environment and to filter chemical and biological contaminants from the water before they reach groundwater, streams, or lakes. The contaminants of greatest concern are biological agents (pathogens), such as certain bacteria in the coliform group, which are hazardous to human health, and nutrients like nitrogen and phosphorus, which accelerate algae growth and cause a decline in aquatic ecosystems.

Tank and drainfield

System Components. Most SAS systems have two components: (1) a holding or *septic tank* where solids settle out and bacteria decompose organic matter, and (2) a *drainfield* through which wastewater is dispersed into the soil. The drainfield is made up of a network of perforated pipes or jointed tiles from which the fluid seeps into the soil (Fig. 6.1). Several layout configurations are commonly employed. The size of the drainfield varies with the rate of wastewater input and the capacity of the soil to absorb it.

Soil Permeability. Critical to the operation of a soil-absorption system is the rate at which the soil can receive wastewater and diffuse it into the soil column. Soils

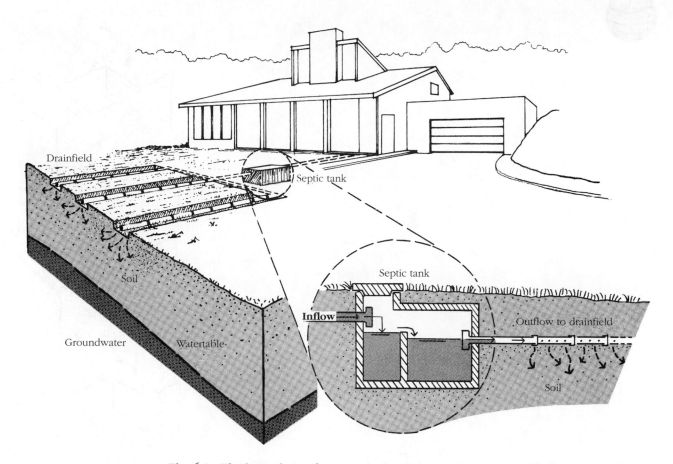

Fig. 6.1 The basic design for a standard soil-absorption system. Solid waste accumulates in the septic tank while effluent water flows into the drainfield, where it seeps into the soil. Tanks may be designed with one or two compartments.

with high permeabilities are clearly preferred over those with low permeabilities. **Permeability** is a measure of the amount of water that will pass through a soil sample per minute or hour. In the health sciences, permeability is measured by the **percolation rate**, the rate at which water is absorbed by soil through the sides of a test pit such as the one shown in Figure 6.2. The so-called *perc test* is usually conducted *in situ*, meaning that it is conducted in the field rather than in the laboratory.

Perc test

    Controls on Percolation. The percolation rate of a soil is controlled by three factors: *soil texture, water content*, and *slope*. Fine-textured soils generally transmit water more slowly than coarse-textured soils, and thus they have lower capacities for wastewater absorption. On the other hand, fine-textured soils are more effective in filtering chemical and bacterial contaminants from wastewater. The ideal soil balances the two attributes and is a textural mix of coarse particles (to transmit water) and fine particles (to act as an effective biochemical filter). In general, loams are the preferred texture, but they vary with compaction and moisture content.

Soil texture

    Soil water tends to reduce permeability in any soil. Soils with high water tables, for example, are limited for wastewater disposal because they are already saturated and therefore reject additional water. This can cause systems to back up and overflow. Soils with very low permeabilities, such as compact clays, can cause the same problem simply because they act as barriers to the intake of wastewater. The areas in Figure 6.3 designated with severe limitations are mainly low-lying areas that are wet much of the year and/or underlain with clayey soils.

Soil drainage

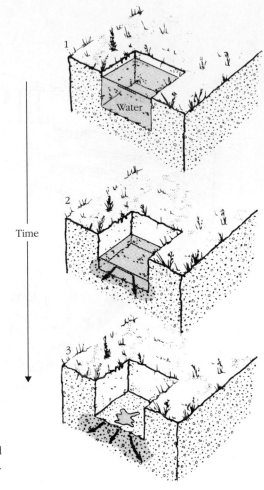

Time

Fig. 6.2   The basic idea of the percolation test. A small pit is excavated, filled with water, allowed to drain, then refilled with water and allowed to drain again. The rate of fall in the water surface is the so-called perc rate.

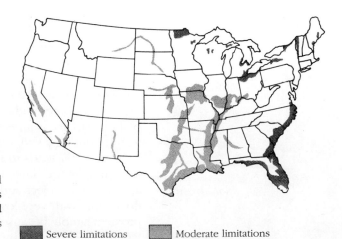

Fig. 6.3   The general distribution of soils with limitations to soil absorption systems in the coterminous United States. Locally, areas with limitations may be found just about anywhere in the United States and Canada, though they are far more common in the areas highlighted.

■ Severe limitations   ▨ Moderate limitations

Slope factor

In addition to soil drainage, *slope* is a consideration. Slope influences percolation inasmuch as it affects the inclination of the drainfield. If the slope of the ground exceeds 3 to 4 feet per 100 feet, it is difficult to arrange the drainfield at an inclination gentle enough (2 to 4 inches per 10 feet is recommended) to prevent wastewater from flowing rapidly down the drainpipes, concentrating at the lower end of the drainfield, and saturating the ground there. For parcels steeper than 4 to 5 percent slope, grading of the drainfield site is required. However, this is not an open-ended option because grading is not feasible for slopes greater than 10 to 12 percent on most sites.

Finally, the thickness of the soil layer must be considered. Where the soil mantle is thin (less than 4 to 6 feet) and bedrock is correspondingly close to the surface, the bedrock may retard the downward percolation of wastewater. The effect varies with the type and condition of the bedrock, but where it has limited permeability, wastewater may build up over it and cause system malfunction and/or surface contamination.

6.3 ENVIRONMENTAL IMPACT AND SYSTEM DESIGN

Soil-absorption systems are a standard means of sewage disposal throughout the world. In the United States, as much as 25 percent of the population relies on these systems, and in Canada the percentage may be somewhat higher. In developing countries such as Mexico, fully 80 to 90 percent of the population probably uses some form of soil absorption for waste disposal, including pit-style privies. Not surprisingly, a high percentage of these systems do not function properly, often resulting in serious health problems and environmental damage.

Causes of System Failure. The causes of system failure are usually tied to one or more of the following:

Sources of failure

- *Improper siting* with respect to slope and soil composition, texture, and drainage.
- *Improper design* of the drainfield.
- *Overloading*, that is, overuse.
- *Inadequate maintenance* of the septic tank and tile system.
- *Loss of soil-percolation capacity* due to the clogging of interparticle spaces or groundwater saturation of the drainfield bed.

Some impacts

Failure often results in the seepage of wastewater into the surface layer of the soil and onto the ground. Here humans are apt to come into contact with it, and that contact may lead to health problems, including, in the extreme, such virulent diseases as cholera. Or the wastewater may enter groundwater and contaminate wells, and it may enter lakes and streams, thus contaminating water supplies, fostering the growth of algae, and degrading the overall quality of the aquatic environment.

Nutrient loading

Impacts on Lakes and Streams. Nutrient enrichment of lakes, ponds, reservoirs, and streams with wastewater seepage is a leading cause of the deterioration of recreation waters in the United States and Canada. For example, where nitrogen and phosphorus, the nutrients of greatest concern, enter a lake, the productivity of algae and other aquatic plants usually rises substantially, resulting in increased organic mass. This mass not only helps fill in the lake bottom, but, as it decays, the consuming bacteria use up available oxygen in the water. In time, the water's oxygen content declines, fish types change to less desirable species, and the lake develops an overgrown (i.e., overfed) character that is often unsightly and smelly (Fig. 6.4). Chapter 11 addresses the problem of nutrient loading in more detail.

Fig. 6.4 An algae bloom on a pond resulting in part from seepage of nutrients into the lake from SAS drainfields.

Design criteria

System management

Biofilm clogging

System Siting and Design. The avoidance of public health and water pollution problems stemming from soil-absorption systems begins with site analysis and soil evaluation. Soils with percolation rates of less than 1 inch per hour are considered unsuitable for standard soil-absorption systems. For soils with acceptable percolation rates, additional criteria must be applied: *slope, soil thickness* (depth to bedrock), and *seasonal high water table.*

If these criteria are satisfied, the system can be designed. The chief design element is drainfield size, and it is based on two factors: (1) the actual soil percolation rate and (2) the loading rate, that is, the rate at which wastewater will be released to the soil. For residential structures, the loading rate is based on the number of bedrooms; each bedroom is proportional to two persons. The higher the loading rate and the lower the percolation rate, the larger the required drainfield must be (Fig. 6.5).

Operational Problems. The successful operation of the system necessitates regular maintenance, especially the removal of sludge from the septic tank, the avoidance of overloading, and the lack of interference from high groundwater. The last-named may occur during unusually wet years or because of the raising of water levels in a lake, wetland, or a nearby reservoir, for example. On the average, the lifetime of a drainfield is 15 to 25 years, depending on local conditions.

By the end of the life cycle, the soil may be too wet to drain properly because the water table has been raised after years of recharge from drainfield water, and/or the soil may have become clogged with minute particles and with the buildup of biofilm (microbe colonies) around and between particles. In addition, the buildup of chemicals such as phosphorus and nitrogen may become excessive, thereby reducing the adsorption capacity of the soil. In any event, old drainfields should be abandoned and new ones constructed in different locations.

6.4 MAPPING SOIL SUITABILITY FOR ON-SITE DISPOSAL

The suitability of an individual site for wastewater disposal is best determined by field inspection and percolation tests. For planning problems involving large areas, however, this approach is not always possible, and planners must resort to less expensive methods to assess soil suitability.

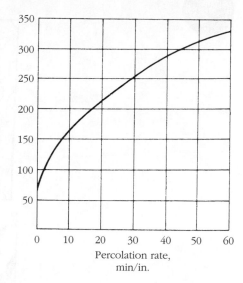

Absorption area required, square ft/bedroom

Percolation rate, min/in.

Fig. 6.5 Size of the absorption area required for soil percolation rates ranging from 0 to 60 minutes per inch of water.

In most cases, the investigator must rely on existing maps and data, and the most prevalent sources in the United States are the county soil reports and maps produced by the Natural Resources Conservation Service (NRCS) (see Fig. 6.9). These maps are suitable for first approximations, but beyond that there may be problems of spatial resolution (mapping polygon size) and geographic accuracy (see Fig. 5.8). In Canada, roughly comparable soil maps are produced by the provinces or by regional environmental agencies.

Building a Classification System. When NRCS or similar studies are not available or when their reliability is in question, the investigator must turn to other sources and synthesize existing maps and data into information meaningful to wastewater disposal planning. The maps in Figure 6.6 are part of a set of maps produced by the

Data sources U.S. Geological Survey (USGS) for a land use capability study in Connecticut. Three of these maps (slope, seasonal high water table, and depth to bedrock) are directly applicable to wastewater disposal problems. However, the fourth map, called unconsolidated materials, requires translation before it can be used in place of soil type or soil texture. The six classes of materials must be translated into their soil counterparts and the soils, in turn, into their potential for wastewater disposal. For example:

Unconsolidated Material* on USGS Map	Approximate Soil Equivalent	Potential for Wastewater Disposal
Till	Loam	Good
Compacted till	Loam, perhaps clayey, that drains poorly	Fair/poor
Clay deposit	Clay; clayey loam	Poor
Sliderock deposit	Rock rubble	Poor
Swamp	Organic: muck or peat	Poor
Sand or sand and gravel deposits	Sand; sandy loam; gravel	Fair/good

* Also called surface deposits.

Synthesis With this classification in hand, we can combine the four maps and produce a suitability map for wastewater disposal. One procedure for this sort of task involves assigning a value or rank to each trait according to its relative importance in wastewater

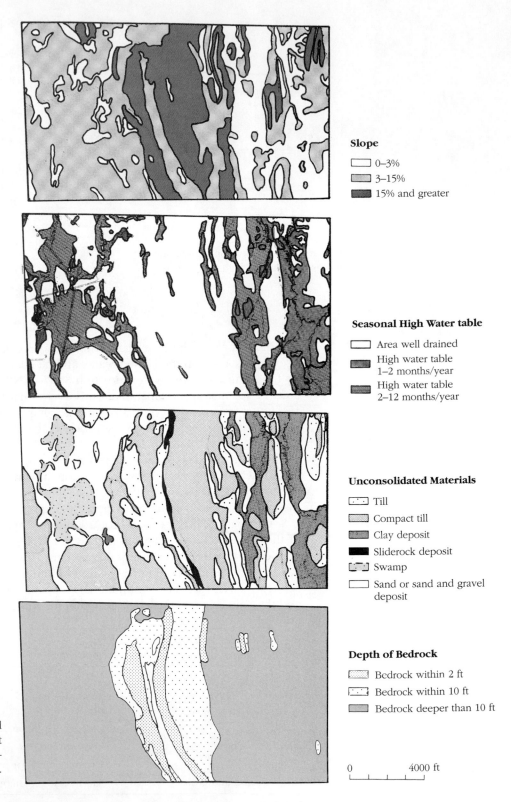

Slope

☐ 0–3%

▨ 3–15%

▨ 15% and greater

Seasonal High Water table

☐ Area well drained

▨ High water table 1–2 months/year

▨ High water table 2–12 months/year

Unconsolidated Materials

▨ Till

▨ Compact till

▨ Clay deposit

■ Sliderock deposit

▨ Swamp

☐ Sand or sand and gravel deposit

Depth of Bedrock

▨ Bedrock within 2 ft

☐ Bedrock within 10 ft

▨ Bedrock deeper than 10 ft

0   4000 ft

Fig. 6.6 A series of maps compiled as part of a land capability study that can be used to assess on-site wastewater disposal suitability. (From U.S. Geological Survey.)

absorption. In this case, three ranks are possible for each of the four traits, and each can be assigned a numerical value: good (3), fair (2), and poor (1) according to the previous translation (Table 6.1).

For any area or site, the relative suitability for wastewater disposal can be determined by simply summing the four values. The final step is to establish a numerical scale

Table 6.1 Criteria and Numerical Values for SAS Suitability for Maps in Figure 6.6

			Trait		
		Material	*Depth to Bedrock (ft)*	*Seasonal Water table*	*Slope (%)*
Rank	3 (good)	Till	>10	Well-drained	0–3
	2 (fair)	Sand, sand and gravel, compacted till	2–10	1–2 mo high water table	3–15
	1 (poor)	Clay, sliderock, swamp	<2	2–12 mo high water table	>15

defining the limits of the various suitability classes. The classes should be expressed in qualitative terms, for example, as high, medium, or low, because numerical values are meaningless and often misleading in terms of their relative significance.

Using NRCS Soil Maps. Modern soil surveys by the U.S. Natural Resources Conservation Service present a somewhat different arrangement. First, all data and information are organized and presented according to soil type. Second, in many areas the NRCS provides information on soil suitability for wastewater disposal by an SAS. In that case, all one needs to do is map the various NRCS classes—taking care, of course, to field-check the NRCS distributions—and record the criteria and rationale for each.

Criteria and rationale   In some cases, however, the criteria and rationale used are not always clear, or a planner may wish to add or delete certain criteria depending on the land use problem and/or local conditions. In such instances, the procedure is largely the same as just outlined, except that the basic mapping unit is already established (Fig. 6.7) and the database may be different. County soil reports often include data on permeability, seasonal high water table, and depth to bedrock, as well as soil texture and slope. When data are available for these criteria, the steps listed in Table 6.2 can be used to build a soil suitability map for wastewater disposal.

6.5 ALTERNATIVES TO STANDARD SAS

The standard soil-absorption system of wastewater disposal has a number of distinct limitations. Site conditions such as steep slopes and wet soils are common to many areas, especially in this era with development increasingly attracted to challenging terrain. Many would also argue that the SAS itself is inherently limited because it is often incapable of fully consuming the organic waste it receives.

Unsustainable SAS. Critics maintain that, as an ecosystem, the SAS is incapable of maintaining a mass balance because its biochemical system cannot keep up with organic inputs. With a proper design—which includes a capacity to foster rich

Life cycle limits   colonies of microbes to gobble up organic matter—the system should be capable of consuming virtually all the organic it receives, leaving only water going to the drainfield. But for a variety of reasons, led principally by inadequate soil oxygen to nurture the necessary levels of microbial activity, large amounts of organic waste are left in the drainpipes and ground.

Water use/loss is another problem with the standard soil-absorption system. Large amounts of water are generated by the average household, upwards of several hun-

Water use   dred gallons a day. Since all this water is fed to the drainfield, it is lost to the ground and atmosphere rather than recycled in some fashion. Given these limitations, it is no surprise that a number of alternative waste disposal systems have been designed and tested for residential-scale development, and we review a few of these now.

Reduced Water Options. The acceptance and use of alternative systems vary from place to place, depending on local health and planning regulations. One of the

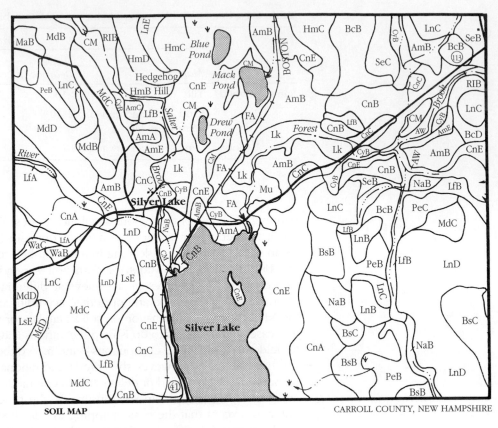

SOIL MAP

CARROLL COUNTY, NEW HAMPSHIRE

Fig. 6.7 An excerpt from a standard U.S. Natural Resources Conservation Service map. The letter codes denote different soil types, and each code is defined in county soil reports.

U.S. DEPARTMENT OF AGRICULTURE
SOIL CONSERVATION SERVICE

Table 6.2 Procedure for Soil Suitability Classification for SAS

Step 1: Select the traits for evaluation and build a matrix listing each soil type and each trait. For example:

	Trait				
Soil	Texture	Depth to Bedrock	Depth to Seasonal Water table	Permeability	Slope
Becket					
Berkshire					
Chocorua					

Step 2: Set up a ranking system such as that in Table 6.1, and complete the matrix by assigning a numerical value to each soil trait.

Step 3: Total the ranking values for each soil type or series.

Step 4: Based on these totals, classify the soils as high, medium, or low.

Step 5: Go to the soils map and color the areas that correspond to each suitability class. For this you need to devise a three-part color scheme and map legend.

Step 6: Check the boundaries of the soil polygons (areas) against other map sources such as detailed topographic maps, and in the field, and adjust accordingly (see Fig. 5.8).

simplest alternatives is the waterless toilet, which eliminates the need for all or part of the drainfield in the standard residence. The **waterless system** is designed to concentrate solid waste and dispose of it in an environmentally safe manner, such as soil compost, after it is free of disease agents.

Waterless option

Many communities view the waterless system very favorably because (1) it eliminates the need to handle and process large amounts of septic tank sludge, and (2) it greatly reduces domestic water use because there is no toilet flushing. Gray water can be directed to a drainfield or stored for recycling. Standard SAS can also be designed or retrofitted to separate toilet water (black water) from gray water to save on water use. The toilet water is directed to the septic tank and drainfield, whereas the gray water, which is not laden with pathogens, can be recycled as irrigation water for lawns and gardens.

Site-Adaptive Options. Where soils are the limitation, several other alternatives are possible. One involves excavating the upper 2 to 3 feet of soil and replacing it with a fill medium of the desired percolation and absorption values. A related system,

Earth mound

called **earth mound**, involves building a drainfield above ground level in a pile of soil medium that contains the release pipes. A pump is used to force the wastewater into the drainfield.

In areas where slopes are the chief limitation, **terraced drainfields** can be used. In this alternative, several benches are designed and the drainfield is laid out in compartments with wastewater distributed according to the size and capacity of each terrace compartment. Another method utilizes a pipe system to transfer wastewater to a suitable disposal site on- or off-site. Because this system usually transfers waste

Pipe and pump

over some distance, pumping is typically required, and it may be designed to serve several residences. Each house is equipped with a holding tank and a pump, often with a grinder to break up the solids, and the disposal field is sized according to the number of houses served. This design has excellent potential with small-scale cluster development (Fig. 6.8).

With the exception of the waterless toilet, construction and maintenance costs are usually greater for alternative systems than they are for the standard soil-absorp-

Drawbacks

tion system. In addition, they require more attention to their design, and permitting by local regulatory agencies in many parts of the United States and Canada is (and should be) more difficult. In many places the waterless toilet, in particular, is not accepted for standard residential units, despite its remarkable success in parts of Europe and North America. However, with rising concern over water use, energy

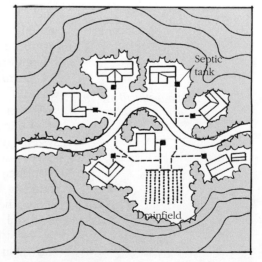

Fig. 6.8 The layout of a wastewater disposal system in an area of hilly terrain designed to serve several residences via a pipe and pump system connected to one large drainfield.

conservation, and/or contamination of groundwater and streams, the waterless toilet and gray water recycling options, among other conservation options, have gained serious attention.

6.6   SMALL-FLOW WETLAND SYSTEMS

Wetlands have also gained serious attention in wastewater disposal planning. It is an established fact that wetlands can serve as effective treatment systems for impure water. They are known to perform several physical and biochemical functions that can reduce suspended sediment, harmful biological agents, and chemical contaminants. Among the wetland's water cleansing processes are:

Wetland services

- *Entrapment* of coarse sediment as runoff slows down and as plant stems, roots, and organic debris capture particles.

- *Filtration* of suspended sediment as water penetrates and percolates through the plant and soil mass.

- *Adsorption* (bonding) of nutrients, organic compounds, and biological agents by organic microparticles (colloids).

- *Plant uptake* of nutrients and other dissolved chemical ions.

- *Breakdown* of certain contaminants by microbial action in organic masses.

Residential Scale. Both natural and constructed wetlands are being used today for the disposal of residential sewage. Although viewed with skepticism by some public health agencies, wetland disposal systems are gaining increased acceptance in the United States and Canada as the list of successful experimental systems grows and the cost of constructing and maintaining mechanical systems becomes more acute. At the top of the list are small-scale, constructed wetland systems designed to serve individual houses or small groups of houses.

System components   These systems consist of a septic tank (with a pump system if necessary), a wetland cell (with a substrate of saturated gravel supporting a cover of wetland plants), and an infiltration bed (to disperse water released from the wetland). The system is equipped with check valves to prevent reverse flows, and the wetland cell may be underlined with plastic to prevent water loss to the ground if that is a concern. The size of the wetland cell varies with climatic conditions and household size, but it typically covers an area of 800 to 1200 square feet. For a household producing 600 gallons of water per day, the effluent detention time in the wetland is two weeks. During this time, suspended solids are filtered out and nutrients are consumed by bacteria and higher plants. Periodically, the wetland needs maintenance, including the thinning of wetland plants and the removal of debris and extraneous vegetation.

Community Scale. Large wetland systems are being used effectively in wastewater treatment in a growing number of communities. They too are often viewed with *Large wetland systems*   skepticism in both the public health and environmental engineering fields. Among other things, these professions tend to be more comfortable with community treatment systems they can control externally, that is, with mechanical devices. But mechanical systems are very expensive and in many respects not consistent with sustainability principles. Wetland-based systems help combat such concerns, for example, by reducing infrastructure and energy use, but they raise other concerns. Among other things, the design and operation of wetland treatment systems must be balanced with local soil, drainage, and climatic conditions.

Design features   Most community-scale systems are designed with a series of wetland basins, as shown in Figure 6.9, through which wastewater is dispersed after the removal of

Fig. 6.9 A small-scale community wastewater treatment system using a series of wetland basins for effluent processing.

solids in a settling basin. The capacity of the wetlands to process effluent water varies seasonally with, among other things, water temperature and supply; therefore, systematic monitoring and day-to-day management are mandatory. Among the communities demonstrating impressive success with wetland wastewater treatment is Arcata, California, where the wetland basins have become prized bird habitats that attract thousands of visitors a year (and substantial revenue to local businesses).

6.7 CASE STUDY

Land Use Implications of On-Site Wastewater Treatment Technologies: Lessons from the Great Lakes Region

James A. LaGro, Jr.

Public policies often are revised in response to new technologies and, in some cases, to unintended consequences. The State of Wisconsin, like other states in the Great Lakes region, has regulated private wastewater treatment for decades. Prior to 1969, when Wisconsin's plumbing code had weak siting standards for private on-site wastewater treatment systems, about 200,000 conventional on-site systems were installed in the state. But septic systems on sites with less than 24 inches (61 centimeters) of suitable soil were failing and polluting nearby lakes and groundwater. After 1969, 36 inches (91.4 centimeters) of soil was required for on-site treatment to mitigate the public health impacts.

Alternative wastewater treatment technologies were developed to replace failing conventional on-site systems in areas not served by municipal treatment facilities. Although intended to replace existing conventional treatment systems, alternative systems eventually facilitated new single-family housing development in previously undevelopable locations. A major revision of Wisconsin's plumbing code—changing it in the 1990s from a prescriptive code to a performance code—was contentious. Environmental and smart growth interests were pitted against property rights advocates and the housing and real estate industries. A lawsuit filed by the League of Wisconsin Municipalities, the Wisconsin Alliance of Cities, and 1,000 Friends of Wisconsin challenged the new plumbing code for its potential land use and ecological impacts. The Wisconsin Realtors Association and other industry interests hired legal counsel to defend against the lawsuit. Ultimately, the State Supreme Court upheld the revised plumbing code, which went into effect in 2001.

Plumbing Code or Stealth Land Use Policy? Most communities in Wisconsin rely on groundwater as their sole source of drinking water. And yet Wisconsin's state plumbing code allows—unless prohibited by local ordinances—the installation of newer alternative treatment systems on sites with just 6 inches (15.2 centimeters) of native soil above bedrock or the seasonal water table. Because more than half of Wisconsin's land area is unsuitable for conventional soil absorption treatment systems, this *de facto* land use policy significantly increases the rural land area in the state that could be converted to residential development.

Coincident with Wisconsin's plumbing code revision, the state enacted the Comprehensive Planning Act (1999). This statute requires Wisconsin towns, cities, villages, and counties to develop comprehensive plans by 2010 to support local zoning and land division decisions. The Comprehensive, or Smart Growth, Planning Act was passed, in part, because of concerns about the land use impacts of the new plumbing code. Yet nearly a decade after the Act's passage, a state inventory of land use planning and policy in the state's 1923 counties, cities, villages, and towns revealed only partial compliance:

> As of April 2008, 740 local governments had adopted comprehensive plans and an additional estimated 660 had a planning process underway. Another 120 units of local government are estimated to be in the preliminary stages of a planning process. Many of the remaining units of local government do not exercise zoning, subdivision regulations, official mapping, or shoreland/wetland zoning.

Land use planning in the United States is highly decentralized, with development decisions made largely at the local level by part-time legislative bodies with limited resources to assess the positive and negative impacts of their land use and infrastructure decisions. A community's geology and the sources, sinks, and water balance of its aquifer system should be assessed to inform land use planning and development policies that protect water supplies, human health, and aquatic ecosystems under present as well as future climate conditions. When rural land use policies do not respond to intrinsic landscape patterns and processes, alternative wastewater technologies—like conventional technologies—can facilitate urban sprawl.

Protecting Environmentally Sensitive Areas. Rural sites with shallow bedrock or a shallow water table are poorly suited for cost-effective, low-impact residential development. These so-called difficult sites increase the excavation and construction costs for basements, building foundations, and site drainage. These sites may also perform valuable ecological and hydrologic functions. The U.S. Environmental Protection Agency cautions:

> In deciding whether to use onsite systems, it is important to consider the risks they might pose to the environment and public health. There may be cases where onsite systems are not appropriate because of the environmental *sensitivity or public health concerns of an area*.

Environmentally sensitive areas, such as the one shown in Figure 6.A, are defined by Wisconsin's Department of Natural Resources as areas that "include but are not limited to wetlands, shorelands, floodways and floodplains, steep slopes, highly erodible soils and other limiting soil types, groundwater recharge areas, and

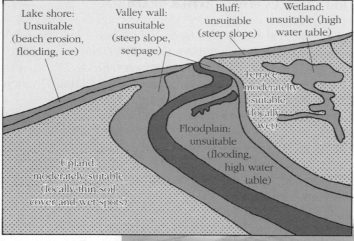

Fig. 6.A An example of an environmentally sensitive area where development is possible but detailed site assessment and strong regulation are necessary to avoid landscape degradation.

other such physical constraints." These areas typically are excluded from municipal sewer service areas because of potential adverse water quality impacts from both point and nonpoint pollution sources. And yet in too many rural communities, alternative wastewater treatment systems serve residential development in environmentally sensitive areas. County and town zoning codes may also contribute to low-density sprawl by requiring large minimum lot sizes (often 2 acres or more) for new single-family residences.

Environmental planners and landscape architects who "design with nature" do not view floodplains, storm surge zones, and other natural site constraints as design challenges to be overcome with engineering technology. Instead, these professionals understand that some locations are better left undeveloped.

James A. LaGro Jr. is a professor of urban planning and landscape architecture at the University of Wisconsin-Madison and author of *Site Analysis: A Contextual Approach to Sustainable Land Use Planning and Design* (Hoboken, NJ: Wiley, 2008).

6.8   SELECTED REFERENCES FOR FURTHER READING

Clark, John W., et al. "Individual Household Septic-Tank Systems." In *Water Supply and Pollution Control*, 3rd ed. New York: IEP/Dun-Donnelley, 1977, pp. 611–621.

Cotteral, J. A., and Norris, D. P. "Septic-Tank Systems." Proceedings American Society of Civil Engineers, Journal of Sanitary Engineering Division 95, No. SA4, 1969, pp. 715–746.

Environmental Protection Agency. "Alternatives for Small Wastewater Treatment Systems." *EPA Technology Transfer Seminar Publication*, EPA-624/5-77-011, 1977.

Gelt, J. "Using Human Ingenuity, Natural Processes to Treat Water, Build Habitat." *Arroyo*. 9, 4, 1997.

Huddleston, J. H., and Olson, G. W. "Soil Survey Interpretation for Subsurface Sewage Disposal." *Soil Science*. 104, 1967, pp. 401–409.

LaGro, J. A. "Designing Without Nature: Unsewered Residential Development in Rural Wisconsin." *Landscape and Urban Planning*. 33, 1996, pp. 1–9.

Last, J. M. *Public Health and Human Ecology*. East Norwalk, CT: Appleton and Lange, 1987.

Public Health Service. "Manual of Septic Tank Practice." United States Public Health Service Publication No. 526. Washington, DC: U. S. Government Printing Office, 1957.

U.S. Environmental Protection Agency. *Voluntary National Guidelines for Management of Onsite and Clustered (Decentralized) Wastewater Treatment Systems*. EPA 832-B-03-001. Cincinnati, OH: EPA Publications Clearinghouse, 2003.

Wilhelm, S., et al. "Biogeochemical Evolution of Domestic Wastewater in Septic Systems: 1. Conceptual Model." *Ground Water*. 32, 1994, pp. 905–916.

Wisconsin Department of Administration. *2008 Wisconsin Local Land Use Regulations and Comprehensive Planning Status Report*. Madison, WI: Division of Intergovernmental Relations, 2008.

World Health Organization. *The International Drinking Water Supply and Sanitation Decade Review of Regional and Global Data*. Geneva: WHO Offset Publication No. 92, 1986.

Related Websites

American Ground Water Trust. Septic Systems for Waste Water Disposal. 2008. http://www.agwt.org/info/septicsystems.htm

Information about soil absorption systems by an organization focused on the importance of groundwater. The site gives detailed descriptions and diagrams of septic systems as well as failure prevention and site selection.

City of Arcata. Arcata Marsh and Wildlife Sanctuary. 2006. http://www.cityofarcata.org/index.php?option=com_content&task=view&id=20&Itemid=47#foam

A citywide wastewater treatment wetland group of concerned citizens who became politically involved. Now a center for education, recreation, and bird habitat, this successful wetland has become a model for other cities.

The Greywater Guerillas. 2007. http://greywaterguerrillas.com/index.html

All things water saving from a collection of educators, designers, and others. This site tells about residential gray water systems and waterless systems—how to make your own composting toilet and reuse your laundry water for irrigation.

GROUNDWATER SYSTEMS, LAND USE PLANNING, AND AQUIFER PROTECTION

7.1 INTRODUCTION

Pollution concerns

Not many years ago, the only consideration given to groundwater in land use planning was as a water supply. Groundwater is still an important source of water for residential, industrial, and agricultural land uses, and locating dependable supplies of usable water is still important in planning. Today, however, there are other problems to contend with. Chief among them is groundwater pollution. Contamination of groundwater is one of the most alarming of today's environmental problems, especially when we consider that groundwater is the single largest reservoir of fresh, liquid water on the planet. At the site scale, groundwater also may pose a serious problem when shallow aquifers interfere with road building, foundation excavations, slope cuts, forestry operations, and agriculture.

Land use connection

Although groundwater is often pictured as a remote and complex part of the environment and traditionally is not the domain of planners, we must keep in mind that virtually all groundwater begins and ends in the landscape. Therefore, changes in the surface environment, especially those involving land use activity, can, and usually do, affect groundwater in some way. For land use planners and developers, the problem is not only knowing the polluting activities, but also finding the proper location for them relative to the local groundwater systems. Indeed, the first line of defense in groundwater protection is judicious site selection for land uses with high-impact potentiality, both in terms of water quality and site conditions.

Buried waste

The responsibility of land use and site planners today goes even further, for they must contend with hidden sources of soil and groundwater contamination. In most states and provinces, thousands of leaking underground storage tanks are releasing oil, gasoline, and other petroleum products into the ground, and tens of thousands of landfills contain hazardous waste. Most of these lie in and around metropolitan regions, and, for a surprisingly large number of them, the locations are unknown. Increasingly, new landowners are uncovering storage tanks and hazardous landfill materials and are being forced to absorb the expense and liability connected with cleanup and environmental restoration. Thus, to the list of considerations in environmental assessment, site selection, and site analysis, we must add that of unrecorded landfills, hazardous waste, and underground storage tanks.

7.2 GROUNDWATER SOURCES AND FEATURES

Groundwater begins with surface water seeping into the ground. Below the surface, the water moves along two paths: (1) Some is taken up by the soil, and (2) some is drawn by gravity to greater depths. The latter, called **gravity water**, eventually reaches a zone where all the open spaces (interparticle voids in soil materials and the cracks in bedrock) are filled with water. This zone is called the *zone of saturation* or the *groundwater zone.*

Water table

Groundwater Features and Measures. The upper surface of the groundwater zone is the **water table** (Fig. 7.1). In some materials the water table is actually a visible boundary line, but in most materials it is more of a transition zone with an irregular configuration. Below the water table, the groundwater zone may extend several miles into the earth, but at great depths the water content is very small. In the upper 5000 feet or so, in the zone of usable groundwater, the actual water content varies significantly, often dramatically, with different soil and rock materials.

Porosity

The total amount of groundwater that can be held in any earth material is controlled by its porosity. **Porosity** is the total volume of void space in a material. It commonly varies from 10 to 40 percent in soils and near-surface bedrock (Fig. 7.2). In general, porosity decreases with depth, and at depths of several miles into the earth's crust, where the enormous pressure of the rock overburden closes out void spaces, it is typically less than 1 percent.

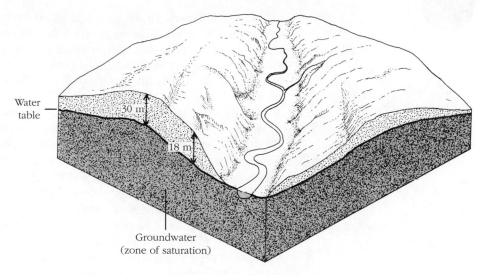

Fig. 7.1 The general relationship between the configuration of the water table and the overlying terrain. The variation in the elevation of the water table is usually less than that of the land surface, resulting in many intersections between groundwater and the surface of the land.

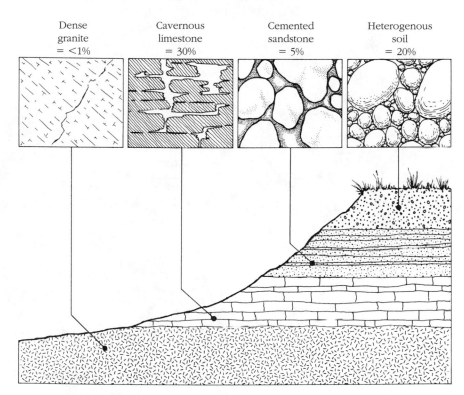

Fig. 7.2 Porosities of four materials near the surface. In all materials, porosity decreases with depth because the great pressure exerted by the soil and rock overburden tends to close the voids. Porosity can be as high as 45 percent in some near-surface bedrock, but it is rarely higher than 2 to 3 percent at depths exceeding 2000 meters.

Aquifers

In most areas, the materials underground are arranged in layers, formations, or zones with different porosities and groundwater capacities. Materials with especially large concentrations of usable groundwater are widely known as **aquifers**. Many different types of materials may form aquifers, but porous material with good permeability, such as beds of sand and fractured rock formations, are usually the best. An aquifer is evaluated or ranked according to (1) how much water can be pumped from it without causing an unacceptable decline in its overall water level, and (2) the quality of its water. With respect to water quality, highly mineralized water, such as saltwater, is not generally usable for agriculture and municipal purposes, and aquifers containing such water are usually not counted among an area's groundwater resources.

Basic classes

Aquifer Types. Aquifers are grouped into two classes depending on the type of material in which they form: *consolidated* (mainly bedrock) and *unconsolidated* (mainly surface deposits). In the central part of North America, where glacial deposits from 50 to 500 feet in thickness lie over sedimentary bedrock, aquifers are found in both the bedrock and the surface deposits. Owing, however, to the extreme diversity of glacial deposits in many areas, aquifers in the unconsolidated class tend to vary greatly in size, depth, and water supply. Bedrock aquifers, on the other hand, tend to be more extensive—especially in areas of sedimentary rocks—often covering hundreds of square miles in area. The largest aquifer in North America is the Ogallala, which stretches more than 1000 miles from South Dakota into northern Texas, and is composed principally of sandstone (Fig. 7.3).

Alluvial aquifers

In the lowlands of large rivers such as the Illinois, Wabash, Platte, Arkansas, Athabasca, and hundreds of other streams, extensive shallow aquifers (at depths of less than 300 feet) lie beneath the valley floors. These are *alluvial aquifers* (i.e., composed of buried stream deposits). They are recharged (i.e., replenished) by river water, and their water supplies fluctuate with the seasonal changes in streamflow. In addition, river aquifers are distinctive for their geographical distributions because they tend to form ribbons following the floors of river valleys and their tributaries.

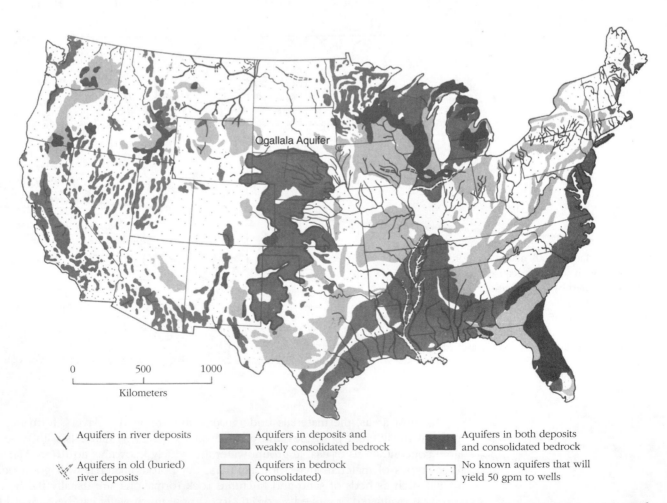

Fig. 7.3 The distribution of major aquifers in the coterminous United States. A major aquifer is defined as one composed of material capable of yielding 50 gallons per minute or more to an individual well of water generally not containing more than 2000 parts per million of dissolved solids.

In the American and Canadian West, aquifers are found in both deposits and bedrock, but many dry areas contain only marginal aquifers with small yields, say, fewer than 50 gallons of water per minute. Locally, however, river aquifers fed by runoff from mountain ranges are very important sources of groundwater in otherwise dry mountain valleys. In general, the diverse mountainous terrain of western North America, with deep valley deposits, alluvial fans on mountain flanks, and fractured bedrock formations, produces a highly varied groundwater environment closely linked to geologic structures, mountain runoff, and related deposits (Fig. 7.3).

Terrain relations

Groundwater Basins. A group of aquifers linked together in a large flow system is called a **groundwater basin**. Groundwater basins are typically complex three-dimensional systems characterized by vertical and horizontal flows among the various groundwater bodies—both those that would qualify as aquifers and those that would not—and between groundwater bodies and the surface (Fig. 7.4). The spatial configuration of a groundwater basin is determined largely by regional geology, that is, by the extent and structure of the deposits and rock formations that house the groundwater bodies. Because these deposits and formations usually differ vastly in their size, composition, and shape, exactly how the various bodies of groundwater in a basin are connected at different depths is seldom clear. This uncertainty is significant not only in planning for water supplies, but also in understanding the spread of contaminants among aquifers.

Aquifer systems

7.3 GROUNDWATER AS A FLOW SYSTEM

If we were to map the elevations of aquifers, we would find that all bodies of groundwater are inclined (tilted) to some degree. This observation can be verified for most shallow aquifers by tracing the elevation of the water table across the landscape. As the diagram in Figure 7.1 illustrates, the water table generally rises and falls with broad changes in surface topography. In deeper aquifers, groundwater commonly slopes with the dip of rock formations. The rate of change in elevation across an aquifer or a segment of the water table is termed the **hydraulic gradient**, and it is calculated in the same fashion as a topographic slope, that is, change in elevation between two points divided by the distance between them.

Hydraulic gradient

Groundwater Movement and Flow Rates. The flow of groundwater is driven by gravity along the hydraulic gradient. For a given material, the steeper the hydraulic gradient, the faster the rate of flow. To determine the flow velocity of groundwater, it

Groundwater flow

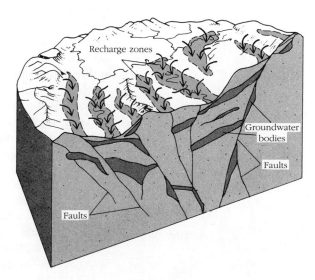

Fig. 7.4 The concept of a groundwater basin: different bodies of groundwater fed by one or more recharge areas in a common geological setting.

is necessary to know not only the hydraulic gradient, but also the resistance imposed on the water by the material it is moving through. *Resistance* is a function of permeability, which is determined by the size and interconnectedness of void spaces. For clays, permeability is low; for sands and gravels, it is high. The basic formula for groundwater velocity, known as **Darcy's Law**, is

Darcy's Law

$$V = I \bullet k$$

where

V = velocity
I = hydraulic gradient
k = permeability

Residence time

Compared to the flow velocities of surface water, groundwater is extremely slow. Typical velocities for large aquifers are only 50 to 75 feet per year. The time it takes water to pass completely through an aquifer, which is called *residence time* or *exchange time*, is measured in decades and centuries. In the case of a typical aquifer 3 or 4 miles in diameter, for example, the exchange time could be as great as 250 to 350 years. This time, of course, may be much shorter for water that has been withdrawn from wells in the aquifer and that has passed only part of the way through the system. In any case, groundwater systems are very slow, and the time required to flush contaminants from polluted aquifers is exceedingly long by human standards.

Recharge and Transmission. The process by which surface water is supplied to a groundwater body is termed **recharge.** Although some aquifers, especially shallow ones, receive recharge water from a broad (nonspecific) surface area, many aquifers are recharged from specific areas, called *recharge zones*. These zones may be places (1) where surface water accumulates, such as in a stream valley, lake, wetland, or a topographic depression; (2) where there is highly permeable soil or rock formation, such as sand deposits or fractured limestone, at or near the surface; or (3) where an aquifer itself is exposed at or near the surface (Fig. 7.5). In all cases, recharge zones are critical to aquifer management because they are the points of most ready access for contaminants from land use activity.

Recharge zones

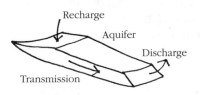

When recharge water enters the aquifer, its continued movement depends on the permeability of the material it encounters and the aquifer's hydraulic gradient. If an aquifer receives rapid recharge, but the material has low permeability (i.e., high resistance), the new groundwater tends to build up. As a result, the hydraulic gradient increases, but as it does, so does flow velocity, according to Darcy's Law. Under such conditions the hydraulic gradient continues to rise until the rate of lateral flow, or *transmission*, is equal to the rate of recharge. Where recharge is low and permeability is high, on the other hand, the hydraulic gradient falls until the rates of transmission and recharge are equal.

Recharge rates

In general, shallow aquifers have a capacity to recharge much faster than deeper ones. Although average recharge rates decrease more or less progressively with depth, as a whole aquifers within 3000 feet of the surface require several hundred years (300-year average) to be completely renewed, whereas those at depths greater than 3000 feet require several thousand years (4600-year average) for complete renewal. This difference is related not only to the distance that recharge water must travel, but also to the fact that percolating water moves more slowly in the smaller spaces found at great depths.

Aquifer output

Although large quantities of groundwater remain in some aquifers for thousands of years, most water remains underground for periods ranging from several years to several centuries. Aside from pumping, groundwater is released to the surface mainly through (1) *capillary rise* into the soil, from which it is evaporated or taken up by plants and released in transpiration, and (2) *discharge*, or seepage into streams, lakes, and

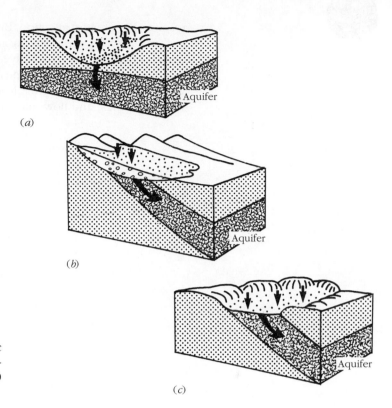

(a)

(b)

(c)

Fig. 7.5 Three types of recharge zones: (*a*) a topographic depression such as a wetland; (*b*) a highly permeable surface material such as an alluvial fan or talus slope; and (*c*) an exposed area of aquifer such as near an escarpment.

Seepage zones

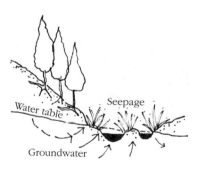

Water features

wetlands where it becomes part of the surface runoff system. Both modes of release involve huge amounts of water, but both are nearly imperceptible in the landscape.

Groundwater Discharge. Many different geologic and topographic conditions can produce seepage. In mountainous areas, for example, it can be traced to fault lines and outcrops of tilted rock formations (Fig. 7.6). In areas where bedrock is buried under deep deposits of soil, **seepage zones** are usually found along the foot of slopes. Steep slopes represent "breaks" in elevation that may be too abrupt to produce a corresponding elevation change in the water table. Under such conditions, the water table may intercept the surface, especially in humid regions where the water table is high. If the resultant seepage is modest, a spring is formed, but if it is strong, a lake, wetland, or stream may be formed. Seepage zones, also called discharge zones, are ecologically significant in any geographic setting because they provide dependable water supplies and temperate thermal conditions. In stream channels, groundwater seeps into the streambed, providing an important long-term source of discharge, called *baseflow*.

The type of water feature that develops from groundwater seepage depends on a host of conditions, including the rate of seepage, the size of the receiving depression, and the rate of water loss to runoff and evaporation. Most of the thousands of inland lakes of Minnesota, Wisconsin, Michigan, Manitoba, and Ontario, for example, are seepage-type lakes whose water levels fluctuate with the seasonal changes in the elevation of the water table. Most of these lakes are either connected to wetlands or evolve into wetlands as they are filled in with organic debris. Where, for example, groundwater is contaminated by nutrients from residential sewer systems, fertilizers, or stormwater, lakes undergo accelerated growth of algae and other aquatic plants, which hastens the filling process.

Site-Scale Considerations. The picture sketched above applies principally to groundwater systems of areawide and regional scales, the scale at which groundwater receives the greatest attention. But groundwater can also play into planning and design at the site scale, especially in areas of varied terrain as the examples in Figure 7.6 demonstrate. Almost invariably site-scale problems involve shallow bodies of groundwater associated with hillslopes, swales, stream valleys, wetlands, lakes, and reservoirs. These bodies come in all sort of shapes and sizes and may vary radically in size, water volumes, and flow rates from season to season.

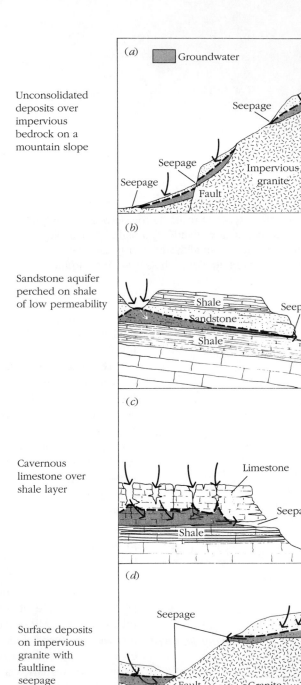

Fig. 7.6 Seepage zones associated with different physiographic settings: (a) unconsolidated sediments over impervious bedrock; (b) inclined sandstone over shale; (c) cavernous limestone over shale; (d) unconsolidated materials broken by a fault.

Problems　Problems usually arise when they are pierced during grading and excavation, creating a low-pressure point in the system, which induces rapid seepage and erosion by sapping, sometimes leading to slope failure (see Fig. 4.1). Stabilizing and remediating such conditions often prove difficult, especially if we fail to realize that we are dealing with a system, not an isolated water body. A common response is to excavate the seepage zone, but in most instances this only exacerbates the problem. Not only does the excavation release more groundwater, but when vegetation is removed, an important dewatering agent is erased. Pumping does not solve the problem and only adds to construction, management, and long-term energy costs.

Responses　The appropriate response is to modify the plan to a more site-adaptive configuration. This may require realigning a road or redesigning a building foundation and backfilling, stabilizing, and replanting the seepage zone. It is also advisable to consider looking for ways of turning a liability into an asset, such as capturing irrigation water, building a wetland, and/or enhancing habitat and landscaping.

7.4 GROUNDWATER WITHDRAWAL AND AQUIFER IMPACT

Groundwater is used for drinking, agriculture, and industry in every part of North America. In rural areas, it provides more than 95 percent of the drinking water and about 70 percent of the water used in farming. Between 25 and 50 percent of the communities in the United States and Canada depend on groundwater for public water supplies.

Aquifer Supply Measures. The best aquifers for water supply are those containing vast amounts of pure water that can be withdrawn without causing an unacceptable decline in the head elevation of the aquifer. Two conditions must exist if an aquifer is to yield a dependable supply of water over many years. First, the rate of withdrawal must not exceed the transmissibility of the aquifer. Otherwise, the water

Safe well yield　is pumped out faster than it can be supplied to the well, and the safe well yield is soon exceeded. *Safe well yield* is defined as the maximum pumping rate that can be sustained by a well without lowering the water level below the pump intake. Second, the rate of total withdrawal should not outrun the aquifer's recharge rate, or else the water level in the aquifer will fall. When an aquifer declines significantly, the *safe aquifer yield* has usually been exceeded, and if the *overdraft* is sustained for many years, the aquifer becomes depleted.

Impacts from Pumping. In arid and semiarid regions where groundwater is pumped for irrigation, the safe aquifer yield is typically exceeded and often greatly

Overdrafts　exceeded. In the American West, many aquifers took on huge quantities of water during and after continental glaciation several thousand years ago, and today these reserves are being withdrawn at rates far exceeding present recharge rates. Unless pumping rates are adjusted to recharge rates, the water level will decline, wells will have to be extended deeper and deeper, and these aquifers will eventually be depleted.

In parts of Arizona, the water level in some aquifers is declining as much as 20 feet a year because of heavy pumping for agriculture and urban uses. In the Great Plains, the huge Ogallala Aquifer is declining as much as 3 feet per year in

Examples　some areas because of heavy irrigation withdrawals for agriculture (Fig. 7.3). With the current shifts in the U.S. population toward the South and Southwest, not only are demands on groundwater reserves rising, but the risk of contamination is rising as well. In Florida, where the population has been growing by about 3 million people a decade and is overwhelmingly dependent on groundwater, a major problem is declining water quality caused by stormwater pollutants transported with recharge water through the sandy soils (Fig. 7.7).

(a)

(b)

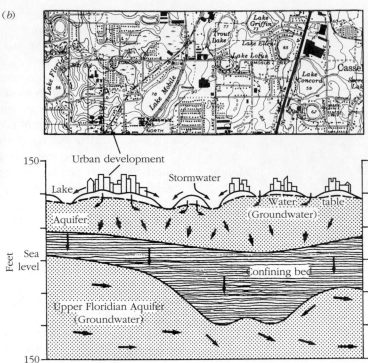

Fig. 7.7 The relationship of the water table aquifer to surface drainage in the urbanized area around Orlando, Florida, shown in the map and photograph.

Cone of depression

In and around urbanized areas (or agricultural areas with high concentrations of large wells), the groundwater level in aquifers may not only be greatly depressed by heavy pumping but also develop a very uneven upper surface. This happens where a high rate of pumping from an individual well is maintained for an extended period of time, causing the level of groundwater around the well to be drawn down by many meters or tens of meters. The groundwater surface takes on a funnel shape, similar to the water surface above an open drain in a bathtub, which is called a **cone of depression** (Fig. 7.8).

As the cone of depression deepens with pumping, the hydraulic gradient increases, causing faster groundwater flow toward the well. If the rate of pumping is fairly steady, the cone usually stabilizes in time. However, if many wells are clustered together, as is often the case in urbanized areas, they may produce an overall lowering of the water table as the tops of neighboring cones widen with drawdown and intersect one other. If the wells are of variable depths, the shallow wells may go dry as the cones of big, deep wells are drawn below the shallower pumping depths.

Surface subsidence

The drawdown of groundwater over a large area can also lead to the loss of volume in the groundwater-bearing materials. Subsidence in the overlying ground can result, as it has in Houston, Texas, for example, where much of the metropolitan area has subsided a meter or more with groundwater depletion. In addition, cones of depression can accelerate the migration of contaminated water because they increase hydraulic gradients and transmission velocities. In coastal areas, groundwater drawdown may also lead to the intrusion of salt groundwater into wells from the bottom up. **Saltwater intrusion** occurs when the mass (weight) of fresh groundwater is reduced by pumping and the underlying salt groundwater rises in response to the reduction in pressure. Once contaminated by saltwater, the wells are effectively lost as a water source.

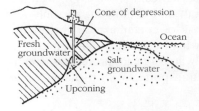

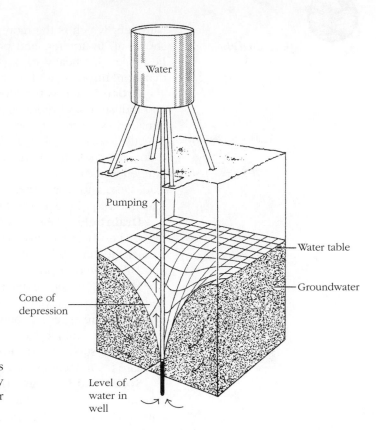

Fig. 7.8 A cone of depression. Groundwater pumping causes a drop in the water table around the well casing, which may in some instances be great enough to cause nearby shallower wells to go dry.

7.5 SOURCES OF GROUNDWATER CONTAMINATION

Range of sources

The sources of groundwater contamination are widespread in the modern landscape. They include all major land uses: industrial, residential, agricultural, and transportational. Planning for groundwater protection, therefore, is not limited to urban and industrial landfills, as is widely thought, but includes agricultural, mining, residential, highway, and railroad activities as well. The level of concern or precaution, however, is not everywhere equal, for several reasons. (1) Groundwater susceptibility to contamination varies widely from place to place. (2) The contaminant loading rate varies with land use type and practices (such as in pesticide applications with different farming methods). (3) The contaminants released to the environment vary in their harmfulness to humans and other organisms. The six principal land use sources of groundwater contamination are described below.

Leachate

Landfills. Buried wastes, both solid and hazardous, discharge contaminated liquids called *leachate* (see Section 5.7). The composition of leachate varies with the composition of the landfill. For urban landfills made up of residential garbage, the leachate may contain organic compounds such as methane and benzene. For agricultural wastes, the leachate may be heavy in nutrients such as phosphorus and nitrogen as well as organic compounds. For industrial wastes, it is commonly heavy in trace elements, that is, metals such as lead, chromium, zinc, and iron, as well as a host of other contaminants including organic compounds, petroleum products, and radioactive materials.

Farmlands. Agricultural fertilizers and pesticides are carried to aquifers with recharge water. Fertilizers are composed principally of nitrogen and phosphorus.

Fugative chemicals

Nitrogen, which is the more mobile of the two in soil and groundwater, commonly shows up in aquifers and poses a serious health problem in public water supplies. Pesticides are heavy in organic compounds, mostly synthetic compounds such as diazinon, fluorene, and benzene.

Urban Stormwater. Runoff from developed areas, especially streets, parking lots, and industrial and residential surfaces, is normally rich in a wide variety of contaminants. Most stormwater is discharged into streams, but a significant share of it goes

Various residues

directly into the soil (Fig. 7.7). Although concentrations of most contaminants are reduced by soil filtration, an appreciable load may be transported through coarse-grained soils to the groundwater zone. This contamination includes metals (e.g., lead, zinc, and iron), organic compounds (mainly insecticides such as diazinon and malathion), petroleum residues, nitrates, and road salt.

Drainfields. Sewage effluent water is released into the soil through seepage beds. The sewage is produced by household drainfields, community drainfields, and

Sewage effluent

animal feedlots. When it is fed to the ground, large amounts of nitrogen, sodium, and chlorinated organic compounds may be discharged into groundwater.

Mining. Mineral extraction and related operations, such as refinement, storage, and waste disposal, yield a variety of contaminants to surface and groundwater. Some operations are especially harmful, such as certain gold mining operations that employ

Leachate

a leaching technique based on the application of cyanide to crushed rock. In most operations, such as coal, iron ore, and phosphate, contaminants are discharged in leachate from decomposing waste rock.

Spills and Leakage. Here the possibilities seem endless. Spills of petroleum products, various organic compounds, fertilizers, metals, and acids are the most common along highways, railroads, and in and around industrial complexes.

Storage tanks

Leakage from underground storage tanks (of which there are more than 10 million in the United States and Canada), pipelines, and chemical stockpiles are also widespread. Residential land uses also contribute: paint, cleaning compounds, car oil, and gasoline, for example. Spills are typically point sources, and where they are associated with a known accident, remedial measures can be applied. However, underground leakages are usually hidden and may go undetected for years (Fig. 7.9).

7.6 APPLICATIONS TO LANDSCAPE PLANNING AND DESIGN

Unlike surface water, which we can see, measure, and map with comparative ease, groundwater is far more elusive as an environmental planning problem. Among other complications, it occupies complex three-dimensional space, with different aquifers and processes operating at different levels. Not surprisingly, determining just how a land use at some location will relate to this vast underground environment is difficult. Thus we approach landscape planning for groundwater protection more or less as a probability problem, that is, by estimating the *likelihood* of a land use's impact on groundwater given different locations, layouts, densities, and management arrangements.

Land uses of concern

Planning for Groundwater Protection. In proposals for new land uses, planning for groundwater protection begins with an understanding of the potential for contaminant production. Among the land uses of special concern are:

- *Industrial facilities*, including manufacturing installations, fuel and chemical storage facilities, railroad yards, and energy plants.
- *Urban complexes*, including highway systems, landfills, utility lines, sewage treatment plants, and automotive repair facilities.
- *Agricultural operations*, including cropland, feedlots, chemical storage facilities, and processing plants.

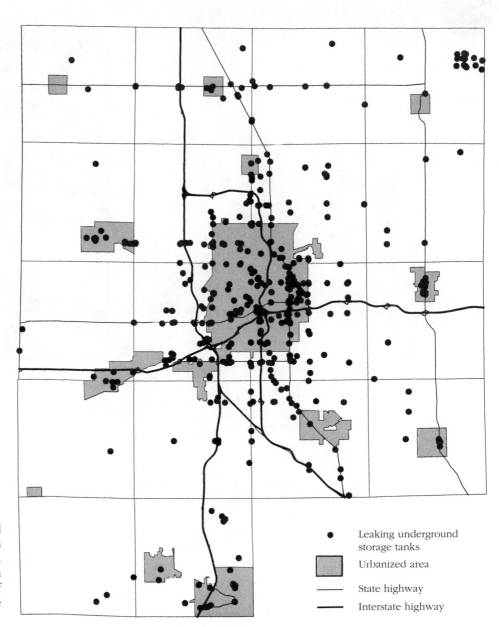

Fig. 7.9 Leaking underground storage tank sites that have been recorded in Genesee County, Michigan, a typical Midwestern urban area about 75 miles north of Detroit. Many more tank sites are unrecorded.

- Leaking underground storage tanks
- Urbanized area
- State highway
- Interstate highway

Land uses of less concern are single-family residential, institutions such as schools and churches, commercial facilities (excluding those with large parking lots), parks, and open spaces.

Once the potential for contaminant production is established, the next step is to assess the proposed sites for groundwater susceptibility to pollution. This entails learning about the aquifer's depths, linkages to the surface, and significance as a drinking water source. Information on water use and well location is usually obtainable from local or state health departments. Most states require that a log be registered with the health department for residential wells that identifies the materials penetrated and the finished depth of the well. In some cases, the results of a water quality test may also be on record.

Aquifer vulnerability

A key question in projects involving land uses with high contaminant production potential is the location of recharge zones. Generally, recharge zones are to be avoided, especially those that feed shallow aquifers such as the one shown in Figure 7.10.

Recharge zones

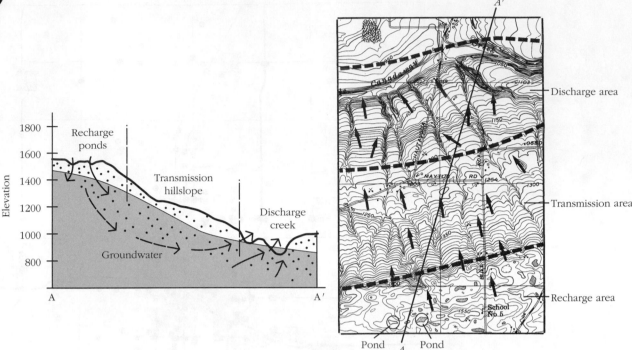

Fig. 7.10   A recharge area upslope from residential wells. Recharge takes place through the small water features at the top of the slope. Such recharge zones are common in glaciated landscapes.

Another important question is the permeability of surface materials because it controls the rate at which contaminated water and leachate from spills can infiltrate the ground.

Shallow aquifers   The shallowest of aquifers is the water table aquifer. It may lie only several meters underground, making it especially prone to contamination. Although it is not widely used as a source of drinking water (because health codes usually require wells to be deeper than 25 feet), the water table aquifer is an important source of water for streams, ponds, and lakes. Pollution plumes can contaminate seepage water discharging into these water features, particularly where they lie within 1000 feet of a pollution source (see Fig. 5.11).

The Planner's Checklist.   The planner's checklist for assessing the likelihood of groundwater contamination from a proposed land use should include the following (also see Table 7.1):

Key questions
■ What are the contaminant production potential and the related risks in handling and storing waste, especially hazardous waste?

■ Does the site lie over an aquifer that (1) presently serves as a source of drinking water; (2) could serve as a future source of drinking water; (3) feeds streams, lakes, and ponds; (4) feeds a bedrock cavern system?

■ Is the aquifer deep or shallow, and is the surface dominated by permeable material such as sand and gravel? Similarly, is the aquifer protected by a confining material, that is, a layer of impervious rock or soil material that limits penetration from sources directly above it?

■ What part of the aquifer system is the site associated with: recharge, transmission, withdrawal, or discharge (seepage)?

- What is the direction of flow in the aquifer: toward or away from areas of concern? (This is often indicated on groundwater maps by a measure called the potentiometric surface, which is an indicator of the overall slope of the aquifer based on readings of well water levels.)

- Will the proposed facilities fall into direct contact with groundwater because of deep footings, foundation work, underground utilities, and/or tunneling?

Three phases

Management Planning. For land uses already in place, the problem is to build a management plan to minimize the risk of groundwater contamination. The plan should address the three phases of the system leading to contamination: (1) *production* of contaminants, (2) *release* from the production and/or disposal site, and (3) *diffusion* into aquifers (Fig. 7.11). Planning for contaminant management and groundwater protection carries the least risk when it is applied early in the system, in the production phase. Three strategies can be employed to lower the risk connected with production: reduce output rates, change technology to less dangerous contaminants, and reduce the risk of accidental spills.

Table 7.1 Suggested Criteria in Land Use Planning for Groundwater Protection

Criteria	Worst	Best
	Desirability	
Contaminant production	High	None
Handling and storage risk	High (e.g., nonsecured storage areas with open soil)	Low
Aquifer use	Drinking water (many wells within a radius of 1000–1500 ft)	None
Aquifer depth	Shallow (less than 200 ft)	Deep (greater than 1000 ft)
Overlying material	Highly permeable (e.g., sands and gravels)	Impermeable (e.g., clayey soil and confining layer)
Aquifer system	Recharge zone	Beyond discharge (seepage) areas
Flow direction	Toward wells	Away from wells

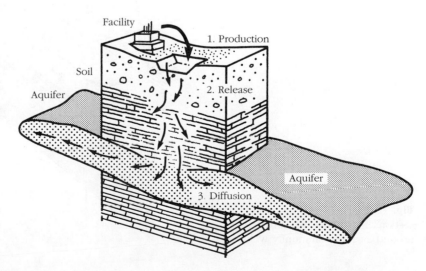

Fig. 7.11 The three-phase system for groundwater management: (1) on-site production, (2) release from the production or disposal site, and (3) diffusion into the aquifer.

Release

At the release phase, the objective is to limit the escape of contaminants from the site. The conventional strategy is safely to contain the contaminants during storage, transportation, and disposal. The U.S. Environmental Protection Agency recommends various methods, including secure landfills of containerized waste (see Fig. 5.12). Soil conditions are extremely important inasmuch as compact clayey soils, with their low permeabilities, retard the advance of leachates, whereas sandy soils, with their high permeabilities, allow the advance of leachates.

Diffusion

By the time fugitive contaminants have reached the diffusion phase, the likelihood of impact on aquifers and human water supplies has risen dramatically. However, the concentrations and chemical composition of the leachate may attenuate in passage as a result of filtration, adsorption, oxidation, and biological decay. Once the contaminants have invaded an aquifer, preventive measures are no longer feasible and the only options left are (1) corrective pumping (heavy pumping to direct the plume away from wells) or (2) abandonment of vulnerable wells. Again, planning efforts for groundwater protection should be made early in the game, at the production and site release phases. Once contaminants escape the site, the environmental risk, techno-

Corrective measures

logical difficulty of recovery, and recovery expense increase dramatically.

7.7 COMMUNITY WELLHEAD PROTECTION

Part of a public awarness program on groundwater protection, Austin, Texas.

To help protect community wells from contamination, the U.S. Environmental Protection Agency (EPA) has mandated states to develop wellhead protection programs. The concept of wellhead protection involves managing the land uses and contamination sources in the contributing or recharge area for a community well (Fig. 7.12). Under this mandate, each state develops its own program plan and submits it to the EPA for approval. Although programs can vary in design, each state's program must include a requirement for the documentation of groundwater levels, flow directions,

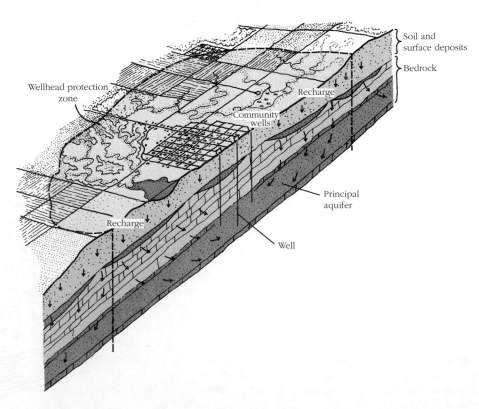

Fig. 7.12 The concept of wellhead protection, illustrating the protection zone around a set of community wells.

transmission rates, recharge/exchange times, recharge areas, and contaminant sources in the recharge area of a community well.

Protection plan

The plan for each community must include a program for testing the public water supply, a management plan for contaminant sources and areas, and a risk management plan for accidents such as chemical spills. Although the Wellhead Protection Program is generally viewed as an important step toward improved management of groundwater resources, the program is not supported by federal funds, and many states and local communities therefore view it with less than enthusiasm.

7.8 CASE STUDY

GIS and the Management of Groundwater Contamination

Martin M. Kaufman, Daniel T. Rogers, and Kent S. Murray

Unlike surface water systems such as watersheds, which are reasonably apparent at a variety of geographic scales, groundwater systems are hidden away underground. This creates a basic difficulty in managing groundwater because it tends to be abstract and remote in the minds of policy makers and the general public. Among other things, it is often difficult to demonstrate the linkage among basic hydrologic processes such as groundwater discharge and stream flow or groundwater recharge and drinking water supply and quality. In our research in Southeastern Michigan, we found Geographic Information Systems (GIS) to be a particularly valuable tool in mapping and documenting groundwater contamination.

For any GIS implementation to be successful, the governing physical systems must be accurately depicted at their proper geographic scales. In the Great Lakes region, the flows of most groundwater contaminants are dictated by the fundamental pathways of the regional hydrologic system: (1) a release point, (2) a soil and/or bedrock medium into which it is released, (3) a migration pathway to and within the groundwater system, and (4) a discharge or withdrawal point somewhere. Different soil types and the near-surface geologic conditions, such as bedrock attitude, play key roles in the pattern and rate of contaminant migration. For example, clayey materials tend to inhibit the downward migration of contaminants, whereas sandy materials provide little resistance. Releases are typically mixtures of contaminants rather than single chemical compounds. Most contain over 100 individual chemical compounds, and each compound behaves and migrates in the near-surface geologic environment differently.

In humid regions like the Great Lakes, massive amounts of groundwater discharge directly into surface waters, most notably streams, lakes, and wetlands. Therefore, it is important to know about landscape, soil, and geological conditions at scales accordant with the geographic coverage of local aquifers. Some lakes and wetlands are geographically isolated with no surface inlets and outlets, and the scale of their groundwater support systems is usually limited. Most streams, on the other hand, are parts of much larger surface networks that may be fed by various aquifers and therefore must be addressed at a much larger scale. Only through this type of scale-specific approach can you include the necessary data for your analysis.

The maps in Figure 7.A show the inferred hydrological and ecological consequences when generalized categories of contaminants are spilled from selected surface locations in the Rouge River watershed in Southeastern Michigan. The contaminant categories include: DNAPLs (dense nonaqueous phase liquids—chlorinated solvents); LNAPLs (light nonaqueous phase liquids—gasoline compounds); PAHs (polycyclic aromatic hydrocarbons—oil compounds); PCBs (polychlorinated biphenyls); and a group of several heavy metals, including chromium and lead.

The first step of the GIS analysis, shown in the upper left map in Figure 7.A, consisted of buffering the contaminated sites. The buffer size represents the average

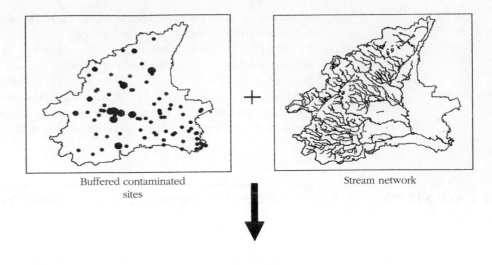

Fig. 7.A A GIS overlay (above) that combines contaminated site locations with the network of stream channels in the Rouge River watershed. Below, the resultant composite map highlighting stream segments directly impacted by contaminants.

distance the contaminants traveled based on measurements taken at their points of release. Therefore, those parts of the watershed with high contaminant migration potentials are represented as larger circles, and they occur in the areas dominated by the highly permeable sand and moraine deposits. Next, the buffered sites were overlaid on the surface stream network map, which includes all first-order streams and higher in the watershed. The composite (lower) map in Figure 7.A indicates the areas where the average contaminant extent and a stream channel intersect.

Field inspection of the contaminant sites and surface stream map layers indicated numerous other sites where contamination within groundwater would migrate to surface water within a few days. This outcome was likely due to the higher flow rates and the high drainage densities in the sand and moraine units in this watershed. At this geographic scale, the runoff system follows this system of pathways: soil to groundwater to low-order streams to higher-order streams to the Great Lakes. Some persistent contaminants such as tetrachloroethene (a common dry cleaning chemical) and Cr^{+6} (the compound used in chrome plating) travel along with the water.

The lesson is this: If groundwater stays abstract and remote in the minds of policy makers, how can they conceive of a link between groundwater and flowing surface water? How can they give adequate consideration to the potential hydrological and ecological impacts on a larger region and system such as the Great Lakes?

Martin M. Kaufman is a professor in the Department of Earth and Resource Science at the University of Michigan—Flint, Daniel T. Rogers is the director of environmental affairs at Amsted Industries, Chicago, and Kent S. Murray is a professor in the Department of Natural Sciences at the University of Michigan—Dearborn.

7.9 SELECTED REFERENCES FOR FURTHER READING

Butler, Kent S. "Managing Growth and Groundwater Quality in the Edwards Aquifer Area, Austin, Texas." *Public Affairs Comment.* 29, 2, 1983.

DiNovo, Frank, and Jaffe, Martin. *Local Groundwater Protection: Midwest Region.* Washington, DC: American Planning Association, 1984.

Hill-Rowley, R., et al. *Groundwater Vulnerability Study for Tyrone Township.* Flint: Regional Groundwater Center, University of Michigan–Flint, 1995.

Josephson, Julian. "Groundwater Strategies." *Environmental Science and Technology.* 14, 9, 1980, pp. 1030–1035.

Moody, D. W. "Groundwater Contamination in the United States." *Journal of Soil and Water Conservation.* 45, 1990, pp. 170–179.

National Research Council. *Groundwater Contamination: Studies in Geophysics.* Washington, DC: National Academy of Sciences Press, 1984.

Page, G. W. *Planning for Groundwater Protection.* New York: Academic Press, 1987.

Pye, V. I., Patrick, Ruth, and Quarles, John. *Groundwater Contamination in the United States.* Philadelphia: University of Pennsylvania Press, 1983.

Rutledge, A. T. "Effects of Land Use on Groundwater Quality in Central Florida— Preliminary Results: U.S. Geological Survey Toxic Waste—Groundwater Contamination Program." *Water-Resources Investigations Report 86–4163.* Washington, DC: U.S. Government Printing Office, 1987.

Tripp, J. T. B., and Jaffe, A. B. "Preventing Groundwater Pollution: Towards a Coordinated Strategy to Protect Critical Recharge Zones." *Harvard Environmental Law Review.* 3, 1, 1979, pp. 1–47.

U.S. Council on Environmental Quality. *Contamination of Groundwater by Toxic Organic Chemicals.* Washington, DC: U.S. Government Printing Office, 1981.

Related Websites

Groundwater Foundation. "Sources of Groundwater Contamination." 2009. http://www.groundwater.org/gi/sourcesofgwcontam.html
  The Groundwater Foundation aims to educate and move people to action on the health of groundwater systems. Read about sources of contamination, watch videos, or find communities in which to get involved.

Ogallala Aquifer Program. 2009. http://ogallala.tamu.edu/index.php
  A collaboration between the USDA, a number of universities, and agricultural research bodies to explore all aspects of water depletion of the Ogallala aquifer. Links lead to water management aspects, an article outlining the issues of the aquifer, and how new irrigation technology can help save the Ogallala.

Province of Nova Scotia. "Groundwater Management." 2009. http://www.gov.ns.ca/ nse/water/groundwater/groundwatermagmt.asp

What Nova Scotia is doing to protect its groundwater supply. This site gives information and links regarding resource evaluation, sustainable development, and monitoring. Follow links to view groundwater maps and publications.

USGS. "Ground-Water Depletion Across the Nation." 2003. http://pubs.usgs.gov/fs/ fs-103-03/

General information about groundwater depletion, including pictures and diagrams. The site reviews the current (2003) condition of groundwater supplies across the United States.

RUNOFF AND STORMWATER MANAGEMENT IN A CHANGING LANDSCAPE

8.1 INTRODUCTION

Rise in surface runoff

One of the most serious problems associated with land development is the change in the rate and amount of runoff reaching streams and rivers. Both urbanization and agricultural development increase surface runoff, resulting foremost in greater magnitudes and frequencies of peak flows in streams. The impacts of this change are serious, both financially and environmentally. Property damage from flooding is increased, water quality is reduced, channel erosion is accelerated, habitat is degraded, and the scenic quality of the riparian environment is compromised.

Corrective measures

Not only that but these problems, especially the flooding and property damage, have traditionally brought on a rash of reactions from the political and engineering communities. The resultant so-called corrective measures they invoke are often more damaging than the problems they are designed to remedy. Among other things, streams have been ditched and piped underground, wetlands drained, and dams erected in an attempt to control stormwater and flooding, all at great expense to the public and the environment.

Management

Responsible land use planning and design require a management—as opposed to a control—approach to runoff from the landscape, the stuff we call stormwater. Stormwater management begins with an assessment of the changes in runoff brought on by land development. In the United States and parts of Canada, this problem has reached epidemic proportions, and in most communities developers are required to provide analytical forecasts of the changes in overland flow and stream discharge resulting from a proposed development. Such forecasts are used not only as the basis for recommending alternatives to traditional stormwater systems in order to reduce environmental impact, but also for evaluating the performance of an entire watershed that is subject to many development proposals.

This chapter is concerned with the stormwater runoff generated from rainstorms, the factors in the landscape that influence how stormwater is produced, and how it can be managed to minimize related environmental impacts. We examine a traditional model for forecasting stormwater from small watersheds, some alternative models, the selection of mitigation measures, and the issue of comprehensive water management.

8.2 RUNOFF AS OVERLAND FLOW

Landscape processes

Most precipitation reaching the landscape is disposed of in four ways (Fig. 8.1). Some is taken up on the surface of vegetation in a process known as **interception**. Some is absorbed directly by the soil in a process known as **infiltration**. Some, called **depression storage**, collects on the ground in small hollows and pockets. The remainder, called **overland flow**, runs off the surface, eventually joining streams and rivers or collecting in low spots. This is **stormwater**.

Disposition of rainfall

Landscape Controls on Overland Flow. Think of rainfall entering the landscape as the beginning of a system and consider all the ways this input of water can be diverted, stored, dissipated, and conducted. The possibilities are enormous, and for decades scientists and engineers have struggled to understand how this system works under different landscape and climatic conditions. In very general terms, forested landscapes retain the most rainfall and yield the least overland flow. Streams are fed mainly from subsurface sources, mainly groundwater and water that moves within the soil, called *interflow*. At the opposite extreme are barren or lightly vegetated landscapes, both natural and human-made, especially those subject to high-intensity rainstorms. These include arid and semiarid landscapes as well as developed landscapes where buildings, streets, parking lots, cropland, and other land uses predominate.

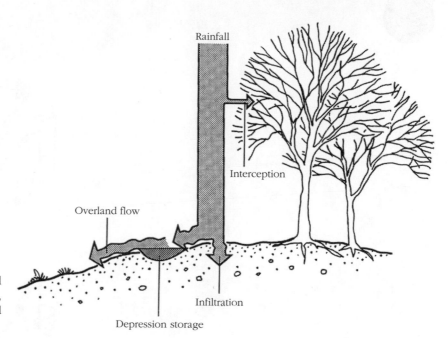

Fig. 8.1 The disposition of rainfall in natural or partially developed landscapes: interception, infiltration, depression storage, and overland flow.

Key variables In conventional stormwater planning and management, the principal concerns are the rate and volume of (1) the rainwater delivered, (2) the rainwater taken up in infiltration, and (3) the overland flow (stormwater) produced. If we put the climatic factor aside for the moment, the controls on infiltration and overland flow take center stage. Three factors are of greatest concern: *land cover* (vegetation and land use),

Controls on infiltration *soil composition and texture*, and *surface inclination* (slope). Figure 8.2 illustrates the relationships among overland flow, soil, and vegetative cover on a hillslope. As a general rule, overland flow *increases* with slope, *decreases* with soil organic content and particle size, *increases* with ground coverage by hard surface material such as concrete and asphalt, and *decreases* with vegetative cover.

The Coefficient of Runoff. For a particular combination of these factors, a **coefficient of runoff** can be assigned to a surface. This is a dimensionless number between 0 and 1.0 that represents the proportion of a rainfall available for overland flow after infiltration has taken place. A coefficient of 0.60, for example, means that

Available rainwater 60 percent of rainfall (or snowmelt) is available for overland flow, whereas 40 percent is lost to infiltration. Table 8.1 lists some standard coefficients of runoff for rural areas based on slope, vegetation, and soil texture. For urban areas, coefficients are mainly a function of the hard (impervious) surface cover, and a number of representative values are given in Table 8.2.

Transient Factors. Several *transient factors* may influence the coefficient of runoff and overland flow that are not standard considerations. For example, as a long rainstorm advances, infiltration may decline (while the coefficient of runoff rises)

Short-term conditions as the soil takes on water. On the other hand, when rainfall intensity rises during a heavy storm and the delivery rate reaches as high as, say, 2 or 3 inches in 30 minutes, a soil's absorptive capacity can be overridden, even in soils with high infiltration capacities.

Prestorm soil moisture is another consideration. If an earlier rainstorm has already loaded the soil with water or if the level of groundwater has risen into the soil column, infiltration capacity falls and the coefficient of runoff rises substantially. In fact, naturally saturated soils such as those in wetlands behave as impermeable surfaces

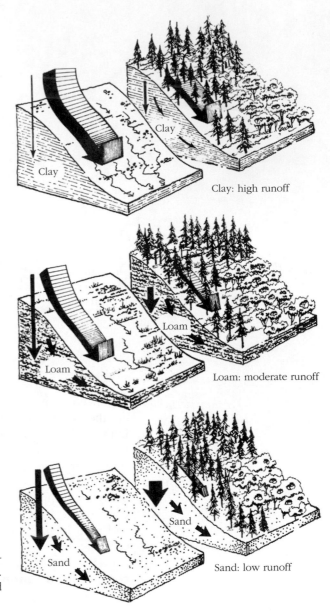

Fig. 8.2 Schematic illustration showing the relative changes in overland flow runoff with soil type and vegetation on sloping ground. The subsurface arrows represent water going to groundwater and interflow.

with coefficients of runoff approaching 0.90 or more. With no medium to take up rainwater, it is converted to overland flow more or less straightaway. These factors understandably complicate the picture and often make the problem of forecasting stormwater runoff under certain circumstances a difficult game.

8.3 COMPUTING RUNOFF FROM A SMALL WATERSHED

Rational method

The runoff generated in the form of overland flow or stormwater from a small watershed can be computed by means of a simple manipulation called the **rational method**. This method is based on a formula that combines the coefficient of runoff with the intensity of rainfall and the area of the watershed. The outcome gives the **peak discharge** (maximum rate of flow) for *one rainstorm* at the *mouth of the watershed*:

$$Q = A \bullet C \bullet I$$

where

Q = discharge in cubic feet per second (or cubic meters per second)
A = area in acres
C = coefficient of runoff
I = intensity of rainfall in inches or feet per hour (or centimeters per hour)

Target rainstorm

Selecting the Design Storm. The rational method requires some essential data. Besides the obvious need to measure the area of the watershed and determine the correct coefficient of runoff, we must select an appropriate rainfall intensity value. Obviously, a wide variety of rainstorms deliver water to the watershed, and one storm that is likely to produce significant runoff, called the **design storm**, has to be selected in order to perform the computation. The design storm is defined according to local rainfall records for intensive storms of short duration (usually 1 hour or less), which occur, for example, on the average of once every 10 years, 25 years, or 100 years. These storms are called, respectively, the 10-year, 25-year, and 100-year storms of 1-hour duration.

Which storm should be used usually depends on the recommendation (or policy) of the local agency responsible for stormwater management. Figure 8.3 gives the values over the coterminous United States and the southern fringe of Canada for the 10-year, 1-hour storm. For more accurate values based on local records, consult local agencies such as the office of the city or county engineer.

Concentration time

Estimating Concentration Time. Once the desired storm has been selected, a second step must be taken to arrive at the actual rainfall value to be used in making the discharge computation. This step is based on two facts: (1) Within the period of a rainstorm, say, 60 minutes, the actual intensity of rainfall initially rises, hits a peak, and then tapers off. (2) The time taken for runoff to move from the perimeter to the mouth of the watershed, called the **concentration time**, varies with the size and conditions of the watershed.

Combining these two facts, we can see that in a small watershed the concentration time may be less than the duration of the storm so that the peak discharge is reached before the storm is over. Therefore, to accurately compute the peak discharge for the

Table 8.1 Coefficients of Runoff for Rural Areas*

Topography and Vegetation	Open Sandy Loam	Clay and Silt Loam	Tight Clay
Woodland			
Flat (0–5% slope)	0.10 or less	0.30	0.40
Rolling (5–10% slope)	0.10–0.20	0.35	0.50
Hilly (10–30% slope)	0.20–0.30	0.50	0.60
Pasture			
Flat	0.10	0.30	0.40
Rolling	0.16	0.36	0.55
Hilly	0.22	0.42	0.60
Cultivated			
Flat	0.30	0.50	0.60
Rolling	0.40	0.60	0.70
Hilly	0.52	0.72	0.82

* These are representative values. Local values typically vary, especially seasonally.

Table 8.2   Coefficients of Runoff for Selected Urban Areas

Commercial	
Downtown	0.70–0.95
Shopping centers	0.70–0.95
Residential	
Single family (5–7 houses/acre)	0.35–0.50
Attached, multifamily	0.60–0.75
Suburban (1–4 houses/acre)	0.20–0.40
Industrial	
Light	0.50–0.80
Heavy	0.60–0.90
Railroad yard	0.20–0.80
Parks, cemetery	0.10–0.25
Playgrounds	0.20–0.40

Fig. 8.3 The amount of rainfall that can be 1.0 expected in the 10-year storm of 1-hour duration. Notice the dramatic differences between the Pacific Coast, the Rocky Mountain-Intermontane Region, and southern Canada on the one hand and the American South on the other.

Storm intensity value

storm in question, we must use the value of rainfall intensity that corresponds to the time of concentration. The graph in Figure 8.4 gives representative curves for various storms and shows how to use the graph based on an example storm intensity of 1.75 inches per hour and a concentration time of 16.5 minutes.

Flow times

Reliable estimates of the concentration time are clearly important in computing the discharges from small watersheds. The approach to this problem generally involves making separate estimates for (1) the time of overland flow and (2) the time of channel flow, and then summing the two:

$$C_t = Of_t + Cf_t$$

where

$$
\begin{aligned}
C_t &= \text{concentration time} \\
Of_t &= \text{overland flow time} \\
Cf_t &= \text{channel flow time}
\end{aligned}
$$

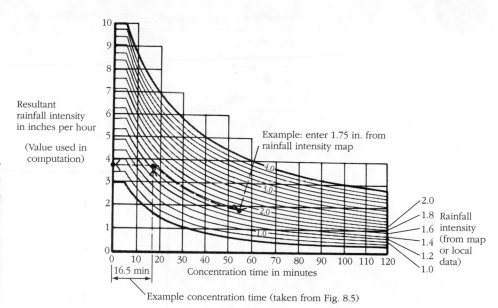

Resultant
rainfall intensity
in inches per hour

(Value used in
computation)

Example: enter 1.75 in. from
rainfall intensity map

Rainfall
intensity
(from map
or local
data)

16.5 min

Concentration time in minutes

Example concentration time (taken from Fig. 8.5)

Fig. 8.4 Rainfall intensity curves. To use the graph, first find the desired rainfall value among the curves, then find the appropriate concentration time on the base of the graph. The rainfall intensity value is read from the intersection of the two on the left vertical scale. In the example, the resultant rainfall value is about 3.75 inches.

Overland flow

The graph in Figure 8.5 can be used to estimate the time of overland flow if three things are known: (1) the length of the runoff path from the outer edge of the watershed to the head of channel flow; (2) the predominant groundcover; and (3) the average slope of the ground to the head of channel flow. If these are not known, we must resort to an approximation based on representative overland flow velocities such as those offered in Table 8.3.

Channel flow

Channel flow velocities can be estimated with the help of the following formula, which is based on the channel's shape, roughness, and gradient. There are numerical values for each parameter, and for most channels they can be approximated from maps and field observations. For most natural channels when they are carrying a discharge at bankfull level, flow velocity falls in the range of 4 to 8 feet per second.

$$v = 1.49 \, \frac{R^{2/3} S^{1/2}}{n}$$

Where

v = velocity, ft or m/sec
R = hydraulic radius
   (wetted perimeter
   of channel divided
   by cross-sectional
   area of channel)
s = slope or grade of
   channel
n = roughness coefficient
   (See p. 295 in Chapter 14
   for roughness values.)

A = Cross-sectional area
P = Wetted perimeter
s = Slope

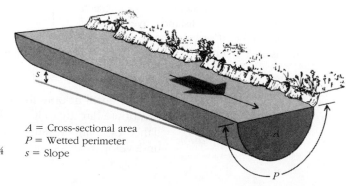

Stormdrains
(not shown on contour maps)

Swale
(not shown on
contour maps)

Channels
(shown on
contour maps)

Because channel flow is so much faster than overland flow (roughly ten times faster), the full channel system, or drainage net, in a watershed must be delineated. This means going beyond the stream channels shown on conventional topographic contour maps and mapping the small natural and human-made channels that carry water only intermittently. These channels include natural swales, road ditches, and

Table 8.3   Representative Overland Flow Velocities

Distance to Channel Head (ft)	Pavement (ft/sec)	Turf (ft/sec)	Barren (ft/sec)	Residential (ft/sec)
Around 100	0.33	0.20	0.26	0.25–0.33
Around 500	0.82	0.25	0.54	0.50–0.65

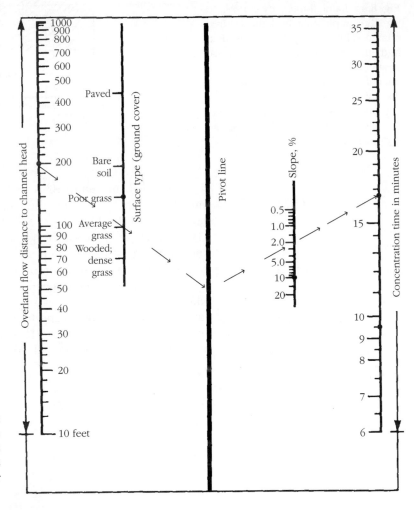

Fig. 8.5 A graph for estimating the concentration time of overland flow from a basin perimeter to a channel head or inlet. The string of arrows represents an example using a 200-foot distance, average grass cover, 2 percent slope. The concentration time for overland flow is 16.5 minutes.

farm drains, but only those that are clearly tied to the main channel system and that contribute directly to its discharge (see Fig. 8.11*b*). As a rule, watersheds with an abundance of channels, such as those in urbanized areas, are much faster than those in undeveloped areas and thus have much shorter concentration times.

Channel delineation

8.4 USING THE RATIONAL METHOD

As a forecasting device, the rational method provides an approximation of stormwater discharge. It is best suited for watersheds that are small [generally less than 1000 acres (400 hectares)] and partially or fully developed. Beyond that, the reliability of the rational method depends on the accuracy of the values used for the coefficient of runoff, the concentration time, the drainage area, and the rainstorm intensity. With

Guidelines

the exception of rainstorm intensity, the necessary data may be generated through field observations and surveys or from secondary sources, in particular topographic maps, soil maps, and aerial photographs. Also, please note that many versions of the rational method have been devised that extend its utility and precision for various situations and geographic locations.

Solving for Qp

Computational Procedure. Given that the pertinent data are in hand, the following *procedure* may be used to compute the **peak discharge (Qp)** resulting from a specified rainstorm. The result gives you the runoff leaving the watershed in a channel at the instant of peak flow. If the watershed or drainage area is not served by a channel, then the peak discharge *cannot* be computed using the rational method, but it may be possible to compute the volume of water *available* as overland flow. (See entry 8 in the following list.)

Steps

1. Define the perimeter of the watershed and measure the watershed area.

2. Subdivide the watershed according to cover types, soils, and slopes. Assign a coefficient of runoff to each subarea, and measure its area.

3. Determine the percentage of the watershed represented by each subarea, and multiply this figure by the coefficient of runoff of each. This gives you a coefficient adjusted according to the size of the subarea.

4. Sum the adjusted coefficients to determine a coefficient of runoff for the watershed as a whole.

5. Determine the concentration time using the graph in Figure 8.5 and the stream velocity formula.

6. Select a rainfall value for the location and rainstorm desired. Using this value and the concentration time, identify the appropriate rainfall intensity from the curves in Figure 8.4 (or a similar one).

7. Multiply the watershed area by the coefficient of runoff and by the rainfall intensity value, to obtain the peak discharge in cubic feet per second. (If you use inches for *I*, the answer is in acre inches per hour; however, these units are so close to cubic feet per second that the two are interchangeable.)

Solving for Qv

8. The total **volume of discharge (Qv)** produced as a result of a rainstorm can also be computed with the rational formula. The formula for volume of storm runoff is

$$Qv = A \bullet C \bullet R$$

where

A = drainage area
C = coefficient of runoff
R = total rainfall

In this case the full one-hour rainfall value is used, and we solve for acre feet, cubic feet, or cubic meters. This is a simpler computation because it does not involve deciphering concentration time and rainfall intensity. It is useful in estimating gross differences in pre- and postdevelopment runoff.

8.5 OTHER RUNOFF MODELS AND CONCEPTS

Scientists recognize a number of shortcomings in the rational method and the concepts on which it is based, and have devised some alternative models that deserve our attention, though none provides an easy computational method for forecasting discharge.

The principal alternative is based on the observation that most undeveloped and many partially developed watersheds do not produce basinwide overland flow in response to intensive precipitation. In other words, when it rains hard, only a fraction of the watershed develops overland flow and yields stormwater to streams. This observation has led to a model called the **partial area** or **variable source concept**.

Small contributing area

The Partial Area Concept. According to this concept, the hydrologically active portion of a watershed in terms of stormwater discharge may be as little as 10 to 30 percent of the total drainage area. Typically, the contributing area is found in collection zones at valley heads, in swales, along footslopes, and around low ground near streams (Fig. 8.6*a*). In addition, the size and shape of the contributing area may be different from event to event depending on soil moisture conditions, groundwater levels, and other factors.

Management implications

The partial area concept has serious implications for stormwater management. First, forecasts of stormwater discharge based on the rationale method may be larger than the actual discharges for many basins, especially forested or partially forested ones, because it assumes that overland flow is generated from a much larger area than is the case. Second, if large parts of a basin do not yield overland flow, then those areas apparently contain *natural buffers* that enable them to retain rainfall and limit its release to streams as stormwater.

Natural buffers

These buffers are usually high-infiltration soils, forested land, wetlands, and/or depression storage areas that take up all the water delivered to them by rainfall and surrounding runoff surfaces. If, in the course of planning and development, these noncontributing areas are equipped with stormwater facilities such as ditches and pipes, as would be the logical assumption using the rational method (as represented in Fig. 8.6*b*), then they are perforce converted into contributing stormwater areas. Not only does stormwater discharge from the basin increase, but an existing stormwater mitigation opportunity is lost.

Interflow Runoff. In many forested watersheds, field research has demonstrated that overland flow is effectively nonexistent. It appears that small streams in such watersheds rely entirely on shallow subsurface water in the form of **interflow**. This water enters the ground at or near the point of rainfall reception and flows laterally within the soil mantle to a swale, valley wall, or streambank.

Response time

The corresponding hydrograph for interflow is slower, lower, and longer than basinwide overland flow systems and most partial area systems that receive overland

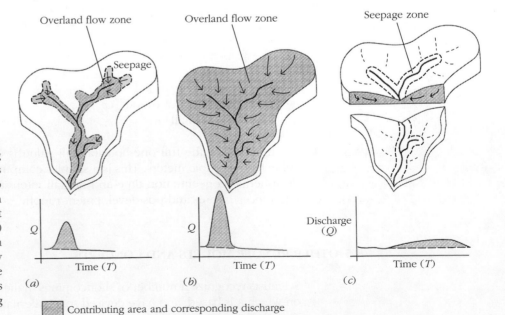

Fig. 8.6 Maps and corresponding hydrographs showing the differences in a basin's response according to (*a*) the partial area concept with contributions from overland flow at valley heads and channel fringes, (*b*) the rational method concept with contributions from overland flow over the entire basin, and (*c*) the interflow concept with contributions limited principally to seepage along channels.

flow (Fig. 8.6c). Unfortunately, no methods comparable in ease of application to the rational method have been devised for computing discharge from an interflow system. The unit hydrograph method, which is described in Chapter 10, has potential in this regard, but it requires stream discharge measurements in response to a rainstorm, and for most small basins, no discharge records are available.

8.6 TRENDS IN STORMWATER DISCHARGE

The clearing of land and the establishment of farms and settlements have been the dominant geographic changes in the North American landscape over the past 200 years. Almost invariably, these changes have led to an increase in both the amount and rate of overland flow, producing larger and more frequent peak flows in streams. With the massive urbanization of the twentieth century, this trend became even stronger and has led to increased flooding and flood hazard, to say nothing of the damage to aquatic environments and the degradation of water quality.

Contributing Factors. From a hydrologic standpoint, the source of the problem can be narrowed down to changes in two parameters:

Key parameters

1. The drastic increase in the *coefficient of runoff* in response to land clearing, deforestation, and the addition of impervious materials to the landscape, as illustrated in Figure 8.7.

2. The corresponding decrease in the *concentration time*, as illustrated in Figure 8.8a.

Land Cover Change. In the extreme, the twentieth-century North American landscape has been driven by the automobile and urbanization. And with these influences has come a massive amount of impervious surface, amounting to an area estimated at 61,000 square miles in the United States. (This is an area larger than the state of Georgia.) In many areas half or more of entire watersheds have been covered, resulting in massive stormwater volumes. Added to the impervious cover are vast unpaved

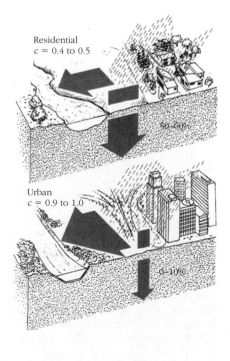

Fig. 8.7 Changes in the coefficient of runoff with land use and land cover: forest, cultivated, residential, and urban. Generally speaking, the coefficient of runoff and water lost to infiltration are inversely related.

areas where coefficients of runoff have been increased by 30 to 50 percent. Chief among these are farmlands where surface runoff is often double that of the landscapes they displaced.

Faster systems   ***Pipes and Ditches.***   The larger volume of runoff in turn demands a more efficient (faster) stormwater removal system. In agricultural areas this comes with the construction of field drains and ditches, as well as with the straightening and deepening of stream channels. And with urbanization the rural ditches are replaced with stormsewers, small streams are piped underground, and gutters are added to streets. All of these changes reduce concentration times, often by more than tenfold. More water gets to streams much faster and the result is a dramatic increase in the magnitude and frequency of peak discharges in these streams, as the hydrographs in Figure 8.8*b* demonstrate.

Climate change   ***Storm Magnitude.***   Changes in a third parameter, rainstorm intensity, may also contribute to discharge increases. Storm magnitudes appear to be on the rise in metropolitan areas in the United States. At the present, however, generalizing about this trend is difficult because it is apparent only in certain metropolitan areas. However, most large urban areas appear to be affected, and, with global warming and urban growth, the effects are apt to intensify and spread. In addition, tropical storm magnitudes appear to be increasing with ocean warming.

Clearing and grading   ***Depression Storage.***   Other changes brought on by development include reduction in depression storage. *Depression storage* is the retention of rainwater and overland

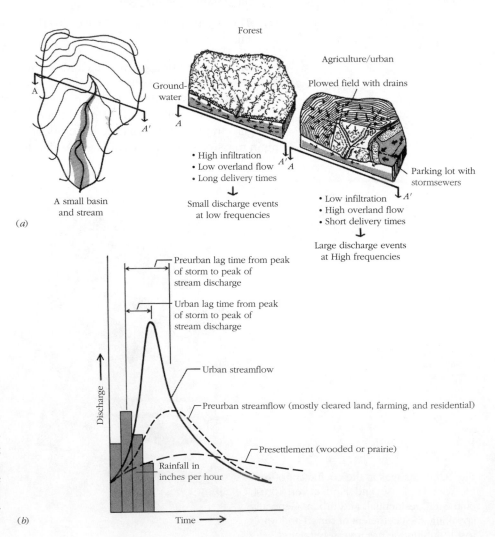

Fig. 8.8 (*a*) Changes in the hydrology of a small watershed as a result of land clearing and development. (*b*) Three stages of landscape change (presettlement, preurban, and urban), leading to a higher and quicker streamflow hydrograph.

Depression storage

Expanding nets

flow in low spots in the microtopography. The water held in depression storage can be considerable, often an entire rainstorm. In old-growth forests the depression storage capacity is typically very high because of the abundance of pits and pockets formed by wind-thrown trees, rotting stumps, burrowing animals, and so on. But when forested land is cleared for agriculture, much of its depression storage capacity is eliminated by plowing away the microtopography (Fig. 8.9). With residential and related development, the ground is not only graded smooth, but often inverted, that is, mounded slightly or substantially to "improve" drainage from yards, and parkways.

Effective Impervious Cover. Roads are almost always built with paralleling drains to carry stormwater away. These drains also serve the land uses along the road, the residences, malls, schools, and so on. Most road drains connect to streams, where they dump their stormwater load. Therefore, when roads and streets are built over a watershed and their drainage ditches and stormsewers are grafted onto natural stream channels, developed areas are given direct access to stream channels (Fig. 8.10).

This is the basic idea behind the term **effective impervious cover**, defined as impervious cover linked by ditch, gutter, or pipe to a stream channel. As the network of roads increases with development and the drainage density (miles of channel per square mile of land) is driven up, effective impervious cover rises until eventually all hard surfaces in a watershed are serviced by stormdrains via roads. The effects are profound, for the natural mitigating effects of the landscape on overland flow are overridden and largely eliminated (Fig. 8.11).

8.7 STORMWATER MITIGATION IN DEVELOPMENT PLANNING AND DESIGN

Most communities have enacted stringent stormwater management standards on new development. Increasingly, these standards call for zero net increase in stormwater discharge from a site as a result of development. In other words, the rate of release across the border of a site, usually measured by the peak discharge for a design storm, can be no greater after development than before. A basic dilemma thus arises

(a) depression storage

(b)

Fig. 8.9 Depression storage before and after development. (*a*) Abundant depression storage (dark spots) before development, but (*b*) very little apparent (or possible) after development.

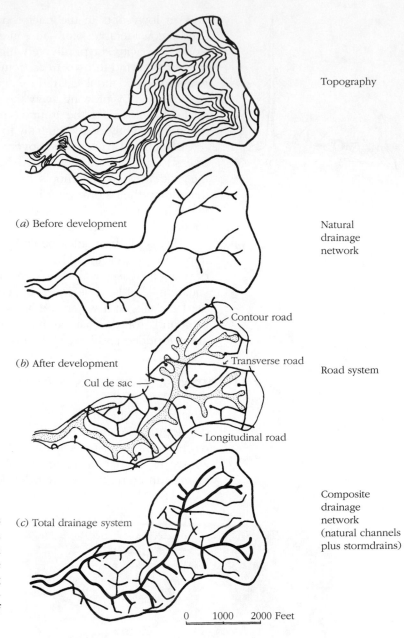

Topography

(*a*) Before development

Natural drainage network

Contour road

(*b*) After development

Transverse road

Cul de sac

Road system

Longitudinal road

(*c*) Total drainage system

Composite drainage network (natural channels plus stormdrains)

0   1000   2000 Feet

Fig. 8.10 The pattern of natural channels and roads (with stormdrains) before and after development in a small (320-acre) basin near Austin, Texas. Roads with curbs, gutters, and stormsewers are grafted onto the natural system of stream channels, more than doubling the drainage density. The increase in drainage density drives up both the magnitude and frequency of stormflows.

because most forms of development effect an increase in runoff but at the same time themselves require relief from stormwater.

Management Strategies. Three strategies may be used to manage stormwater: (1) *plan* the development so that it produces little or no increase in stormwater dis-

Plan, return, store...   charge; (2) *return* the excess water (stormwater) to the ground, where it would have gone before development; (3) *store* the excess water on or near the site, releasing it slowly over a period of time beyond the duration of the runoff event.

Site-Adaptive Planning. The *first strategy* is aimed at eliminating or reducing the need for corrective measures, such as storage basins or gravel beds for infiltration, by practicing **site adaptive planning**. Site adaptive planning begins at the local watershed scale by defining sites with inherently good hydrologic performance,

Finding preferred sites   that is, places where the land is capable of holding and absorbing rainwater and

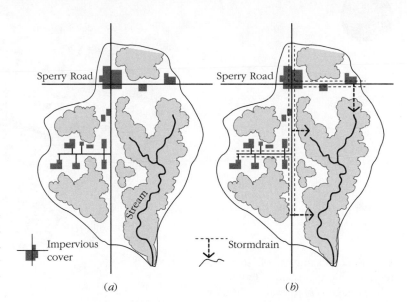

Fig. 8.11 An illustration of the concept of effective impervious cover: (*a*) low-effective impervious cover—no direct linkage between impervious cover and streams; (*b*) high-effective impervious cover—all impervious surfaces linked directly to streams via stormsewers and road ditches.

stormwater generated by land use facilities. These places or parcels should be classed as preferred sites, some with the potential to support higher land use densities (see Fig. 13.2).

Next, attention is paid to surface materials, avoiding the introduction of impervious surfaces wherever possible. *Clustering* is one means of reducing impervious areas by grouping buildings and related facilities to improve the per-person impervious cover ratio and to allow for more comprehensive approaches to stormwater mitigation (Fig. 8.12*c*). Although it is possible to achieve zero net increase in runoff solely through careful site planning, especially when previously cleared land is involved, most successful stormwater management requires some combination of site planning, source control, and storage strategies, especially in large and complex developments. (These ideas are developed further in Chapter 13.)

Reducing cover ratio

Source Control. The *second strategy* relies on **source control** of stormwater, that is, disposing of it at or very close to its point of origin. This strategy usually involves some form of soil infiltration and is usually accomplished onsite. Stormwater is directed into vegetated areas, shallow depressions, troughs, or pits, where it percolates into the ground (Fig. 8.12*b*). Since infiltration rates are often slower than rainfall rates, this strategy is usually most effective for small, low-intensity, or long-duration storms or for the water produced by the first part of large rainstorms. It is particularly useful on the North American West Coast north of California where the majority of rain comes in events producing 1 cm or less.

Point-of-origin rule

The lower diagram in Fig. 8.12 illustrates a source-control concept based on depression storage in residential development. For a 0.25-acre lot (11,000 square feet), the entire volume of a 0.5-inch rainfall can be held in an area of 1400 square feet (13 percent of the lot area) at a depth of 4 inches. In dry regions, this method may prove to be an effective means of recharging local soil moisture supplies and reducing yard and parkway irrigation. But to use depression storage effectively, many communities, their engineers, and their landscape architects would have to change their standards and practices that now dictate smooth and/or mounded grading of yards and parkways, which rule out any possibility for depression storage.

Example

Basin Storage. The *third strategy* involves directing stormwater to a **holding basin** and then releasing it slowly over an extended period of time. The main objective is to "shave" the peak discharge, thereby reducing the otherwise rapid rate of stormwater delivery to streams from developed land. This strategy usually involves

Shaving Qp

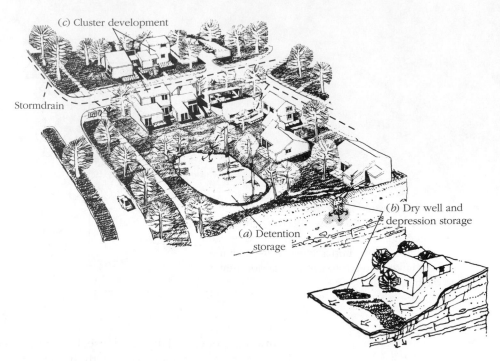

Fig. 8.12 Examples of stormwater mitigation measures: (*a*) detention storage, (*b*) on-site infiltration (source control), and (*c*) cluster development. The lower drawing illustrates a version of (*b*) based on depression storage.

the construction of storage facilities such as detention basins to catch the runoff delivered from a network of stormdrains, and has come to be known as the "pipe and pond" method (Fig. 8.12*a*).

Detention basins are ponds sized to store the design storm and then allow it to discharge at a specified rate. For instance, if a design storm generates a total runoff of 4.2 acre feet (182,952 cubic feet) in one hour, and the allowable release rate is 12 cubic feet per second (43,200 cubic feet per hour), then about 3.2 acre feet of gross storage are needed. Basins that store water on a long-term basis with the objective of releasing it to the ground and to evaporation are called *retention basins*, and they are generally used in areawide stormwater management.

Basin types

Increasingly, the pipe and pond strategy is falling into disfavor. Among other things, receiving streams often suffer a significant decline in summer low-flows. In addition, the absence of channel-shaping peak flows has been shown to reduce normal stress on aquatic and riparian habitat, thereby allowing invasion by less tolerant alien plant species. Moreover, the ponds are often unsightly, expensive (especially in terms of land costs), and dangerous in residential areas.

Selecting a Strategy. Which strategies and mitigation measures should be used depends on many factors, including construction cost, local topographic and soil conditions, and the character and design of the development program. In some areas there is no choice in the matter because local policy dictates what is to be used, and many of these regulations are based strictly on traditional engineering methods. Where there is a choice, however, several criteria can be used to guide the selection of stormwater mitigation measures.

Where in the system...?

Discharge Magnitude. The first criterion is discharge magnitude. The question is where within the flow system to focus mitigation efforts: (1) on small flows in and around planning sites or (2) on large flows downstream? This is significant because the force of running water, that is, the stress it exerts on the channel and riparian environment, increases substantially with discharge. As a rule, therefore, large flows are inherently more difficult to contain and regulate than small ones. It follows that large flows usually require substantial engineering treatments, whereas small flows can often be handled as a part of landscape design treatments.

Resistance

Unfortunately, the engineering community has tended to resist landscape design treatments, because, among other things, these treatments are usually built into privately held lands, which are beyond their control. Thus stormwater management in most North American communities is limited to traditional engineering approaches involving gutters, pipes, and basins patterned according to street and road networks and tied to local streams. Any mitigation of this system usually involves neighborhood or regional stormwater basins that inevitably lead to more engineering and higher mitigation costs.

Finding opportunities

Environmental Compatibility. The environmental compatibility criterion addresses the questions of (1) environmental sacrifices and tradeoffs, such as wetland impacts and habitat loss, and (2) compatibility with landscape design as an aesthetic and land use issue (Fig. 8.13). The second question includes the potential for achieving multiple outcomes, such as recreation and habitat enhancement, as well as improved neighborhood character and social efficiency. Among these is rainwater harvesting and recycling for food production and landscape gardens.

Reducing costs

Cost Considerations. The third criterion is the cost of both construction and maintenance. The costs of conventional infrastructure systems, which feature curbs, gutters, piping, and ponds, are substantially higher than source control and site adaptive measures. In community development the latter usually includes some form of clustering, reduced road widths, grass-lined swales, on-site infiltration, and related measures. Test cases reveal that planning and source control approaches together may reduce the cost of stormwater management by as much as 40 to 50 percent over conventional engineering.

Critical issue

Liability Concerns. The fourth criterion is liability. The risks of children around stormwater ponds and the hazards associated with downstream flooding are the principal liability concerns in stormwater management. However, with the rise of interest in alternative stormwater management measures such as source controls, contractors and communities have expressed concern over performance expectations and the risk of water damage to facilities. On balance, risk management and liability are becoming colossal issues on many fronts in environmental planning, often demanding inordinate attention from engineers, planners, and facility managers. Unfortunately, these issues are hampering implementation of improved management strategies.

Fig. 8.13 Retention ponds in Iowa farm country built for flood control. While compatible with agricultural needs, such measures may conflict with other objectives, such as habitat conservation, downstream water supply, and open space linkage.

8.8 THE CONCEPT OF PERFORMANCE

A values issue

Any effort to plan and manage the environment rests on a concept about how the environment should perform. When plans are formulated, the objective is either to guide and structure future change in order to avoid undesirable performance and/ or to improve performance in an environment, setting, or system whose existing performance is judged to be inadequate. The phrase "judged to be inadequate" is important because any judgment on performance is based on human values. "That stream floods too often and it poses a danger to local inhabitants" is a value judgment by someone about environmental performance. Similarly, an ordinance restricting development from wetlands reflects a societal value about either the performance of wetlands, the performance of land use, or both.

In watershed planning and management, **performance goals** must be set early in the program to determine the *best management practices* (BMPs) to be used in the planning process. Generally, the larger the watershed, the more difficult is the task because large watersheds usually involve many communities and interest groups with different views and policies. The task is usually less complex for a small watershed, not only because fewer players are involved, but also because the watershed itself is usually less complex.

Setting Performance Goals. The process of formulating performance goals and defining a BMP strategy usually begins with the definition of local (i.e., watershed land users') values, attitudes, and policies. Local values are especially critical and often the key to setting priorities. Next, regional factors are considered, in particular, policies pertaining to development density, stormwater retention, wetlands, open space, and the like. In addition, the values and needs, both articulated and apparent, of downstream riparians must be given serious consideration. For instance, upstream users may depend on a watershed and its streams to carry stormwater away, and downstream riparians may have set a goal to maintain habitat and quality residential land near water courses. In such a case, the two goals are in conflict, and unless modified, the downstream group stands little chance of success.

System perspective

Watershed Carrying Capacity. A third consideration in setting performance goals is the *carrying capacity* of the watershed based on its composite character. It is important to recognize what we might call the functional diversity of watersheds, even small ones. A basin often contains a wide range of land types and roles, and this diversity calls for different types, densities, and configurations of land use. In other words, watersheds cannot be treated as uniform entities, ignoring fundamental differences in terrain, soils, and water processes from place to place, as is usually the case with zoning maps.

Assigning capacity

Each watershed must be evaluated as a biophysical system, including existing land use, and weighed against the proposed goals based on local and regional values and policies. A watershed and streams that cannot support a trout population under predevelopment conditions certainly cannot support one after development. Therefore, a performance goal of habitat quality suitable to sustain trout would be unrealistic. Usually, an environmental specialist in geomorphology, hydrology, or ecology, for example, would examine the watershed, determine its condition and potentialities, and recommend appropriate modifications in proposed performance goals.

Performance Standards and Controls. Once performance goals have been formulated, performance standards and controls have to be defined. **Performance standards** are the specific levels of performance that must be met if goals are to be achieved. For example, a site-scale standard might require that 90 percent of the rainwater received by a site has to be retained on-site and disposed of by infiltration and in depression storage or recycled as irrigation water. At a community scale, a stormwater discharge standard might call for zero net increase in peak discharges on first-order streams after a particular date or development density has been reached.

Setting standards

This means that in planning new development, it is necessary to take into account all changes in land use activities in the watershed at this time and to determine from the balance of both positive and negative changes whether special measures or strategies are needed. If, for instance, cropland is being converted to woodland at the same time woodland is being cleared for new residential development, analysis may show that one change offsets the other, thereby maintaining the performance standard of zero net change in stormwater discharge.

Enabling means

Performance controls are the rules and regulations used to enforce the standards and goals. These may be specific regulations such as limits on the percentage of impervious surface and on effective impervious cover, that is, limits on pipe, ditch, and road linkages to natural water features. Or they may be broader, such as requirements to integrate stormwater management with open space planning and design or provisions for tax breaks to property owners who install stormwater-reducing measures on their land. Performance controls are necessary because without them the plan has no real teeth and thus no regulatory strength.

8.9 COMPREHENSIVE WATER MANAGEMENT PLANNING

The tendency in modern land planning and engineering is to isolate the various components of the environment and address them individually and separately. This approach is fostered by community regulations and development guidelines that typically address stormwater, wetlands, grading, landscaping, roads, parking, and utilities as though they were more or less mutually independent. As a result, not only do the rules and guidelines often lack balance and coordination, but some are actually

Incompatible policy

incompatible. In many communities, for example, grading specifications for development projects call for mounding lawn areas in parkways, a practice that tends to increase stormwater runoff even while stormwater management guidelines for the same site call for reducing runoff. In one California community, residential property owners are urged and sometimes required to practice water conservation but are restricted by ordinance from storing and recycling stormwater for garden and lawn irrigation.

The common thread

Watershed as the Unifying Model. Water is the one theme that is central to virtually all landscape planning and through which site planning can be unified. Water management planning provides an opportunity to weave a common thread through a land use plan, integrating landscape design, architecture, and engineering, leading to reduced construction and landscape management costs, to improved risk management, and to more sustainable landscapes and facilities in the long run. Water management planning must be framed within the system that is universal to all terrestrial environments: the *watershed* and its *drainage net.*

A place in the system

Every planning site is nested within a drainage system, and it is important to remember that the principles that govern watersheds also apply at smaller scales, including individual sites. A first consideration in building a water management plan is simply to pinpoint the project location in the drainage system and then to define what that location means in terms of upstream and downstream opportunities and constraints, such as flood risk and water supply, and what these determinations, in turn, imply for stormwater management needs and responsibilities. A second consideration is the search for complementary water systems and how to integrate them in a compatible way (Fig. 8.14). Wetland and stormwater systems are examples.

Although wetland mapping is standard practice in planning projects today, usually little or no attention is paid to wetland watersheds as a part of the exercise. As a result, land use projects in wetland watersheds rarely make allowances for the maintenance of wetland water supplies as a part of stormwater management. Consequently, some wetlands are overloaded with redirected stormwater, whereas others are deprived of natural runoff supplies. If stormwater systems can be coordinated with wetland water

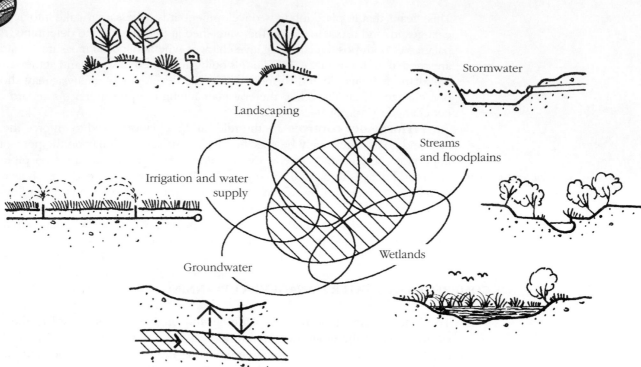

Fig. 8.14 The principal systems involved in comprehensive water management for land use development projects. Watershed concepts also apply at the site scale.

An example

supply needs, there is an opportunity, with proper allowances for changes in discharge rates and water quality, to achieve better performance in both systems at lower costs. The same sorts of complementary relationships can be found between wetlands and groundwater systems, between stormwater and irrigation systems, and between landscaping (grading, soil treatment, ground covers, etc.) and stormwater runoff.

8.10 CASE STUDY

A Case for Green Infrastructure in Stormwater Management, Surrey, British Columbia

Patrick Condon

In 1998, farmers sued the City of Surrey, British Columbia, for flood damages on their agricultural lands and received a settlement of an undisclosed amount in the tens of millions. Farmers blamed the flooding on suburban development upstream, which had cleared forests and grasslands, added impervious cover, funneled runoff into stormdrains, which in turn ran unmitigated and unabsorbed to streams, and flooded farmland downstream.

Low-density, highly paved residential communities are a major threat to natural drainage systems. Conventional stormwater management requires that rainwater be drained along street curbs into grates and then conveyed by pipes directly into waterways. This stormwater surges to streams at velocities and volumes many times greater than predevelopment rates, scouring streambeds, and sometimes causing flooding.

In addition to being environmentally damaging, building and maintaining standard infrastructures with stormdrains and arterial roads is costly to communities, developers, and ultimately homeowners. The high price of new homes in low-density development reflects the cost of the inefficient and overbuilt street networks that are part and parcel of the stormwater infrastructure. Furthermore, the replacement and upkeep costs of infrastructure in areas of conventional low-density residential development are consuming a large and growing share of municipal budgets.

The City of Surrey recognized the need to change its basic approach to stormwater management. In 1998, the city council voted to collaborate with the Landscape Architecture Program at the University of British Columbia to build a model community that would apply sustainable planning principles and alternative development standards "on the ground." The preservation and promotion of stream corridors and their drainage regimes was a key sustainability goal. The city chose a 250-hectare site called East Clayton for a demonstration project (Fig. 8.A).

Our approach employed a green infrastructure strategy. We saw this not only as a way of managing stormwater and reducing the community's liability to landowners downstream, but as a way to combat unsustainable conventional development practices. Green infrastructure relies on source control of stormwater; that is, it encourages infiltration of rainwater at or close to its point of origin, where it can be filtered into the soil before either being taken up by trees, recharging groundwater, or flowing slowly to streams as interflow. This approach helps support stream baseflows, reduces stream peak flows, and eliminates flooding downstream, ultimately bringing watershed performance close to its predevelopment levels (Fig. 8.B).

In addition, organizing development around green infrastructure creates more walkable communities. The excessive paving and hierarchical pattern of suburban streets (typically five times the amount of paving per capita than in traditional, higher-density, urban neighborhoods) creates an unfriendly and dysfunctional pedestrian environment. A green infrastructure system, by contrast, uses narrow streets arranged in an interconnected network, which makes walking and cycling easier while reducing stormwater runoff. Conversely, the need for automobile travel is reduced, especially high-frequency short trips, thereby reducing per-capita energy use and lowering greenhouse gas emissions. Street networks that encourage walking and cycling also provide an opportunity to improve the health of residents, for researchers have linked obesity with the sedentary, automobile-dependent lifestyles of suburban development.

In 2003, the first phase of the East Clayton neighborhood was completed on a 20-hectare parcel. Integral to the East Clayton design is a green infrastructure system that builds on existing waterways to create an integrated and multifaceted

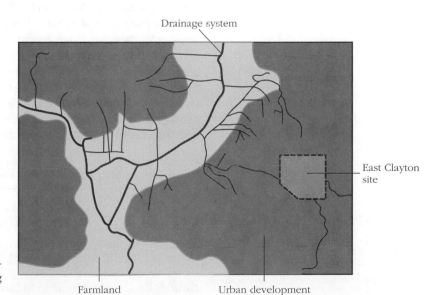

Fig. 8.A The lowland corridor of farmland surrounded by Surrey's urban development, including the East Clayton site.

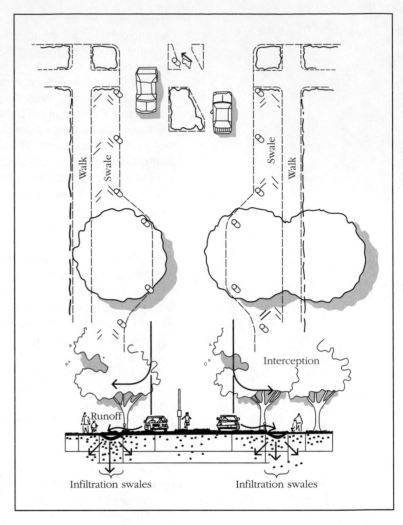

Fig. 8.B A street design featuring green stormwater infrastructure that includes reduced road width, street trees, and infiltration swales.

network of streams, green streets, greenways, self-mitigating parcels, parks, and riparian areas. The stormwater system at East Clayton has the capacity to retain 2.4 centimeters (1 inch) of rainfall per day, which accounts for 90 percent of the 60-plus inches of annual rain falling on the site.

The integration of green infrastructure into our built environments is a viable solution for maintaining the ecological health of watersheds and building affordable and sustainable communities. The cost of green infrastructure can be less than half the cost of conventional infrastructure; the savings in Surrey were more than $12,000 per household. Lower infrastructure costs in turn lead to lower housing and long-term maintenance costs. Finally, green infrastructure creates more walkable neighborhoods, giving us safe and more socially sustainable communities. East Clayton is a demonstration of the links between affordability, sustainability, and walkability.

Patrick Condon is professor of landscape architecture at the University of British Columbia. He holds the James Taylor Chair in Landscape and Liveable Environments.

8.11 SELECTED REFERENCES FOR FURTHER READING

Burchell, Robert, and Lostokin, David. *Land, Infrastructure, Housing Costs and Fiscal Impacts Associated with Growth: The Literature on the Impacts of Sprawl Versus Managed Growth.* Cambridge, MA: Lincoln Institute of Land Policy Research Paper, 1995.

Condon, Patrick, and Gonyea, Angela. "Status Quo Versus an Alternative Standard, East Clayton Two Alternative Development Standards Compared." *Technical Bulletin No. 2.* James Taylor Chair, University of British Columbia, 2000.

Dunne, Thomas, and Black, R. D. "Partial-Area Contributions to Storm Runoff in a New England Watershed." *Water Resource Research.* 1970, pp. 1296–1311.

Ferguson, Bruce K. "The Failure of Detention and the Future of Stormwater Design." *Landscape Architecture.* 81, 1991, pp. 76–79.

Hewlett, J. D., and Hibbert, A. R. "Factors Affecting the Response of Small Watersheds to Precipitation in Humid Regions." In *Forest Hydrology.* Oxford: Pergamon Press, 1967, pp. 275–290.

Horton, Robert E. "The Role of Infiltration in the Hydrologic Cycle." *American Geophysical Union Transactions.* 14, 1933, pp. 446–460.

Marsh, W. M., and Marsh, N. L. "Hydrogeomorphic Considerations in Development Planning and Stormwater Management, Central Texas Hill Country, USA." *Environmental Management.* 19, 5, 1995, pp. 693–702.

Poertner, Herbert G. *Practices in Detention of Urban Stormwater Runoff.* Chicago: American Public Works Association, 1974.

Seaburn, G. E. "Effects of Urban Development on Direct Runoff to East Meadow Brook, Nassau County, Long Island, New York." *U.S. Geological Survey Professional Paper 627-B*, 1969.

U.S. Natural Resources Conservation Service. *Urban Hydrology for Small Watersheds.* Technical Release No. 55. Washington, DC: U.S. Department of Agriculture, 1975.

Whipple, W., et al. *Stormwater Management in Urbanizing Areas.* Upper Saddle River, NJ: Prentice-Hall, 1983.

Related Websites

Center for Watershed Protection. 2008. http://www.cwp.org/

An organization whose mission is to protect and restore watersheds provides professionals with best management practices, retro-fitting guides, and model ordinances. The site has large resource sections for stormwater management and run-off controls.

City of Seattle, Seattle Public Utilities. 2009. http://www.seattle.gov/util/About_SPU/Drainage_&_Sewer_System/Natural_Drainage_Systems/index.asp

Seattle is taking a nature-mimicking approach to its stormwater management. Follow the links to the virtual tour of the Street Edge Alternatives (SEA streets), which have reduced runoff by 99 percent.

Natural Resources Defense Council (NRDC). Issues: Water. http://www.nrdc.org/water/default.asp

As a nonprofit committed to protecting the earth's natural resources, one of NRDC's major issues is water. The "City Stormwater Solutions" section provides several case studies and pictures of what cities in various states are doing to reduce runoff. Readers can also take a look at the water and global change section for a forecast of worldwide water supply.

Waterbucket. http://www.waterbucket.ca/

Related to the Water Sustainability Action Plan for British Columbia, Waterbucket tells about water resources and management in BC. Content covers watercentric planning, rainwater management (as they try to get away from stormwater management), and green infrastructure.

WATERSHEDS, DRAINAGE NETS, AND LAND USE PLANNING

9.1 INTRODUCTION

The drainage net

Think of overland flow moving slowly over the ground and at some point gathering together to form minute threads of water. These threads merge with one another, forming rivulets capable of eroding soil and shaping a small channel. The rivulets in turn join to form larger channels and some point streams take shape, and they in turn join to form larger streams, and so on. This system of channels, characterized by streams linked together like the branches of a tree, is called a **drainage network**, and it represents nature's most effective means of getting liquid water off the land. The area feeding water to the drainage network is the **drainage basin**, or **watershed**, and, for a given set of geographic conditions, the size of the main channel and its flows increase with the size of the drainage basin.

Net organization

Early in the history of civilization, humans learned about the advantages of channel networks, for both distributing water to and removing water from the land. The earliest sewers were actually designed to carry stormwater, and they were constructed in networks similar to those of natural streams. Whether or not they knew it, the ancients often applied the *principle of stream orders* in the construction of both storm-sewer and field irrigation systems. This principle describes the relative position, called the *order*, of a stream in a drainage network and helps us to understand the relationships among streams in a complex flow system.

Net alternatives

Modern land development habitually alters drainage networks by obliterating natural channels, adding artificial channels, and changing the size and shape of channel systems. Such alterations can have serious environmental and land use consequences, including increased flooding, loss of aquatic habitats, reduced water supplies during low-flow periods, and degraded water quality. In land use planning, generally little attention is paid to drainage networks as geographic entities, especially as systems. Site planners and designers typically ignore questions related to a project's location in a drainage network and its possible implication in terms of risk, liability, and water management.

Likewise, community planners often ignore the implications of retention basins and other stormwater facilities in terms of streamflow, water quality, and flood management at the regional scale. In this chapter we examine the drainage network as a system of channels that service a watershed and its land uses. We are interested in how to map and analyze them and how small drainage basins function and respond to land use change.

9.2 THE ORGANIZATION OF NETWORKS AND BASINS

Stream orders

Networks in nature have a fundamental logic. The **principle of stream orders** defines that logic for drainage nets, apple trees, the human lung, and many other branching phenomena. It is a simple model built on a classification system based on the rank of streams in the drainage network. Streams ranked as *first-order* are channelized flows with no tributaries. *Second-order* streams are those with at least two first-order tributaries. *Third-order* streams are formed when at least two second-order streams join together, and so on (Fig. 9.1).

Streams per order

Network Parameters. The number of streams of a given order that combine to form the next higher order generally averages around 3.0 and is called the **bifurcation ratio**. From a functional standpoint, this ratio tells us that the size of the receiving channel must be at least three times the average size of the tributaries. For drainage networks in general, a comparison of the total number of streams in each order to the order itself reveals a remarkably consistent relationship in which stream numbers decline progressively, with increasing order (see the graph in Fig. 9.1). Therefore, first-order streams are the most abundant streams in every drainage network.

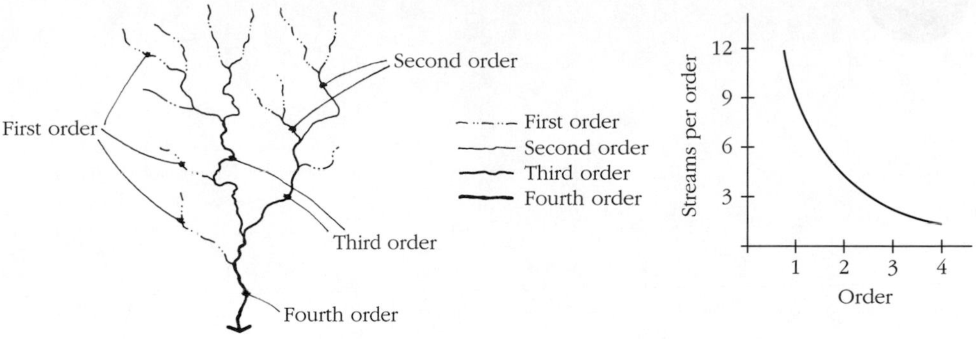

Fig. 9.1 Stream order classification according to rank in the drainage network. The number of streams in each order declines as order rises.

Given this classification by order of the streams in a drainage net, we can examine the relationship between orders and other hydrologic characteristics of the river system such as drainage area, stream discharge, and stream lengths. The system of classification provides a basis for:

■ Comparing drainage nets under different climatic, geologic, and land use conditions.
■ Analyzing selected aspects of streamflow, such as changes due to urbanization.
■ Defining zones with different land use potentials.

Table 9.1 lists and defines several factors involved in drainage network and basin analysis.

Table 9.1 Factors Important to the Analysis of Drainage Networks

---

Number of Streams—The total number of streams in each order.
Bifurcation (branching) ratio—Ratio of the number of streams in one order to the number in the next higher order.

$$BR = \frac{N}{N_u}$$

where

BR = bifurcation ratio
N = the number of streams of a given order
N_u = the number of streams in the next highest order

Drainage basin order—Designated by the highest-order (trunk) stream draining a basin.
Drainage area—Total number of square miles or square kilometers within the perimeter (divide) of a basin.
Drainage density—Total length of streams per square mile or square kilometer of drainage area.

$$\text{drainage density} = \frac{\text{sum total length of streams (mi or km)}}{\text{drainage area (mi}^2 \text{ or km}^2)}$$

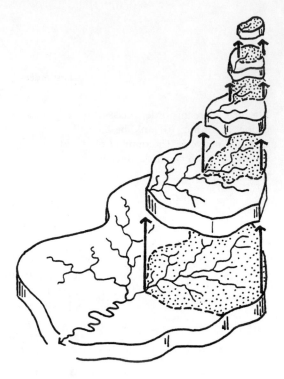

Fig. 9.2 Illustration of the nested hierachy of lower-order basins within a large drainage basin.

Basin Order. *Drainage basins* themselves can also be ranked according to the stream order principle. First-order drainage basins are those emptied via first-order *Basin nesting* streams; second-order basins are those in which the main channel is of the second order; and in a fifth-order basin, the trunk stream is of the fifth order. Just as all high-order streams are products of a complete series of lower streams, high-order (large) basins are comprised of a complete series of lower-order basins, each set inside the other. This is sometimes referred to as a *nested hierarchy*, as illustrated in Figure 9.2.

Nonbasin Drainage Area. In accounting for the combined areas of the basins that make up a large basin, however, not all the land area is taken up by the lower-order basins. Invariably a small percentage of the land drains directly into higher-*Direct input area* order streams without first passing through the numerical progression of lower-order basins (Fig. 9.3). This land is called the *nonbasin drainage area*, and it generally constitutes 15 to 20 percent of the total drainage area in basins of second order or larger. When nonbasin areas are developed and engineered to discharge stormwater, this water passes directly into a higher-order channel without being subject to the processes associated with the sequence of lower channels.

Drainage Density. *Drainage density* is the total length of channels per unit area (square miles or square kilometers) of land. It is an important parameter in watershed management because it indicates how efficiently land is served by channel systems. Because open channel (or pipe) flow is the quickest way of removing water from the land, the more channels there are per square mile, the faster the runoff system operates and the more stormwater reaches stream channels.

Virtually everywhere drainage density increases with land clearing and development as drainage ditches are cut along roads and farmfields; as cropland is laced *Channels everywhere* with drain tiles; and as curbs, gutters and stormsewers are added to communities (Fig. 9.4). Unless counterbalanced with water management measures such as holding basins and source controls, stormwater loading and delivery rates in such landscapes increase with drainage density while water quality decreases, often dramatically.

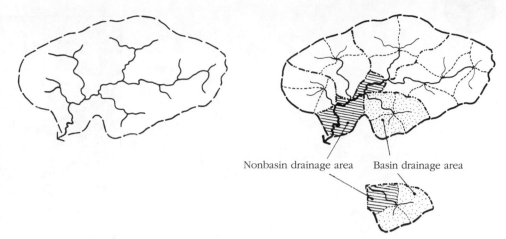

Fig. 9.3 Two types of drainage areas can be defined in basins larger than the first order: basin and nonbasin areas. Nonbasin area borders the trunk stream(s) and releases its water directly to the main channel.

Nonbasin drainage area   Basin drainage area

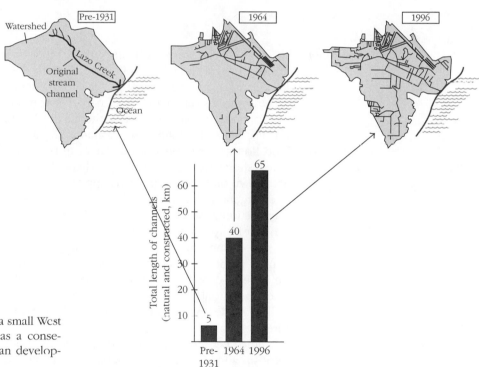

Fig. 9.4 Drainage density change in a small West Coast watershed from 1931 to 1996 as a consequence of road, agricultural, and urban development, a 13-fold increase.

9.3 MAPPING THE DRAINAGE BASIN

The process of mapping the individual drainage basins in a large drainage net requires finding the **drainage divides** between channels of different orders. This is best accomplished with the use of a topographic contour map and the assistance of a Geographic Information System (GIS). However, field inspection is advisable (1) in developed areas in order to find culverts, diversions, and other drainage alterations and (2) in heavily forested areas where the technicians responsible for contour map preparation were unable to see the ground on aerial photographs.

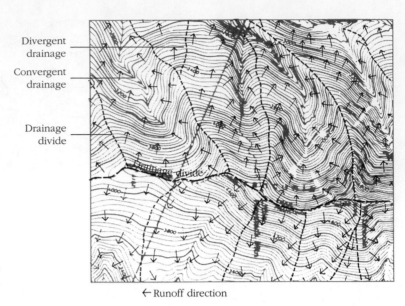

Divergent drainage

Convergent drainage

Drainage divide

← Runoff direction

Fig. 9.5 Mapping and partitioning the watershed using vectors of overland flow. Flow is divergent on divides and convergent in the basins.

Channels first

Some Basic Steps. As a first step, channels should be traced and ranked, taking care to note the scale and the level of hydrographic detail provided by the base map. This is important because maps of large scales show more first-order streams, owing to the fact that they are usually drawn at a finer level of resolution than smaller-scale maps.

In areas of complex terrain, the task of locating basin perimeters can be tedious, but it can be streamlined somewhat if the assumed pattern of surface runoff (overland flow) is first demarcated. This can be done by mapping the direction of runoff using short arrows drawn perpendicularly to the contours over the entire drainage area. Two basic runoff patterns appear: divergent and convergent. Where the pattern is divergent, a drainage divide (a basin perimeter) is located; where it is convergent, a collection area and drainageway form (Fig. 9.5). These patterns are also illustrated schematically in Figure 9.11.

Flow vectors

First-order basins

By delineating all the drainage divides between first-order streams, a watershed can be partitioned into first-order basins. From this pattern, second-order basins can be traced. Second-order basins are comprised of two or more first-order basins plus a fraction of nonbasin area. Roughly 80 to 85 percent of the second-order drainage area will be made up of first-order basins. The remainder will be nonbasin area that drains directly into the second-order stream.

Subsurface variations

Since perennial streams are also fed by groundwater emanating from subsurface deposits and bedrock, we must be aware that the drainage area for groundwater runoff may be different from that for surface runoff. In other words, an aquifer may extend beyond the topographic perimeter of a basin. In most instances, however, identifying such an arrangement without the aid of an areawide groundwater study is impossible. A clue to the existence of a significant difference in the two can sometimes be found in first- or second-order streams that have exceptionally large or small baseflows for their surface drainage area relative to similar streams in the same region.

9.4 LAND USE ALTERATIONS OF SMALL DRAINAGE BASINS

Land clearing and development can have pronounced effects on drainage networks and basins. Deforestation, overgrazing, and crop farming may initiate soil erosion and gully formation. As gullies advance, they expand the drainage network, thereby

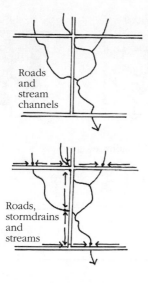

Roads
and
stream
channels

Roads,
stormdrains
and
streams

Urban hydrograph

Stormsewers

increasing the number of first-order streams and the drainage density. Road building, lumbering, and commercial agriculture can have the same effect because these uses require the construction of drainage ditches that form new first-order streams or lengthen existing ones (Fig. 9.4).

The main hydrologic consequence of higher drainage density is shortened concentration times—that is, faster response times—because the distance that water must travel as overland flow or interflow is reduced (Fig. 9.6). In its place comes channel flow, which is much faster. Discharges are in turn larger because, for small basins, concentration times fall closer to the peaks of rainstorm intensity. Therefore, large flows occur with higher frequencies, floods are larger and more frequent, channel erosion is greater, more sediment is moved, and water quality can be expected to decline.

Urban Development. Urbanization is particularly effective in changing the shape and density of drainage networks. One of the first changes that takes place is a pruning of natural channels, that is, the removal of parts of the channel network. These channels are often replaced by ditches, usually where no channel or swale existed previously, and by underground channels in the form of stormsewers. Although the natural network may be pruned, the total drainage network is usually enlarged by channel grafting and intensified by adding pipes and ditches.

The *net effect* of urbanization is usually an increase in total channels and in turn an increase in the overall drainage density (Fig. 9.7; also see Fig. 8.10). Coupled with the lowered infiltration rates and extensive effective impervious cover in urbanized areas, these changes lead to increased amounts of runoff and shorter concentration times for the drainage basin, both of which produce larger peak discharges. As a result, both the magnitude and frequency of peak discharges are increased for receiving streams and rivers (see Fig. 8.8).

Stormsewers are pipe systems that conduct surface water by gravity flow from streets, buildings, parking lots, and related facilities to streams and rivers. The pipes are usually made of concrete, sized for the area they serve, and capable of conducting stormwater at a very rapid rate, up to four times that of natural channels. Studies show that stormsewers and associated impervious surfaces increase the frequency of floodflows on streams in fully urbanized areas by as much as sixfold (Fig. 9.8).

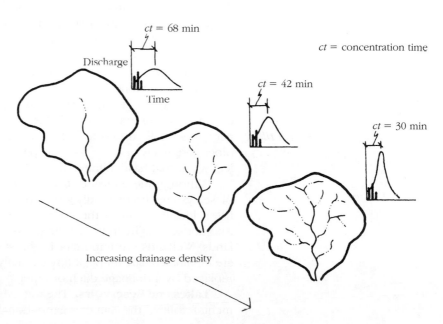

Fig. 9.6 Hydrographs showing the changes brought about by increased drainage density: concentration time is reduced, causing greater peak discharges for a given rainstorm.

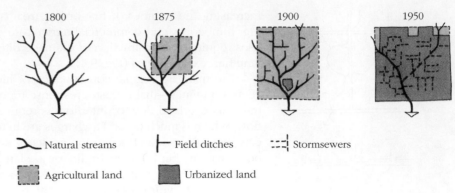

Fig. 9.7 Pruning, grafting, and intensification of a drainage network with community development.

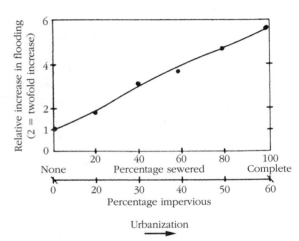

Fig. 9.8 Increased frequency of floodflows related to stormsewering and impervious surface.

Basin grafting

The graph in Figure 9.8 assumes that the size of the drainage basin has remained unchanged as sewering advanced. In many instances, however, sewering increases the size of the basin, as illustrated for the Reeds Lake watershed in Figure 9.9. The additional drainage area here is completely urbanized, is laced with stormsewers, and produces peak discharges perhaps four to five times larger than drainage areas of comparable size without stormsewers and such heavy development.

Agriculture. Modern agriculture is also responsible for altering drainage basins and stream networks. In humid regions, farmers often find it necessary to improve field drainage to facilitate early spring plowing and planting. In addition to cutting ditches through and around fields and deepening and straightening small streams, networks of drain tiles are often installed in fields. Drain tiles are small perforated pipes (originally ceramic but now plastic) buried just below the plow layer. They collect water that has infiltrated the soil and conduct it rapidly to an open channel.

Drain tiles

Although the effects of tile systems on streamflow have not been documented, these systems undoubtedly serve to increase the magnitude and frequency of peak flows locally in much the same way as yard drains and stormsewers do in cities. Another serious hydrologic change associated with agriculture is the draining of wetlands. Wetlands are important in the storage of runoff and floodwater. Where they are eliminated, stream peak flows usually increase, especially if the wetland has been replaced by a drainage ditch or pipe.

Lakes and Reservoirs. The watersheds that serve lakes and reservoirs (impoundments) follow the same organizational principles as river watersheds, with the

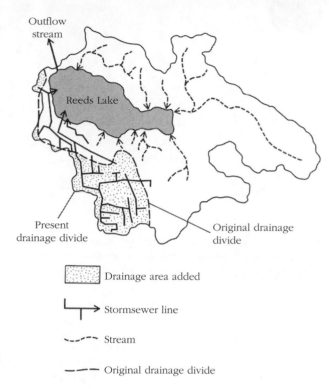

Outflow
stream

Reeds Lake

Present
drainage divide

Original drainage
divide

Drainage area added

Stormsewer line

Stream

Original drainage divide

Fig. 9.9 The Reeds Lake watershed showing the drainage added
(in tint) through stormsewering with urban development.

impoundment itself representing a segment of some order in the drainage network or
a node linking streams of lower orders. On the other hand, the land use patterns are
different in one important respect: The heaviest development is usually found in the
Nonbasin area   nonbasin drainage area (see Fig. 11.11).

The nonbasin drainage area forms a discontinuous belt of shoreland that encir-
cles the waterbody. As water frontage it is highly attractive to residential, recreational,
and commercial development. And when development takes place, the need for
storm drainage usually arises and yard drains, ditches, and stormlines are constructed
where, under natural conditions, channels never existed. In addition, the shoreland
zone is often expanded inland with stormsewer construction, thereby capturing additional runoff and rerouting it directly to the waterbody.

9.5 WATERSHED PLANNING AND MANAGEMENT CONSIDERATIONS

Since small drainage basins are the building blocks of large drainage systems, watershed planning and management programs must address the small basin. Most small
basins (primarily first, second, and third orders) are comprised of three interrelated
parts which are illustrated in Figure 9.10:

Watershed zones
1. An outer, *contributing zone* that receives most of the basin's water and generates runoff in the form of groundwater, interflow, and overland flow.

2. A low area or *collection zone* in the upper basin where runoff from the contributing zone accumulates.

3. A central *conveyance zone* represented by a valley and stream channel through
which water is transferred from the collection zone to higher-order channels.

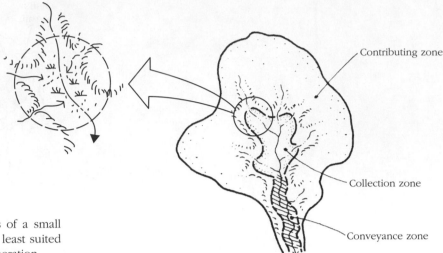

Contributing zone

Collection zone

Conveyance zone

Fig. 9.10 The three main hydrologic zones of a small drainage basin. The lower two zones are the least suited to development but essential to the basin's operation.

The hydrologic behavior of each zone is different, and each in turn calls for different planning and management strategies.

Organizational Framework. The **contributing zone** is uppermost in the watershed and generally the least susceptible to drainage problems. It provides the greatest opportunity for site-scale stormwater management because most sites in this zone have little upslope drainage to contend with and because surface flows are generally small and diffused.

By contrast, the **collection zone** in the upper basin is subject to serious drainage problems. Seepage is common along the perimeter, and groundwater saturation can be expected in the lower central areas during much of the year. During periods of *Headwaters* runoff, this zone is prone to inflooding, caused by interflow and stormwater loading superimposed on an elevated groundwater surface. Inflooding is very common in rural areas of modest local relief; in fact, it is probably the most common source of local flood damage today in much of the Midwest and southern Ontario.

Edge Forms. The edge separating the contributing (upland) zone and the collection zone is one of the most critical borders in landscape planning. This border is made of two edge forms: convex and concave (Fig. 9.11). The *convex forms* create *Dry and wet edges* divergent patterns of runoff and therefore tend to be dry along their axes—indeed very dry in some instances. By contrast, *concave forms* create convergent (funnel-like) patterns of runoff, concentrating flows downslope, and they therefore tend to be wet along their axes.

The concave edges are the functional links between the upland surface, where runoff is generated, and the lowland collection areas, where water accumulates and streams head up. As such they are vital management points in the drainage basin, serving as the gateways to the lowland and riparian environments of the collection zone. It follows that the concave edges also represent important habitat corridors between upland and lowland surfaces.

Lower Basin Drainage. The central **conveyance zone** contains the main stream channel and valley, including a small floodplain. The flows in this zone are derived *Conveyance zone* from the upper two zones as well as from groundwater inflow directly to the channel. Groundwater contributions provide the stream baseflow and constitute the vast majority of the stream discharge over the year. Stormwater, on the other hand, is derived mainly from the upper zones, and, though relatively small in total, it constitutes the bulk of the largest peak flows in developed or partially developed basins.

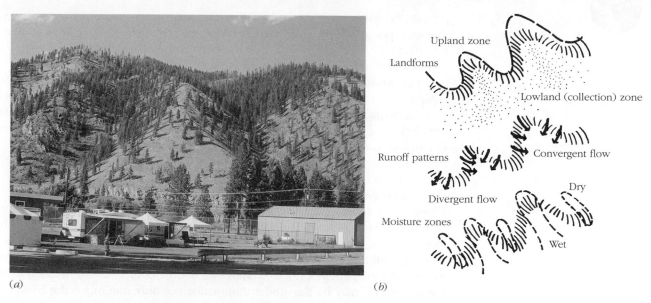

Fig. 9.11 Concave and convex edges and associated runoff patterns. Vegetation (photograph) reveals the contrast in moisture conditions.

The central conveyance zone is also subject to flooding, but in this case it is outflooding caused by the stream overtopping its banks. Because this zone is prone to comparatively large floods, especially if the upper basin is developed, it is generally the least suited to development and the most difficult to manage hydrologically. In hilly and mountainous terrain, such as coastal California, where slopes tend to be unstable during wet periods, the upper fringe of the conveyance zone and the upper collection zone receive the landslide and mudflow debris from slope failures.

Planning Implications. Recognition of the constraints and opportunities associated with each of these drainage zones is an important step toward forming development guidelines for small basins, including buildable land, open space, and special-use areas. What it does not provide, however, is a rationale for establishing density guidelines. *Density* is a measure of the intensity of development, defined, for example, as the percentage of impervious surface, total houses per acre, or popu-

The density question

lation density. For water and land management, the percentage of impervious surface or of effective impervious cover is commonly used to define density. As density rises, both the magnitude and frequency of stormflows and the pollution load delivered by runoff increase. In the Pacific Northwest, researchers have found that fish habitat declines measurably when effective impervious cover reaches only 8 percent of a basin's drainage area.

Watershed Carrying Capacity. Establishing the appropriate level of development for a drainage basin should be based on performance goals and standards (see Chapter 8 for a discussion of performance concepts) and basin carrying capacity. The

Sustainable load

carrying capacity of a drainage basin is a measure of the amount and type of development it is able to sustain without suffering degradation of water features, water quality, biota, soils, and land use facilities. Although this concept is easy to envision, the determination of a basin's carrying capacity may be difficult to derive. Just what percentage of a basin should be the allowable development limit?

Design Considerations. Thirty percent has been suggested as a rule of thumb, but this number is highly questionable. Among other things, development limits vary with the type and style of development as well as with the character of the basin, especially its hydrologic versatility. For residential development, for example, much

Site-adaptive design

depends on site selection and design practices. If *site-adaptive design* is employed that incorporates sensitive road alignments, source control of stormwater, and little or no effective impervious cover, then the percentage of basin area devoted to development can be higher than with conventional residential development.

Also important is thinking in terms of development forms that minimize roads, piping, and ditching because such infrastructure relates directly to stormwater loading and water quality impacts, among other effects. For residential land use, infrastructure development, when measured on a per-capita basis (either per person or per dwelling unit), reaches its highest ratio for large-lot single-family development. Cluster development, on the other hand, supports a larger population with smaller infrastructure and less impact on the drainage basin.

Physiographic Considerations. But this is only part of the picture because watershed *physiography* and *hydrologic versatility* also exert a strong influence on carrying capacity. Basins in hilly or mountainous terrain, similar to that shown in Figure 9.12*a*, usually have the lowest carrying capacity because of the abundance of steep, unbuildable slopes, complex drainage networks, rapid rates of stormwater runoff, and the paucity of opportunities for source control of stormwater. In addition, the soil mantle may be thin and a thin mantle not only minimizes the basin's capacity to filter and retain infiltration water, but often increases the tendency for slope failure during wet periods (see Section 4.6).

Basin inequality

For basins with less relief and gentler slopes, along the lines of the one shown in Figure 9.12*b*, where soil mantles are likely to be deeper and drainage less complex, the carrying capacity may be considerably higher. Curiously, few communities take basin physiography and hydrologic versatility into account when formulating land use planning and engineering policy. Basins with steep terrain, complex drainage systems, and few opportunities for stormwater management are, within certain limits, usually treated no differently in terms of zoning, land use design, road alignments, and stormwater engineering than low-relief basins with simple drainage systems and lots of management opportunities.

Systems Considerations. On balance, to achieve the best performance of a drainage basin, we must begin with an understanding of the system—not just its

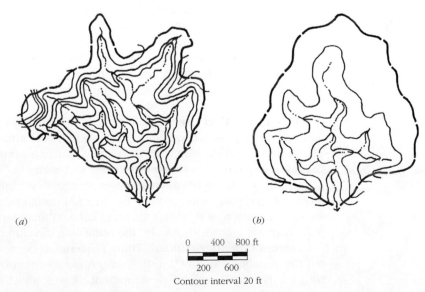

Fig. 9.12 Two watersheds of similar size but contrasting carrying capacities based on topography and drainage complexity. Basin (*a*) has a lower carrying capacity than basin (*b*). Among other things (*a*) is more topographically rugged with steeper slopes and a higher drainage density.

Think like a system

forms and features, but how it functions and how these functions relate to development. This understanding must be grounded in the perspective that there are better and poorer places in a watershed to site different facilities, that there are better and poorer ways of designing land use facilities, and that design by prescription (that is, without a site-adaptive perspective) is a seriously flawed approach to development planning that commonly results in unsustainable land use arrangements. Above all, we need to remember that we are dealing with natural drainage systems, which are fraught with complexity and irregularity. Rarely do they fit the engineer's notion of system efficiency. Therefore it is best to avoid invoking efficiencies where they do not exist and do not belong.

9.6 FRAMING A PLAN FOR THE SMALL DRAINAGE BASIN

First steps

The procedure for building land use plans for small drainage basins begins with a definition of the stream system, patterns of runoff, and the three hydrologic zones: contributing, collection, and conveyance (Fig. 9.13*a*). Each zone should be analyzed not only for soils, slopes, vegetation, and existing land use but for evidence of its hydrologic behavior, that is, for signs of overland flow, groundwater seepage, and flooding. Steep slopes, runoff collection areas, unstable and poorly drained soils, forested areas in critical runoff zones, and areas prone to flooding and seepage should be designated as nonbuildable. The remaining area is more or less open to consideration for development, and the bulk of it usually lies in the upland zone of the basin separated from the lowlands by the concave and convex edges described earlier.

Nonbuildable land

Next step: land units

The next step is to define land use units. These are physical entities of land, with the fewest constraints to development in general, that set the spatial scale and general configuration of the development program (Fig. 9.13*b*). Within this framework, development schemes may be evaluated and either discarded or assigned to the appropriate land unit. To achieve the desired performance, different options for the siting of different activities should be exercised; for instance, high-impact activities may be assigned to land units that are buffered by forest and permeable soils from steep slopes and runoff collection areas.

Siting options

Site Selection and Design Guidelines. In general, the recommended guidelines for site selection and site planning and design in small drainage basins are to:

Guidelines

1. Choose sites and design schemes that lend themselves to source control of stormwater.

2. Mimic predevelopment runoff systems and processes in designing source control schemes.

3. Wherever possible, disconnect human-made channels (and pipes) from natural channels.

4. Where source controls and disconnects are not possible:

    a. Maximize the distance of stormwater travel from the site to a collection area or stream.

    b. Maximize the concentration time by slowing the rate of stormwater runoff.

    c. Minimize the volume of overland flow per unit area of land.

    d. Utilize or provide buffers, such as forests and wetlands, to protect collection areas and streams from development zones.

    e. Divert stormwater away from or around critical features such as steep slopes, unstable soils, and valued habitats (Fig. 9.13*c*).

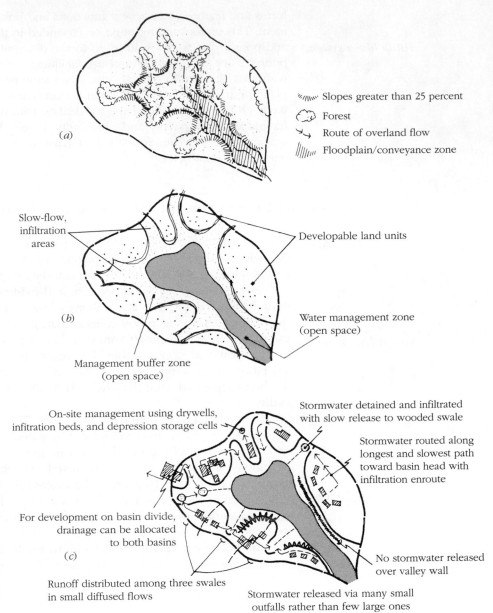

Fig. 9.13 Maps produced as a part of a procedure for land use planning in a small drainage basin: (*a*) the three hydrologic zones and the significant features of each; (*b*) the land use units; (*c*) some strategies and guidelines in site selection and site planning.

9.7 CASE STUDY

Watershed Concepts Applied to Industrial Site Management, Saltillo, Mexico

Jack Goodnoe

With the shift in industrial operations from the United States to Mexico, many corporations are facing the challenge of siting and designing new manufacturing facilities. But unlike the industrial site planning and design practiced during most of the twentieth century, the exercise today calls for more than muscling engineered facilities onto a parcel of ground. This was the case with an industrial site near Saltillo in northeastern Mexico where General Motors built a set of automobile plants. The program not only called for state-of-the-art manufacturing systems, but for water and landscape management systems quite unlike those of the humid, temperate settings where most GM facilities were situated in the United States.

Rimmed by the Sierra Madre Mountains, the Saltillo Valley is extremely arid. Average annual rainfall amounts to about 10 inches (25 centimeters) per year, and most of it is given up to evaporation. On the face of it, stormwater management would seem to be a fairly insignificant part of planning an industrial site in such a location. But that is far from the case because most of the rain each year comes in a dozen or so intensive downpours, and when it hits the desert landscape, there is practically nothing to hold it back. As a result, rainstorms can produce not only massive stormwater runoff, but dramatic erosion of soil unprotected by vegetation.

Consequently, stormwater management is a major concern in such arid regions because facilities must, among other things, be protected from flooding and sedimentation. Using conventional stormwater methods of collecting runoff in pipes and ditches, such as shown in Figure 9.A, has serious limitations because (1) such systems only concentrate the runoff into even bigger flows that produce even greater erosion and flooding downstream; (2) the pipes and ditches become clogged with sediment, and (3) water essential to the maintenance of vegetation is discarded from the site.

Given a development program that called for an engine plant, an assembly plant, and an administrative building (together covering about 150 acres including parking and storage areas), a strategy for stormwater management and landscape development was needed that would serve multiple purposes. These included the reduction of:

■ Stormwater runoff and downstream flooding.
■ Catastrophic on-site flooding that would impair manufacturing operation and wash away spill debris.
■ Soil erosion and sedimentation of pipes, ditches, and stream channels.
■ Airborne dust, a threat not only to human health, but to air supplies used for industrial cooling and automotive painting operations.
■ Infrastructure construction and maintenance costs.
■ The harsh industrial character of the site, making it more attractive and functional as a working landscape.

Our approach to water management and landscape stabilization, shown in Figure 9.B, began by viewing the site as several small watersheds rather than one large one. Because upslope drainage flowed toward the operations area, we needed to separate runoff into subbasins and invoke source controls as the main line of defense. With this approach the stormwater system is designed to hold runoff as close to its point of origin as possible and thereby prohibit rapid transfer and concentration downslope. It has the advantage of reducing runoff and soil erosion by keeping flows small and promoting local infiltration, which in turn reduces the need for stormwater piping and irrigation infrastructure.

Because of the size of the industrial site (550 acres) and the availability of soil material from building excavations, we seized the opportunity to create several small subbasins to capture upslope drainage before it could reach the central operations area. Within each subbasin, we created a broad depression storage cell. Each subbasin and storage cell formed a water management unit, a parcel of land designed to integrate water, soils, plants, and landscape aesthetics into a manageable and ultimately sustainable subsite or landscape cell. In this manner, the site was organized into a number of individual water management units that were, in turn, framed by a larger landscape design scheme organized around a micro watershed theme.

Each unit functioned more or less independently, and the need for piping upslope stormwater through the center of the site was virtually eliminated. Exceptionally large storms were allowed to overflow, with the discharge redirected laterally to the arroyos on the site perimeter, but only after substantial quantities of water had already been detained in the depression storage cells and introduced to the soil.

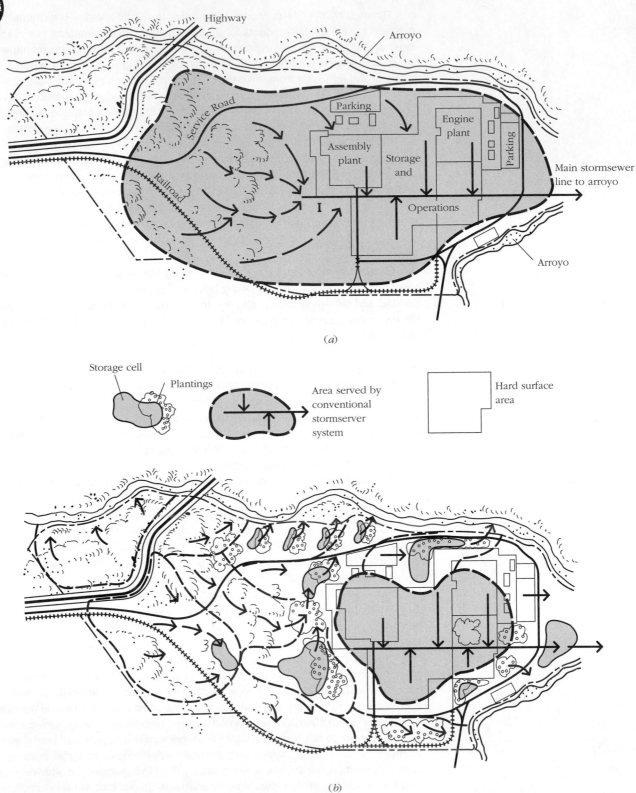

Fig. 9.A (*a*) Original stormwater drainage scheme; (*b*) alternative water management scheme designed to protect the central operations area.

Fig. 9.B The effects of runoff from one rainstorm on the piping system proposed in Alternative A. This inlet was located near the point marked I in Figure 9.A*a*.

Risks of disruption to manufacturing operations and facilities were significantly reduced (1) by breaking down the stormwater system into smaller drainage units (rather than concentrating stormwater into a single large system, as is customary in conventional stormwater engineering), (2) by reducing the need to move stormwater over long distances, (3) by diverting stormwater away from manufacturing facilities at the heart of the site, and (4) by reducing the magnitude and frequency of damaging flows in the lower site and beyond.

The depression storage cells took the form of shallow basins (averaging 0.5 to 1.5 meters in depth). Their small scale and simple design allowed construction with uncomplicated engineering, using readily available excavators and untrained labor. The cells were planted with grasses and groves of trees using species, such as *Pinus resinosa*, adapted to sporadic water supplies. The cells were designed with groves of trees placed where they would help to screen and soften views of the industrial buildings and storage yards. In addition, worker recreation fields were designed to retain stormwater and serve as a no-cost water management feature with self-irrigating capacity.

This strategy of integrated stormwater management and landscape design helped create a multifunctional industrial landscape in which water, soil, vegetation, and land use are not only complementary but mutually supportive. It improved the performance and the success of each of the individual landscape systems while sharing costs, enhancing aesthetic quality, and reducing environmental and economic risks. Among other things, the plan demonstrated the appropriate application of an alternative approach to industrial site management, one that does not rely solely on conventional stormwater and site engineering. Finally, the concept of water management units was easily translated into a site management plan. Because of their simple design and operation, the water management units can be serviced and maintained by conventional grounds crews that find it easy to grasp the purpose behind the system and how it functions for the site as a whole.

Jack Goodnoe, a landscape architect, is president of Land Planning and Design, Ann Arbor, Michigan. He specializes in strategic approaches to institutional planning and landscape design.

9.8 SELECTED REFERENCES FOR FURTHER READING

Copeland, O. L. "Land Use and Ecological Factors in Relation to Sediment Yield." *Misc. Publication 970.* Washington, DC: U.S. Department of Agriculture, 1965, pp. 72–84.

Dietrich, W. E., et al. "Analysis of Erosion Thresholds, Channel Networks, and Landscape Morphology Using a Digital Terrain Model." *The Journal of Geology.* 101, 1993, pp. 259–278.

Dunne, Thomas, and Leopold, L. B. "Drainage Basins." In *Water in Environmental Planning.* San Francisco: Freeman, 1978, pp. 493–505.

Dunne, Thomas, Moore, T. R., and Taylor, C. H. "Recognition and Prediction of Runoff-Producing Zones in Humid Regions." *Hydrological Sciences Bulletin.* 20, 3, 1975, pp. 305–327.

Gregory, K. J., and Welling, D. E. *Drainage Basin Form and Process.* New York: Halsted Press, 1973.

Horton, R. E. "Erosional Development of Streams and Their Drainage Basins: Hydrophysical Approach to Quantitative Morphology." *Geological Society of America Bulletin.* 56, 1945, pp. 275–370.

Leopold, L. B. "Hydrology for Urban Land Planning—A Guidebook on the Effects of Urban Land Use." *U.S. Geological Survey Circular 554.* Washington, DC: U.S. Department of the Interior, 1968.

Leopold, L. B., and Miller, J. P. "Ephemeral Streams: Hydraulic Factors and Their Relation to the Drainage Net." *U.S. Geological Survey Professional Paper 282-A.* Washington, DC: U.S. Department of the Interior, 1956.

Meganck, Richard. Multiple Use Management Plan, San Lorenzo Canyon (summary). Organization of American States and Universidad Autonoma Agraria, 1981.

Riggins, R. E., et al. (eds.). *Watershed Planning and Analysis in Action.* New York: American Society of Civil Engineers, 1990.

Strahler, A. N. "Quantitative Geomorphology of Drainage Basins and Channel Networks." In *Handbook of Applied Hydrology* (Van te Chow, ed.). New York: McGraw–Hill, 1964.

Related Websites

Lake Superior Conservancy and Watershed Council. 2009. http://www.lscwc.org/
   An organization committed to protecting the Lake Superior watershed. Issues of that area include dams and mines, among others. The site gives resource, protection, and rehabilitation information, suggesting a so-called medical model for the latter.

Ontario Ministry of Natural Resources. Watershed Management and Planning. 2009. http://www.mnr.gov.on.ca/en/Business/Water/2ColumnSubPage/STEL02_163404.html
   Government page for watershed management and planning in Ontario. Links lead to watershed planning information, implications, and goals. Further reading on elements of watershed planning is available.

Saskatchewan Watershed Authority. http://www.swa.ca/
   Water resource management to meet economic, environmental, and social needs for Saskatchewan. The stewardship division links several case studies of local watershed management plans.

State of Washington Department of Ecology. Watershed Planning. 1994-2009. http://www.ecy.wa.gov/watershed/index.html
   The process of watershed planning and a publication helping users manage watersheds under the Watershed Management Act. The site also looks at the role of Native Americans in the process.

STREAMFLOW, FLOODPLAINS, FLOOD HAZARD, AND LAND USE PLANNING

10.1 INTRODUCTION

Early advances

Prior to the nineteenth century, it was widely believed that the water carried by streams could not possibly be produced by precipitation. After all, if streams were the result of rainfall, then how could they continue flowing when no rain was falling? Moreover, it was argued, there was not nearly enough rainfall to account for the flows seen in streams, especially large ones like the Seine and the Rhine. In the 1700s, a Frenchman named Pierre Perrault proposed to test this supposition by comparing rainfall and stream discharge in the Seine Basin of France. He found that the annual volume of precipitation exceeded streamflow by sixfold.

Runoff sources

But how rainwater got to the river and what controlled its rate of delivery remained largely unanswered until the twentieth century. We now understand that streams are supplied water from various sources—some large, some small; some slow, some fast; some variable, some steady. But all the sources begin as precipitation and many factors—including numerous human factors as we saw in previous chapters—influence the disposition of this water and the path it takes on its way to stream channels. In some instances, more water is delivered to a channel than it can carry, and the result is a flood.

Flooding

Floods have been part of the human experience for thousands of years. Our understanding of the nature and causes of flooding has increased dramatically over the past century or so, yet property damage and the loss of life continue to increase. Much of this trend is related to global population growth and to the shift of people toward river valleys, coastal areas, and cities throughout the developing world. But flood damage is also rising in the developed world.

In the United States, flooding ranks near the top of the national environmental agenda. About 7 percent of the country lies within a 100-year floodplain, and these areas have been favorite places for rural and urban development throughout the twentieth century. By 1980, between 3.5 and 5.5 million acres of floodplain had been developed for urban land uses, including more than 6000 communities with populations of 2500 or more. Since then floodplain development has been growing by about 2 percent a year. Today the total list of vulnerable land uses is massive, and disasters like the Katrina-related flood bear this out in spades.

Time for Change. The traditional response to the flood problem has been to engineer the threat away. In the United States, the federal government's major involvement in flood projects began with the Flood Control Acts of 1928 and 1936. Congress assigned the U.S. Army Corps of Engineers the responsibility of constructing reservoirs, levees, channels, and diversions to prevent flood damage along the Mississippi and its major tributaries. Since the late 1930s, the Corps has built 76 reservoirs and built or improved 2200 miles of levees in the upper Mississippi Basin alone.

Traditional response

Added to this are an estimated 5800 miles of nonfederal levees in the upper basin, built mainly by state and local governments. In addition, more than 3000 reservoirs on the smaller tributaries of the upper Mississippi Basin have been constructed under the auspices of another federal agency, the U.S. Natural Resources Conservation Service. But despite all these efforts, flood damage in the Mississippi and other watersheds has climbed since 1900. We now realize that the problem requires strategies beyond those offered by engineering alone, a point driven home by the shocking flood disasters of the past few decades.

10.2 SOURCES OF STREAMFLOW

Discharge

A logical place to begin is with streamflow and its sources. The quantity of water carried by a stream is termed **discharge**. It is measured in volumetric units of water passing a point on a stream, such as a bridge, over time. The conventional units in the United States are *cubic feet per second* (cfs). In Canada and most other countries, *cubic meters per second* are used. One cubic meter is equal to 35.3 cubic feet.

Surface and Subsurface Sources. Stream discharge is derived from four sources: channel precipitation, overland flow, interflow, and groundwater (Fig. 10.1). *Channel precipitation* is the moisture falling directly on the water surface, and in most streams, it adds very little to discharge. *Groundwater*, on the other hand, is a major source of discharge, and in large streams, it accounts for the bulk of the average daily flow.

Baseflow

Groundwater. Groundwater enters the streambed where the channel intersects the water table, providing a steady supply of water, termed **baseflow**, during both dry and rainy periods. Owing to the large supply of groundwater available to streams and the slowness of the response of groundwater to precipitation events, baseflow changes only gradually over time, and it is rarely the main cause of flooding. However, it does contribute to flooding by providing a stage onto which runoff from other sources is superimposed.

Interflow. *Interflow* is water that infiltrates the soil and then moves laterally to the stream channel in the zone above the water table. Much of this water is transmitted within the soil itself, some of it moving within the soil horizons. Until recently, the role of interflow in streamflow was not understood, but field research has begun to show that next to baseflow, it is the most important source of discharge for streams in forested lands. This research also reveals that overland flow in heavily forested areas makes negligible contributions to streamflow (see Fig. 8.6).

Overland Flow. In dry regions, cultivated lands, and urbanized areas, however, overland flow is usually a major source of streamflow, particularly in small streams.

Surface runoff

Overland flow is stormwater (or snowmelt) runoff that begins as a thin layer of water that moves very slowly (typically less than 0.25 feet per second) over the ground. Under intensive rainfall and in the absence of barriers such as rough ground, vegetation, and absorbing soil, however, it can mount up, rapidly reaching stream channels in minutes and causing sudden rises in discharge. The quickest response times between rainfall and streamflow occur in urbanized areas where yard drains, street gutters, and stormsewers collect overland flow and route it to streams straightaway. Runoff velocities in stormsewer pipes can reach 10 to 15 feet per second.

Relative Runoff Contributions. It is important to understand that the relative balance of contributions from overland flow, interflow, and groundwater changes dramatically with (1) the nature of rainfall and (2) the conditions of the drainage basin.

Landscape change

As explained in Chapter 8, land clearing and development tend to reverse the natural

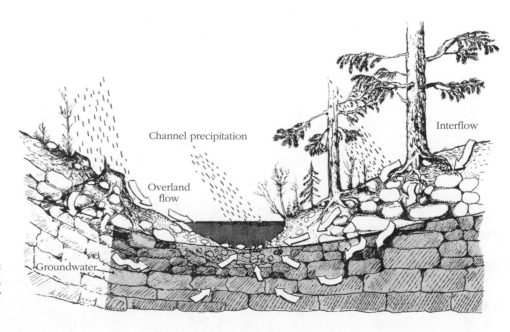

Channel precipitation

Interflow

Overland flow

Groundwater

Fig. 10.1 Sources of streamflow: channel precipitation, overland flow, interflow, and groundwater, in order of their arrival at the channel.

balance between overland flow and infiltration. In addition, the rate of stormwater delivery changes with land use development as the area served by slow-moving overland flow is reduced and replaced by a much faster delivery system of gullies, gutters, ditches, and stormsewers. Converted into one or more of these engineered forms of channel flow, overland flow increases in volume and quickens in travel time with land use development, resulting in much greater magnitudes and frequencies of peak discharges in receiving streams.

Rainfall delivery

The amount and rate of water delivered to streams also vary with rainfall intensity and duration. Moderate rainfalls generally produce little or no overland flow but, if extended over many days, they may produce enough percolation through the soil to cause an increase in baseflow. This concept is illustrated by the graph in Figure 10.2*a*. Thus discharge may rise without appreciable contributions from surface runoff. In striking contrast to this situation is the streamflow from a short, heavy rainstorm in which rainfall intensity exceeds the soil's infiltration capacity. The result is immediate overland flow and a sudden rise in stream discharge. If coupled with a so-called fast watershed, that is, one that has been developed and laced with engineered drains, the stormwater load amassed in the stream may be extreme indeed (Fig. 10.2*b*).

Massive Flooding. The 1993 Mississippi River flood, the largest ever recorded on the river, was a response to heavy, long-duration spring and summer rainfalls.

A great flood

Early rains saturated the soil over more than 300,000 square miles of the upper watershed, greatly reducing infiltration and leaving soils with little or no storage capacity. As rains continued, surface depressions, wetlands, ponds, ditches, and farmfields filled with overland flow and rainwater.

Contributing factors

With no remaining capacity to hold water, additional rainfall was forced from the land into tributary channels and thence to the Mississippi. For more than a month, the total load of water from hundreds of tributaries exceeded the Mississippi's channel capacity, causing it to spill over its banks onto adjacent floodplains (Fig. 10.3). Where the floodwaters were artificially constricted by an engineered channel bordered by constructed levees and unable to spill onto the large sections of floodplain, the flood levels were forced even higher.

Lessons

This flood event taught us a number of lessons. Among them is the realization that, no matter how big an event we prepare for in a watershed, there is always a bigger one waiting in the wings. It is only a matter of probability of how things, such as rainfall events, line up to produce a great flow. But there is also another lesson related

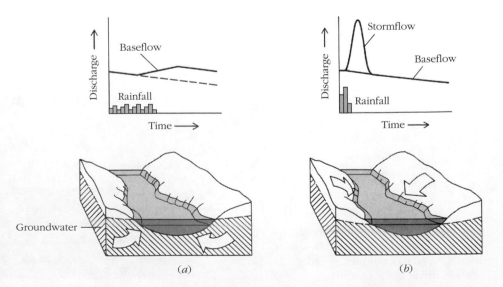

Fig. 10.2 (*a*) Stream response to a long steady rainfall that produces groundwater recharge and a rise in baseflow. (*b*) Stream response to an intensive rainstorm that produces only overland flow.

Preflood channel

Fig. 10.3 The 1993 Mississippi flood, the largest in the recorded history of the river, was a response to prolonged spring and summer rainfall.

to our own actions in the watershed. Undoubtedly, the vast areas of cleared, ditched, and piped farmland, towns, and cities contributed to greater and faster runoff, and in the river valleys our efforts to reduce flood risk by building levees and related structures not only interfered with the river's ability to redistribute excess waters, but helped give people a false sense of security about safe places to live and work.

10.3 METHODS OF FORECASTING STREAMFLOW

Seemingly endless debate, even professional and political wrangling, goes into making forecasts of streamflows, and we can only outline the basic concepts in this book. For most streams, especially those with small watersheds, no record of discharge is available. In that case, as described in Chapter 9, we must make discharge estimates using the rational method or some modified version of it. However, if chronological records of discharge are available for a stream, a short-term forecast of discharge can be made for a given rainstorm using a hydrograph.

Rainfall vs. runoff

Unit Hydrograph Method. This method involves building a graph in which the discharge generated by a rainstorm of a given size is plotted over time, usually hours or days (Fig. 10.4a) It is called the **unit hydrograph method** because it addresses only the runoff produced by a particular rainstorm in a specified period of time—the time taken for a river to rise, peak, and fall in response to the storm. Once

Flow period	Total flow	Storm-flow
0–12 (hr)	63 (m³/s)	63 (m³/s)
12–24	192	127
24–36	1065	991
36–48	1101	1019
48–60	714	623
60–72	453	354
72–84	275	170
84–96	194	85
96–108	144	28

Peak = 1253; 1175 at hour 36

Flow period	Percentage of total flow
0–12 (hr)	0.2 (%)
12–24	3.6
24–36	29.0
36–48	29.8
48–60	18.3
60–72	10.3
72–84	5.0
84–96	2.4
96–100	0.8

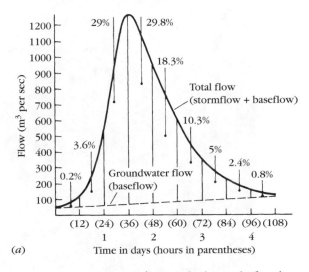

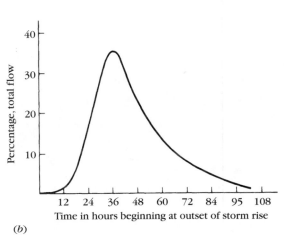

Fig. 10.4 Unit hydrograph for the uppermost part of the Youghiogheny River near the Pennsylvania–West Virginia border. Streamflow (*a*) is expressed as a percentage of total flow (*b*).

the rainfall–runoff relationship is established, then subsequent rainfall data can be used to forecast streamflow for selected storms, called standard storms. A *standard rainstorm* is a high-intensity storm of some known magnitude and frequency. One method of unit hydrograph analysis involves expressing the hour-by-hour or day-by-day increase in streamflow as a percentage of total runoff (Fig. 10.4*b*). Plotted on a graph, these data form the unit hydrograph for that storm, which represents the runoff added to the prestorm baseflow.

To forecast the flows in a large drainage basin using the unit hydrograph method would be difficult because in a large basin, such as the Ohio, Hudson, Fraser, or Columbia, geographic conditions may vary significantly from one part of the basin to another. This is especially so with the distribution of rainfall because an individual rainstorm rarely covers the basin evenly. As a result, the basin does not respond as a unit to a given storm, making it difficult to construct a reliable hydrograph.

Flow records

Magnitude and Frequency Method. In such cases, we turn to the **magnitude and frequency method** and calculate the probability of the recurrence of large flows based on records of past years' flows. In the United States, these records are maintained by the Hydrological Division of the U.S. Geological Survey for most rivers and large streams. For a basin with an area of 5000 square miles or more, the river system is typically gauged at five to ten places. The data from each gauging (gaging) station apply to the part of the basin upstream of that location.

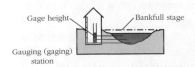

Because we are concerned mainly with the largest flows, only *peak annual flows* are recorded for most rivers. With the aid of some statistical techniques, we can use these flow data to determine the probability of recurrence of a given flow on a river. Key data available for each gauging station include:

Data
- *Peak annual discharge:* The single largest yearly flow in cubic feet per second.
- *Date of peak annual discharge:* The month and year in which this flow occurred.
- *Gage height:* The height of water above channel bottom.
- *Bankfull stage:* Gage height when the channel is filled to bank level.

The following steps outline a procedure for calculating the recurrence intervals and probabilities of various peak annual discharges:

Steps
- First, *rank* the flows in order from highest to lowest; that is, list them in order from the biggest to the smallest values. For two of the same size, list the oldest first.
- Next, determine the *recurrence interval* of each flow:

$$t_r = \frac{n + 1}{m}$$

where

t_r = the recurrence interval in years
n = the total number of flows
m = the rank of flow in question

- Third, determine the probability (p) of any flow:

$$p = \frac{1}{t_r}$$

- This calculation yields a decimal, for instance 0.5, that can be converted to a percentage (50 percent) by multiplying by 100.
- In addition, it is also useful to know *when* peak flows can be expected; accordingly, the monthly frequency of peak flows can be determined based on the dates of the peak annual discharges. Furthermore, the percentage of peak annual flows that actually produce floods can be determined by comparing the stage (elevation) represented by each discharge with the bankfull elevation of the gauging station; that is, any flow that exceeds bankfull elevation is a flood.

Making Projections. Given several decades of peak annual discharges for a river, limited *projections* can be made to estimate the size of some large flow that has not been experienced during the period of record. The technique most commonly *Projecting a curve* used involves projecting the curve (graph line) formed when peak annual discharges are plotted against their respective recurrence intervals. In most cases, however, the curve bends rather strongly, making it difficult to plot a projection accurately using all the points, as Figure 10.5 illustrates.

This problem can be overcome by plotting the discharge and/or the recurrence interval data on logarithmic graph paper or logarithmic/probability paper. Once the plot is straightened, a line can be ruled drawn through the points. A projection can then be made merely by extending the line beyond the points and then reading the appropriate discharge for the recurrence interval in question.

In Figure 10.6 the curve was straightened by using a logarithmic scale on the recurrence interval (horizontal) axis. The period of record covers 30 years, but we would

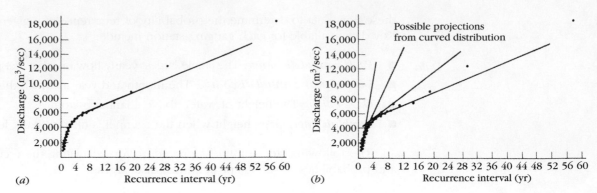

Fig. 10.5 (*a*) Graph based on recurrence interval and discharge of peak annual flows for the Eel River, California, during the period of 1911–1969. (*b*) This graph shows why it is impossible to plot one straight graph line from a curved distribution.

like to know the magnitude of the 100-year flow. Assuming that the 100-year (or larger) flow is not represented in the 30 years of record, we can project the curve beyond the 30-year event and read the magnitude of the 100-year event, which is at 97,000 cfs, as shown in Figure 10.6. This number represents our best approximation of the 100-year flow.

Forecasting Limitations. Making forecasts of riverflows based on discharge records is extremely helpful in planning land use and in engineering bridges, buildings, and highways in and around river valleys. On the other hand, this sort of forecasting technique has several distinct limitations.

Notes of caution

First, the period of record is very short compared to the total time the river has been flowing; therefore, forecasts are based only on a glimpse of the river's behavior. For a stream with records covering, say, the period 1900–2000, there is no guarantee that

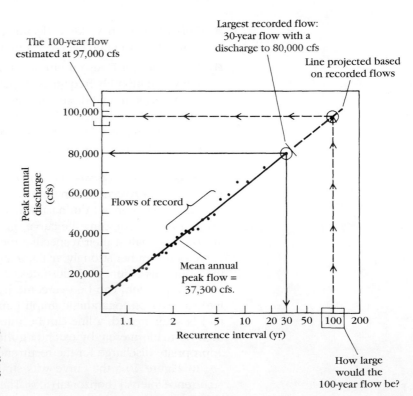

Fig. 10.6 Straightened curve of the distribution of discharges for the Skykomish River at Gold Bar, Washington. An estimate of the magnitude of flows at intervals beyond 30 years is now possible, as shown by the broken line.

the true 100-year event actually occurred in the period of record. Second, throughout much of North America, watersheds have undergone such extensive changes because of urban and agricultural development, forestry, and mining that discharge records for many streams today have a different meaning than those of several decades ago.

Climate change Third, climatic change in some areas, such as near major metropolitan regions, may be great enough over 50 or 100 years to produce measurable changes in runoff. In addition, regional climate change as a part of global warming has now become a concern. As a result, some regions will experience larger and more frequent storms than only one or two decades ago. Together, these factors suggest that forecasts based on projections of past flows should be taken only as approximations of future flows, and for greatest reliability, they should be limited to events not too far beyond the years of record, say, 50-year or 100-year flows. Indeed, climate change may render even relatively short-term forecasts unreliable.

10.4 THE SIZE AND SHAPE OF VALLEYS AND FLOODPLAINS

Flood analysis and forecasting also require an understanding of the geomorphology of stream valleys because valley size, shape, and topography influence the distribution, movement, and force of floodwaters. As a rule, stream valleys are carved by the streams that occupy them. It follows that large streams flow in large valleys and small ones in small valleys.

Streams and valleys

Misfitness. There are exceptions, however, called **misfit streams.** These streams are either too large or too small for their valleys. In the midlatitudes, for example, are some streams that carried huge discharges and carved broad valleys during the last glaciation but today carry much smaller flows. But most of the streams we encounter, especially those in the rural and predeveloped landscape, more or less fit the scale of their valleys.

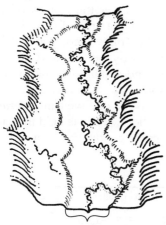

Original channel

In the developed landscape, particularly in and around urban areas, however, many streams have become misfit. Stormwater loading in urban streams (and some agricultural ones as well) has so changed the magnitude and frequency of peak flows that these streams often behave as much larger streams. Not only is flooding more frequent, but because it occurs in relatively small valleys, its stage is higher. That is, floodwaters are much deeper over the valley floor. When streamflow exceeds the channel capacity and spills onto the valley floor, the valley more or less takes on the function of a huge channel.

The elevation over the valley floor reached by the floodwaters for a given discharge depends mainly on the valley's size and shape. If the valley is wide, the water spreads over a large area in a thin layer. However, if it is narrow, it builds up into a deep layer. As we noted earlier, the levees along the Mississippi reduced the size of the river's effective valley, restricting and elevating floodwaters in the 1993 flood. At St. Louis the width of the Mississippi's channel in 1849 was 4300 feet; today it is only 1900 feet.

Valley Formation and Features. Valley formation involves two essential geomorphic processes: (1) downcutting or deepening and (2) lateral cutting or widening.

V-shapes Downcutting generally leads to *V-shaped valleys* with little or no flat ground or floodplain between the stream channel and the valley walls. V-shaped valleys are characteristic of many headwater streams and streams in mountainous terrain. In extreme cases, the valleys are so sharp that they merge with the channel without a topographic break (Fig. 10.7*a*).

Most stream valleys, however, even those in deep gorges, have an area of flat ground, or **floodplain** (or **alluvial plain**), on the valley floor (Fig. 10.7*b*). This land-

Floodplains form contains the stream channel, and within it the stream shifts its course, erodes and deposits sediments, shapes soils and habitats, and overflows with floodwaters. These processes and related features, which are important to understanding the local dynamics of flooding, are part of the topographically diverse landscape of floodplains, and they are discussed in detail in Chapter 14.

U-shapes

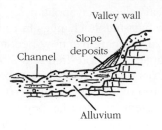

Besides V-shaped valleys and gorges with small floodplains, there are five other basic valley forms. If we widen a V-shaped valley, the next form is a *U-shape* (Fig. 10.7*c*). This valley has a central corridor of flat ground bordered by gentle slopes leading up to steeper valley walls. The gentle slopes are comprised of various deposits, such as alluvial fans, built by small tributary streams and debris from slides and mudflows. Where the valley walls are lower and/or more resistant, or where the stream has been in contact with them, these gentle side slopes may be absent and the valley has more of a *rectangular shape*, with straighter and often steeper valley walls (Fig. 10.7*d*).

Terraced valleys contain benchlike areas situated above the valley floor (Fig. 10.7*e*). They are usually remnants of former floodplains when the channel and valley floor were both at a higher elevation in the valley. Terraces are typically found along the sides of valleys, next to the valley walls, but occasionally they occur as isolated parcels (or islands) in the modern floodplain. They are, of course, less prone to flooding, and if large enough, are attractive ground for settlement.

Terraced

Broad, Flat Valleys. *Broad, flat floodplains* without distinct valley walls are characteristic of many streams crossing lowlands such as coastal plains (Fig. 10.7*f*). These valleys are best described as hydrologic rather than geomorphic because their geographic limits are defined more or less by the areal extent of various flood events rather than by landforms. In fact their internal topography, represented by levees, terraces, backswamps, and other features (see Fig. 3.3), is more important to understanding flood management than topography on the valley margins. The lower third or so of the Mississippi (the area called the Embayment) is the strongest example of this kind of valley in North America. Many other streams, including the Trinity, Red, Tombigbee, and Savannah, also exhibit hydrologic valleys where they cross the Gulf and Atlantic Coastal Plain; to the north in North Dakota, Minnesota, and southern Manitoba the Red River has a hydrologic valley of enormous dimensions.

Broad, flat . . .

Decidedly the most dangerous valley/floodplain type is the one we call *inverted*. In these valleys, the streambed lies at or above the valley floor, and it is held in place by natural or constructed levees (Fig. 10.7*g*). The inverted condition is caused by

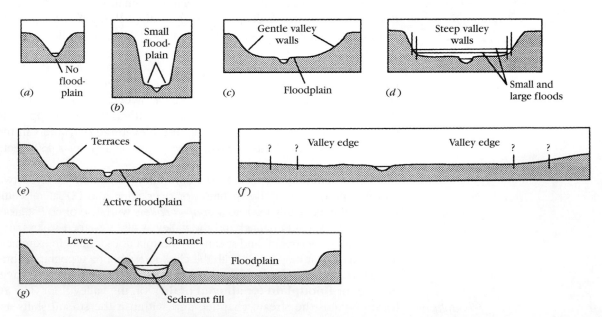

Fig. 10.7 Common shapes of stream valleys and floodplains and their implications for the depth and extent of flooding.

Inverted

sediment buildup in the channel, which can be caused by various conditions including excessive soil erosion higher in the watershed. Not many streams fall into this class; the lower Mississippi in the Embayment section is in this category where tributary basins parallel the main channel. New Orleans lies behind high levees, which failed in the Katrina disaster. But perhaps the most extreme case is the Yellow River of China, whose elevated channel is dangerously high above the floodplain and poses a threat to millions of people living on the valley floor should the confining levees fail under the force of a major event.

10.5 APPLICATIONS TO LAND PLANNING AND DESIGN

Property damage from floods in the United States and Canada increased appreciably in the twentieth century (Fig. 10.8). The numbers are still rising, and the trend is similar beyond North America as populations push into river valleys and flooding increases with watershed alteration related to land uses changes.

Stage and topography

Analysis of a river's flood potential involves translating a particular discharge into a level (stage or elevation) of flow and then relating that flow level to the topography in the river valley (Fig. 10.9a). In this manner we can determine what land is inundated and what is not. This procedure requires accurate discharge and detailed topographic data, as well as knowledge of channel conditions. The latter is needed to compute flow velocity, which is the basis for determining the water elevation in different reaches of the channel, that is, just how high a given discharge will reach above the channel banks.

Best approximation

Approximating Flood Potential. For planning purposes, however, neither the time nor money is usually available for such detailed analysis. We must, instead, turn to existing topographic maps and existing discharge data and make a best estimate of the extent of various frequencies of flow. The accuracy of such estimates depends not only on the reliability of the data, but also on the shape of the river valley. Where the walls of a river valley rise sharply from the floodplain, the limits of flooding are often easy to define, as the profile in Figure 10.7d suggests. And as Figure 10.9b shows, there is not much difference in the extent of the 25- and 100-year floods because of the

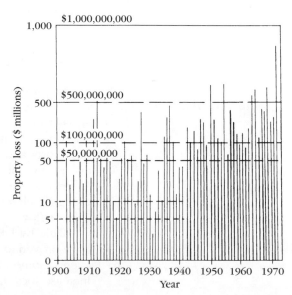

Fig. 10.8 Property loss (in millions of dollars) as a result of floods in the United States over 70 years. (Data from the U.S. National Weather Service.)

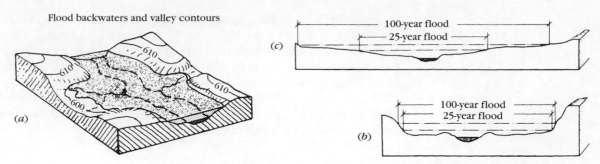

Flood backwaters and valley contours

Fig. 10.9   (*a*) Flood backwaters in relation to valley contours. (*b*) The influence of valley shape on the extent of the 25-year and 100-year floods.

stream valley's box-shape. However, where a river flows through broad lowlands, the problem can be very difficult, and the differences in the extents of the 25-, 50-, or 100-year event might be very hazy indeed (Fig. 10.9*c*).

Key zones

U.S. National Flood Insurance. The **U.S. National Flood Insurance Program** is based on a definition of the 100-year floodplain. In most areas the boundaries of this zone are delineated by relating discharge data and flow elevations to the topography of the stream valley. According to this program, two zones are defined: (1) the regulatory *floodway*, the lowest part of the floodplain where the deepest and most frequent floodflows are conducted, and (2) the *floodway fringe*, on the margin of the regulatory floodway, an area that would be lightly inundated by the 100-year flood. Buildings located in the regulatory floodway are not eligible for flood insurance, whereas those in the floodway fringe are eligible provided that a certain amount of floodproofing is done (Fig. 10.10).

Flood insurance

The flood insurance program is intended to serve as one of the controls to limit development in floodplains. It can, however, be argued that it has actually had the opposite effect and encouraged development in the floodway fringe by offering government insurance in exchange for adding certain floodproofing measures to buildings, such as elevated foundations and waterproofing. Experience has shown that these floodproofing measures are ineffective in curbing damage from truly large floods, which are, of course, inevitable in the areas labeled as floodway fringe. Moreover, the geographic definitions of the 100-year flood coverage are often suspect because many stream valleys were mapped so long ago (in the 1970s and 1980s or before) that the magnitude of the 100-year event has changed with land development, engineering projects, and climate change.

Concerns

Zoning, information, and...

Other Approaches. Other controls designed to prevent floodplain development include zoning restrictions against vulnerable land uses and educational programs to inform prospective settlers of the hazards posed by river valleys. In the case of river valleys where development is already substantial, the only means of reducing damage is to retrofit vulnerable facilities into more site-adaptive configurations, relocate flood-prone land uses, or reduce the size of hazardous river flows. Because of the high costs and the sociological problems associated with relocation, decision makers have traditionally turned to the flow-reduction and flood-control options. These alternatives call for structural (engineering) interventions such as building reservoirs, dredging channels, diverting flows, and/or constructing embankments (levees) to confine flow.

A paradigm shift?

Little by little the ineffectiveness of structural measures is coming to light. They have proven to be expensive and deleterious to the environment, and, for large flood events, some measures have been shown not only to be ineffective but actually to increase local flood magnitudes and damage. During the 1993 Mississippi flood, for

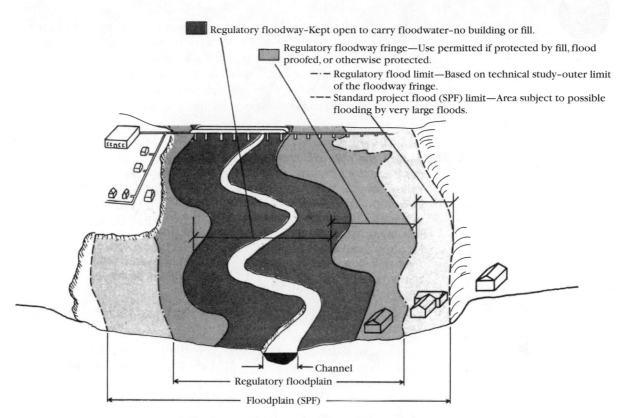

Fig. 10.10 Illustrations of the regulator floodway and the floodway fringe, according to the U.S. National Flood Insurance Program.

example, engineers concluded that the levees they had constructed along the channel to keep floodwater out of the floodplain had in fact forced more water downstream, thereby creating higher flood levels there. Along with escalating construction costs, tighter environmental regulation, and the growing specter of public liability, public agencies are now giving more consideration to various sorts of nonstructural pro grams in floodplain management.

Risk management   Increasingly, **risk management** programs offer a promising alternative or a sup-plement to traditional approaches. This approach recognizes the inevitability of floods and the risks they pose in floodplains and coastal areas. Risk management planning endeavors to build response plans to minimize disruption, property damage, and loss of life. Such plans include public information programs on disaster preparation, early warning systems, evacuation plans, and plans and programs for disaster recovery.

Site-Adaptive Planning. Unfortunately we have also become victims of rigid and prescriptive development practices in stream valleys. Sensitivity to local variations in valley landscapes and the ways they function have for decades been replaced by blanket zoning practices, simple subdivision layouts, and heavy-handed manipulation of development sites. *Site-adaptive planning* is based on the observation that stream

Reading the land   valleys are inherently diverse physiographically and thus embody a rich assortment of opportunities and constraints for dealing with drainage problems. If facilities are sited and designed according to the character of the local landscape, using not only a varied design palette but an appropriate set of performance standards, many local flooding problems can be avoided.

Of course, this approach necessitates actually going into the field with maps in hand and finding and translating landscape features such as terraces, scour chan-nels, and backswamps in terms of land use potential. It also necessitates thinking in

Site-adaptive design

terms of *site-adaptive design*, that is, giving consideration to alternative site layouts, unconventional building forms such as houses on piers, and landscaping schemes that use water-resistant plant types. It also necessitates building realistic management programs that spell out how sites can be expected to behave in valley and floodplain settings in response to changing seasonal conditions and runoff events, as well as how to respond to flood events of various magnitudes and durations.

10.6 SMALL BASIN FLOODPLAIN MAPPING

Local concerns

Small streams are far more abundant than large ones, and for most small stream valleys, little or no information is available on their drainage. Most streams in Canada and the United States (and elsewhere in the world) do not qualify for an official gauging station or floodplain mapping, yet they merit serious consideration in local land use planning. Communities bordering on such streams are often pressed for information not only on flood hazards, but also on many other features of the stream valley that are important to planning decisions.

Typical concerns include soil suitability for foundations and basements, drainfield siting and design, road layouts and infrastructure design, habitat protection, and wetland conservation. Much of this information can be generated from published sources, in particular, topographic contour maps, aerial photographs, and soil maps, and when coupled with a little field work, a reasonably intelligent picture of the valley environment can be sketched.

Why the floodplain?

Of all the features of the stream valley, the floodplain is the most important from a planning standpoint. First, excluding the stream channel itself, the floodplain is generally the lowest part of the stream valley and thus the most prone to flooding. Second, floodplain soils are often poorly drained because of the nearness of the water table to the surface and saturation by floodwaters. Third, floodplains are formed by incremental erosion and deposition associated with the lateral migration of streams in their valleys. Therefore, the borders of the floodplain can be taken as a good indicator of the extent of alluvial soil. These features are the key indicators for floodplain mapping (Fig. 10.11).

Floodplain Delineation. Using secondary sources, floodplains can be delineated in four ways, according to (1) topography, (2) vegetation, (3) soils, and (4) the extent of past floodflows.

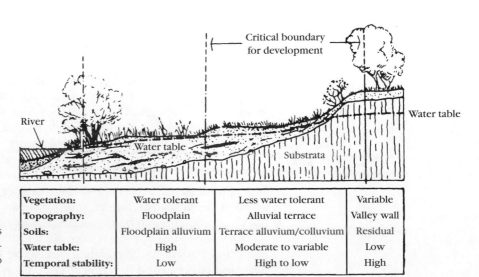

Fig. 10.11 An illustration of the changes in topography, soils, drainage, and vegetation in the transition from the floodplain to the valley walls.

Vegetation:	Water tolerant	Less water tolerant	Variable
Topography:	Floodplain	Alluvial terrace	Valley wall
Soils:	Floodplain alluvium	Terrace alluvium/colluvium	Residual
Water table:	High	Moderate to variable	Low
Temporal stability:	Low	High to low	High

■ *Topography:* Topographic contour maps can be used to delineate floodplains if the valley walls are high enough to be marked by two or more contour lines. However, valley walls with only a single contour or those that fall within the contour interval cannot be safely delineated, even though they may be topographically distinct to the field observer. In the latter case, aerial photographs can be helpful in mapping this feature. Viewed with a standard stereoscope, topographic features appear exaggerated, thereby making identification of the valley wall and other features a relatively easy task in many instances.

Contour lines

■ *Vegetation and land use:* Aerial photographs also enable us to examine vegetation and land use patterns for evidence of floodplains and wet areas. In the North American Corn Belt, for example, stream valleys stand out as forest corridors that escaped land clearing because they are poorly suited for crop farming. The same pattern is often found in the prairies where trees are limited to valley floors by dry conditions (and farming) on adjacent upland surfaces. In areas where forest covers both lowlands and uplands, floodplains are often distinguishable by the tree species growing there. Indeed, some floodplain corridors are neatly delineated by distinct tree associations, and their canopies can usually be identified on aerial photographs. (Floodplain habitat and vegetation are discussed in Section 14.7.)

Aerial photographs

■ *Soil:* Because river floodplains are built from river deposits, the soils of floodplains often possess characteristics distinctively different from those of the neighboring uplands. In the United States, these soils are generally described in the U.S. Natural Resources Conservation Service (NRCS) reports as having high water tables, seasonal flooding, and varied composition. Therefore, soil maps prepared by the NRCS (or similar agencies in Canada) can also be used in the definition of floodplains, and an excerpt from one such map is shown in Figure 10.12. *A note of caution:* Part—and often a large part—of the decision for drawing the boundaries of different soil types is based on topography, vegetation, and land use patterns. Therefore, we must be aware that a good correlation between floodplains based on NRCS soil boundaries and floodplains based on observable vegetation, topography, and land use may be just circular reasoning.

Soil maps

■ *Past floodflows:* Finally, floodplains may be defined according to the extent of *past floodflows.* Evidence of past flows may be gained from firsthand observers who are able to pinpoint the position of the water surface in the landscape and from features such as organic debris on fences and deposits on roads that can be tied to the flood. In addition, geobotanical indicators such as scars on tree trunks from ice damage or flood-toppled and partially buried trees are helpful evidence (see Fig. 14.14). Understandably, this method usually demands extensive fieldwork, including interviews with local residents and tramping through brushy bottomlands.

Telltale

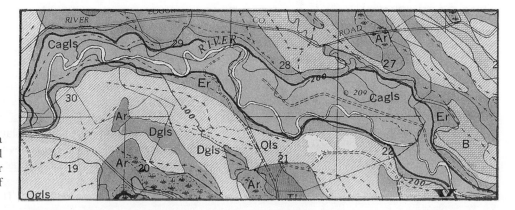

Fig. 10.12 An excerpt from a soil map showing a belt of alluvial soil (coded Cagls) marking a river floodplain. The belt is about a half mile wide.

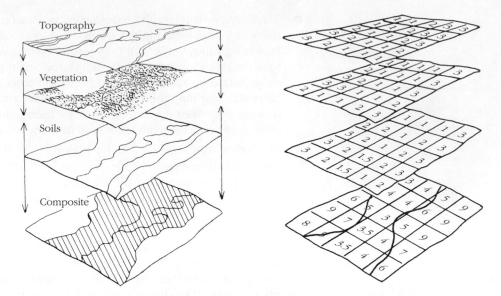

Fig. 10.13   A schematic portrayal of the map overlay technique used in floodplain mapping. *Left*, maps delineating the pertinent features; *right*, these features codified into a numerical scheme based on land use suitability.

Map overlays

Data Synthesis. The map overlay technique is most commonly employed in synthesizing all these data. Although it has been long used by geographers and others, the application of this technique has been advanced and greatly improved with the use of Geographic Information Systems (GIS), and today is employed in all sorts of landscape planning, design, and engineering problems involving complex spatial arrangements among multiple sets of data. One overlay scheme involves assigning numerical values to the various classes of each component of the landscape (vegetation, drainage, soils, and the like) according to their land use suitability. The values are then summed and the results used to delineate areas or zones of different development potentials (Fig. 10.13). The values themselves are, of course, arbitrary; in other words they carry no intrinsic value and should be interpreted accordingly. Although GIS techniques are enormously helpful in floodplain mapping, we face the same problems as in manual techniques, including data acquisition, data reliability, and the meaning of weighted overlays.

10.7 CASE STUDY

Flood Risk and the Impacts of Fire in a Small Forested Watershed, Northern Arizona

Charlie Schlinger, Cory Helton, and Jim Janecek

Communities downstream of large earthen dams are potentially at risk of disastrous flooding. Indeed, catastrophic floods associated with earthen dam failures—such as the 1889 "rainy day" Jonestown, Pennsylvania, flood or the 1976 "sunny day" Teton Dam failure—collectively have moved legislators and regulators to focus on worst-case scenarios. Although such scenarios have an exceedingly low probability of occurring, no one involved is willing to suffer the consequences. In many locales, dam owners must deal with so-called hazard creep, which occurs when development takes place downstream of existing dams. This is an issue for many dams because they were not designed or constructed with the expectation of a significant future downstream population.

In addition to flood hazards posed by dams, many areas experience infrequent high-magnitude storm events with the potential to cause severe flooding. "Flashy" ephemeral watercourses in arid and semiarid regions generate stormflows that rapidly peak following brief, intense storms. This effect is counterintuitive to many unfamiliar with hydrologic processes. To make matters more interesting, the effects of fire on a watershed can magnify the flashiness of these systems and greatly increase peak flows, particularly for low- to moderate-intensity precipitation events. High-intensity fires not only destroy water-absorbing vegetation, but they also change the soil structure, creating a phenomenon called hydrophobicity, a highly impermeable surface condition that promotes increased runoff. The successful management of a watershed with an earthen dam, flashflood potential, and forest fire impacts requires a strategy that considers watershed and stream channel conditions as well as the performance and reliability of engineered structures.

The watershed feeding the Black Canyon Lake reservoir covers 5.6 square miles in remote semiarid ponderosa pine country above the Mogollon Rim in north-central Arizona. The dam is nearly 80 feet tall, was built in the early 1960s, and has a capacity of 1580 acre feet. Downstream lies the community of Heber, Arizona, with a population of nearly 2800, which partially occupies the floodplain (Fig. 10.A). Failure of the Black Canyon Dam could be catastrophic to Heber. Therefore, in the summer of 2002, when the Rodeo-Chediski fire burned much of the Black Canyon watershed, regulatory agency staff became concerned. Approximately 84 percent of the watershed was affected by the fire; nearly 25 percent was severely burned.

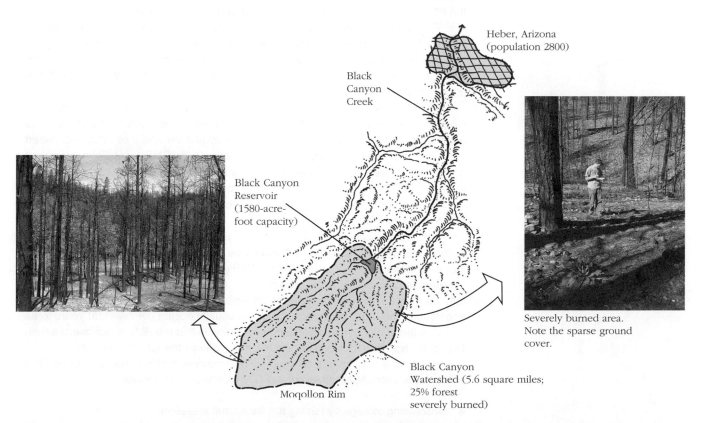

Heber, Arizona
(population 2800)

Black
Canyon
Creek

Black Canyon
Reservoir
(1580-acre-
foot capacity)

Mogollon Rim

Black Canyon
Watershed (5.6 square miles;
25% forest
severely burned)

Severely burned area.
Note the sparse ground
cover.

Fig. 10.A A schematic drawing of the Black Canyon Watershed, reservoir, and creek connecting to Heber, Arizona. Some of the effects of the fire are shown in the photographs.

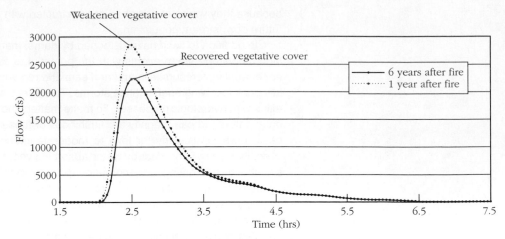

Fig. 10.B Hydrographs illustrating the change in discharge from the watershed largely in response to vegetation conditions one year after the fire and six years after the fire. Vegetation is credited with reducing the probable maximum flood event by more than 5000 cfs.

The Black Canyon Dam is regulated by the Arizona State Department of Water Resources. Given the downstream risk to Heber from a dam failure at Black Canyon, the department stipulated that the reservoir and spillway must accommodate the probable maximum flood (PMF). The PMF arises from a hypothetical worst-case rainfall event, known as the probable maximum precipitation (PMP). The concern is that overtopping of the earthen dam during a PMF event could lead to catastrophic failure of the dam and downstream propagation of a flood wave not only driven by the PMF but also augmented by the nearly 1600 acre feet of reservoir water.

The 6-hour PMP is estimated at around 10.4 inches, with nearly 5.5 inches falling in a 15-minute period midway through the storm. This is important because the spillway at Black Canyon was originally designed based on a 6-hour 100-year storm with *total* precipitation of only 4.5 inches, less than half of the estimated probable maximum precipitation (PMP)!

Using a unit hydrograph method developed by the U.S. Natural Resources Conservation Service, and a multibasin model with numerous subcatchments to account for varying degrees of the burn and spatial variability of other watershed parameters, discharge projections were calculated for the PMP event. For a watershed condition representing recovery one year after the fire, peak discharge was estimated at 28,000 cfs. For a watershed condition representing recovery six years after the fire, the estimated discharge had dropped 5500 cfs to 22,500 cfs (see the lower curve in Fig. 10.B).

For events such as the 2- or 25-year storm, the hydrological model should be detailed. That is, a watershed should be subdivided into multiple catchments. However, detail is less of an issue for a PMP event. The impact of fire on runoff from a watershed can be extreme, especially for small watersheds and low-recurrence-interval storms. Yet the recovery of understory vegetation (grasses and shrubby vegetation) and topsoil conditions from fire usually occurs over several years to as many as ten or more years. A final consideration is that the PMF is not overly sensitive to soil type or vegetation conditions; nonetheless the runoff is extreme.

Handling a flow generated by the PMP requires making changes in the dam and/or watershed. Alternatives for the Black Canyon Dam include:

- Increasing storage by raising the dam crest elevation.
- Increasing discharge capacity by widening the spillway channel or lowering the spillway crest elevation.

■ Hardening the earthen embankment so that an overtopping flow can pass over it without causing significant damage.

■ Or some combination of these three alternatives.

Selecting an alternative, however, has been delayed because of concerns over the reliability of the PMP value used in runoff calculations. This doubt has led to a multiagency effort to re-estimate PMP values across the entire state. In the meantime, the state has decided to hold off on raising the dam crest elevation while moving forward with improvements to the spillway.

Finally, although PMF study results often lead to a focus on structural remedies, such as those listed above, watershed and forest conditions deserve equal consideration. The management of forested watersheds to mitigate risks associated with catastrophic burns can alleviate the need for conservatism in PMF evaluations and thus reduce the required magnitude, and cost, of structural remedies. Presently, much of the U.S. Southwest, including Arizona, is in the grip of a decade-long drought, which has, in many places, degraded vegetative and hydrologic conditions in forests. Furthermore, forecasts suggest that long-term climate change will be accompanied by increased drought magnitude, duration, and frequency. Given these recent, ongoing, and anticipated changes in the southwestern climate, the adaptive environmental assessment and management (adaptive management) of forested watersheds is an essential strategy for mitigating the hazards associated with high-magnitude, low-frequency, short-duration floods.

Charlie Schlinger is associate professor of civil and environmental engineering, and Jim Janecek is a research hydrologist at Northern Arizona University, where they work on water management problems. Cory Helton is an engineer with Jon Fuller Hydrology & Geomorphology, Inc.

10.8 SELECTED REFERENCES FOR FURTHER READING

Bhowmik, N. G., et al. *The 1993 Flood on the Mississippi River in Illinois* (Miscellaneous Publication 151). Champaign: Illinois State Water Supply Survey, 1994.

Burby, Raymond J., and French, S. P. "Coping with Floods: The Land Use Management Paradox." *Journal of the American Planning Association*. 47, 3, 1981, pp. 289–300.

Dunne, Thomas, and Leopold, Luna B. "Calculation of Flood Hazard." In *Water in Environmental Planning*. San Francisco: Freeman, 1978, pp. 279–391.

Glassheim, Eliot, "Fear and Loathing in North Dakota." *Natural Hazards Observer*. 21, 6, 1997, pp. 1–4.

Holling, C.S. (ed). *Adaptive Resource Assessment and Management*. Chichester: Wiley, 1978.

Interagency Floodplain Management Review Committee. *Sharing the Challenge: Floodplain Management into the 21st Century*. Washington, DC: U.S. Government Printing Office, 1994.

Kochel, R. Craig, and Baker, Victor R. "Paleoflood Hydrology." *Science*. 215, 4531, 1982, pp. 353–361.

Linsley, R. K. "Flood Estimates: How Good Are They?" *Water Resources Research*. 22, 9, 1986, pp. 1595–1645.

Natural Hazards Research and Applications Information Center. *Floodplain Management in the United States: An Assessment Report*. Boulder: University of Colorado, 1992.

Platt, R. H. "Metropolitan Flood Loss Reduction Through Regional Special Districts." *Journal of the American Planning Association*. 52, 4, 1986, pp. 467–479.

Rahn, Perry H. "Lessons Learned from the June 9, 1972, Flood in Rapid City, South Dakota." *Bulletin of the Association of Engineering Geologists.* 12, 2, 1975, pp. 83–97.

Schneider, William J., and Goddard, J. E. "Extent of Development of Urban Flood Plains." *U.S. Geological Survey Circular 601-J,* 1974.

Thompson, A., and Clayton, J. "The Role of Geomorphology in Flood Risk Assessment." *Proceedings of the Institution of Civil Engineers.* Paper 12771, 2002, pp. 25–29.

White, Gilbert F. *Flood Hazard in the United States: A Research Reassessment.* Boulder: University of Colorado, Institute of Behavior Science, 1975.

Wolman, M. Gordon, "Evaluating Alternative Techniques for Floodplain Mapping." *Water Resources Research.* 7, 1971, pp. 1383–1392.

Related Websites

Environmental Protection Agency. Stream Corridor Structure. 2008. http://www.epa.gov/watertrain/stream/index.html

Part of the Watershed Academy Web, a thorough description of stream channels with excellent diagrams and pictures. Also included is associated vegetation and vegetation forms.

Federal Interagency Stream Restoration Working Group. 2001. http://www.nrcs.usda.gov/technical/stream_restoration/newgra.html#Anchor1

An interagency group to document stream corridor restoration practices. Links lead to slide shows of natural and constructed steam corridors, case studies of rivers altered to reduce peak flood damage, and additional reading about stream restoration.

FEMA. National Flood Insurance Program. 2009. http://www.fema.gov/about/programs/nfip/index.shtm

The U.S. flood insurance program and floodplain management for communities.

National Weather Service. Northwest River Forecast Center. 2009. http://www.nwrfc.noaa.gov/

An up-to-date record of current and past floods in the Northwest region. See how different inputs impact the rivers and read more about the hydrologic cycle.

U.S. Global Change Research Program. "Impacts of Climate Change and Land Use in the Southwestern United States." 2004. http://geochange.er.usgs.gov/sw/index.html

Follow the links from this workshop to realize the link between flooding and climate change. This website looks specifically at the Southwestern United States but has general information useful to all regions.

11

WATER QUALITY AND RUNOFF CONSIDERATIONS IN LANDSCAPE MANAGEMENT

11.1 INTRODUCTION

National trends

The quality of water in lakes and streams has been a serious national issue in the United States and Canada for nearly a half century. Both countries have enacted complex bodies of law calling for nationwide pollution control programs. In the United States the past 30 years have produced several significant trends. The Water Pollution Control Act of 1972 (now amended and known as the Clean Water Act) provided federal funding for new and improved sewage treatment facilities, which has resulted in reduced fecal bacteria and nutrient loading and in decreased BOD (biological oxygen demand) in the nation's inland and coastal waters. Lead levels also declined with the change to unleaded gasoline for automobiles.

Despite these strides, the overall quality of the nation's surface waters has not met forecasts based on the programs implemented under the Clean Water Act. Sewage from many municipalities is still inadequately treated. More than 1000 communities, including a number of large cities, rely on combined sewer overflow systems that discharge raw sewage directly into streams, lakes, and the ocean via stormsewers when treatment plants become overloaded during wet weather. Even more alarming is the pollution from agricultural runoff, which has escaped most of the abatement regulations that apply to other land uses.

Land use relationships

All land uses contribute to stormwater pollution, and although many communities and industries have invoked measures to control stormwater, contaminant loading of runoff remains one of our most serious environmental problems. And the problem is growing with economic development and population growth. BOD loadings from both agriculture and urban stormwater sources have risen in most parts of North America and stormwater is now considered to be the major source of nitrogen, phosphorus, sediment, and chloride pollution. The nitrogen from agricultural fertilizers and feed lots in the U.S. Midwest now reaches (via the Mississippi River) into the Gulf Coast, where it is causing widespread damage to aquatic ecosystems. Agriculture is also responsible for most of the increase in sediment loading, whereas road deicing is mainly responsible for the increase in chloride.

Managing quality of runoff is not only a legitimate part of landscape planning and design but clearly a critical part of it. For landscape planners, the issue goes hand in hand with stormwater management. Land use activities are the sources of pollution, and the drainage systems we design deliver the pollutants to water features, not just lakes, streams, and harbors, but to groundwater as well. This chapter begins with a brief description of the main classes of water pollutants and their delivery systems and then goes on to land use relations and management considerations.

11.2 POLLUTION TYPES, SOURCES, AND MEASUREMENT

Water pollutants can be classified in different ways according to various criteria, including environmental effects, influence on human health, types of sources, and pollutant composition. Based on their influences on the environment and human health, pollutants can, for our purposes, be grouped into eight classes.

■ *Oxygen-demanding wastes:* As organic compounds in sewage and other organic wastes are decomposed by chemical and biological processes, they use up available oxygen in water that is essential to fish and other aquatic animals. Biological oxygen demand (BOD) is the most common measure of this form of water pollution.

■ *Plant nutrients:* Water-soluble nutrients such as phosphorus and nitrogen accelerate the growth of aquatic plants, causing the buildup of organic debris from these plants that can lead to increased BOD and the elimination of certain organisms in streams, lakes, and other water features.

Major pollutants

■ *Sediments:* Particles of soil and dirt eroded from agricultural, urban, and other land uses cloud water, cover bottom organisms, eliminate certain aquatic life forms, clog stream channels, and foul water supply systems.

■ *Disease-causing organisms:* Parasites such as bacteria, viruses, protozoa, and worms associated mainly with animal and human wastes enter drinking water and cause diseases such as dysentery, hepatitis, and cholera.

■ *Toxic minerals and inorganic compounds:* Substances such as heavy metals (e.g., lead and mercury), fibers (e.g., asbestos), and acids from industry or various technological processes are harmful to aquatic animals and cause many diseases in humans, including some forms of cancer.

■ *Synthetic organic compounds:* Substances, some water-soluble (e.g., cleaning compounds and insecticides) and some insoluble (e.g., plastics and petroleum residues manufactured from organic chemicals), cause various conditions in animals and humans, including kidney disorders, birth defects, and probably cancer.

■ *Radioactive wastes:* By-products of nuclear energy manufacturing from both commercial and military sources emit toxic radiation, a cause of cancer.

■ *Thermal discharges:* Heated water discharged mainly from power plants and some industrial facilities causes changes in species and increased growth rates in many aquatic organisms.

Point and Nonpoint Sources. There are two conventional ways of describing water pollution sources. One is based on the activity that produces the pollutant, and the other on the way the pollution is discharged into the environment. The latter uses only two categories, generally called point sources and nonpoint sources. **Point source pollution** is from a specific source, usually a facility, and is released at a known discharge point or outfall, usually a pipe or ditch. Chief among point sources are municipal sewer systems, industry, and power plants. Although point source pollutants are very important, the majority of stream pollution in the United States and Canada today comes from nonpoint sources.

Point and nonpoint

Nonpoint source pollution represents spatially dispersed, usually nonspecific, sources that are released in various ways at many points in the environment. Most nonpoint pollution is delivered by stormwater from a wide array of land uses, both urban and rural, and is discharged into streams, lakes, and other water features at countless places along channels and shores. In most regions, agriculture is the chief contributor of nonpoint source pollution and sediment is the main contaminant. However, the most geographically extensive source of water pollution is fallout (in both wet and dry forms) from the atmosphere.

System linkage

The Production-Delivery System. In general, stormwater pollution systems consist of three essential parts: (1) on-site production; followed by (2) removal from the production site; and ending with (3) delivery to a stream, lake, wetland, or harbor (Fig. 11.1).

■ *Production* represents the amounts and types of contaminants generated by a land use. For example, on streets and highways, cars and trucks are the principal contaminant producers (e.g., oil, gasoline, exhausts, and tire residues), and the rate of output is directly tied to traffic levels.

■ *Removal* is accomplished mainly by runoff. Roads are built with crowns in the middle so that runoff moves efficiently to the shoulders and collects with its load of contaminants in the drains along the road edge.

■ *Delivery* entails routing the stormwater in a pipe or ditch from the road edge to a stream or waterbody.

This three-part system applies to all land uses and provides a framework in which planning and management programs can be designed and implemented.

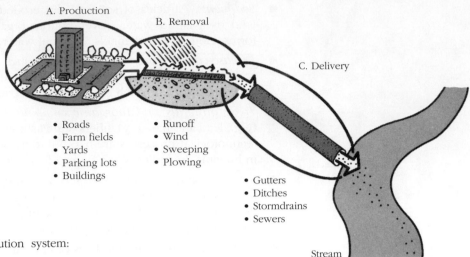

Fig. 11.1 Three-part stormwater pollution system: production, removal, and delivery.

ppm = mg/l

Measuring Water Pollution. To describe and analyze water pollution problems, we must be able to measure pollution levels. For most pollutants, such as sediments and plant nutrients, measurement involves separating the pollution from the water and measuring the total units or parts of pollutant per parts of water, usually given in *parts per million* (ppm). A more widely used measure is *milligrams per liter (mgl)*, which is the weight of the pollutants in a liter of water. Although they are different units of measurement, parts per million and milligrams per liter are equivalent and both are standard measures of pollution concentration.

Concentration and loading

Important to the understanding of pollution data is the distinction between pollution concentration and total pollution load or discharge. **Pollution concentration** is the amount of contaminants in water at a single moment, such as during a pollution spill. It is measured in parts per million or milligrams per liter and varies with both changes in streamflow and with changes in the amount of pollutants discharged. When there is a lot of water, pollution is diluted and the concentration is low. When there is a lot of pollution and little water, concentration is high. The **pollutant load**, on the other hand, is measured by adding up the total pollution discharge into a stream over an extended period of time, such as a year. Unlike pollution concentration, load is measured irrespective of the amount of water involved.

Background levels

Setting Performance Standards. In most cases, measuring water pollution usually involves first determining background levels of the contaminant in question because many substances labeled as pollutants (e.g., sediment, nitrogen, and certain heavy metals) are found naturally in water. In a stream where sediment is the issue, this measurement can be difficult because sediment levels naturally fluctuate with changes in discharge; so defining background levels may require considerable sampling. But what if sediment from pollution sources is already present? Then there is really no way to define background levels precisely. As a result, background levels of naturally occurring pollutants are often estimates, but for manufactured pollutants such as synthetic organic compounds, which are *not* found in the environment, detection at any level constitutes pollution, strictly speaking.

U.S. permitting

Stormwater quality is getting serious attention as part of environmental regulation. Most municipalities recognize that the stormwater detention basins required for new development also help reduce sediment and some other contaminants in runoff. The U.S. Environmental Protection Agency (EPA) has implemented stormwater management rules and guidelines, including a permitting requirement under the **National Pollutant Discharge Elimination System (NPDES)**. This system addresses mainly point sources of water pollution but can also include contaminated stormwater, such as that from industrial areas, which is released from pipes. In 1987 this requirement was extended to communities with populations over 100,000, mandating them to obtain an NPDES permit for their stormwater management systems. Unfortunately, agriculture remains untouched by stormwater regulations despite its enormous contributions to water pollution.

11.3 STORMWATER, LAND USE, AND WATER QUALITY

A significant question concerning the relationship between water quality and land use is the balance between the pollution contributions of urban and agricultural land uses. This is not an easy question to answer, and the general response is that stormwater pollution is very large in both urban and agricultural lands. Based on one measure, BOD, it is estimated that (1) in 35 percent of the watersheds in the United States, urban nonpoint sources (principally stormwater) are greater contributors than point sources, and (2) in 54 percent of the watersheds, agricultural nonpoint sources (also principally stormwater) are greater contributors than point sources. The concentration of BOD pollutants in stormwater is generally equal to or greater than that in effluents *Stormwater contributions* discharged from modern sewage treatment facilities. However, the total loading from stormwater may be significantly greater than that from point sources because of the huge volumes of stormwater discharged from the land.

The Automobile Connection. Our concern is mainly with urban stormwater, which brings us first to the automobile. The rise of **urban stormwater pollution** more or less parallels the growth of cities and the use of automobiles since 1920 or so. In the United States the number of operating motor vehicles has grown to 200 million or more. Between 1970 and 1995, the total miles traveled annually by Americans has more than *The car connection* doubled, from 1.1 trillion to 2.25 trillion miles, and the number is continuing to rise. There are now more than a million miles of paved roads and countless acres of parking lots, driveways, garages, and other hard surfaces for automobiles (Fig. 11.2). In most

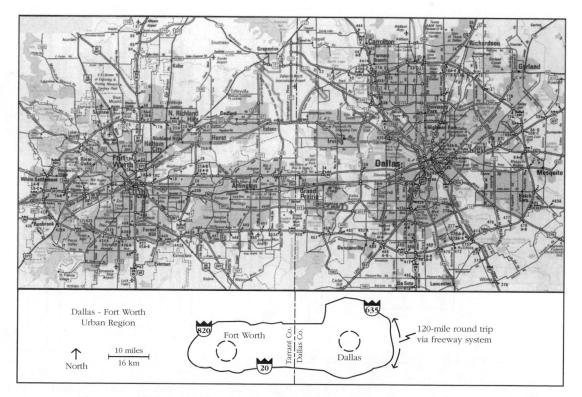

Fig. 11.2 The system of highways in the Dallas–Fort Worth urban region is similar to that in most large urban areas. Add city streets to this system, and the road density more than doubles.

metropolitan areas, the total length of paved streets is equivalent to a two-lane highway stretching from Michigan to Florida. And it bears repeating that, in total, impervious surfaces in the United States cover an area larger than the state of Georgia.

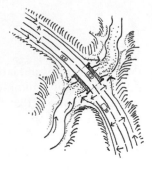

As impervious surfaces, these automobile-based facilities generate huge amounts of stormwater containing a host of contaminants—oil, paint, lead, organic compounds, and many other residues. Today the quantity of petroleum residue washed off streets, highways, parking lots, and industrial sites each year exceeds the total spillage from oil tankers and barges worldwide. Between 1950 and 1980, highway salt applications in the United States increased 12 times. The salt content in expressway drains during spring runoff may be a hundred times greater than natural levels in freshwater. Unfortunately, we have not demanded of transportation planners the same standards demanded of development (land use) planners, and it remains conventional engineering practice to release contaminated stormwater directly to streams and other water features in the quickest and cheapest way.

Density and pollution

Land Use Density Relations. In urban areas, the pollutant loading of stormwater increases with *land use density* (Fig. 11.3). Land use density is measured by the percentage of land covered by impervious materials, and it reflects the level of pollution-producing activities such as automobile traffic, spills, leakage, atmospheric fallout, and garbage production per acre or hectare. It also reflects the efficiency of surface flushing by stormwater runoff, which is nearly 100 percent in heavily built-up areas. The heaviest pollution loads are typically carried by the so-called *first-flush flow*, representing the first half inch or so of a rainstorm.

Although impervious cover is an important indicator of pollution loading, it, of course, is not a direct cause of stormwater pollution. Rather, the amount and types of land use activity associated with impervious cover are the sources of pollution—commercial activity, automobile traffic, air pollution, and garbage production, for example. Other measures of land use may also be used as indicators of pollution

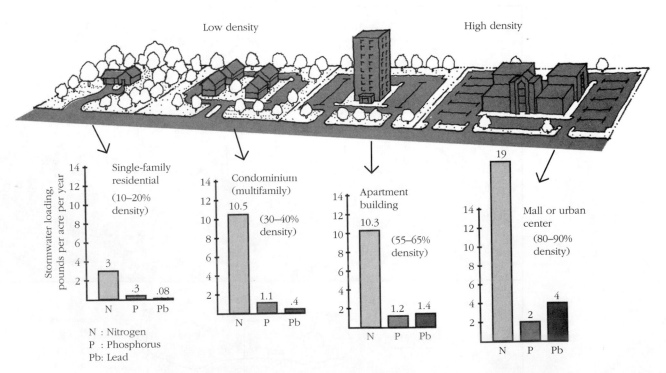

Fig. 11.3 The relationship of stormwater pollution and land use density in an urban region. Pollution loading increases with density from the suburban fringe toward the city center.

Table 11.1 Annual Stormwater Pollution Loading for Residential Development

Density	Phosphorus[a] (per person rate)	Nitrogen[a] (per person rate)	Lead[a] (per person rate)	Zinc[a] (per person rate)	Sediment[b] (per person rate)
0.5 unit/ac (1.25 person/acre)	0.8 (total) (0.64)	6.2 (4.96)	0.14 (0.11)	0.17 (0.14)	0.09 (0.07)
1.0 unit/ac (2.5 persons/acre)	0.8 (0.32)	6.7 (2.68)	0.17 (0.07)	0.20 (0.08)	0.11 (0.04)
2.0 units/ac (5 persons/acre)	0.9 (0.18)	7.7 (1.54)	0.25 (0.05)	0.25 (0.05)	0.14 (0.03)
10.0 units/ac (25 persons/acre)	1.5 (0.06)	12.1 (0.48)	0.88 (0.04)	0.50 (0.06)	0.27 (0.01)

[a] Pounds per acre per year.
[b] Tons per acre per year.
The number in parentheses is the per person rate

Source: Northern Virginia Planning District Commission, *Guidebook for Screening Urban Nonpoint Pollution Management Strategies* (Washington, DC: Prepared for the Metropolitan Washington Council of Governments, 1980).

loading, such as dwelling units per acre and people per acre. As the data in Table 11.1 indicate, pollution loading in residential areas increases with both dwelling units and persons per acre. This is very significant, but equally significant for planning purposes is the *rate* of loading per person or dwelling units at different density levels.

The *per-capita loading* rate, that is, the amount of stormwater pollution per person, actually *decreases* with higher residential densities. By way of example, consider residential densities of 0.5, 2.0, and 10 dwelling units per acre. At a density of 0.5 house per acre (2-acre lots), which is equivalent to about 1.25 persons per acre, the per-person loading rate for phosphorus is 0.64 pound per year (i.e., 0.8/1.25) and

Per-capita loading   lead's rate is 0.11 pound per year (i.e., 0.14/1.25). At a density of 2 houses per acre (0.5-acre lots), the per-person loading rate is 0.18 pound phosphorus and 0.05 pound lead. For town house apartments at 10 units per acre, the person loadings are 0.06 pound phosphorus and 0.035 pound lead.

The trend is clear. Measured on a per-capita basis, large residential lots in the range of 1 to 2 acres tend to be the most damaging to water quality. The reasons are related to the large size of lawns, the large numbers of automobiles per household, and the relatively great lengths of roads and drives, among other things. The fact that the per-capita loading rate decreases with higher densities suggests that communities should discourage large-lot development on the urban fringe and promote cluster

Important trend   development as a means of reducing water pollution. For communities faced with the dilemma of how to accommodate growth with minimal water quality degradation, the answer is clearly not large-lot single-family residential development.

11.4 WATER QUALITY MITIGATION ON LAND

Broadly speaking, the first objective in mitigating stormwater quality is to *slow the overall rate of response* of the stormwater flow system (Fig. 11.4). Decelerating the system induces infiltration and the settling of contaminants and reduces rates of surface flushing and erosion. One measure of system response is concentration time. If the concentration times of stormflow, especially those that flush surfaces frequently—for instance, flows from storms in the range of a 0.5- to 1.0-inch rainfall—can be extended rather than

Hobbling the system   decreased with development, an important step can be taken toward water quality control. Slower flow systems encourage infiltration, settling, and sequestering of nutrients

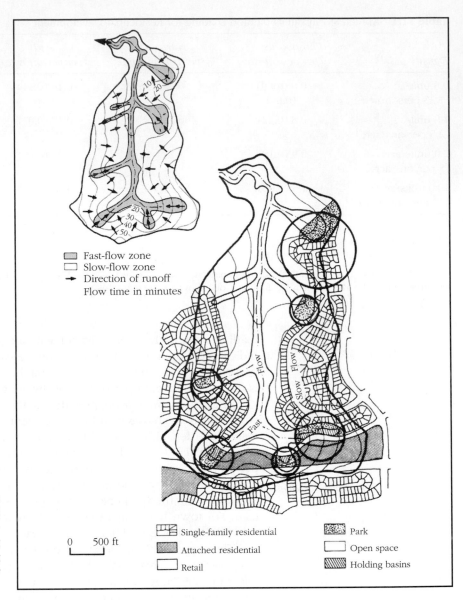

Fig. 11.4   The pattern of runoff travel times in a small basin (inset) and a proposed land use plan designed to minimize acceleration of runoff in fast-flow zones by limiting development near channel heads and using stormwater basins and parkland to slow delivery rates to channels.

and other contaminants. In addition, slow-flows lower the risk of channel erosion and the sediment it releases into the delivery system itself.

Mitigation Strategies and Facilities. We can approach the mitigation problem at three levels: (1) pollutant production, (2) pollutant removal from the site, and (3) pollutant transfer through the delivery system. The first level includes the regulation of land use types, development density, and land use practices such as lawn fertilizing and garbage burning. Considering the wide range of per-capita pollutant loading rates associated with different residential densities (Table 11.1), cluster forms of development are clearly an important planning strategy for minimizing the pollution loading of stormwater in residential development. In fact, the tighter the cluster, the better the performance. (These and other approaches are developed further in Chapter 13, which discusses best management practices.)

Limiting production

Absorption and Filtering. Measures for limiting the transfer of pollutants from the site are aimed largely at regulating the volume of runoff. The most common strategy relies on increasing **soil absorption** by, for example, increasing the ratio of vegetated-to-impervious groundcover, using porous pavers, grading for depression storage, and diverting runoff into infiltration beds or dry wells. According to a major study

Soil efficiency

conducted in the Washington, D.C., region, soil-absorption measures are the most effective means of removing pollutants from stormwater. This study found that, for soil with average permeability, expected removal capacities are in the range of 35 to 65 percent for total annual phosphorus, 40 to 85 percent for annual biochemical oxygen demand (BOD), and 80 to 90 percent for annual lead, depending on land use type.

Wells and cells

There is a wide variety of soil-absorption/filter systems in use today and several are described in this section. We have already mentioned *dry wells* and *depression storage cells* (see Fig. 8.12*b*). Dry wells are vertical shafts filled with stones or coarse aggregate, designed to store and infiltrate stormwater at subsoil depths. They have the advantage of not taking up much surface area and aiding in groundwater recharge.

Infiltration beds

Infiltration beds, like the one shown in Figure 11.5, come in various forms and names, with and without plants. Those that come with plants (sometimes called *bioremediation beds*) lend themselves to ready integration into landscape design schemes. Among the most interesting are *stormwater gardens* (or rain gardens) similar to the one shown in the photograph in Figure 11.5*b*.

Some communities have experimented with the use of soil material as a filter medium for stormwater. Austin, Texas, for example, has favored two such measures in new developments: filter berms and filtration basins. **Filter berms** are elongated earth mounds constructed along the contour of a slope. They are usually constructed of soil containing different grades of sand and a filter fabric, and they are designed to function basically in the same fashion as soil infiltration trenches, which have been shown to be highly effective in contaminant removal (Fig. 11.5). With berms and related measures, treatment is limited to small flows, either those from individual lots, small groups of lots, and/or residential streets and sidewalks. Soil-filtering efficiency is very high according to tests based on the application of treatment plant effluent to soil in various parts of the United States.

Filter berms

Filtration basins (also called water quality basins or filtering ponds) are concrete structures floored with several grades of sand and a filter fabric through which the stormwater is conducted (Fig. 11.6). Filtration basins are generally used for higher-density land uses, such as shopping centers, than would be appropriate for filter berms. They are designed to filter the first 0.5 inch of runoff, the so-called first flush, which is heaviest in contaminants. The performance of filtration basins based on the city of Austin's experience is good for small stormflows (less than 0.5 inch runoff). For first-flush

Filtration basins

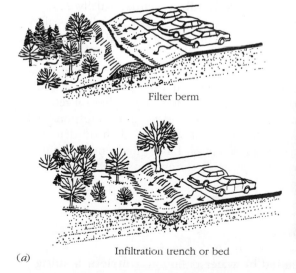

Filter berm

Infiltration trench or bed

(*a*)

(*b*)

Fig. 11.5 (*a*) A filter berm and an infiltration trench, two simple, small-scale mitigation devices to reduce contaminant levels in stormwater. (*b*) An elaborate infiltration swale, or stormwater garden.

Fig. 11.6   A filter basin used to reduce contaminant loads in stormwater. Water is filtered through layers of sand between the baffles.

flows, Austin has reported the following filtering rates: Fecal coliform 76 percent, total suspended solids 70 percent, total nitrogen 21 percent, total Kjeldahl nitrogen 46 percent, nitrate nitrogen 0, total phosphorus 33 percent, BOD 70 percent, total organic carbon 48 percent, iron 45 percent, lead 45 percent, and zinc 45 percent.

Another filtering measure strongly favored by some communities is the **vegetated buffer**. Several experimental studies show that vegetative buffers can be extremely efficient in sediment removal (up to 90 percent or more) if they meet the following design criteria: (1) continuous grass/turf cover; (2) buffer widths generally greater than 50 to 100 feet; (3) gentle gradients, generally less than 10 percent; and (4) shallow *Plant buffers* runoff depths, generally not exceeding the height of low grass. In hilly terrain, vegetative buffers should, to the greatest extent possible, be located on upland surfaces and integrated with depression storage and soil-filtration measures such as berms and dry wells (see Fig. 8.12*b*). It is doubtful that vegetated buffers alone are effective in reducing contaminants such as nitrogen and phosphorus, especially where stormwater is discharged over wooded hillslopes.

Holding basins   ***Holding and Settling.***   The final class of mitigation measures are those placed in the delivery system. These measures are **holding basins**, usually detention ponds and retention basins, designed to store stormwater in order to reduce peak discharge. By holding stormwater, water quality, especially that of the early runoff from a storm, can also be improved. Investigators generally point to the importance of sediment settling in stormwater basins and its role in the overall removal of pollutants from *Removal estimates* the water. Table 11.2 offers representative removal values reported by various studies from retention and detention basins. Values tend to be higher for retention basins because they hold water on a permanent basis and have a correspondingly higher potential for sediment settling and biochemical synthesis than do detention basins. Retention basins are often larger than detention basins and employ infiltration and evaporation rather than outflow to dispose of stormwater, except in the case of massive runoff events that overflow the basin.

11.5 EUTROPHICATION OF WATERBODIES

Among the many problems caused by water pollution, **nutrient loading** is one of the most serious and widespread in North America. Nutrients are dissolved minerals that nurture growth in aquatic plants such as algae and bacteria. Among the many

Table 11.2 Representative Removal Efficiencies for Stormwater Holding Basins

Pollutant	Percentage Removal (%)
Suspended sediment	40–75
Total phosphorus	20–50
Total nitrogen	15–30
BOD	30–65
Lead	40–90
Zinc	20–30

Source: Based on a compilation from various sources by Michael Sullivan Associates, Austin, Texas.

Lake loading

nutrients found in natural waters, nitrogen and phosphorus are usually recognized as the most critical ones because, when both are present in large quantities, they can induce accelerated rates of biological activity. Massive growths of aquatic plants in a lake or reservoir will lead to (1) a change in the balance of dissolved oxygen, carbon dioxide, and microorganisms and (2) an increase in the production of total organic matter. These changes lead to further alterations in the aquatic environment, most of which are decidedly undesirable from a human use standpoint:

Outcomes

- Increased rate of basin infilling by dead organic matter.
- Decreased water clarity.
- Shift in fish species to rougher types, such as carp.
- Decline in aesthetic quality, such as an increase in unpleasant odor.
- Increased cost of water treatment by municipalities and industry.
- Decline in recreational value.

The Eutrophication Process. Together, the processes of nutrient loading, accelerated biological activity, and the buildup of organic deposits are known as **eutrophication**. Often described as the process of aging a waterbody, eutrophication is a natural biochemical process that works hand in hand with geomorphic processes to close out waterbodies. Driven by natural forces alone, an inland lake in the midlatitudes may be consumed by eutrophication in several thousand years, but the rate varies widely with the size, depth, and bioclimatic conditions of the lake. In practically every instance, however, land development accelerates the rate by adding a surcharge of nutrients and sediment to the lake. So pronounced is this increase that scientists refer to two eutrophication rates for waterbodies in developed areas: natural and cultural. The graph in Figure 11.7 illustrates this concept for three lakes at different eutrophic states.

For terrestrial and aquatic plants to achieve high rates of productivity, the environment must supply them with large and dependable quantities of five basic resources: heat, light, carbon dioxide, water, and nutrients. According to the biological *principle of limiting factors*, plant productivity can be controlled by limiting the supply of any one of these resources. In inland and coastal waters, heat, light, carbon dioxide, and water are abundant on either a year-round or seasonal basis. Nutrients, on the other hand, often tend to be the most limited resource, not just in terms of quantities but also in terms of types. If nitrogen and phosphorus are available in ample quantities, productivity is usually high; however, if either one is scarce, productivity may be retarded and eutrophication can be held in check.

Limiting factors principle

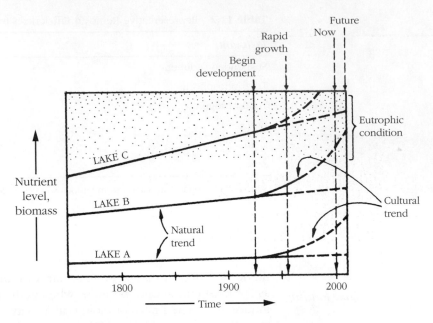

Fig. 11.7 Graph illustrating the concepts of natural and cultural eutrophication. The three sets of curves represent waterbodies at three stages of natural eutrophication as they enter the cultural stage of their development in the twentieth century.

Nitrogen and Phosphorus Systems. When introduced to the landscape in dissolved form, phosphorus and nitrogen show different responses to runoff processes. **Nitrogen**, which is generally the more abundant of the two, tends to be highly *Nutrient mobility* mobile in the soil and subsoil, moving with the flow of soil water and groundwater to receiving waterbodies. If introduced to a field as fertilizer, for example, most of it may pass through the soil in the time it takes infiltration water to percolate through the soil column—as little as weeks in humid climates.

In contrast, **phosphorus** tends to be retained in the soil, being released to the groundwater very slowly. As a result, under natural conditions most waters tend to be phosphorus limited, and when a surcharge of phosphorus is directly introduced to a waterbody, accelerated rates of productivity can be triggered. Accordingly, in water management programs aimed at limiting eutrophication, phosphorus control is often the primary goal. In the study of inland lakes, a classification scheme for levels or stages of eutrophication has been devised based on the total phosphorus content (both organic and inorganic forms) of lake water (Table 11.3).

Landscape planners may draw an important lesson here. Soil provides a critical service in controlling eutrophication by retaining phosphorus, and when this function is bypassed by routing surface water overland, this service is lost or greatly reduced. *Lesson for planners* Inland lakes and ponds are especially vulnerable because they are such ready receptacles for stormdrains in the watersheds that surround them. Because stormdrains make no provision for phosphorus reduction, management programs should rule against the routing of stormdrains directly into waterbodies. Road and street drains are the main culprits. In watersheds where such drains already exist, they should be rerouted and terminated well before reaching the lake. Wetlands, natural or constructed, are often appropriate receptacles for their discharge. For new development, it is advisable to use source control techniques for residential sites, coupled with narrow streets and low-gradient grass-lined swales.

Table 11.3 Levels of Eutrophication Based on Dissolved Phosphorus

Level or Stage	Total Phosphorus (mg/l)[a]	Water Characteristics
Oligotrophic (pre-eutrophic)	Less than 0.025	No algal blooms or nuisance weeds; clear water; abundant dissolved oxygen
Early eutrophic	0.025–0.045	
Middle eutrophic	0.045–0.065	
Eutrophic	0.065–0.085	
Advanced eutrophic	Greater than 0.085	Algal blooms and nuisance aquatic weeds throughout growing season; poor light penetration; limited dissolved oxygen

[a] Representative mean annual values of phosphorus in phosphorus-limited waterbodies.

Oligotrophic

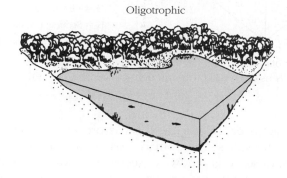

Advanced eutrophic

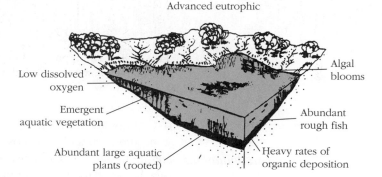

11.6 LAKE NUTRIENT LOADING AND LAND USE

Inputs and outputs

The Nutrient Budget Concept. For any body of water, it is possible to compute the **nutrient budget** by tabulating the inputs, outputs, and storage of phosphorus and nitrogen over some time period. *Inputs* come from four main sources:

Inputs

- Surface runoff (streamflow, stormwater, and point sources).
- Subsurface runoff (chiefly groundwater).
- The atmosphere.

Outputs take mainly three forms:

Outputs

- Streamflow.
- Seepage into the groundwater system.
- Burial of organic sediments containing nutrients (Fig. 11.8).

The *storage* of nutrients is represented by plants and animals, both living and dead, which, if not buried by sediment, release synthesized nutrients upon decomposition. A formula for the nutrient budget may be written as follows:

$$P + R + O + G + A + Q - S - B = 0$$

where

P = point source contributions
R = surface runoff contributions
O = organic sediment contributions
G = groundwater contributions

A = atmospheric contributions
Q = losses to streamflow
S = losses to groundwater
B = losses to organisms and sediment burial

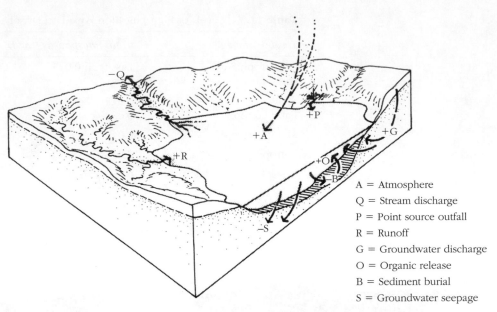

A = Atmosphere
Q = Stream discharge
P = Point source outfall
R = Runoff
G = Groundwater discharge
O = Organic release
B = Sediment burial
S = Groundwater seepage

Fig. 11.8 Main inputs and outputs of nutrients to a waterbody. Computation of a nutrient budget over some time period tells us the trend of the nutrient concentrations in the water.

Data limitations

Although the nutrient budget is easily described, it has proven very difficult to compute accurately for most waterbodies. The primary reasons are: (1) Some of the pertinent data, such as nutrient losses to groundwater seepage, are difficult to generate. (2) Exchanges of nutrients among water, organisms, and organic sediments are difficult to gauge. As a result, most nutrient budgets are based on a limited set of data, usually those representing streamflow, stormdrain discharge, septic drainfield seepage, point sources, and atmospheric fallout. In the case of streamflow and stormwater, data have been produced from field studies relating land use and surface cover to the nutrient content of runoff.

Land use relationships

In a study of the nitrogen and phosphorus contents of streams throughout the United States, the U.S. EPA and various state water quality programs made several interesting findings. First, the export of these nutrients from the land by streams tends to vary widely for different runoff events and for different watersheds with similar land uses. Second, although regional variations in nutrient export appear in the United States for small drainage areas, they do not correlate very well with rock and soil type. Instead, nutrient export by small streams tends to correlate best with land use and cover, in particular with the proportion of agricultural and urban land in a watershed. Nitrogen and phosphorus concentrations in streams draining agricultural land, for example, are typically five to ten times higher than those draining forested land.

Loading by land use

Estimating Nutrient Loading. Given these findings, it is possible to estimate the nitrogen and phosphorus loading of streams and lakes in most areas. The loading values are applicable to surface runoff, principally channel flow. The values in Table 11.4 are given in kilograms per square kilometer for seven basic land use/cover types. (The definition of each land use/cover type is given at the bottom of page 243.) Figure 11.9 gives the loading values in milligrams per liter of water (runoff) for agriculture and forest use/cover types for the eastern, middle, and western regions of the coterminous United States. For comparative purposes, examine these values with those representative of the water systems listed in Table 11.5.

Procedure

To determine the nutrient contributions to a waterbody, first delineate the drainage system and define its drainage areas. This may be a difficult task in developed areas because of the complexity of the land uses and the artificial drainage patterns that are superimposed on the natural drainage system. As a first step, it is instructive to identify

Table 11.4 Nutrient Loading Rates for Six Land Use/Cover Types

Use/Cover	Nitrogen (kg/km²/yr)	Phosphorus (kg/km²/yr)
Forest	440	8.5
Mostly forest	450	17.5
Mostly urban	788	30.0
Mostly agriculture	631	28.0
Agriculture	982	31.0
Mixed	552	18.5
Golf Course	1500	41.0

Source: J. Omernik, *The Influence of Land Use on Stream Nutrient Levels* (Washington, DC: U.S. Environmental Protection Agency, 1977).

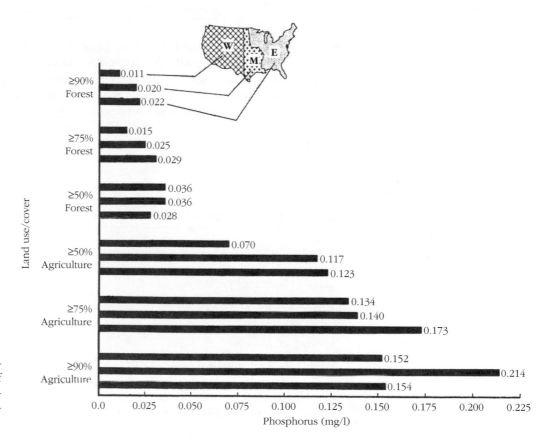

Fig. 11.9 Phosphorus loading values per liter of runoff based on agricultural and forest land use/cover for the coterminous United States.

the various land uses and major cover types in the area, noting their relationships to the drainage system and water features, what kinds of pollutants they are apt to contribute (both nutrients and other types), and the locations of critical entry points.

The remaining steps are as follows:

- Determine the percentage of each drainage area occupied by forest, agriculture, and urban development.

Steps   ■ Classify each area according to its relative percentages of forest, agriculture, and urban land uses based on the following percentages:

Forest: >75% forested.

Mostly forest: 50–75% forested.

Agriculture: >75% active farmland.

Mostly agriculture 50–75% active farmland.

Mostly urban: >40% urban development (residential, commercial, industrial, institutional).

Mixed: Does not fall into one of the preceding classes; for example, 25% urban, 30% agriculture, and 45% forest.

■ Using the nutrient-loading values given in Table 11.4, multiply the appropriate value for each area by its total area in square kilometers.

■ To calculate the loading potential from septic drainfield seepage, count the number of homes within 100 yards of the shore or streambank for each drainage area, and multiply this number by the following nutrient loading rates:

Phosphorus/Year	Nitrogen/Year
0.28 kg/home	10.66 kg/home

Total input

■ Combine the two totals for each drainage area for the total input from the watershed. If the *grand* total input from all sources is called for, then atmospheric contributions must also be considered. Atmospheric input needs to be considered only for water surfaces because it is already included in the values given for land surfaces. Given a fallout rate for the area in question, this value (in milligrams per square meter) should be multiplied by the area of the waterbody. Add this quantity to those from the watershed to obtain the grand total input (less groundwater contributions, if any) to the waterbody.

Table 11.5 Representative Levels of Phosphorus and Nitrogen in Various Waters

Water	Total P, mg/l	Total N, mg/l
Rainfall	0.01–0.03	0.1–2.0
Lakes without algal problems	Less than 0.025	Less than 0.35
Lakes with serious algal problems	More than 0.10	More than 0.80
Urban stormwater	1.0–2.0	2.0–10
Agricultural runoff	0.05–1.1	5.0–70
Sewage plant effluent (secondary treatment)	5–10	More than 20

Source: John W. Clark et al., *Water Supply and Pollution Control*, 3rd ed. (New York: IEP/Dun-Donnelley, 1977); and American Water Works Association, "Sources of Nitrogen and Phosphorus in Water Supplies," *Journal of the American Water Works Association*, 59, 1967, pp. 344–366.

11.7 PLANNING FOR WATER QUALITY MANAGEMENT IN SMALL WATERSHEDS

On the waterfront

Planning for residential and related land uses near water features presents a fundamental dilemma. People are attracted to water, yet the closer to it they build and live, the greater the impact they are apt to have on it. As impacts in the form of water pollution and scenic blight rise, the quality of the environment declines, and in turn land values usually decline (Fig. 11.10). Moreover, the greater the proximity of development to water features, generally the greater the threat to property from floods, erosion, and storms. In spite of these well-known problems, the pressure to develop near water features has not waned. In fact, it has increased, and in areas where there are few natural water features, developers are often inclined to build artificial ones to attract buyers.

In planning and management programs aimed at water quality, basically two approaches may be employed: corrective and preventative. The *corrective approach* is

Fig. 11.10 Examples of water quality degradation that can lead to deterioration of land values: (*a*) Long Beach, California, at the mouth of the Los Angeles River; (*b*) Montreal, Canada, along the St. Lawrence Seaway.

(*a*) (*b*)

Corrective approach

used to address an unsatisfactory condition that has already developed—for example, cleaning beaches and surface waters after an oil spill. On small bodies of water with weed growth problems, corrective measures may involve treating the water with chemicals that inhibit aquatic plants, harvesting the weeds, and/or dredging the lake basin to remove organic sediments. Such measures are generally used as a last resort for waterbodies with serious nutrient problems and/or advanced states of eutrophication.

The Preventative Approach. A *preventative approach* is generally preferred for most bodies of water, though it may actually be more difficult to employ successfully. This approach involves limiting or reducing the contributions of nutrients and other pollutants from the watershed by controlling on-site sources of pollution, limiting the transport of pollutants from the watershed to the lake, or both. Measures to control nutrient sources include improving the performance of septic drainfields for sewage disposal, replacing septic drainfields with community sewage treatment systems, reducing fertilizer applications to cropland and lawns, controlling soil erosion, eliminating the burning of leaves and garbage, and controlling animal manure. Measures to limit nutrient transport to the lake include filtering water through the soil or a soil medium, source control of stormwater, eliminating stormsewers in site planning and design, disconnecting farmfields from the drainage network, maintaining wetlands, floodplains, and natural stream channels, and constructing wetlands such as the one described in the case study at the end of this chapter.

Preventative approach

Building a Management Program. The formulation of a water quality *management program* for a waterbody must begin with an accurate map of the watershed as a drainage system (that is, the map should show where water comes from and how it gets to the waterbody) and as a land use/land cover system. Next, an effort should be made to estimate the nutrient budget. Nutrient contributions are placed in two categories: (1) those that can be managed using available technology and funding and (2) those that cannot. The latter usually includes groundwater and atmospheric contributions, whereas the former includes primarily surface and near-surface runoff, that is, stormwater, septic drainfield seepage, and the like.

First step

Second step

The second step entails defining management goals, such as "to slow the rate of eutrophication" or "to improve on the visual character" of a certain part of the waterbody. This process is important because the goals must be realistic and attainable. Generally speaking, the more ambitious the goals are (for example, "to reverse the trend in eutrophication"), the more difficult and expensive they are to achieve.

Third step

Once goals are established, a plan (or set of plans) is formulated identifying the actions and measures that need to be implemented. Some actions, such as prohibiting the use of phosphorus-rich fertilizers, may apply to the entire watershed, whereas others may apply only to a specific subarea that the nutrient budget data show to be a large contributor. In addition, measures should be identified to limit the nutrient production and delivery related to future development by defining appropriate development zones and enacting guidelines on sediment control, stormwater drainage, and sewage disposal.

Fourth step

Finally, after the plan is implemented, the waterbody must be monitored and the results weighed against the original goals, financial costs, and public support. Evaluation is a difficult task because it often requires comparing the results with forecasts about what conditions would have been without the plan. Uncertainty often arises relating to the validity of the original forecasts and to the possibility that natural perturbations in climatic, hydrologic, or biotic systems may have masked or diminished the efforts of the plan. And in this era of complex changes in foundation systems like climate and groundwater, planning and management efforts are all the more uncertain.

The Essential System. In water quality planning for lakes and ponds, the logical starting point is the drainage system. As noted in Chapter 9, every waterbody is supported by a watershed, which is comprised of many systems or subsystems, both natural and manmade, including stream networks, stormdrains, groundwater, and precipitation and evaporation. The watershed runoff system, particularly that on the surface, provides the spatial framework within which land use planning related to water quality management takes place.

The watershed illustrated in Figure 11.11 is typical of that around most impoundments. The bulk of the area is taken up by *subbasins* that drain all but the narrow belt

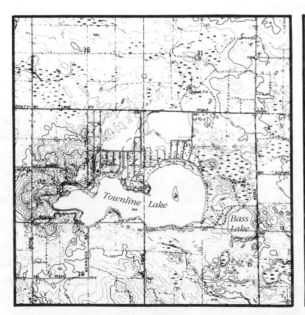

Fig. 11.11   The spatial organization of a small lake watershed showing the two types of drainage subareas: shoreland and subbasin.

Drainage subareas

of land along the shore, called *shoreland*. Subbasins are small watersheds with integrated drainage systems that usually discharge via a single stream. Shorelands, being too narrow to develop streams, lack integrated drainage systems, and drain directly to the waterbody by overland flow, interflow, and groundwater.

Land uses in the watersheds of impoundments tend to correlate with these two types of drainage areas. Water-oriented development (recreational and residential uses) is concentrated in the shoreland, whereas nonwater-oriented development (agriculture, suburban residential, commercial, and so on) is located mainly in the subbasins. The proportion of the watershed occupied by these two groups is usually very lopsided in favor of the subbasin users, which in turn can create a stakeholder's dilemma. Occupants of both shorelands and subbasins are watershed stakeholders,

Land use relationships

but only shoreland occupants with lake frontage property are usually recognized as such. It is necessary to remember, however, that it may take only one major polluter in a distant subbasin, such as a cattle feedlot or an industrial-scale poultry farm, to bring a waterbody to its knees (Fig. 11.12).

In the shoreland zone, management efforts usually involve dealing with individual property owners because each site drains directly into the waterbody (see the graphics in Case Study 21.9). In the subbasins, on the other hand, many land uses are linked by a runoff system that connects to the waterbody at a single point. The management of a subbasin, therefore, is in three respects a more difficult task. First,

Subbasin problems

because a great many players may be involved; second, the system is often larger and more complex; and third, land users lack a sense of connection and hence stakeholder responsibility for the lake and its management problems. Subbasins do, however, provide management opportunities that shorelands do not. Among other things, the runoff system is more diverse and may provide more management options, including sites where runoff can be collected and processed by natural or artificial means to reduce nutrients, sediment, and other impurities, as the following case study illustrates.

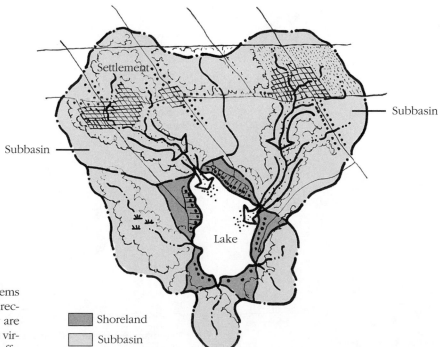

Fig. 11.12 Lake management requires systems thinking. Distant land users who are not usually recognized as stakeholders by the lake community are often the biggest players in lake management by virtue of their linkage to the lake via subbasin runoff.

11.8 CASE STUDY

Sediment and Nutrient Trapping Efficiency of a Constructed Wetland Near Delavan Lake, Wisconsin

John F. Elder and Gerald L. Goddard

Delavan Lake is a highly valued and heavily used sport fishery and recreational lake in southeastern Wisconsin, but for many years it has been plagued by water quality degradation and algal growth. By the late 1970s, the 2072-acre lake had reached a eutrophic state. Despite efforts in 1981 to reduce nutrient loading by diverting sewage and septic tank effluent from the lake basin, a severe algal bloom occurred in 1983 that signaled the need for additional action and that led to hydrologic and water quality studies and a detailed rehabilitation plan. Among other recommendations, the plan called for the construction of a wetland at the confluence of Jackson Creek and its two tributaries, which combine to form the principal inflow to Delavan Lake. The purpose of this wetland was to reduce sediment and nutrient inflow to the lake and thus contribute to its long-term protection and maintenance (Fig. 11.A).

Jackson Creek Wetland was constructed in the fall of 1992 when a 15-acre wetland was enlarged to 95 acres. Nearly all surface water inflow to the wetland is through three streams that drain an area of mixed land use dominated by agriculture with increasing residential and commercial development. In addition to areas of sedge meadow, wet prairie, and shallow-water marsh, the wetland contains three retention ponds along its upstream edge at the mouths of the inflowing streams. The surface areas of these ponds are about 3.4, 1.2, and 1.2 acres, respectively. Each pond has one to four outlet swales that distribute runoff into the wetland.

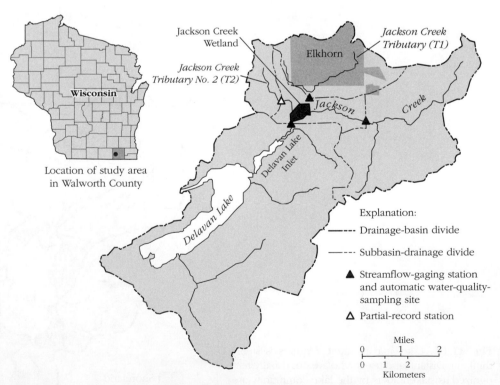

Fig. 11.A The management target, Delavan Lake, its watershed, and the constructed wetland, Jackson Creek Wetland. The lake faces declining water quality, which has reduced its recreation value.

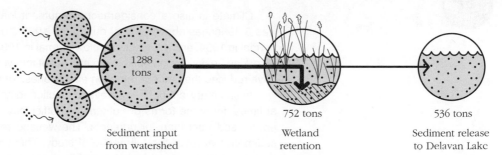

1288 tons	752 tons	536 tons
Sediment input from watershed	Wetland retention	Sediment release to Delavan Lake

Fig. 11.B The constructed wetland system covers 95 acres and contains three ponds that collect and redistribute runoff. Over a two-year period, it retained 58 percent of the suspended sediment and reduced most nutrients.

The primary objective of this study was to assess the effectiveness of the wetland as a retention system for suspended sediment and nutrients. The study was conducted between 1993 and 1995.

Sediment accumulated in the wetland ponds each year of the study. The total accumulation over two years was 752 tons in all three ponds. This represents a retention efficiency of 58 percent of the suspended sediment input to the ponds from stream discharge. Retention was greater than 20 percent at all times of the year except during the winter/early spring period. During the growing season, retention was frequently greater than 80 percent. For nutrients, however, the trapping effectiveness of the wetland was much less consistent. With the exception of ammonia, total nutrient loads generally decreased substantially between the inflows and the outflow over the course of a year. The inflow/outflow ratio, however, was highly variable seasonally and was frequently less than 1, an indication that more nutrients were given up by the ponds than retained.

In 1994, the season of greatest transport by a large margin, was the winter/early spring period. That period was also the time of greatest fractional retention of total and dissolved phosphorus. This combination of high transport and uptake accounted for nearly all of the annual net retention; it overshadowed later small-volume releases during low-flow periods. In fact, virtually all retention in 1994 occurred during a single month: February. Nearly all other months were actually periods of net phosphorus release, reflected by the large positive changes during the third and fourth quarters. Phosphorus mobilization during spring and summer might be due to increased solubility in anaerobic conditions that could, in turn, be caused by higher rates of microbial respiration in warmer temperatures.

What do these seasonal variations imply for aquatic ecosystems downstream? Although no direct evidence of their significance is available, the fact that phosphorus releases tended to occur in late spring and summer makes them coincident with the likely timing of algal blooms downstream in Delavan Lake Inlet and Delavan Lake. Further indication of a potential problem is the fact that the proportional release of dissolved orthophosphate—the fraction that is available for algal uptake—was even greater than that of total phosphorus.

The lack of coupling between suspended sediment and nutrient transport dynamics is also notable. The suspended material presumably is retained because of the reduced velocity of water flow, which allows the material to settle out. One might expect that the nutrient loads would be largely associated with oxidized suspended sediments and therefore would be entrained together with the sediments. However, this was clearly not the case; the net release of nutrients often coincided with substantial retention of sediments. This result indicates that biogeochemical processes mobilized the sediment-associated nutrients and thus prevented their effective retention in the wetland, at least periodically.

Climate is also a consideration on nutrient load characteristics. Precipitation was 3.49 inches above normal during the 1993 study period, 12.19 inches below normal in 1994, and 4.74 inches below normal in 1995. The high water flows in 1993 produced heavy nutrient fluxes into and out of the wetland. The large flows may also have reduced the nutrient retention effectiveness of the wetland.

In summary, although the wetland functioned as a sink for both sediments and, at times, for some forms of nutrients, its efficiency was greater and more consistent for sediment than for nutrients. The wetland retained 46 percent of the input sediment load over the full period of study. This finding, though illustrating some limitation of the wetland as nutrient sink, does not necessarily conflict with the general concept of nutrient retention in the wetland. As observed in other wetlands, the variability reflects the capacity of the system to function not only as a sink but also, periodically, as a facilitator of nutrient transformation and transport. The results of this study thus illustrate the complexities of biogeochemical cycles in any ecosystem, wetlands included. Understanding these complexities may help us to avoid placing unrealistic expectations on natural or constructed wetlands as systems with unlimited filtering capacity.

John F. Elder (retired) is a research limnologist and Gerald L. Goddard is a hydrologic technician. Both are with the U.S. Geological Survey, Wisconsin District Lake-Studies Team. ∎

11.9 SELECTED REFERENCES FOR FURTHER READING

Clark, John W., Viessman, Warren, Jr., and Hammer, Mark J. *Water Supply and Pollution Control*, 3rd ed. New York: IEP/Dun-Donnelley, 1977.

Dillon, P. J., and Vollenweider, R. A. *The Application of the Phosphorus Loading Concept to Eutrophication Research*. Burlington, ON: Environment Canada, Center for Inland Waters, 1974.

Elder, J. F. "Factors Affecting Wetland Retention of Nutrients, Metals, and Organic Materials." In *Wetland Hydrology* (J. H. Kusler and G. Brooks, eds.). Chicago: Association of State Wetland Managers, 1988.

Marsh, William M., and Hill-Rowley, Richard. "Water Quality, Stormwater Management, and Development Planning on the Urban Fringe." *Journal of Urban and Contemporary Law*. 35, 1989, pp. 3–36.

Naiman, R. J., and Turner, M. G. "A Future Perspective on North America's Freshwater Ecosystems." *Ecological Applications*. 10, 4, 2000, pp. 958–970.

National Academy of Sciences. *Eutrophication: Causes, Consequences, and Correctives*. Washington, DC: National Academy of Science Press and the National Research Council, 1969.

OECD. *Environmental Impact Assessment of Roads*. Paris: Organization for Economic Cooperation and Development, 1994.

Omernik, James M. *Nonpoint Source–Stream Nutrient Level Relationships: A Nationwide Study*. Corvallis, OR: U.S. Environmental Protection Agency, 1977.

Smith, R. A., Alexander, R. B., and Wolman, M. G. "Water Quality Trends in the Nation's Rivers." *Science*. 235, 1987, pp. 1607–1615.

Tilton, Donald L., and Kadlec, R. H. "The Utilization of a Fresh-Water Wetland for Nutrient Removal from Secondarily Treated Waste Water Effluent." *Journal of Environmental Quality*. 8, 3, 1979, pp. 328–334.

U.S. Soil Conservation Service. *Ponds for Water Supply and Recreation*. Washington, DC: U.S. Department of Agriculture, 1971.

Vallentyne, John R. *The Algal Bowl: Lakes and Man.* Ottawa: Environment Canada, Fisheries and Marine Service, 1974.

Wolman, M. Gordon, and Chamberlin, C. E. "Nonpoint Sources." *Proceedings of the National Water Conference.* Philadelphia: Philadelphia Academy of Sciences, 1982, pp. 87–100.

Related Websites

Environment Canada. "Water Pollution." 2009. http://www.ec.gc.ca/default. asp?lang=En&n=9CA329A6-1

Canadian government site giving information on sources and environmental effects of water pollution. The water quality section has information about remediation projects in Canada.

Environment Protection Authority–Victoria. "Stormwater." 2007. http://www.epa.vic. gov.au/water/stormwater/default.asp

Governmental site focuses on urban stormwater runoff. The types of pollution, their effects, and sources. Also gives residents and businesses ways to reduce their run-off.

Environmental Protection Agency. "National Pollutant Discharge Elimination System." 2009. http://cfpub.epa.gov/NPDES/

The NPDES permit system explained. The "Permit Program Basics" link gives the fundamental information about water permitting and the strategic plan. Water quality trading is also explained.

U.S. Geological Survey. "Oregon District Hydrologic Studies. PN381 Assessment of Nutrient Loading to Upper Klamath Lake, Oregon." 2002. http://or.water.usgs. gov/projs_dir/pn381/pn381.html

A case study and its resulting publications of the nutrient loading in Upper Klamath Lake due to changes in land use. Links have been made to draining wetlands, grazing, timber harvesting, and agriculture.

12

SOIL EROSION, STREAM SEDIMENTATION, AND LANDSCAPE MANAGEMENT

12.1 INTRODUCTION

Soil erosion may be the most serious land management problem facing humanity today. Simple estimates based on sediment discharged to the sea indicate that since the origin of agriculture some 12,000 years ago, the annual sediment loss worldwide has more than doubled—from 9 billion tons to more than 20 billion tons. This dramatic increase is even more alarming when we consider that only a fraction of the soil eroded from the land actually ends up in the sea. Most of the sediment from soil erosion is left in terrestrial and freshwater environments such as woodlands, stream valleys, wetlands, and reservoirs, where it is a major source of habitat and water quality degradation.

Land degradation

Virtually all land uses contribute directly or indirectly to soil erosion, but agriculture is decidedly the single greatest contributor. Crop farming is the main cause, but grazing (cattle, sheep, and goats) also contributes to soil loss by causing extensive damage to rangeland. Broadly speaking, we can say that soil loss from erosion has followed the trend of world population. Today it is estimated that 4.5 million square miles of land have been seriously degraded by crop farming, deforestation, overgrazing, and other factors; this is an area larger than all of Canada and soil erosion is the principal culprit.

Erosion costs

The costs of erosion are enormous and rising. In 1995 it was estimated that soil erosion cost $44 billion a year in the United States and $400 billion worldwide. These estimates included direct costs represented by loss in soil fertility (productivity) and indirect costs represented by damage to waterways, infrastructure, and human health. Indirect costs included sediment clogging of channels, which increases flooding and damage to land use facilities, and the reduction of water quality. Direct costs were calculated according to the value of fertilizers needed to replace soil fertility losses. Neither the direct nor indirect costs calculated for these estimates include damage to ecosystems.

Erosion controls

For most soils, vegetative cover is the most important control on erosion. When vegetation—no matter whether it is grass or trees—is disturbed or removed, erosion inevitably increases. Other factors influencing erosion by runoff include rainfall magnitude and frequency, the slope of the land, cropland management, and soil erodibility. The most erodible soils are loams, sands, and silts situated on sloping ground without a protective plant cover. If this condition is coupled with heavy rainfall and no runoff controls, such as berming, terracing, or contour plowing, then soil loss may be as high as 1 to 2 inches a year. This is equivalent to more than 100 tons of soil loss per acre. At this rate, it is only a matter of a few years before the topsoil is entirely wiped out, leaving only the underlying and much less fertile mineral soil at the surface.

Our agenda

What is our particular interest in soil erosion? Although nonagricultural land uses are minor contributors to soil loss worldwide, they are major sources of the problem locally, especially on the urban fringe where the environments most affected—streams, lakes, and wetlands—are among the most prized and heavily used. The agricultural problem has long held the attention of soil scientists and geomorphologists, and it is on their research and management recommendations that we build our approach to soil erosion and sedimentation related to residential land use and similar forms of development. In this chapter we are interested in understanding the process of soil erosion and sediment transport, methods for estimating soil loss rates, techniques for controlling erosion and sedimentation, and the implications of sediment loss and transport in watershed management.

12.2 SOIL EROSION, BIOCLIMATE, AND LAND USE

From field studies conducted in different parts of North America, we are able to sketch a brief picture of the general relationship between bioclimatic conditions and soil erosion. Barring the influence of land use, it appears that erosion is relatively low

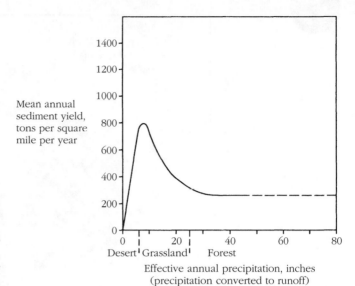

Fig. 12.1 Sediment yield related to bioclimatic conditions less the influences of land use. Rates are highest in the semiarid zone, where short grasses provide poor protection against occasionally heavy rainstorms and runoff.

in the bioclimatically extreme environments, namely, arid and humid regions, but relatively high in semiarid regions, especially in their drier zones (Fig. 12.1).

Bioclimate and Soil Erosion. The explanation for this phenomenon is related to the effects of precipitation and plant cover. In the semiarid zones, precipitation is modest, averaging 10 to 20 inches a year in the drier sections, but is capable of producing substantial runoff events in most years. But evapotranspiration is also high, and soil moisture is generally insufficient to support more than a weak and discontinuous cover of short grasses, which provides poor protection against erosion by runoff. In addition, rainfall is quite variable in these areas, further promoting erosion because drought years weaken plant covers and wet years produce much higher-than-average runoff.

Vegetation and runoff

Humid regions have significantly more rainfall and thus a greater potential for erosion. However, forest is the predominant vegetative cover and provides strong defense against the forces of erosion. This is reflected in the relatively low sediment loads in streams in the eastern and northwestern sections of the United States (Fig. 12.2). At the other bioclimatic extreme, in arid lands receiving annual runoff-producing precipitation of less than 7 to 8 inches, runoff rates are so low that, despite the high erodibility of poorly protected desert soils, erosion is very limited.

Land Use Impacts. When land use is added to the picture, erosion rates change appreciably. In the semiarid grasslands, erosion by both wind and runoff usually rises dramatically with plowing, cropping, and grazing. Historically, droughts have been especially damaging to farmland, but today's widespread use of irrigation reduces the risk of massive soil loss from drought-related events, such as during the Dust Bowl days of the 1930s. Nevertheless, wherever prairie grasses are removed for farming, with or without irrigation, erosion by runoff invariably rises, especially on sloping ground. The same situation holds in humid lands where forests are cleared for farming.

Clearing and plowing

Land clearing breaks down the protective vegetative cover, and, if not replaced by a permanent substitute cover, topsoil is subject to rapid erosion. On most cropland, fields are left barren for the winter months when soil moisture levels and the potential for runoff are high. In Canada and much of the northern United States, fields are often cropless for six months a year. Today, improved soil protection methods, such as no-till farming, which leaves crop stubble on fields over the winter, helps

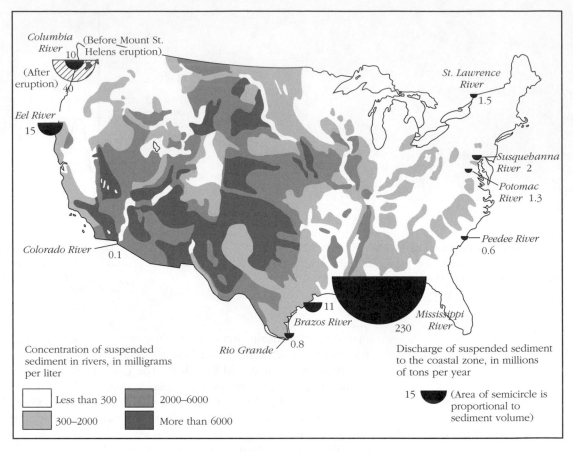

Fig. 12.2 The average concentration of suspended sediment in streams in the contiguous United States with figures on the average discharge of suspended sediment by major rivers. Suspended sediment is made up of fine soil particles, mainly clay, eroded by runoff and discharged into streams. It is a key indicator of soil erosion rates.

Sediment and land use

Traditional and modern land clearing

reduce erosion. However, as a defense against soil erosion it is a poor substitute for the original plant cover.

In the past century or so, land use change leading to urbanization has also become important in the soil erosion issue, especially as it is practiced in North America. The general sequence of land use change that ends in urban development and the corresponding rates of sediment yield (from soil erosion) are given in Figure 12.3. A description of the process begins with forest clearing and agricultural development sometime in the 1800s for most areas. Erosion rises with the destruction of natural vegetation and the establishment of cropland and pasture. This trend continues until the first half of the twentieth century, when agriculture declines with the shift of rural population to urban areas. As abandoned farmland is taken over by a volunteer vegetation, and soil erosion probably declines somewhat.

Urbanization, Sediment, and Stream Channels. With the growth of cities in the second half of the twentieth century comes urban sprawl and the development of suburbia. Abandoned farmland and woodland, as well as active farmland on the urban fringe and beyond, are cleared for development. Unlike the slow and laborious process of early land clearing, the process is now expedited by earth-moving machinery. What once took years to clear now takes only days, and the process is thorough and devastating. The wholesale exposure of soil gives rise to massive erosion rates, particularly for large construction sites left open for many months. Although this phase of landscape change is short-lived, its effects—which today are usually mitigated

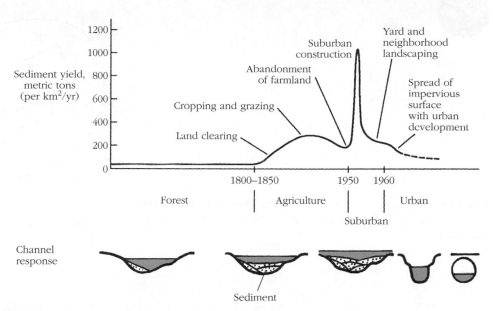

Fig. 12.3 The trend of soil erosion related to land use/cover change beginning with pre-settlement forest and ending with twentieth-century urban development. The corresponding changes in local stream channels are sketched below the chart. (Based on Wolman, 1967.)

Suburban phase

in some fashion in most parts of North America—were often profound and lasting. Chief among them were sediment-glutted stream channels, silted lakes and reservoirs, degraded aquatic habitat, and increased flooding due to reduced channel capacities.

Urban phase

As urban sprawl advances and the suburban landscape takes shape, soil exposure to erosion declines because the land is covered by buildings, roads, and landscaped surfaces and runoff is diverted into stormdrains. Soil erosion and stream sedimentation rates decline. With full urbanization and impervious cover approaching 75 percent or more, soil erosion rates decline further, probably approaching early agricultural or even lower levels. On the other hand, stormwater runoff rates increase dramatically. For streams that remain open (that is, not piped underground), the heavy stormwater discharges lead to channel scouring, downcutting, and the degradation of aquatic habitat. For streams whose channels are replaced by large pipes—a destiny suffered by most small streams in urban areas—the "channel" becomes little more than a stormwater conduit carrying sullied runoff (see the diagrams below the graph in Fig. 12.3).

12.3 THE SOIL EROSION–SEDIMENT TRANSPORT SYSTEM

As with any environmental problem, the path to effective planning and management begins with understanding the nature of the system and its drivers. For our purposes, a soil erosion–sediment transport system is made up of four essential components: erosion, transport, storage, and export. These components are interconnected, but not necessarily sequential, especially the sediment storage and export components.

System components

In addition, the system operates within a relatively small spatial framework, the local drainage basin, with an area ranging from several square miles to 100 square miles or more (Fig. 12.4).

Erosion is the process by which particles are dislodged from the soil. *Transport* is the transfer of particles by runoff from the point of erosion to a storage site or point of export. For most sediment, especially sand and gravel, transport distances are usually short. Small sediments (clays and silts) may go for long rides but only if they become

Definitions

entrained in a stream's discharge and drainage net.

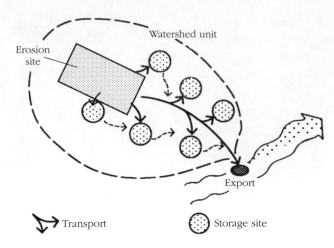

Fig. 12.4 A graphic model of a soil erosion–sediment transport system made up of four components: erosion, transport, storage, and export.

Splash and wash

Storage is merely the deposition of sediment. This process begins with larger particles on and around an erosion site and continues as sediment moves down the drainage basin and is trapped in various places. Storage may be brief, lasting only until the next runoff event, or it may be long term if it is immobilized in standing water or by burial or plant overgrowth. The sediment discharged after losses to storage within the basin constitutes the *export* component of the system.

Soil Erosion Processes. On exposed soil, the erosional process itself begins with *rainsplash* from the impact of raindrops. When the drops hit the ground, they explode, driving small soil particles downslope. If the rainfall rate (intensity) exceeds the soil's infiltration capacity, then runoff in the form of overland flow results. Overland flow moves in thin sheets, called *sheetwash*, and in tiny threads, called *rivulets*. Both processes are capable of moving sediment, and rivulets can cut microchannels, termed *rills*, into exposed soil.

Gullying

The soil erosion from sheetwash and rivulet runoff is also known as **rainwash**. At some point in the midsection of a slope, this runoff begins to concentrate in small channels, and, as its volume and velocity increase, its energy rises dramatically, giving it the power to erode gullies into the slope face (see Fig. 4.10). Known as **gullying**, this process is one of the most severe forms of soil erosion. In erodible soils such as silt and loose sand in exposed farmfields or construction sites, gullying may carve channels several feet deep and tens of yards long in response to several, heavy rainfalls over a year or less. Entire hillsides may be gutted in a matter of several years (Fig. 12.5).

Sediment storage

Sediment Storage and Export. The vast majority of the sediment released from an erosion site is stored in the local drainage basin relatively close to its place of origin. Most of this sediment is deposited in various places, called **sinks**, near the erosion site, such as wooded hillslopes, swales, stream channels, wetlands, and floodplains (Fig. 12.6). Based on field studies in several different places, it appears that only 10 to 20 percent or so of the sediment released through soil erosion is actually exported from small rural watersheds on an annual basis.

Controlling linkage

There is an important lesson here. Beyond the loss of soil itself, the greatest impacts of soil erosion occur from sedimentation close to the erosion site and within the local watershed. A key management consideration, then, is controlling the *linkage* between sediment sources and transport processes, mainly channel flows in swales, ditches, and streams that can distribute the sediment down the system and over large areas. This explains why cropland erosion has such great impacts on aquatic habitats

Fig. 12.5 Gullies in California rangeland. In wet years, gullies such as these can advance upslope several meters while cutting deeper and wider cavities.

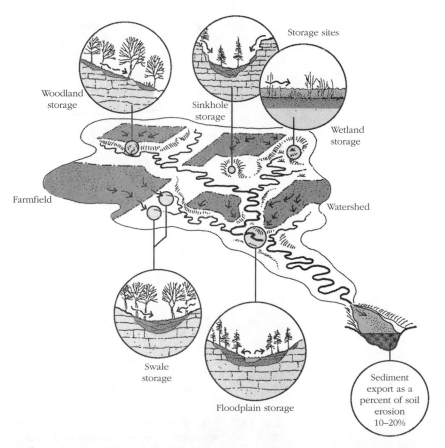

Fig. 12.6 Sediment sinks in a small drainage basin. Only 10 to 20 percent of sediment released in soil erosion may be exported from the basin.

in humid regions. To expedite water removal for spring plowing and planting, soil must be drained via tile and ditch systems, which discharge directly to local streams. The efficient movement of field runoff means efficient movement of sediment. This arrangement serves to expand the zone of sediment impact from field runoff rather than contain it on or near cultivated fields.

12.4 PRINCIPAL FACTORS INFLUENCING SOIL EROSION

Any attempt to forecast soil erosion for a site must take into account four basic factors: (1) vegetation, (2) soil type, (3) slope size and inclination, and (4) the frequency and intensity of rainfall. Rainfall and the runoff it generates are the driving factors, whereas vegetation, soil texture, and slope are the resisting factors in the system.

Rainfall. Tests show that, in general, heavy rainfalls such as those produced by thunderstorms promote the highest rates of erosion. Accordingly, the incidence of such storms, together with the total annual rainfall, can be taken as a reliable measure of the effectiveness of rainfall in promoting soil erosion. The U.S. Natural Resources Conservation Service has translated this into a **rainfall erosion index** that *Erosive energy* represents the erosive energy delivered to the soil surface annually by rainfall in an average year. The index values vary appreciably over the United States in response to climate with the highest values in the humid South and much lower values to the north and west (Fig. 12.7). And in some regions they vary appreciably from one side of a state to another. For instance, values decline from 250 to less than 100 from the southeastern to the northwestern corner of Kansas; in other words, on the average, erosion on comparable sites should be more than 2.5 times greater in the southeast.

Vegetation and Soil. On most surfaces, vegetation appears to be the single most important factor regulating soil erosion. Foliage intercepts raindrops, reducing

Fig. 12.7 Rainfall erosion index for the United States east of the Rocky Mountains. In the West, index values are less reliable and best calculated from local rainfall records.

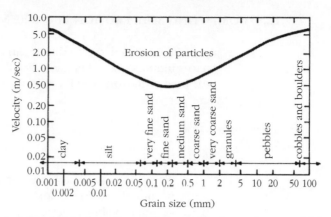

Fig. 12.8 Erosion thresholds for soil particles ranging from clay to cobbles and boulders. Sand has the lowest threshold or highest erodibility. Conditions in nature may vary with compaction, cementing, and other factors.

Cover density   the force with which they strike the soil surface. Organic litter on the ground further reduces the impact of raindrops, and plant roots bind together aggregates of soil particles, increasing the soil's resistance to the force of running water. The one feature of vegetation that appears to have the greatest influence on erosion is **cover density**: the heavier the cover, either in the form of groundcover or tree canopy, the lower the soil loss to runoff.

If running water is applied to soils of different textures, sand usually yields (erodes) first. To erode clay, the velocity of the runoff has to be increased to create sufficient stress to overcome cohesive forces that bind the particles together. Similarly, high velocities are needed to move pebbles and larger particles because their masses

Soil erodibility   are so much greater than those of sand particles. Thus, in considering the role of soil type in erosion problems, it appears that **erodibility** is greatest with intermediate textures, whereas clay and particles coarser than sand are measurably less erodible (Fig. 12.8). Other soil characteristics, such as compactness and structure, also influence erodibility, but in general *texture* can be taken as the leading soil parameter in assessing the potential for soil erosion.

Slope. The velocity that runoff is able to attain is closely related to the **slope** of the ground over which it flows. Slope also influences the quantity of runoff inasmuch as long slopes collect more rainfall and thus generate a relatively large volume of

Slope and runoff   runoff, other things being equal. In general, then, slopes that are both steep and long tend to produce the greatest erosion because they generate runoff that is high in both velocity and mass. But this is true only for slopes up to about 50 degrees (110–115 percent) because at steeper angles the exposure of the slope face to rainfall grows rapidly smaller, vanishing altogether for vertical cliffs.

In land use problems, however, the greatest consideration is normally given to slopes of less than 50 degrees, particularly insofar as urban development, residential

Slope value   development, and agriculture are concerned. Table 12.1 gives relative values for soil erosion or the potential for it based on slope steepness and length for slopes up to 50 percent inclination and 1000 feet length. In using this table, note that the rate of change with slope steepness is not linear, whereas it is with slope length. For example, for a 500-foot slope of 10 percent, the value is 3.1, whereas for the same length at 20 percent, the value is 9.3.

12.5 COMPUTING SOIL EROSION FROM RUNOFF

An estimate of soil loss to runoff can be computed by combining all four of the major factors influencing soil erosion. For this we use a simple formula called the **universal soil loss equation** (USLE or Revised USLE), which gives us soil erosion in tons per acre per year:

Soil loss equation

$$A = R \bullet K \bullet S \bullet C$$

where

A = soil loss, tons per acre per year
R = rainfall erosion index
K = soil erodibility factor
S = slope factor, steepness and length
C = plant cover factor

In problems involving agricultural land, a fifth factor, *cropping management* (P), is included, but for problems involving nonagricultural land, abandoned farmland, and urban land types, it is not generally applicable.

Interpretation. To interpret the universal soil loss equation reliably, we need to understand three points: (1) The computed quantity of soil erosion represents only the displacement of particles from their original positions. (2) Only sheetwash and rill *Prerequisites* erosion are covered by the equation, whereas gullying is not. (3) The soil loss equation is designed for agricultural land, and applications to other situations may or may not be appropriate. Cleared land, former cropland, and construction sites can generally be considered appropriate for application of this method.

Erosion versus loss What constitutes soil loss depends on the study objectives. Strictly speaking, soil erosion takes place when particles are displaced from their place of origin no matter how far or little they are moved. But soil erosion may be different than soil loss. For example, much erosion can take place on a construction site, but if the resultant sediment is contained within the site by, say, perimeter berms, it can be argued that no soil loss has taken place even though erosion has degraded the soil.

Table 12.1 Slope Factor Based on Slope Steepness and Length

Slope Length in Feet	Slope Steepness (%)														
	4	6	8	10	12	14	16	18	20	25	30	35	40	45	50
50	0.3	0.5	0.7	1.0	1.3	1.6	2.0	2.4	3.0	4.3	6.0	7.9	10.1	12.6	15.4
100	0.4	0.7	1.0	1.4	1.8	2.3	2.8	3.4	4.2	6.1	8.5	11.2	14.4	17.9	21.7
150	0.5	0.8	1.2	1.6	2.2	2.8	3.5	4.2	5.1	7.5	10.4	13.8	17.6	21.9	26.6
200	0.6	0.9	1.4	1.9	2.6	3.3	4.1	4.8	5.9	8.7	12.0	15.9	20.3	25.2	30.7
250	0.7	1.0	1.6	2.2	2.9	3.7	4.5	5.4	6.6	9.7	13.4	17.8	22.7	28.2	34.4
300	0.7	1.2	1.7	2.4	3.1	4.0	5.0	5.9	7.2	10.7	14.7	19.5	24.9	30.9	37.6
350	0.8	1.2	1.8	2.6	3.4	4.3	5.4	6.4	7.8	11.5	15.9	21.0	26.9	33.4	40.6
400	0.8	1.3	2.0	2.7	3.6	4.6	5.7	6.8	8.3	12.3	17.0	22.5	28.7	35.7	43.5
450	0.9	1.4	2.1	2.9	3.8	4.9	6.1	7.2	8.9	13.1	18.0	23.8	30.5	37.9	46.1
500	0.9	1.5	2.2	3.1	4.0	5.2	6.4	7.6	9.3	13.7	19.0	25.1	32.1	39.9	48.6
550	1.0	1.6	2.3	3.2	4.2	5.4	6.7	8.0	9.8	14.4	19.9	26.4	33.7	41.9	50.9
600	1.0	1.6	2.4	3.3	4.4	5.7	7.0	8.3	10.2	15.1	20.8	27.5	35.2	43.7	53.2
650	1.1	1.7	2.5	3.5	4.6	5.9	7.3	8.7	10.6	15.7	21.7	28.7	36.6	45.5	55.4
700	1.1	1.8	2.6	3.6	4.8	6.1	7.6	9.0	11.1	16.3	22.5	29.7	38.0	47.2	57.5
750	1.1	1.8	2.7	3.7	4.9	6.3	7.9	9.3	11.4	16.8	23.3	30.8	39.3	48.9	59.5
800	1.2	1.9	2.8	3.8	5.1	6.5	8.1	9.6	11.8	17.4	24.1	31.8	40.6	50.5	61.4
900	1.2	2.0	3.0	4.1	5.4	6.9	8.6	10.2	12.5	18.5	25.5	33.7	43.1	53.5	65.2
1000	1.3	2.1	3.1	4.3	5.7	7.3	9.1	10.8	13.2	19.5	26.9	35.5	45.4	56.4	68.7

For a site such as the one shown in Figure 12.9, much of the soil lost from the upper surface is deposited near the foot of the slope. Therefore, in computing the soil loss for the lower surface, we should take into account sediment added from upslope and stored on-site, and then we should solve for the net change in the soil mass. For purposes of environmental planning and management, however, the key consideration is often soil loss from the site and its impacts on neighboring environments.

Maps, table, field

Data Generation. The data needed to make a soil erosion computation can usually be obtained from topographic sheets, soil reports, and a few additional maps and tables. Field inspection of the site is recommended to determine plant cover and patterns of erosion and deposition. If that is not possible, then ground and aerial photographs may be used instead, but the resultant data are approximations at best. Forecasting where soil lost to erosion goes beyond the site requires delineation of drainage lines (gullies, swales, ditches, stream channels) in the field with the aid of topographic maps and aerial photographs.

Soil erodibility

K Factor. Soil type is expressed in terms of an erodibility factor, or *K factor*, which is a measure of a soil's susceptibility to erosion by runoff. This factor was derived from tests conducted by soil scientists on field plots for each soil series in various states of the United States and is given on a dimensionless scale between 0 and 1.0, though actual values range between 0.02 and 0.7. The higher the number is, the greater the susceptibility there is to erosion. In some cases, two numbers are given for a soil: one for disturbed ground and one for undisturbed ground. Disturbed ground includes filled and rough-graded ground. *K* factors are usually available from county or state offices of the U.S. Natural Resources Conservation Service. If *K* values are not available, then an alternative method can be used based on laboratory analysis of a soil's particle size composition.

Rainfall, slope, cover

R, S, and C Factors. The rainfall erosion (*R*) values for most states may be read from the map in Figure 12.7. Reliable *R* values for the western states have not been generated because rainfall is highly variable owing to the diverse topography. For locations in the West, *R* values can be calculated from local rainfall records with best results using the 2-year, 6-hour rainfall figures. The slope factor (*S*) can be read from Table 12.1, and the plant cover factor (*C*) can be approximated from Table 12.2 based on groundcover and canopy density.

Setup: subareas

The reliability of soil erosion computations based on the universal soil loss equation depends not only on the accuracy of the data used, but also on how the problem is set up. For sites in areas of diverse terrain, the parcel should be divided into subareas in which soil, slope, and vegetation are reasonably uniform. Soil loss should be computed for each subarea and then adjusted for any deposition that may be received from adjacent subareas. In making the adjustment for deposition, consider not only

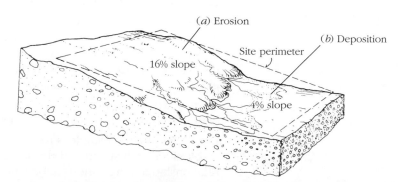

Fig. 12.9 A site comprised of two classes of slope, one that yields sediment and one that accumulates it. Erosion is significant on the upper slope, yet there may be little or no net loss of sediment from the site.

Table 12.2   Plant Cover Factors

Canopy Cover (%) — *Ground Cover (%)*

Canopy Cover (%)	0	20	40	60	80	95-100
0	0.45 / 0.45	0.20 / 0.20	0.10 / 0.15	0.042 / 0.09	0.013 / 0.043	0.003 / 0.011
25	0.39 / 0.39	0.18 / 0.22	0.09 / 0.14	0.039 / 0.085	0.013 / 0.042	0.003 / 0.011
50	0.39 / 0.39	0.16 / 0.19	0.08 / 0.13	0.038 / 0.080	0.012 / 0.040	0.003 / 0.011
75	0.27 / 0.32	0.10 / 0.18	0.08 / 0.12	0.035 / 0.080	0.012 / 0.040	0.003 / 0.011

1: Grassy surface cover with turf or litter at least 2 in. deep.

2: Broadleaf herbs, such as weeds, with little top soil development.

the potential contribution from upslope areas, but also the patterns of gullies, streams, and swales through which runoff and sediment can be funneled across flat areas and discharged onto low ground or into streams, lakes, and wetlands.

   Computational Procedure. To summarize, the procedure for computing soil loss from a site takes seven steps:

Suggested steps

1. Define the site (problem area) boundaries on a large-scale base map, preferably a topographic contour sheet.

2. Using aerial photographs, soil maps, topographic maps, field inspection, and whatever sources are available, divide the site into subareas. If the site is essentially uniform throughout, this step is not necessary.

3. For each subarea, assign a *soil erodibility factor* (based on NRCS soil series *K*-factor designation), a *slope factor* (from Table 12.1 based on steepness and length data taken from topographic map) and a *plant cover factor* (from Table 12.2 based on aerial photographs or field observation).

4. Multiply the appropriate value from the *rainfall erosion* map in Figure 12.7 times the three factors assigned to the subarea in step 3. The answer is in tons of soil erosion per acre per year. If agricultural land is involved and one of the cropping management practices given in Table 12.3 is used, apply the appropriate *P* factor.

5. Examine the relations between slopes and drainage patterns in each subarea, identify where sediment is expected to accumulate, and, if possible, adjust the quantity of soil erosion computed in step 4 and derive an estimate of soil loss.

6. Determine the total soil loss for the entire site by summing the net amounts of soil loss for each subarea. If this is not possible because of uncertainty over the relationship among various subareas, then the gross soil loss from the subareas should be summed for a gross site total.

7. Based on slope and runoff patterns, both within the site and on neighboring land, identify the points and areas to which sediment is transported and where sedimentation (storage) is likely to take place.

Table 12.3   Cropping Management (P) Factor

Land Slope, Percentage	P Values		
	Contouring	Contour Stripcropping	Contour Irrigated Furrows
2.0 to 7	0.50	0.25	0.25
8.0–12	0.60	0.30	0.30
13.0–18	0.80	0.40	0.40
19.0–24	0.90	0.45	0.45

12.6 APPLICATIONS TO LANDSCAPE PLANNING AND MANAGEMENT

Rationale

An important task of most local planning agencies is the review and evaluation of land development proposals for housing, industrial, and commercial projects. Proposals are evaluated according to a host of criteria, including soil erosion or the potential for it. Consideration of soil erosion stems not only from a concern over the loss of topsoil and depletion of the soil resource in general, but also from the impact of sediment on terrestrial vegetation, wetlands, stream channels, lakes, reservoirs, water supply systems, and drainage facilities such as stormsewers. To gain the necessary information, planners are often asked to prepare a site plan and respond to critical questions like these:

Key questions

■ What is the watershed address of the site and what is its position in the drainage net (such as headwaters, conveyance zone, confluence point)?

■ What are the minimum setback distances between the proposed development and (a) water features (streams, ponds, reservoirs, and wetlands) and (b) existing drainage facilities (such as water intakes and stormdrains)?

■ What percentage of the site lies in slopes of 15 percent or greater, and of this area how much (*a*) is proposed for development, and (*b*), if developed, will be affected by construction?

■ What percentage of this site is forested, grass covered, and shrub covered? How much of each of these covers will be destroyed as a result of development?

■ What erosion and sedimentation control measures are proposed during (*a*) the construction phase and (*b*) the operational phase of the proposed project?

■ What is the anticipated length of the construction period, and which months in the year are proposed for (*a*) land clearing, (*b*) excavation and grading, (*c*) construction of facilities, and (*d*) regrading and landscaping? How do these phases relate to the seasonal pattern of rainfall, especially the months of heaviest rainstorms (Fig. 12.10*a*)?

■ Finally, how will these activities be phased so that not all the site will be opened at once (Fig. 12.10*b*)?

Mitigation Measures. Control of erosion and sedimentation during the construction phase of development is viewed very seriously in many communities—so much so that many have legislated erosion control policies. At the heart of such regulations are the control measures, or **mitigation measures,** to be employed. These include a wide variety of techniques and approaches, and there is a very active dialogue among permitting agencies, scientists, landscape architects, civil engineers, construction management professionals, and others on this topic.

The approaches employed vary widely, depending mainly on local requirements and the level of monitoring and enforcement employed by public agencies. Generally,

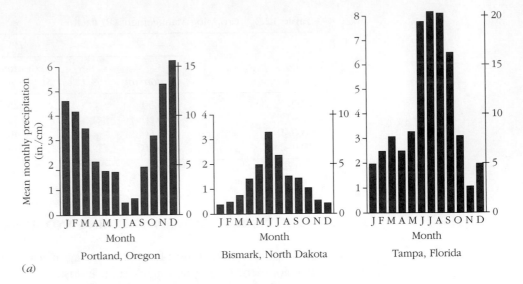

(a)

Portland, Oregon   Bismark, North Dakota   Tampa, Florida

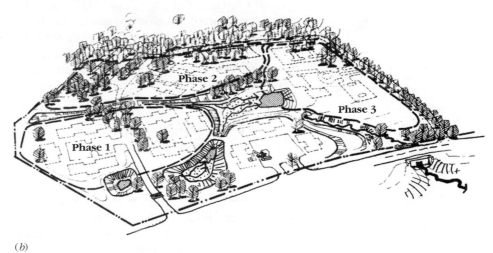

(b)

Fig. 12.10 (*a*) Mean monthly precipitation for three locations in the United States. Both the amount and seasonal distribution of precipitation are important considerations in project scheduling. (*b*) Phasing of development is necessary in large projects to minimize land exposure and to hold down soil erosion and sedimentation of water features on and near the site. In this project, only 25 to 30 percent of the site is open at any time.

Strategies

however, the mitigation measures used are simple and small scale, and most are aimed at (1) holding an exposed soil in place; (2) blocking sediment-laden runoff from draining into water features such as streams, ponds, or wetlands; or (3) filtering muddy water through fabric or straw. Measures include the placement of silt (cloth) fences, straw bales, and berms around construction zones, the use of fiber nets and fabric such as burlap on slopes, and the use of sedimentation (sump) basins to collect and isolate muddy water. (For additional comments on water quality mitigation techniques, refer to Chapter 11, Section 11.4.)

Performance

Despite the requirements of ordinances/bylaws and related efforts to control erosion and sedimentation, mitigation programs on many, if not most, construction sites are inadequate. The reasons for this include too great a reliance on silt fences, inadequate attention to sites as parts of larger drainage systems, inattention to local runoff patterns and volumes, and poor understanding of rainstorm magnitudes and frequencies. Especially critical is the use of silt fences around construction sites where

Fig. 12.11 Silt fence overloaded by fill and sediment. A typical scene on constructions sites in the United States and Canada.

potentially large volumes of runoff are involved (Fig. 12.11). In general, silt fences can retain only small amounts of runoff and sediment such as that from exposed strips along roadways or embankments.

   Beyond Silt Fences. For open areas larger than 3 acres or so, it is advisable to berm the perimeter, particularly on the downslope side, to contain both runoff and sediment. The height of the berm can be set according to the 10- or 25-year storm magnitudes and the site acreage. (In this situation, the proper location of a silt fence is along the toe of the outer slope of the berm itself.) For steeply sloping ground, perimeter berms may have to be supplemented by terraces to slow runoff and to capture sediment before it reaches the edge of the site, as the photograph in Figure 12.12 suggests.

Fig. 12.12 Simple soil terraces are sometimes a helpful first-phase measure to slow runoff and check soil loss.

In addition, several traditional measures, using stakes and woody cuttings, can be employed to help arrest erosion. Some of these date from the 1930s when they were used in public works projects to stabilize banks and gullies. One measure, called *contour wattling* or *wattle fencing*, involves placing live cuttings such as willows along the contour of the slope and staking them in place, as shown in Figure 12.13*a*. The willow bundles may be partially buried to form small terraces that catch runoff and sediment. Properly done, the cuttings should sprout roots and stems, strengthening the slope and dissipating runoff. Figure 12.13*b* shows a similar concept, called a *wattle check dam*, applied to an active gully. The objective is to slow runoff, capture sediment, induce infiltration, and arrest channel degradation. Such facilities are best suited to small and moderate flows; large flows can blow them out. To avoid this, it is advisable to carefully consider upslope drainage areas and the sources and sizes of storm flows.

Old standbys

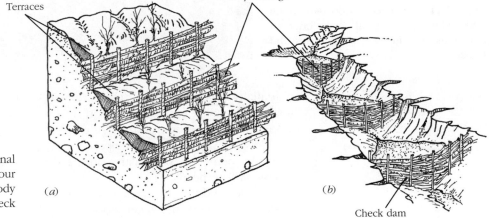

Fig. 12.13 Examples of traditional methods of erosion control: (*a*) contour wattling or wattle fencing using woody cuttings and stakes and (*b*) wattle check dams in a gully.

12.7 WATERSHED MANAGEMENT CONSIDERATIONS

Our perspective on soil erosion and stream sedimentation must always extend beyond the site scale of observation. As we have stressed throughout this book, virtually every site is situated in a drainage basin where runoff is generated and where sediment is not only produced and entrained in the flow system but deposited in various types of sinks along the way.

The Sediment Transport System. This flow system is made up of a network of drainage channels that together function as the principal conduits in the sediment transport system. In most natural drainage systems, the size of an individual channel is adjusted to the size of the flows it conducts, in particular to certain flows of larger-than-average magnitudes. We reason, therefore, that when two streams join in the network, the size of the resultant channel should approximate the sum of the two tributary channels. Actually, the *capacity* of the channel, defined by the magnitude of discharge that it can accommodate, increases with the merger of tributaries.

Channel system

Sediment transport capacity also increases with the magnitude of stream discharge. As magnitude rises toward bankfull stage, channel scouring mobilizes sediment, and there is a rapid rise in the rate of sediment transport (see Fig. 14.4). And because large discharge events occur infrequently, streams tend to move most sediment in infrequent surges. Therefore, in assessing the disposition of sediment released to a stream from an erosion site, it is important to consider first the stream's discharge regime, then the size of the receiving channel and its relative position in the drainage network. Massive loading of small streams (such as the one shown in Fig. 12.14) located in the upper parts of a watershed often results in channel glutting, which

Sediment transport

reduces the capacities of channels to handle large flows. Flooding increases, carrying additional sediment onto the floodplain, into wetlands, and to other areas. Much of the sediment deposited in the channel and on the valley floor is relegated to storage, especially after vegetation has become established on it, and then not even large floodflows can displace it.

Sedimentation Sites. In a watershed where several sites are releasing sediment to a stream network at the same time, sediment becomes differentiated into bodies or waves that move through the stream system at different rates, making it difficult to assess the impact on the system as a whole. Clays and silts that are carried in suspension in stream water move through the entire system rapidly, whereas sand and coarser materials not only move more slowly but also build up at selected points (Fig. 12.15).

Often we can estimate where in the drainage system sediment is likely to accumulate based on sediment size and channel size, shape, and gradient. Broad reaches of slow moving water are the favored deposition sites for sands, pebbles, and other large particles. Among these, wetlands and impoundments such as reservoirs and *Sediment sinks* lakes are generally considered the most critical from a watershed management standpoint, and it is often necessary to estimate the amount of sediment that would be added to them based on land use practices in the drainage basin. On the other hand, knowledge of where and why sediment accumulates in a drainage system gives us clues about how to design facilities to capture fugitive sediment, as illustrated in Case Study 12.8.

For purposes of watershed management, *sediment loading* of a drainage basin requires a yearly calculation of sediment production on a site-by-site basis for the entire watershed. The locations represented by each of these quantities must be pinpointed in the watershed and the drainage network, taking special care to determine where sediment would actually enter the network. In cases where a site falls on a drainage divide within the watershed, it may contribute sediment to two or more channels. Once these relationships are known, the receiving waterbodies downstream can be identified and a loading rate can be calculated.

Fig. 12.14 A small channel glutted with sediment from upstream soil erosion. The sediment reduces channel capacity to conduct discharge and destroys most of the aquatic plants and animals.

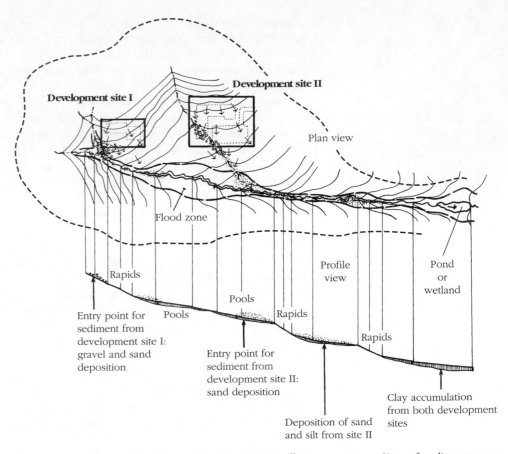

Fig. 12.15 Location of sedimentation sites in a small stream system. Sites of sediment accumulation vary with sediment size; loading rate; proximity to sediment sources; and channel size, shape, gradient, and stream discharge regime.

Finally, the sediment loads must be converted to volumetric units to determine how much space in the channel or impoundment the sediments will actually take up.

Impoundment loading The following sediment densities are recommended:

■ Clay: 60–80 pounds per cubic foot
■ Clay/silt/sand mixture: 80–100 pounds per cubic foot
■ Sand and gravel: 95–130 pounds per cubic foot

Using an intermediate value of 90 pounds per cubic foot, we find that a reservoir receiving 10,000 tons of sediment per year would lose 222,000 cubic feet (8200 cubic yards) of volume each year. Given a reservoir capacity of 2,000,000 cubic feet (one, for example, with dimensions of 200 feet wide, 1000 feet long, and 10 feet deep), the life of the reservoir would be only 9 years. Thousands of reservoirs in North America were lost to sediment in the twentieth century. Especially vulnerable were those overtaken by urban sprawl whose watersheds were torn open during the construction phase of development.

For impoundments with low resident times, that is, those that exchange water in only a matter of hours or several days rather than months or years, most of the clays do not settle out. Owing to their small sizes, clay particles remain in suspension, allowing most to pass through small impoundments. Impoundments with long residence times, which may be as great as 10 to 12 years in some lakes, retain fine sediments and deposit them over the lake floor, burying or coating bottom organisms.

12.8 CASE STUDY

Erosion and Sediment Control on a Creek Restoration Project, South Lake Tahoe, California

Steve Goldman

Erosion and sediment control methods must be tailored to site conditions. No one plan will work for every site. In this case study, which involved creek restoration in the Trout Creek watershed of Lake Tahoe, conventional erosion and sediment control measures did not and could not do the job (Fig. 12.A). The solution came from understanding the nature of the problem and taking advantage of the natural filtering system that existed on this site.

This project involved the construction of 6000 feet of meandering stream channel. The stream, known as Cold Creek, is located in El Dorado County, California, a few miles south of Lake Tahoe. It is a tributary to Trout Creek and has a watershed of 13 square miles. Above Cold Creek's confluence with Trout Creek is a 3000-foot-long meadow, through which Cold Creek originally meandered. During the 1950s, a dam was placed across the narrow part of the meadow, the creek was rerouted into a ditch along the northeast side of the meadow, and a small feeder channel was constructed from the creek to the meadow to create a lake there. The meadow-turned-lake, though originally constructed for agricultural purposes, became known as Lake Christopher. The creek was restored in 1994 by the City of South Lake Tahoe, with a grant from the California Tahoe Conservancy.

The objectives of the project were to restore a naturally functioning creek in the meadow, to reduce erosion, to protect Lake Tahoe water quality, and to enhance wildlife and fish habitat. The restored creek was connected, via a small side channel, to two waterfowl ponds, which were constructed in the meadow as part of an earlier project. A major challenge of the project was the desilting of the 6000 linear feet of newly constructed stream channel without allowing sediment or other pollutants to be discharged into Trout Creek.

The contractor hired to construct the project was also responsible for erosion and sediment control during construction. Because there are no standard methods for erosion control for stream restoration projects, the contractor chose to use conventional erosion control devices for construction sites, namely, silt fences and straw bales. Unfortunately, these devices proved to be ineffective at trapping sediment because the flow greatly exceeded the ability of the fences to conduct the water. As a result the area behind each fence quickly filled with sediment and most of the water poured over the top, discharging its sediment load downstream. Turbid water was soon observed entering Trout Creek at levels significantly exceeding state standards.

State water quality officials then convened a meeting of the city, the contractor, and the project team. It was agreed that the city would submit an erosion control plan for the final phase of the project, with the intent that the plan would become a model for future stream restoration projects at Lake Tahoe. A plan was developed that mimicked nature's way of trapping sediment. It involved discharging overbank flows into the meadow—similar to floodflows on a floodplain—to desilt the creek. First, a small wooden dam was constructed at the lower end of the newly constructed creek channel, adjacent to a low beaver dam. The inflow to the new creek channel was controlled by placing sandbags at the junction point of the new creek channel and the old outlet channel from Lake Christopher. The sandbags allowed the creek flow to be routed in either channel at variable rates. From the new channel, creek water then gradually overflowed into the meadow. Workers were stationed at various points to observe where the water was spreading and to make adjustment in the flow released to the new channel.

Fig. 12.A   The Lake Tahoe Basin. The watershed covers 314 square miles and reaches high into the Sierra Nevada Mountains, the lake's principal source of water.

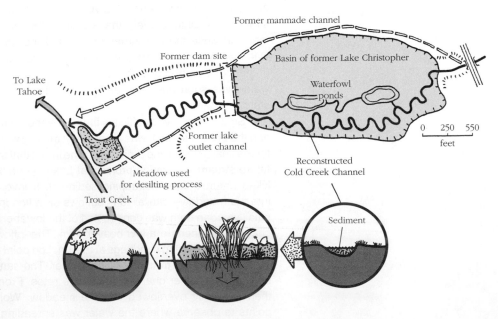

Fig. 12.B   The settings for the Cold Creek restoration project. Sediment-laden water from the reconstructed channel was spread over a meadow before reaching Trout Creek, which trapped most of the sediment.

Based on estimates from photographs of the meadow, the spreading area covered 72,000 square feet. Using this figure, the inflow rate, and standard particle-settling velocities, the sediment trapping efficiency of the treatment area was calculated, and it was revealed that all sand- and silt-sized particles would be trapped in the meadow. Only clay-sized particles could possibly escape the meadow trap. However, the equation for this calculation does not include the effect of vegetation, and, because the meadow was densely covered with sedges, the actual trap efficiency was probably greater than the calculation indicated. Because the silty water had to flow through thousands of plant stems on the way to Trout Creek, much of the sediment load was caught along the way. Resuspension of trapped fine sediments was unlikely because of the low velocities across the meadow and because much of the settled soil tended to fall into the myriad crevices among the plant stems and roots. Thus, the water entering Trout Creek from the flooded meadow was clean, a sharp contrast to the highly turbid water that poured over the silt fences during the first phase of construction (Fig. 12.B).

After spreading the water across the meadow for a week, the boards in the dam at the bottom end of the project were then removed one at a time over a 5-day period. Each time a board was removed, a small amount of turbidity was observed, which typically lasted less than 30 minutes. When the last board was removed, the turbidity lasted about 2 hours, but it was at a much lower level than the turbidity that occurred during the first phase of the project. This method of controlling sediment proved to be a success, and no further regulatory actions were necessary.

We learned a lot from this project. As with so many aspects of environmental planning, if we model our approaches after natural processes, we can often achieve highly satisfactory results at far lower costs with less environmental disturbance than would be possible through more conventional technical approaches.

Steve Goldman is an erosion control and stream restoration specialist formerly with the California Tahoe Conservancy, South Lake Tahoe, California. He is senior author of *Erosion and Sediment Control Handbook* (McGraw-Hill).

12.9 SELECTED REFERENCES FOR FURTHER READING

Ferguson, Bruce K. "Erosion and Sedimentation in Regional and Site Planning." *Journal of Soil and Water Conservation*. 36, 4, 1981, pp. 199–204.

Glanz, James. "New Soil Erosion Model Erodes Farmers' Patience." *Science*. 264, 1994, pp. 1661–1662.

Goldman, S. J., et al. *Erosion and Sediment Control Handbook*. New York: McGraw–Hill, 1986.

Heede, B. H., "Designing Gully Control Systems for Eroding Watersheds." *Environmental Management*, 2, 6, 1978, pp. 509–522.

Herweg, K., and Ludi, E. "The Performance of Selected Soil and Water Conservation Measures—Case Studies from Ethiopia and Eritrea." *Catena*. 36, 1999, pp. 1–2, 1999.

Hjulström, F. *"Transport of Detritus by Moving Water."* In *Recent Marine Sediments: A Symposium* (P. Trask, ed.). Tulsa, OK: American Association of Petroleum Geologists, 1939.

Meade, R. H., and Parker, R. S. "Sediment in Rivers of the United States." In *National Water Summary 1984*. U.S. Geological Survey Water-Supply Paper 2275, 1985, pp. 49–60.

Patterson, R. G., et al. "Costs and Benefits of Urban Erosion and Sediment Control: The North Carolina Experience." *Environmental Management.* 17, 2, 1993, pp. 167–178.

Pimentel, David, et al. "Environmental and Economic Costs of Soil Erosion and Conservation Benefits." *Science.* 267, 1995.

Trimble, S. W. "A Sediment Budget for Coon Creek Basin in the Driftless Area, Wisconsin, 1853–1977." *American Journal of Science.* 283, 1983, pp. 454–474.

U.S. Environmental Protection Agency. *Erosion and Sediment Control: Surface Mining in the Eastern United States.* Washington, DC: U.S. EPA Technology Transfer Seminar Publication, 1976.

Wischmeier, W. H., et al. *Procedure for Computing Sheet and Rill Erosion on Project Areas.* Washington, DC: U.S. Natural Resources Conservation Service, Technical Release No. 51, 1975.

Wolman, M. G. "A Cycle of Sedimentation and Erosion in Urban River Channels." *Géografiska Annaler.* 49A, 1967, pp. 385–395.

Related Websites

Huron River Watershed Council. "Impacts of Land Use: Soil Erosion and Sedimentation." 2005. http://www.hrwc.org/text/wqsedimentation.htm
Watershed council that successfully educates the public and acts to protect the Huron River in Michigan. Focuses on land use planning and water quality, including soil erosion and sedimentation.

Landplan Engineering Supplies. 2006. http://www.landplan.com.au/index.htm
A company producing soil erosion options. See pictures and information about gabions, geotextiles, grasspavers, and the like.

Natural Resource Conservation Service, U.S. Department of Agriculture. http://soils.usda.gov/
Governmental division protecting soil from erosion. Searching "soil erosion" brings up an article that tells you all about erosion, its effects, and mitigation and that includes several pictures.

Soil Conservation Council of Canada. 2009. http://www.soilcc.ca/about-us.htm
A nongovernmental organization of parties concerned about soil conservation. The site focuses on the agriculture sector, specifically no-till practices and greenhouse gas emissions in farming and ranching. See the feature article, "Good News and Bad News of Soil Conservation."

Vetiver Network International. http://www.vetiver.org/
Vetiver grass as a way of controlling soil erosion. Links are available to learn more about the plant, particularly its application and installation worldwide.

BEST MANAGEMENT PRACTICES, LOCAL WATERSHEDS, AND DEVELOPMENT SITES

13.1 INTRODUCTION

In previous chapters we examined loading of the runoff system with stormwater, sediment, and nutrients and other contaminants. In each of Chapters 8, 9, 11 and 12, we highlighted and briefly discussed various management concepts and measures. In this chapter we focus exclusively on best management practices (BMPs) and present a general approach or strategy that can help build BMP plans to mitigate the effects of land use on the runoff system. Through this approach we hope to broaden the conventional concept of BMPs and introduce a multitiered concept of BMP planning.

Definition

BMPs can be defined as measures taken to prevent or reduce the detrimental impacts of land use development and practices on the environment. They may address any part of the environment, but their most common use is associated with runoff systems, especially stormwater. In this context, BMPs usually take the form of structural devices such as detention ponds, which are installed with or after buildout and are designed to regulate stormwater discharge and to help reduce soil loss and water pollution. But BMPs can also include other measures, often nonstructural and micro in scale, such as special elements in landscape and building design, watershed-based planning, policy guidelines, landscape design, information programs, and ecological restoration.

Our approach

Our approach to BMP planning begins with the watershed and calls for measures best described as preventative, that is, measures that are introduced before development as part of the planning process. This is followed by site-scale BMP planning that deals with corrective measures in the postdevelopment landscape. At the watershed scale, we are interested in understanding how the runoff system functions and how land uses can be designed to conform with these functions and capitalize on mitigation opportunities that already exist in the watershed. At the site-scale, stormwater and its contaminant load are addressed in terms of the three-part system outlined in Chapter 11 and illustrated in Figure 11.1: (1) production, (2) removal or release from the site, and (3) delivery to a receiving water feature. The overall goal of the BMP package at both the watershed and site scales is to mimic the predevelopment performance of areas subject to development and to reduce the need for large, structural mitigation measures.

13.2 THE WATERSHED RUNOFF SYSTEM AND BMP OPPORTUNITIES

BMPs have traditionally been the domain of civil engineering, and their application has been mainly corrective rather than preventative. Where development is already in place, engineered BMPs remain an appropriate tool, but for undeveloped and partially developed lands, they should not be our first choice. Instead, *BMP planning* should take place before development, as part of land use planning and design, and

Proactive planning

it should focus on finding where the watershed has the capacity to take on development without the assistance of structural devices to mitigate stormwater. In other words, BMP planning should be proactive rather than reactive in the planning and development process.

If we were able to plan and design land uses to properly fit the land and its carrying capacity, new development would not need to be retrofitted with structural BMPs. Unfortunately, we have gotten into the habit of planning less for land and more for real estate and zoning interests. Therefore, we commonly miss opportunities to reduce or even eliminate elaborate and costly structural measures in attempting to mitigate stormwater runoff. In short, if we were better problem solvers in landscape planning and design—using land use densities, mixes, and configurations appropriate for various terrains and drainage systems—we would probably have little need for pipes and ponds to take care of stormwater.

Watershed context

System Considerations. Virtually every development site, no matter where it is located, belongs to a watershed and is inherently part of a drainage system. For undeveloped lands, the first step in the BMP process is to define where you are in the drainage system, what runoff processes are operating there, and what conditions are associated with that location and its processes. In small watersheds (where most development takes place), development sites fall within one of the *three hydrologic zones* we discussed in Chapter 9: contributing, collection, and conveyance (Fig. 13.1).

Each zone in the drainage system handles water differently, and it is necessary to understand these functions to make BMP decisions at the watershed scale. Once the essential processes and related conditions have been identified, it is often helpful to model the subject area to estimate predevelopment hydrologic performance. The model should include not only estimating how much stormwater will be generated but also how that stormwater moves over the area, that is, the patterns and processes involved in its movement downslope.

Partial area concept

Most watersheds function as *partial area systems*, meaning that only part of the basin actually contributes surface runoff to streamflow in response to a rainstorm (see Fig. 8.6). The total area contributing runoff, such as stormwater, ranges roughly from 10 to 30 percent, depending on the magnitude of the rainstorm and prestorm basin conditions such as soil moisture, ground frost, and snow cover. Because most of the basin never or rarely contributes stormwater, it is extremely important to define the distribution of noncontributing areas in order to figure out, first, how land uses should be assigned to the watershed and, second, where there are opportunities to take advantage of built-in BMP features such as wetlands and sandy soils.

Field evidence

Noncontributing Drainage Area. Finding *noncontributing areas* begins with a basic mapping of slopes, soils, vegetation, water features, and land uses. Areas with permeable soils and substantial vegetation are good candidates, as are wetlands, especially closed or partially closed ones (Fig. 13.2). Once candidate areas have been defined, a field examination should be conducted to verify the absence or paucity of surface runoff. The main features to look for are small channels such as swales, gullies, and rills. Where there is no evidence that the ground is serviced by ephemeral channels, even microchannels such as rills, then it is doubtful that the site generates stormwater runoff in the conventional sense (that is, as overland flow). In other

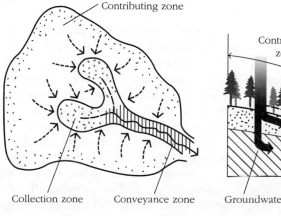
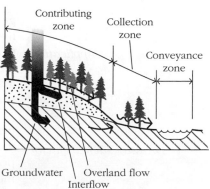

Fig. 13.1 The three main hydrological zones of a small watershed: contributing, collection, and conveyance. The contributing zone represents the large part of the watershed and offers the greatest opportunity for stormwater mitigation.

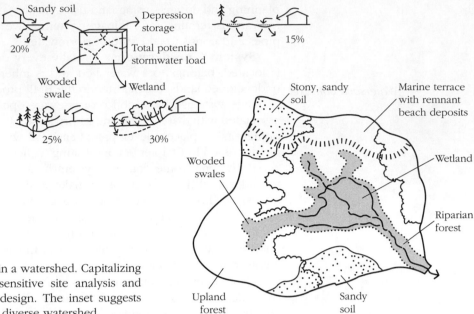

Fig. 13.2 Potential BMP opportunities in a watershed. Capitalizing on these opportunities requires both sensitive site analysis and thoughtful land use planning and site design. The inset suggests how stormwater might be allocated in a diverse watershed.

words, surface runoff is so weak and/or infrequent that it is incapable of mounting flows large enough to etch the surface.

In such areas runoff is moved in other ways, as interflow and/or groundwater after infiltration has taken place, or it is taken up in depression storage, as soil moisture, or in interception. This is often the case in forested watersheds where previous land uses have not erased the microtopography and topsoil; have not cleared and extended swales; and have not built roads, ditches, field drains, and the like (see Fig. 8.11). Tree roots, ground cover, and organic litter, coupled with tree throws, rotting stumps, and burrowing animals, all impart roughness, permeability, and porosity to the forest floor. Unless the floor has been graded in some fashion as a part of earlier land use activity, the ground is usually incapable of generating overland flow.

Landscape attributes

Identifying such conditions is important in BMP planning. They are clues that an opportunity may exist to incorporate natural mitigation measures into the land use plan, thereby reducing or even eliminating the need to construct BMP facilities with or after development. And where there are ample opportunities for natural mitigation, we need to take our thinking to the field, define and map the significant features, and assess how development plans and site design schemes might be adapted to best take advantage of them.

Seize the day

Source Control Planning. This approach is fundamental to the application of **green infrastructure** concepts because green approaches to stormwater management rely heavily on **source control** of runoff. To employ source controls, the site planner must either use existing means to hold and infiltrate stormwater or create the means. The most successful source control measures, such as infiltration beds or grass-lined swales, combine existing opportunities like permeable soils with grading, soil treatments and planting schemes as a part of site planning and design.

Application

Another attribute of watersheds that is important in making assessments for BMP planning at the watershed scale is drainage density. Low drainage densities indicate that rainwater is being taken up and stored, returning to the atmosphere via evapotranspiration, and/or running off by means other than surface flow, that is, by interflow and groundwater. High drainage densities, on the other hand, indicate that stormwater and channel flows are abundant and that few, if any, sites in the watershed are not serviced by a stormwater channel, either natural or human-made (see Fig. 8.10).

Drainage-density factor

Where a watershed is laced with stormdrains, the opportunities to use natural opportunities for BMP planning are greatly diminished, especially at the source control level, because the stormdrain system taps into most sites and in many instances overrides the watershed's built-in opportunities. In some cases, a site can be detached from this network of stormdrains, but, short of ripping out the stormdrain system or abandoning it, it is advisable under such conditions to look to the site itself as the place to apply BMPs.

13.3 THE SITE-SCALE STORMWATER SYSTEM

Components

Within the watershed, each development site has its own stormwater system. This system consists of three components: (1) the on-site *production* of stormwater and contaminants, (2) the *removal* of stormwater from the production site, and (3) the *delivery* of stormwater and its contaminant load to a receiving waterbody (Fig. 13.3).

From Here to There. *Production* is defined as surface water (and its contaminants) generated from cleared and developed ground that is available for stormwater runoff if it is released from the site. *Removal* is the means by which stormwater is released from the site and discharged into a delivery system. *Delivery* is the means by which stormwater is conducted to a receiving waterbody.

The *production system* consists of all the facilities and land use practices added to a site that result in increased amounts of stormwater and contaminants. Production begins with land clearing and commonly includes soil compaction, construction of impervious cover, lawn fertilization, and garbage burning. The *removal system* is usually made up of gutters, downspouts, yard drains, and field tiles, whereas the

System overview

delivery system is usually made of curbs, gutters, ditches, and stormsewers. Once in place, the removal and delivery systems function as bypass devices that override the site's natural hydrological system, making it "more efficient" as a flow system. In conventional land use development, stormwater is produced quickly, removed from the site quickly, dumped quickly into a delivery system, and quickly discharged into nearby streams.

Levels of Control. The choice of BMPs for the site stormwater system is based on how the development and associated drainage system functions to produce, convey, and deliver stormwater from the site to the receiving waterbody. Three categories

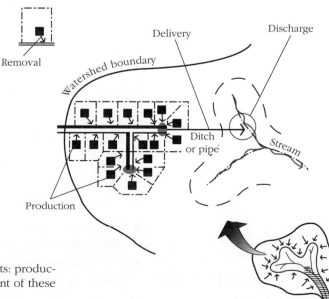

Fig. 13.3 The site stormwater system consists of three components: production, removal, and delivery. This diagram illustrates the arrangement of these three components in a conventional stormwater system.

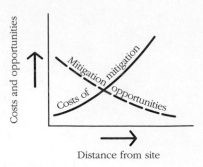

Fig. 13.4   The general trends in mitigation (BMP) opportunities and costs with distance from the site.

Application levels

of BMPs are defined, corresponding to each phase of this three-part system: planning, design, and engineering. Various combinations of BMPs representing these three levels of control may be selected for different settings and problems. Once stormwater passes through the system and is delivered to the receiving waterbody, BMP opportunities are greatly diminished or lost altogether. As a rule of thumb, we can expect BMP opportunities not only to decline with distance from the site, but also to rise in cost, usually significantly (Fig. 13.4). At the delivery phase, engineering solutions are usually the only options.

13.4 PRODUCTION BMPs

The basic idea

Production BMPs are used to decrease the on-site production of stormwater runoff and contaminants. This class of BMPs is typically ignored in conventional development in preference to catch-all BMPs, such as stormdrains and detention basins designed to capture large amounts of stormwater runoff from one large site or many small ones. Production BMPs utilize planning strategies, planning and development policies, and educational programs to help reduce runoff and pollution production on individual sites. This includes not only siting facilities in appropriate locations based on geographic variations in landscape carrying capacity, but also following site-adaptive design principles by designing facilities and connecting systems to optimize site conditions, thereby advancing sustainable design principles at both the watershed and site scales (Table 13.1).

Policy, design, management

Types of Production BMPs. In general, production BMPs fall under three major headings: (1) planning policy, (2) architectural and landscape design, and (3) site management practices. Planning policies that affect the production of stormwater and related pollutant loading include zoning regulations addressing land use type and density; lot coverage restrictions; wetland protection regulations; and slope limitations on street, road, and building placement. Architectural and landscape design BMPs include limits on roof-to-floor-area ratios; limits on site coverage by outbuildings, walks, and drives; rules on foundation and basement design; and guidelines on grading and the use of plant and soil materials in landscaping. Site management practices address BMPs related to, among other things, waste disposal, fertilizer and pesticide application, irrigation and water recycling, and the number and type of domestic animals housed on a site.

Production BMPs also include:

■ Limits on impervious cover, including decreased road widths, smaller driveways, and reduced roof areas.

■ Site management guidelines that encourage retaining woodland and wetland areas.

Table 13.1 Planning Options and Production BMP Measures at the Watershed and Site Scales

Planning Options That Affect Stormwater and Contaminant Production	Production							
	Watershed Scale			Site Scale				
	Preservation of Natural Wetlands	Preservation of Permeable Soils	Preservation of Natural Depression Storage Areas	Density-Site Carrying Capacity	Vegetation Conservation and Riparian Buffers	Road Configuration and Design	Production-Based Policies (e.g., hazardous waste disposal)	Education Programs
Land cover	✓	✓	✓		✓	✓		✓
Impervious area				✓		✓		✓
Runoff storage	✓		✓					✓
Runoff treatment	✓	✓			✓			✓
Runoff infiltration	✓	✓			✓			✓
Pollution production				✓		✓	✓	✓

Additional measures

- Road and neighborhood configurations designed to reduce automobile travel distances and encourage sustainable modes of transportation.
- Roof water storage and recycling through the use of cisterns coupled with garden and lawn applications.
- Backyard burning, pet feces cleanup, and household and industrial hazardous waste disposal practices to reduce on-site pollution production.
- Education and information programs to encourage home and business owners to adopt more sustainable waste disposal and landscape maintenance practices.
- Incentive programs, including property tax abatements, for installing stormwater reduction measures such as porous pavers, dry wells, and woodlands.

Implementation. Production BMPs can be implemented in various ways. At the planning policy level, they can be implemented through:

Planning policy

- *Community planning:* For example, designating land use types.
- *Zoning ordinances/bylaws:* For example, restricting lot coverage and density.
- *Subdivision review:* For example, directing road location and size.
- *Development permits:* For example, designating environmentally sensitive areas as development permit areas.
- *The building permit process:* For example, requiring onsite stormwater measures.

The same implementation recommendations hold for architectural and landscape design BMPs. At the site management level, however, implementation demands an approach that combines public policy enforcement (e.g., garbage disposal), incentives (e.g., property tax abatement for stormwater recycling), and public information and education (e.g., fertilizer applications and car washing).

Developed sites

Production BMPs can be used for both undeveloped and developed areas, but developed areas offer less flexibility and fewer options. Not only are buildings already in place but stormwater facilities usually exist as well. Nevertheless, mitigation opportunities may be available in neighborhood parks, greenbelts, street parkways, remnant wetlands, and residential yards. These areas need to be evaluated to determine whether they have stormwater management potential and can be integrated into a larger management plan.

Retrofitting

Developed sites can also be retrofitted in various ways to reduce stormwater and contaminant production. A simple example is replacing worn-out driveways and sidewalks with porous pavers. Another is capturing part of roof drainage in cisterns with overflow quantities going to dry wells or infiltration beds on the yard perimeter (Fig. 13.5). The resulting reduction in stormwater production not only lowers environmental impacts but also reduces wear and tear on stromdrains, thereby increasing their life cycle periods.

Plans and designs must maximize opportunities to reduce stormwater production and maximize on-site infiltration. To test the efficacy of a proposed plan, runoff simulation models can be applied to determine whether it meets predevelopment performance standards. In certain situations, management programs that utilize only production-level BMPs are not sufficient to effectively treat and manage stormwater production from a developed site.

Testing the Site-Scale BMP Plan. A simple way of testing a BMP plan is to use a spreadsheet approach to stormwater accounting. This approach involves calculating the volume of stormwater (Q_v) produced from each surface within the site before and after development based on its coverage and coefficient of runoff. The procedure begins with the predevelopment (or presettlement) site and its performance. If, for example, we are dealing with a proposed residential lot of 0.25 acre (11,000 square feet) and a predevelopment C value of 0.17 (old pasture and woodland), then the volume of water available from a 1-inch rainfall (the storm size is arbitrary) is 156 cubic feet. Using this as the performance standard, the next step is to lay out a design scheme whose performance meets this standard.

Application and testing

The design process usually requires several rounds of juggling among the program, the facility and landscape layout, the building materials, and the BMP measures

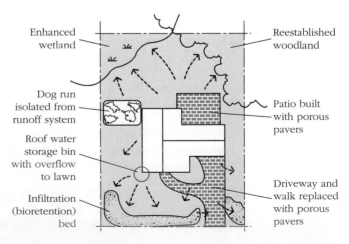

Fig. 13.5 Measures that can be used to retrofit a residential parcel to reduce the on-site production of stormwater and contaminants.

Q$_v$ accounting

until the postdevelopment Q_v value is brought to or below the predevelopment value. Because most residential landscapes are highly varied (that is, represent a wide range of hydrologic performance), the accounting process can become tedious, and here is where a spreadsheet program, aided by a Geographic Information System (for manipulating and measuring surface areas), comes in handy. These tools make it relatively easy to assess, for instance, the advantage of changing a patio from concrete to gravel, of adding vegetation to a roof, or of decreasing a building footprint. From this exercise a set of models can be formulated to assist in programming and design on similar sites in the project area.

13.5 SITE REMOVAL (RELEASE) BMPs

On-site controls

The next level of BMP application is intended to prevent rainwater from leaving the site and from contributing to stormwater surcharging of streams. Site removal, or release, BMPs are designed to disconnect the site as a source of stormwater from the larger watershed drainage system. These BMPs reduce stormwater volume and minimize the site's access to drainage facilities, such as street gutters and stormdrains. Detaching downspouts and diverting yard drains are, for example, important lines of defense at this level. Such measures are well suited to handle low- to moderate-magnitude runoff events because these events, which are the most commonly sized events, produce the most stormwater and contaminants in total. On the other hand, stormwater runoff from larger, infrequent events can usually be handled by the delivery BMPs, which are taken up in Section 13.6.

Disconnecting the Site. In conventional designs, stormwater runoff produced on-site is quickly released from the site and dumped into stormdrain systems. Removal BMPs, however, intercept this runoff and direct it to spots such as infiltration beds that are designed to detain and absorb it. The objective of these BMPs is to isolate the site as a source of stormwater. The treatment and infiltration of runoff from rainfall events producing less than an inch of water, for example, go a long way toward improving the hydrological performance of the entire watershed.

Lethargy by design

Slowing the release of runoff and disconnecting the site from the watershed can be achieved by:

1. Increasing travel time through the use of longer and slower routing schemes (for example, see Fig. 9.13c).

2. Increasing surface roughness to slow down overland flow and delay the formation and advance of channel flows.

3. Disconnecting impervious surfaces from drainage systems to reduce flow continuity and stormwater drainage area.

4. Utilizing grading and planting designs to diffuse runoff and promote infiltration (for example, see Fig. 8.12b).

The key principle of site removal BMPs is to treat stormwater runoff at the site scale, that is, at its source, not at the neighborhood, community, or watershed scale. Studies have found that site-level treatment is more effective and considerably cheaper than broadly integrative stormwater BMPS. Site removal BMPs include:

Key measures

- Disconnecting downspouts and yard drains from the stormwater system.
- Bioretention features such as wetlands and tree canopies that intercept rainfall and take up stormwater runoff.

- Infiltration facilities such as permeable swales, trenches, and dry wells that enhance soil intake of surface water.

- Diversion channels that direct stormwater away from stormwater delivery systems.

- Depression storage cells and other grading features that slow the movement of runoff and promote infiltration.

- Special planting schemes such as stormwater gardens designed to slow runoff, enhance infiltration, and intercept and treat stormwater pollutants.

13.6 DELIVERY BMPs

Rationale

Delivery BMPs are the final set of practices that can be used to manage stormwater runoff. In conventional land use planning, they are engineering measures, such as curbs, gutters, and stormdrains, used to move stormwater quickly and efficiently from developed areas to streams, lakes, harbors, and constructed ponds. Their main purpose has been to reduce the risk of flooding and property damage caused by large, infrequent runoff events. Unfortunately, they themselves have become serious problems in the landscape mainly because they damage water features, and the capital costs involved have increasingly proven to be excessive.

Beyond convention

Paradigm Shift. With the introduction of stormwater source controls and site-based BMPs, the need for massive stormwater delivery systems can be reduced and often eliminated. However, because of liability concerns and other issues—some justifiable, some not—delivery-scale BMPs are still called for in most areas of North America. However, they need not be designed in the traditional manner, that is, as high-efficiency conduits, especially in areas of low-density land uses and ample source control opportunities. In such areas, delivery BMPs, like site removal BMPs, should be designed to slow stormwater delivery rates, induce infiltration, minimize construction costs, and, above all, avoid prized water features (Fig. 13.6). Delivery BMPs remain the domain of civil engineers, and most engineers favor structural measures, especially in connection with streets and roads. Not surprisingly, structural measures, including detention and retention basins, are decidedly the most widely used BMPs in stormwater management.

Delivery BMPs of the alternative (green) variety include the following options:

Green measures

- Open (swale) drains without curbs and gutters.

- Diversion channels that direct stormwater away from valued habitat and water features.

- Rerouted flow patterns that lengthen travel times and slow delivery rates.

- Storage basins designed to lengthen travel times, promote groundwater recharge, and reduce pollution loads.

- Very low-gradient delivery systems such as wide swales with roughened beds.

- Constructed wetlands.

- Infiltration trenches.

13.7 AN APPROACH TO BMP PLANNING

Think like a system

Framing a BMP plan requires one to think like a system. In other words, we must envision a parcel of land as part of a watershed, how a particle of water in that parcel is converted into runoff, and how the particle "makes a decision" about its destiny. Landscape planners and designers control that destiny depending on where they place the parcel, what they program for it, how they design the program, and how they tie the design to the larger watershed system. In the end, what they add to the watershed in the way of facilities and landscapes must become integral working parts of the system, not alien appendages taped onto it.

(a)

(b)

(c)

Fig. 13.6 Not all open swales are appropriate BMPs in delivery systems: (*a*) an example of one that probably performs more poorly than a pipe; (*b*) a better solution, grass-lined; (*c*) a fully vegetated road drain that is better yet.

Beginning at the watershed scale, here are some steps that might be followed in formulating a BMP plan:

Suggested steps

1. Delineate the watershed and its subbasins, and locate existing and proposed development in this area.

2. Define the watershed's drainage system, showing the network of channels and other drainage features such as lakes, wetlands, floodplains, and seepage zones, and delineate the flow system. Highlight key habitats, especially fish-bearing streams and connecting features.

3. Build an inventory of biophysical features that have stormwater management potential for the watershed as a whole and/or around the proposal development area, and highlight critical environmental features, such as streams.

4. Formulate a watershed plan that:

   a. Defines buildable land units;

   b. Relates land units to watershed BMP opportunities, and

   c. Provides a site design model that:

      i. Meets site performance standards that minimize stormwater (and contaminant) production; and

      ii. Includes covenants and measures that minimize stormwater release from the site.

5. Calculate predevelopment and postdevelopment discharges for a modest design storm and allocate excess stormwater to existing biophysical features on and near the site, such as parkways, woodland buffers, and wetlands, taking care not to exceed their loading capacities.

For any stormwater that remains after biophysical uptake, reexamine the site design model (step 4), and modify it to accommodate excess stormwater. If a sizable surplus still exists, consider appropriate delivery-based BMPs.

13.8 TOWARD GREEN INFRASTRUCTURE

Why?

Stormwater and its contaminants constitute a major problem throughout the world. The conventional ditch, gutter, and pipe approach to stormwater management, which is actually more control than management, is inadequate for the settings and scales at which most of today's development is carried out. Among other things, this approach largely ignores the natural functions of drainage systems and the abundant opportunities they provide to reduce infrastructure costs and create sustainable landscapes by utilizing the natural services available in most watersheds.

Begin on site

The Rationale. As a rule, stormwater BMP efforts should focus on source control because the site offers the greatest number and the least expensive of management opportunities. In the production-removal-delivery system, the emphasis should be on production and removal, not on delivery. Given reduced stormwater output from developed areas, delivery systems can be downscaled and simplified, and they can function more as backup measures to be used selectively rather than habitually.

Question storage basins

Detention and retention basins do not solve the problem either because, among other things, they too are fed by expensive pipe systems. The argument that these basins are needed everywhere to curtail large flows is untenable for most small watersheds in the face of evidence that (1) they deprive streams of discharge during summer low-flow periods; (2) many, perhaps most, small watersheds function as partial area systems, often producing much less stormwater than we forecast; (3) stream systems should not be deprived of large flows because these flows are essential to the movement of sediment, shaping of channels, and maintenance of habitats; and (4) with proper planning, large, infrequent flows can be handled as risk management events.

Although land use should be integrated into the watershed, we need to avoid transforming watersheds into systems that are more hydrologically efficient than they were before development. Indeed, advances in the efficiency of stormwater drainage have brought us to the depressed state of affairs we face today in watershed management. The alternative is to figure out how landscapes function as hydrologic environments before land use and infrastructure plans are implemented—and not impose surface runoff and channels where none existed before development. And where a surcharge in surface runoff is unavoidable, the resultant stormwater should be forced to find slower, more natural means of getting to a stream channel.

Question imposed efficiency

This argument, of course, points to **green infrastructure** as the preferred approach to stormwater management, an approach that endeavors to integrate low-impact architecture, landscape design, and engineering into a balanced whole. In this way we may be able to build watershed arrangements that include people, their facilities, and natural features functioning together in a sustainable manner that is not measurably different from nature operating alone.

Integrate design

Finally, BMP programs should not be narrowly focused and limited to the perspectives and services of one professional discipline. Rather, they should include a wide range of measures and multiple professions. At a minimum, the balanced BMP program should include planning, landscape design, and engineering components, and it should draw on the services of scientists, landscape architects, planners, architects, and engineers, as Figure 13.7 suggests.

Seek diversity

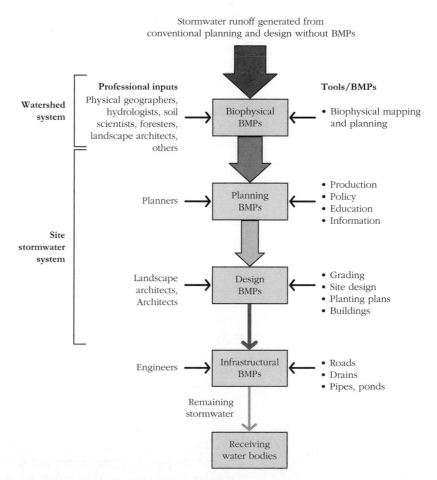

Fig. 13.7 Best management practices for the planning and design process utilize a variety of planning, design, and engineering tools, with different professions in the lead at each level.

13.9 CASE STUDY

Making the Shift to On-Site BMPs: Rainwater Management on a Small Urban Site

Scott Murdoch and Paul deGreeff

The application of landscape-based stormwater management BMPs in urban areas has long been a tough sell. Despite the numerous arguments in favor of using a softer approach to managing stormwater, the application of landscape-based BMPs calls for a major paradigm shift that threatens conventional authority and raises the liability specter of increasing local flood risk and soil contamination. It has been all too common for regulators and design professionals alike to dismiss landscape BMPs as risky, untested, unreliable, and short-lived. However, if a keen eye for identifying site-specific opportunities is applied and if an understanding of natural systems is brought to bear on design problems, the practicality, cost-benefit ratio, and resilience of landscape-based stormwater BMPs are difficult to dismiss. In our local experience over the last decade, the surmounting measurable benefits demonstrated by successful installations are beginning to change the tide on the acceptance and application of landscape-based stormwater BMPs.

In Victoria, British Columbia, as in most West Coast urban areas, extensive pipe systems move stormwater and its pollutants quickly and efficiently from buildings and roads to local creeks and harbors. Once entrained, pollutants are extremely expensive to remove and so are simply accepted as an unavoidable burden of urban stormwater relief, despite public outcries over the contamination of local beaches. Furthermore, the replacement cost of these at-capacity old piped systems is exorbitant, and the damage to local streams caused by increased magnitudes and frequencies of large flows is undeniable. So why not go with a lower-cost alternative that begins to address some of the ills of piped systems and that provides a better fit with the existing urban fabric?

Oak Bay Avenue encompasses an affluent area of Victoria with large picturesque street trees, beautiful homes, and quaint shops. Redevelopment here has been deliberate and slow, but it has resulted in discernable improvements, including wider pedestrian walkways, updated streetscape amenities, and traffic-calming features. Environmental systems, on the other hand, have largely been ignored as part of the redevelopment process. This case involves the redevelopment of a gas station and garage site. The buildings, parking area, and underground gas storage tanks were removed, and in their place a small hardware store was proposed with a small parking lot in the rear. Upon seeking approval for redevelopment, the applicant owners and their architect were looking to dress up the character of the site as a whole and create a better tie to the character and curb appeal of the neighboring shops. Because the site would need soil remediation, we saw an opportunity to introduce some alternative stormwater management measures to the site that would not only add to the visual appeal of the street, but set an example for the community on the use of source control BMPs.

During site preparation, a considerable amount of granular aggregate material was brought in to fill the void left by the removal of the gas storage tanks and contaminated soil (that was remediated off-site). The aggregate fill created an opportunity for stormwater storage in a landscape otherwise dominated by heavy marine clays and hard glacial tills. We responded with a simple design that featured a linear rain garden along one edge of the building that would receive all the runoff generated by the roof (Fig. 13.A). Any drainage infiltrating through the rain garden was directed to the 3-meter-deep aggregate bed beneath it (Fig. 13.B). In addition, permeable pavement was incorporated into the parking lot design, so that all parking stalls drained

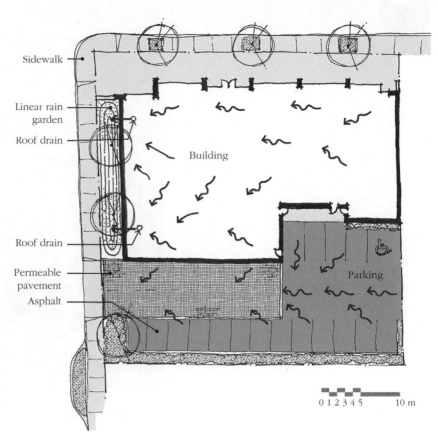

Fig. 13.A Site plan for water management. Roof water is directed to the rain garden, and parking lot runoff is infiltrated through permeable pavers. These measures reduced stormwater runoff by 80 percent.

to the permeable paving and thence into the aggregate reservoir below. On balance, of the total annual stormwater produced on this site, nearly 80 percent was infiltrated at its source, thereby introducing a lag time and significantly reducing the loading of the stormwater system over the original gas station's loading/output. This approach is well suited to the climate of the Victoria region, which is characterized by frequent low-intensity, long-duration rainfall events. For the occasional extreme storm event, the existing pipe system served as a backup to handle the excess runoff.

This caught the attention of the city engineering department and triggered new interest in pursuing source control BMPs. Among other things, the engineers realized such BMPs offer low-cost solutions for dealing with aged infrastructure while improving environmental quality. In addition, they pointed out that reductions in runoff reduce peak demand on infrastructure and reduce sedimentation in pipes, thereby contributing to increased longevity of existing infrastructure. As a consequence, the city is now building rain gardens and other green BMPs into the public streetscape, as well as being more receptive to private land owners who are seeking approvals for source control measures as part of redevelopment projects. For Victoria, a broad-based application of stormwater BMPs offers long-term hope not just for cleaning up local public beaches and creeks, but for also vastly improving the experiential quality of the urban landscape.

In follow-up with the hardware store owners, two key outcomes are noteworthy. First, the vegetation in the rain garden is robust and vigorous, and it has developed into a wonderful visual buffer between the building and the sidewalk. The irrigation system has been turned off because roof drainage is proving adequate to sustain this garden of mostly native plants. This effect is significant considering that nonirrigated

Runoff
from roof

Roof drain to
rain garden

Linear rain
garden
planted with
native and site
adapted non-
native plants

Pooled water

Growing
medium

3-m deep
aggregate
bed

Parent
material

Road
base

Fig. 13.B A section showing the rain garden design. Excess roof water is directed to an aggregate bed under the garden.

vegetation typically tends to brown off during Victoria's normally dry summers, to say nothing of the savings in water costs. Second, considerable positive feedback from neighbors and walkers has been directed to the store owners, expressing an interest in the rain garden and their design. Although the stormwater management component is not entirely obvious, interest in the landscape has resulted in casual education of the general public about the link between stormwater management and public health and about designing at the site scale with larger systems in mind. It has become a small model of the future: how to keep a community beautiful, healthy, and more effectively connected to the larger landscape while saving money.

Scott Murdoch and Paul deGreeff are landscape architects who own a small consulting firm in Victoria, British Columbia, that specializes in environmental analysis, site planning, and rainwater management.

13.10 SELECTED REFERENCES FOR FURTHER READING

Department of Environmental Resources. *Low-Impact Development: An Integrated Design Approach.* Prince George's County, MD, 1999.

Dreiseitl, H., and Grau, D. *New Waterscapes: Planning, Building and Designing with Water.* Basel: Birkhäuser, 2005.

Dunnett, N., and Clatden, A. *Rain Gardens: Managing Water Sustainably in the Garden and Designed Landscape.* Portland, OR: Timber Press Inc., 2007.

Marsh, William M., "Toward a Management Plan for Lazo Watershed and Queen's Ditch." Courtenay, BC: Regional District of Comox-Strathcona, 2002.

Marsh, William M., and Hill-Rowley, Richard. "Water Quality, Stormwater Management, and Development Planning on the Urban Fringe." *Journal of Urban and Contemporary Law.* 35, 1989, pp. 3–36.

Murdoch, Scott P. *The End of the Pipe: Integrated Stormwater Management and Urban Design in the Queen's Ditch.* Master of landscape architecture thesis. Vancouver: University of British Columbia.

Richman, Tom, et al. *Start of the Source: Residential Site Planning and Design Guidance Manual for Stormwater Quality Protection.* Palo Alto, CA: Bay Area Stormwater Management Agencies Association, 1997.

Sabourin, J. F., et al. *Evaluation of Roadside Ditches and Other Related Stormwater Management Practices.* Toronto: Metropolitan Toronto and Region Conservation Authority, 1997.

U.S. Environmental Protection Agency. *Urbanization and Streams: Studies of Hydrologic Impacts.* Washington, DC: Office of Water, 1997.

Related Websites

Environmental Protection Agency. Watershed Academy Web. "Forestry Best Management Practices in Watersheds." 2008. http://www.epa.gov/watertrain/forestry/
An interactive slide show from the EPA about forestry BMPs in watersheds. The links/pictures teach about planning, forest wetland protection, revegetation, and road construction.

International Stormwater BMP database. http://www.bmpdatabasc.org/index.htm.
A U.S. government and nongovernmental organization (NGO) collaborative initiative studying BMPs. The website contains case studies and performance analysis of hundreds of BMPs regarding stormwater management.

Low Impact Development Center, Inc. 1998–2008. http://www.lowimpactdevelopment .org/index.html
An NGO educating professionals and promoting low-impact development. Information on permeable surfaces and many case studies using low-impact BMPs is given under publications/projects.

Natural Resources Conservation Service. USDA. "Water Related Best Management Practices in the Landscape." http://www.wsi.nrcs.usda.gov/products/UrbanBMPs/
Governmental website for BMPs to protect water quality in the urban area. Links suggest methods of reducing impact during construction, ways to manage water volume to reduce erosion and sedimentation, and many other protection measures.

THE RIPARIAN LANDSCAPE: STREAMS, CHANNEL FORMS, AND VALLEY FLOORS

14.1 INTRODUCTION

Sad legacy

Stream channels and their valleys are among the truly spectacular environments on the planet—wonderfully diverse, aesthetically pleasing, scientifically challenging, economically valuable, and, above all, ecologically rich. Yet for all these qualities, streams are one of our most maligned, degraded, and mismanaged natural phenomena. In North America and Eurasia, only 20 percent of the major rivers remain without structural barriers such as dams and locks. The great channel of the Missouri between Kansas and Montana, which Lewis and Clark traversed in 1804 and 1805, today is little more than a long string of dams and reservoirs.

The channels of 70 percent of the streams in the United States have been altered by land use activity. In agricultural and urban areas, countless streams are routinely deepened, widened, straightened, piped, and diverted. As a result, these streams are effectively destroyed as natural entities and converted into conduits to carry water, waste residues, and sediment. Although no recent data are available, more than 300,000 miles of streams in the United States and Canada appear to have been treated in this manner.

Bit of hope

As land use pushes farther into watersheds and marginal and submarginal lands are taken up by sprawling cities, new settlements, roads, and farms, stream systems will suffer further from engineering schemes in the form of dredging and channelization, road construction, irrigation projects, flood control programs, and navigation projects. At the other end of the spectrum are efforts to recover stream channels and to restore them as habitats, sources of clean water, and scenic environments. Driven by a variety of interests led increasingly by streamkeeper and watershed organizations, much work is being done in erosion and sediment control, aquatic habitat conservation, and riparian corridor ecology and restoration. These activities all require a basic understanding of how streams operate in the landscape, including the processes that move water and sediment, the forms and features they create, and how it all comes together in the form of these wonderful fluvial landscapes.

14.2 HYDRAULIC BEHAVIOR OF STREAMS

As with any part of the landscape, planners and designers need to do more than merely map and describe the forms and features of stream valleys. They need also to understand them as systems because the systems drive the processes responsible for change. And since fluvial landscapes are places where change is the norm rather than the exception, land planners must read these places as dynamic game boards in order to find the right fit between site and land use.

Flow Velocity. To understand how streams erode, deposit material, shape channels, and build valleys, let us first examine a few principles about running water, beginning with velocity. **Flow velocity** in stream channels is apparently related to three factors: the slope of the channel, the depth of the water, and the roughness of the channel. These variables form the basis of the *Manning equation*, an empirical formula commonly used to calculate the mean velocity of a stream:

Calculating velocity

$$v = 1.49 \frac{R^{2/3} s^{1/2}}{n}$$

where

A = width × depth

v	=	velocity in feet or meters per second
R	=	hydraulic radius, which represents depth and is equal to the wetted perimeter (*P*) of the channel divided by cross-sectional area (*A*) of the channel
s	=	slope or gradient of the channel
n	=	roughness coefficient:

n Value	Channel Material
0.020–0.025	Soil material (e.g., loam)
0.040–0.050	Gravel with cobbles and boulders
0.050–0.070	Cobbles and boulders
0.100–0.150	Brush, stumps, boulders

Velocity controls

Increases in slope or depth (or hydraulic radius) cause increases in velocity, whereas an increase in the channel roughness causes a decrease in velocity. The effect of an increase in depth is slightly greater than that of an increase in slope of equal proportion. Hence the fastest velocities in stream systems may be found not in the mountainous headwaters, but farther downstream, where the depth and discharge are greater. For example, the average flow velocity near the mouth of the Amazon, where the slope is only a few inches per mile and the depth is 150 to 200 feet, is about 8 feet per second, whereas the average velocity in the Colorado River of Yellowstone National Park, where the slope is 200 feet per mile and the depth is 3 to 6 feet, is only about 3 feet per second. The point is simply this: Water in deep, low-gradient channels that *looks* slow may actually be relatively fast, and water that *looks* fast may not be so fast.

In-channel velocity

Within a stream channel, the velocity is lowest near the bottom and the sides of the channel and highest near the top in the middle. The highest velocity is generally slightly below the surface because of friction between the water and the overlying air. At channel bends, the faster-moving water responds more to centrifugal force than slower water does, and so it slides toward the outside of the bend (Fig. 14.1). In addition, this may also produce super elevation of the stream, in which the water surface near the outside of the bend is higher than the water surface near the inside. Where channels meander, the belt of fast moving water is thrown from one side of the channel to the other in successive bends, much as a bobsled team does in negotiating a run.

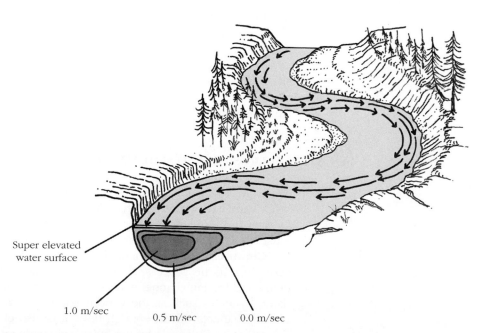

Super elevated water surface

1.0 m/sec   0.5 m/sec   0.0 m/sec

Fig. 14.1 The distribution of streamflow velocity in a meandering channel. The zone of fastest flow, called the thalweg, hugs the outside of the meanders just below the surface.

Turbulent motion

Motion and Force. Streamflow is always characterized by *turbulent motion*, that is, by the rolling and swirling action of eddies. This mixing motion is a primary source of flow resistance between the slower moving water molecules on the streambed and the faster moving water above the bed. Although some streams, especially those with smooth water surfaces, appear to be dominated by laminar, or sheetlike, flow, none in fact are. Laminar flow occurs only in a very thin sublayer immediately at the streambed where velocity is effectively zero. Above this sublayer, that is, above the streambed, the intensity of turbulence in streams increases with flow velocity.

Bed shear stress

Any inferences we make about stream behavior based on average flow velocities are limited by an important observation. Flowing water very quickly establishes an equilibrium with the prevailing frictional environment of the channel. Unless channel shape or roughness changes, or unless the discharge is increased or decreased, the water does not accelerate or decelerate. Because the downhill force of the moving water mass does not result in any acceleration of the stream, an equal and opposite force must be holding the water back. This opposing force is the **bed shear stress**, which is the amount of energy loss in the movement of water over the streambed. Bed shear stress can generally be given as a quantity equal to water density times gravitational acceleration times water depth times channel slope:

$$\text{Bed shear stress} = \rho g D S$$

The density of water (ρ) varies slightly with temperature, but for our purposes we can consider it constant at 1000 kilograms per cubic meter. Similarly, gravitational acceleration (g) varies slightly with latitude and altitude, but we can consider it constant at 9.8 meters per second squared. Therefore, the only variables are depth (D) and slope (S), and thus the mean bed shear stress is proportional to their product.

Energy balance

The product of bed shear stress and discharge is the *rate* at which the stream is losing energy as it runs its course. Variations in energy, in turn, can be related to changes in discharge, slope, and/or water depth. Thus, a combination of low discharge, slight gradient, and shallow water means low stream energy. These three variables are at the root of most problems in managing open channel flow and in stabilizing and restoring channel environments.

From headwaters to mouth, the stream system involves a great deal of energy. It begins in the upper watershed with potential energy (an elevated mass of water), which is converted to kinetic energy as the water is put into motion. But only a small amount of this energy is devoted to the important work of dislodging particles and moving sediment. Most of the stream's kinetic energy is converted into heat, which explains why streams do not freeze out in winter.

14.3 STREAM EROSION AND SEDIMENT TRANSPORT

Nevertheless stream systems are capable of performing immense amounts of work. Worldwide each year more than 20 billion tons of sediment are brought to the oceans by rivers and streams. But for running water to pick up (erode) and move particles, it must overcome the particles' weight and other resisting forces.

Scouring

Channel Erosion. Channel erosion occurs mainly by means of a process called **scouring**. In this process, heavy particles bump and skid along the streambed, loosening and freeing channel material. Scouring is highly effective in eroding unconsolidated materials such as the various soil and sediment materials that make up most stream channels. Given enough time, it is also effective in eroding bedrock, as evidenced at a grand scale by the huge cuts made into the earth's crust by streams such as the Colorado River at the Grand Canyon.

Sediment sources

But streams like the Colorado are far larger than those we work with in most land use problems. The streams we face tend to flow in valleys made up of various kinds of deposits, typically loose materials deposited by the stream system itself. Therefore, most of their work consists of the erosion and re-erosion of the sediment already on the valley floor, mainly old channel deposits mixed with flood deposits, material that readily gives way to scouring. Added to this is the sediment brought to the channel by tributary streams and by hillslope processes on the valley margins (Fig. 14.2).

Load types

Sediment Loads and Transport. The total mass of material entrained and transported by a stream is called **sediment load**. Streams carry three types of loads: bed load, suspended load, and dissolved load. **Bed load** consists of larger particles (sand, pebbles, cobbles, and boulders) that roll and bounce along in virtually continuous contact with the streambed. **Suspended load** consists of small particles, principally clay and silt, that are held aloft in the stream by turbulent flow. The amount of the suspended load carried by a stream is determined mainly by sediment supply rather than by the force of the stream's flow, and most streams are capable of carrying a large suspended load if the sediment is available. On the other hand, the size and amount of bed load transport are determined by the level of bed shear stress.

Bed load transport

The mechanism of bed load movement involves two forces: *drag*, a force along the bedstream, and *lift*, an upward force. The lift force is of primary importance in initiating motion; laboratory flume experiments have demonstrated that a lift force of about 70 percent of the submerged weight of a particle is sufficient to pivot it upward on its axis, whereupon the drag force can push it downstream. For any size class of particles there is a *threshold of force*, or critical level of bed shear stress, below which no movement takes place.

Above this threshold value, the rate of sediment movement increases as some power of bed shear stress. In simplified form, this concept can be expressed in terms of threshold velocities required to dislodge and move particles of different sizes (Table 14.1). These data have practical value in channel management because they tell us the size particles that need to be placed on stream banks, in ditches, and in gullies to stabilize them against erosion under different flow velocities.

Erodibility

Sand particles are the most erodible sediment in stream channels. Clays are less erodible because of particle cohesion (see Fig. 12.8). A relatively large bed shear stress is required to overcome this cohesive force, and the smaller the clay particle, the greater the force required, especially if the material is compacted. Clay particles

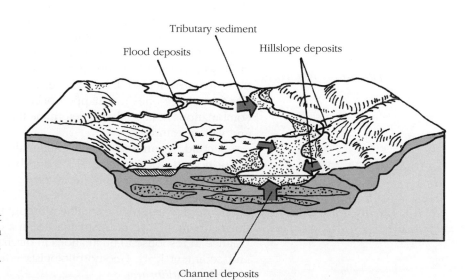

Fig. 14.2 The four main sources of sediment in stream systems. Most sediment comes from the alluvium on the streambed and banks. Tributary input is also a major source, especially for large streams.

Table 14.1   Relative Resistance of Channel Materials to Stream Erosion

Material	Relative Resistance[a]
Fine sand	1.0
Sandy loam	1.13
Grass-covered sandy loam	1.7–4.0
Fine gravel	1.7
Stiff clay	2.5
Grass-covered fine gravel	2.3–5.3
Coarse gravel (pebbles)	2.7
Cobbles	3.3
Shale	4.0

[a] Resistance relative to fine sand, the least resistant channel material.

are so small, however, that once dislodged, they will remain in suspension even under the slowest of stream flows. Accordingly, clay sediments, which make up most suspended load, are extremely mobile in stream systems and, once entrained, tend to be moved great distances (see Fig. 12.2).

Dissolved load

Dissolved load is that material carried in solution. Ions of minerals produced in weathering are released into streams mainly through groundwater inflow. Total ionic concentrations in stream water are generally on the order of 200 to 300 milligrams per liter, but in areas with low relief and high rates of soil leaching, as in parts of the U.S. East and South, concentrations may reach several thousand milligrams per liter. In dry areas, such as the Southwest, dissolved load is often much lower, typically less than 10 percent of the total stream load. This is explained by the arid and semiarid climates that limit the rates of chemical weathering and soil leaching.

14.4 CHANNEL DYNAMICS AND PATTERNS

Discharge and work

Stream systems are very much unlike manufactured water systems. For one thing, variability is the norm rather than the exception in the behavior of streams, and this characteristic applies to how they do work. Understanding this is important in interpreting the features we see in and around stream channels because the magnitudes of flows that can be observed on most days usually have little or nothing to do with these features. Rather, the processes associated with a limited number of flows, usually fairly large ones that happen once or twice a year or every other year, or some combinations of large flows, determine the size, character, and distribution of channel and valley features.

Equilibrium trend

Second, stream systems are continually adjusting themselves through erosion and deposition. If, for example, slope and bed shear stress are low at one point in the stream profile, deposition takes place there. But deposition, in turn, increases the slope and with it the bed shear stress, eventually leading to the restoration of equilibrium, whereby the sediment brought to that point continues to be transported downstream. In fact, most streams are in a constant state of flux as they trend toward balance in response to fluctuations in discharge and changes in channel forms, slope, and sediment load. Geomorphologists refer to this conditional state of balance as *dynamic equilibrium*.

Riffle-Pool Sequences. One feature that illustrates the complexity of the processes involved in channel formation is the **riffle-pool sequence**. During low flow in many streams, the channel becomes partitioned into a distinctive sequence of segments called pools and riffles. The pools are quiet segments where the water is deeper and slower and where the streambed is lined with fine sediment; the riffles are rough segments where the water is faster and shallower and the streambed is lined with gravel and cobbles.

Spacing ratio

The spacing between riffles and pools is surprisingly regular. As a rule, the pool to pool or riffle to riffle spacing is five to seven times the channel width (Fig. 14.3). As habitats, riffles, with their shallow, turbulent flows, are distinctly different from pools, with their deeper, less turbulent flows. Among other things, trout lay their eggs in the riffles where the newly hatched fish can escape through the gravels into the stream, but adult trout occupy the pools and feed from the sediment that collects there.

Why pools?

This raises a question: If the pools collect sediment, do they not eventually fill in and become obliterated? The answer lies in the variable nature of streamflow, for the infrequent high flows are responsible for the maintenance of the pools. When the stream is at a high-flow stage, the entire riffle-pool sequence is drowned, and from the water surface the distinction between pools and riffles is not apparent (see the broken line in the profile in Fig. 14.3). Underwater inspection reveals that the pools and riffles are still there, but the former difference in the slope (gradient) of the water surface between pool and riffle is no longer present. In other words, the slope of the stream surface now is almost the same over both pool and riffles. But the water depth

Pool maintenance

is greater over the pools, therefore bed shear stresses are maximized at these points. Hence, at high flow the stream exerts greater force over the pools, they are scoured and deepened, and the larger particles are moved downstream to the next riffle area.

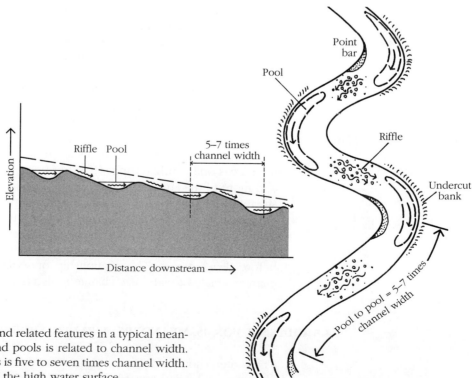

Fig. 14.3 The pattern of riffles, pools, and related features in a typical meandering channel. The spacing of riffles and pools is related to channel width. The pool-to-pool spacing in most streams is five to seven times channel width. The broken line in the profile represents the high water surface.

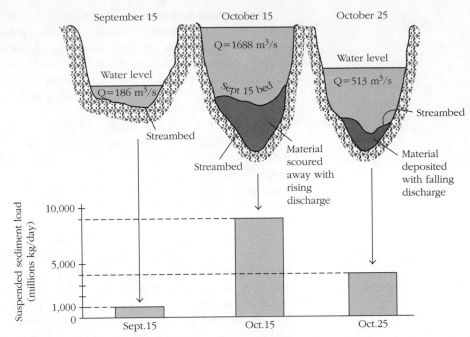

Fig. 14.4   Channel scouring and changes in suspended sediment load with changes in discharge in the San Juan River near Bluff, Utah. Notice that both increase as discharge rises. (From Leopold et al., 1964.)

Thus the modest flows we are apt to see during a visit to a stream appear to have little to do with the formation of the riffles and pools that produce the sequence of fast-water and slow-water segments that are so important to the character of stream channels.

Sediment-Discharge Relations. If we examine a stream channel in cross section, we find that the geometry of the channel changes substantially in response to changes in discharge and sediment supply. When discharge rises, both velocity and water depth increase, producing scouring of the streambed and increased sediment load (Fig. 14.4). Conversely, much of the channel bed fills in with sediment as the discharge falls.

These channel changes, referred to as **degradation** and **aggradation**, respectively, are also associated with changes in sediment supplied to the stream as a result of land use changes (Fig. 14.5). Aggradation is initiated with the rise in runoff and soil erosion brought on by early land clearing and cultivation, but then it peaks years later with massive sediment inputs as suburban construction spills over into agricultural areas on the urban fringe (Fig. 14.5*b* and *c*). As urban sprawl advances and more and more of the land is covered up by pavement and buildings, sediment loading of streams declines while stormwater discharge rises.

Channel responses

The end result, shown in Fig. 14.5*d*, is the degradation of stream channels. Such conditions are manifested not only by the absence of typical amounts of channel sediments but also in severely eroded channels, which are deeper and often wider than before development. The relationship of sediment production and stream channel changes related to land use change is also depicted in Figure 12.3.

14.5   CHANNEL FORMS, MEANDERS, AND RELATED PROCESSES

The writer Thomas McGuane reminds us that "Our history is understood by riverine chapters, and it is impossible to know American life without constant reference to rivers. . . . It is still to them we look for restoring some poetry to our national life."

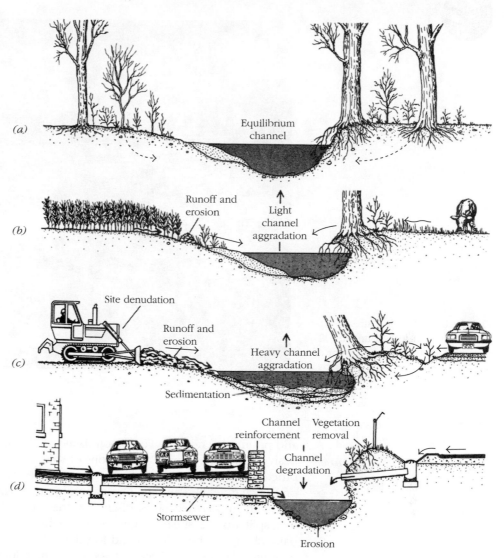

Fig. 14.5 Stream channel changes associated with land use: (*a*) uncleared land; (*b*) agricultural development; (*c*) suburban residential development; (*d*) urban development. Channel aggradation begins with agriculture and historically peaks with suburban development. Conversely, channel degradation increases with urbanization and rising stormwater flows.

And this is a poetry fraught with river symbolism, and among these symbols, none is more apparent than channel forms and patterns.

Channel Patterns. Two distinctive channel patterns occur in natural streams: braided and single-thread. In **single-thread** channels water is confined to one conduit. The channel form is relatively stable and characterized by one or two steep banks held in place by bedrock, soil materials, and plant roots. The main axis of the

Single-thread channels

channel is marked by a zone of relatively deep, fast moving water known as the *thalweg* (see Fig. 14.1). Most single-thread channels are sinuous or curving, within which the thalweg swings from middle channel to one bank and then the other. Many are also conditionally stable or metastable. This means that they owe their stability to a few key factors, such as the bank vegetation, and the loss of which (due, for example, to excessive sediment loading or construction activity) can cause the channel form to break down and become braided.

Braided channels are often found in geomorphically active environments, such as near the front of a glacier or at an unmanaged construction site, where massive amounts of sediment are being moved and/or the stream banks are highly erodible

25

Fig. 14.6 A braided stream channel in the Rocky Mountains. The woody debris marks the extent of a recent large flow.

Braided channels

and poorly defended by vegetation. Braided channels usually form in coarse, noncohesive bed material under conditions of sharply fluctuating discharge (Fig. 14.6). The process begins with falling discharge, which leads to the deposition of coarse debris on the channel bed and the formation of gravel bars.

With further decline in discharge, the bars becomes exposed, forming barriers that split up the stream's flow. Subsequent high discharges may erode the gravel bars, changing their shapes and locations, but the bars that survive several seasons may grow and stabilize as driftwood becomes lodged on them and plants become established. Multiplied many times, this process, in all its varied forms, can lead to incredibly diverse channels.

Meandering channels occur in a wide variety of environments—in arid and humid regions and in all kinds of surface materials. They also occur in a wide variety of patterns, from slightly bending channels to those that are excessively sinuous, called *tortuous* channels. Analysis of meanders usually begins with a description of the meander geometry, including meander belt width, wavelength, and sinuosity. The *meander belt* is the corridor containing the meander system as illustrated in Figure 14.7. The width of the meander belt varies with the size of the stream and appears to be related to the river's mean annual discharge: the larger the discharge, the greater the width.

Meander forms

Channel Sinuosity. A line drawn down the center of the meander belt is the *meander belt axis.* **Sinuosity** is the ratio of channel length along the curve of the meanders to the length of the meander belt axis. Geomorphologists class channels with sinuosities of less than 1.5 as straight or relatively straight, and those with sinuosities greater than 1.5 as meandering. At sinuosities approaching 4.0 or more, a channel is tortuous. The most extreme channels, called *anastomosing,* are streams with multiple single-thread channels, all with meandering and even tortuous, patterns. Table 14.2 summarizes the forms and conditions of the various channel forms and patterns.

Sinuosity classes

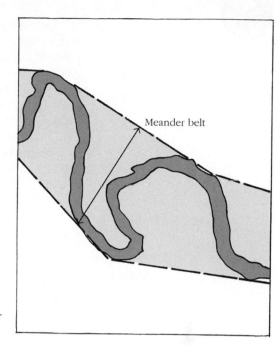

Fig. 14.7 A meandering, single-thread channel with the meander belt defined by the area between the broken lines.

Understanding the dynamics of meanders is essential to managing the channel environment, including facility planning and maintenance, habitat restoration, and sediment control. Where pools and riffles occur, we often find successive riffles favoring opposite sides of the stream. In a meander sequence, the pools are generally found at the bends, with the riffles between them. Hence there are two sets of riffle-pool sequences in a full meander (Fig. 14.3).

Table 14.2 Stream Channel Types and Characteristics

Channel Type		Morphology	Sinuosity	Load Type	Behavior
	Meandering	Single-thread channel	>1.5	Suspended or mixed	Lateral erosion and point bar formation
	Straight	Single-thread channel	<1.5	Suspended, mixed, and/or bed load	Minor downcutting and widening
	Braided	Multiple, intertwined subchannels	<1.3	Bed load	Aggradation
	Anastomosing	Two or more single-thread channel with large islands	>2.0	Suspended load	Slow lateral migration

Source: Adapted from Selby, M. J., 1985.

Why meanders?

Detailed investigations of the distribution of bed shear stresses along meandering and straight reaches of the same rivers show that the meandering reaches have less variable energy distributions. Thus, as a river trends toward a meandering course, the distribution of energy along the channel grows less variable. This fact indicates that meandering flow represents a more stable state than does flow in a straight channel. It also suggests that road ditches, straightened streams, and earthen stormdrains are fundamentally unstable channel forms that eventually will cut into banks, roads, and field margins as the flows in them drive toward equilibrium.

Channel Shifting. Active meanders in most streams are continuously changing with lateral channel erosion and deposition. Erosion is concentrated on the outsides of bends and slightly downstream from them where it cuts back the bank, forming an *undercut bank*. Deposition occurs on the insides of bends, forming features called *point bars* (Fig. 14.8a). Each year or so, a new increment is added to the point bar while the river erodes away a comparable amount on the opposite bank, the undercut bank. In this way the river shifts laterally, gradually changing its location in the valley and, at selected sites, cutting back the land on the edge of the valley.

Processes and landforms

But the river can also undergo sudden changes in location when it erodes new segments of channel and abandons old ones. This is especially commonplace where

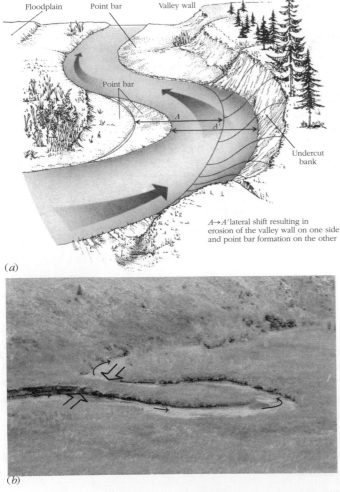

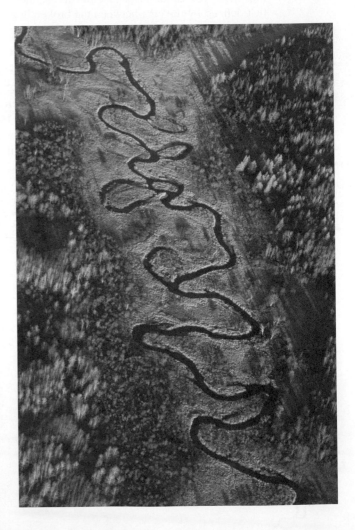

Fig. 14.8 (*a*) A schematic diagram showing the lateral migration of a stream channel in a meander bend. (*b*) Left, a meander loop in a small stream nearing the point where the neck will be breached as bank undercutting advances. Right, a complex sequence of loops and breaches.

Floodplain features

a meander forms a large loop and the river erodes toward itself from opposite sides of the loop, eventually breaching the meander (Fig. 14.8*b*). The old channel is abandoned because the new route is steeper and thus more efficient. The old channel forms a small lake, called an *oxbow*, but in time it fills with sediments and organic debris, becoming a wetland. Such features are often vividly defined by the patterns of vegetation on the valley floor, and the aerial photographs in Fig. 3.3 show typical examples.

14.6 FLOODPLAIN FORMATION AND FEATURES

If we step out of the channel in most stream valleys, we enter an area of fairly flat ground made up of alluvial deposits. This is river plain, or **alluvial plain,** and in most quarters it is referred to as **floodplain** because it is the ground that first receives floodwaters when the stream overtops its banks.

Floodplain Formation. Floodplains form mainly by the channel processes just described, that is, the lateral shifting of streams on their valley floors. The process works as follows. When the river flows against the high ground at the edge of its valley, called the *valley wall*, it undercuts the wall, which fails and thereby retreats a short distance. At the same time, new ground is formed on the opposite bank in the form of a point bar, as illustrated in Figure 14.8*a*. The new ground forms at a low elevation, near that of the stream. In time the valley walls are cut back so far that a continuous ribbon of low ground is formed along the valley floor.

Valley floor

This ribbon of land is the floodplain. It is composed principally of channel deposits of sands, gravels and other materials that were parts of former point bars, riffles, pools, and other channel features. (See Fig. 10.7 for examples of different valley and floodplain forms.) Although floods can alter the surface of the floodplain by eroding it and leaving deposits on it, they are clearly not the main cause of its formation, and in this regard the term *floodplain* is a little misleading.

Floodplain

In addition to lateral cutting by the channel in widening the valley and building the floodplain, other processes also contribute to valley formation and composition. These are various slope processes such as slumping, overland flow, and gullying that work on the valley walls, wearing them back and adding debris deposits to the floodplain. Especially significant are aprons of *talus* that develop along steep bedrock walls and *alluvial fans* and *colluvial deposits* where runoff processes are active (Fig. 14.9). Colluvial deposits are mixes of alluvium and mass wasting debris (such as talus) and are very common on valley margins in mountain areas.

Other factors

Understandably, the composition of the floodplain is often very diverse indeed. Various types of channel deposits are laced together with deposits from floods, organic materials from resident forests and wetlands, and hillslope deposits from the valley walls. Floodplain soils, it follows, are diverse, often extremely so, and require detailed investigation in planning and engineering projects. But it is also necessary to recognize that soil materials are not completely random in their distribution in the valley. At least three soil (or parent material) zones can be identified where different concentrations of floodplain, alluvial fan, and valley wall materials can be expected and these are illustrated in Fig. 14.9.

Floodplain composition

Floodplain Features. Floodplains abound with distinctive topographic features. *Meander scars* and *oxbows* are the most salient features in many floodplains. As the name implies, meander scars are the imprints left by former channel locations. They often occur in regular series or sequences called *scrolls*, marking the progressive lateral shift of a channel (again see Fig. 3.3). Oxbows are the most abundant in the valleys of streams with tortuous meander patterns. Although meander breaching is a natural process in meandering streams, breaching is also performed by engineers to make streams straighter and more efficient as navigation channels.

Meander features

Natural levees, scour channels, backswamps, and terraces are also common floodplain features. *Levees* are mounds of sediment deposited along the river bank by

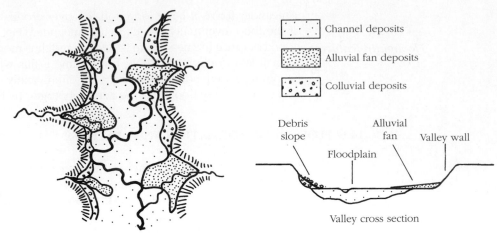

	Channel deposits
	Alluvial fan deposits
	Colluvial deposits

Fig. 14.9 The three main types of deposits that form the soil parent material in stream valleys.

floodwaters. They occur on the bank because this is where flow velocity declines sharply as the water leaves the channel, which causes it to drop part of its sediment load. In the low areas behind levees, water may pond for long periods, forming *back-swamps*. *Scour channels* are shallow channels etched into the floodplain by floodwaters. They often form across the neck of a meander loop and carry flow only when floodwaters are available. *Terraces* are elevated parts of a floodplain that form when a river downcuts and begins to establish a new floodplain elevation (see Fig. 10.7*e*).

Floodplain features

Channel Elevation Changes. Changes in the elevation of stream channels do not necessarily require long (e.g., geologic) periods of time. Significant changes can take place in a matter of years, and such changes are often great enough to pose serious management problems. Two examples of changes arising from channel degradation are noteworthy. The first is found in urban areas where streams are overloaded with stormwater and deprived of much of their sediment load resulting in severe channel scouring (see Fig. 14.5*d*). The second is found in streams that have been dammed and lost their sediment load to a reservoir. The channel below the spillway generates a new sediment load by severely eroding its channel. Under large spillway discharges, the channel degrades, thus rapidly reducing it to large boulders and exposed bedrock.

Induced degradation

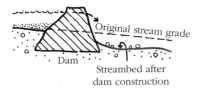

At the opposite extreme is channel aggradation from excessive sediment loading. We have already covered one example of this phenomenon related to land use change involving agriculture and suburban construction. Another example is found in eastern North America where streams were dammed in the eighteenth and nineteenth centuries to create mill ponds. These became sites of heavy sedimentation, and today the streams flow in elevated channels. At a much larger scale is the Yellow River of China, which suffers from such massive aggradation that the channel elevation now lies well above large sections of floodplain and its waters are held in only by constructed levees (see Fig. 10.7*g*). Heavy erosion in tributary basins upstream is the cause, and, when the levees eventually break, the flooding and sediment burial of farmland and settlements will be disastrous.

Induced aggradation

14.7 RIPARIAN AND CHANNEL HABITATS

Stream channels and valleys are complex and dynamic environments. As habitats, they are both attractive and risky places. Their diverse forms, compositions, and scales offer opportunities for a huge number of organisms and their communities, while presenting constraints to others. Life is nurtured and sustained by the stream of water, energy, and nutrients and at the same time is threatened, rearranged, and even destroyed by the same flow system. Stream valleys are places of fresh starts for both humans and other organisms.

Adjusting to Site Conditions. Although humans have struggled with adaptive land use practices—preferring instead to manipulate the environment with dams, channels, levees, and the like—many organisms have shown great success in adjusting to the radical behavior of streamflow and the related geomorphic changes. Floodplain and riparian vegetation is distinctive, among other things, for both its resilience and sensitivity to variations in drainage processes and conditions on the valley floor.

Biodiversity

Floodplains contain some of the most diverse vegetation in North America. In Canada and the United States more than 5000 species of wetland plants are found in floodplains. Most of these plants are adapted to a particular *hydrologic habitat* governed by two factors: (1) the frequency and duration of flooding and (2) groundwater conditions. For most floodplains the influence of flooding and groundwater can be approximated on the basis of elevation changes associated with landforms such as levees, oxbows, scour channels, and terraces.

Floodplain associations

Vegetation Patterns. This influence is illustrated in Figure 14.10. In this typical floodplain in the American South, bald cypress, tupelo, and water elm occupy wetlands immediately along the channel, and a different association, led by willow oak and pin oak, occupies slightly higher ground where flooding can be expected every one to two years. On the margin of the floodplain next to the valley wall, where the flood frequency is greater than 10 years, are the least water-tolerant species. A broadly similar pattern holds for stream valleys in the North and West, but the species are different and the biodiversity generally lower.

Point bar patterns

Another distinctive vegetation pattern is related to point bar deposits on migrating channel meanders, such as the one in Figure 14.8a. As point bars are formed on the insides of meanders, riparian plants such as cottonwood and poplar seedlings become established there. The channel subsequently shifts, creating a new point bar and new tree habitat. In time a belted pattern of vegetation is established within meander loops that is characterized by a sequence of tree stands trending from young to old along the meander axis.

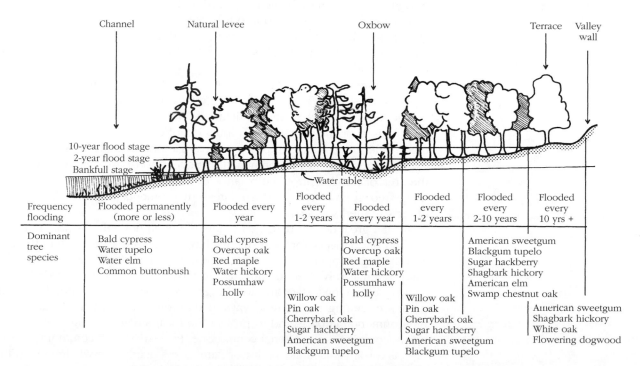

Fig. 14.10 A representative distribution of floodplain trees for stream valleys in the American South and the related flood frequency of their habitat.

Tolerant species

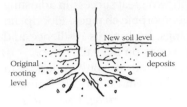

Adjusting to Extreme Events. Another notable trait of riparian and floodplain vegetation is its **tolerance** to extreme levels of disturbance from powerful floodflows. Many shrub and tree species are able to regenerate after heavy damage from flooding, ice ramming, and sediment burial. Trees such as willow, poplar, cottonwood, and sycamore, for example, can generate new stems after severe disturbances that include complete uprooting and burial.

The same species, among others, also have the capacity to raise their rooting level when new layers of sediment are added to the floodplain surface. Furthermore, they have the ability to develop dense root systems and spread by cloning, creating thickets of stems along banks, point bars, and levees. These species are true survivors and understandably are excellent ground stabilizers as well as important agents in shaping and sustaining riparian habitats.

Riparian Forest. The forests and woodlands that line stream channels also play an important role in shaping the stream's aquatic habitat. Root masses and fallen trunks alter water flow patterns that in turn diversify channel topography and improve fish habitat. Parenthetically, we should add that these are the very features engineers *Habitat and food* remove from channels to decrease roughness and to quicken flow as a part of traditional flood control programs. For small streams, forest canopies shade the summer channel, thereby helping to maintain cool water temperatures essential to trout habitat. And riparian vegetation is the principal source of food for aquatic organisms. In forested watersheds it is estimated that well over 90 percent of the aquatic food supply originates in the riparian forest as insects and plant matter are dropped, blown, and washed into streams.

14.8 MANAGEMENT PROBLEMS AND GUIDELINES

Stream environments are natural sinks in the landscape where the excess water, sediment, land use debris, sewage effluent, and many other things collect. At the same time, they are the environments on which we depend for water supply, navigation, farmland, habitat, recreation land, and open space. Fluvial environments are deeply embedded in our history and mythology, so much so that streams and their valleys are central to our geographic ethos. Not surprisingly, we devote tremendous social, political, scientific, and economic energy learning how to use, manage, and reuse them.

The Need for New Thinking. Much of the energy and ingenuity, however, has been misdirected over the years. Until relatively recently, our approach to river management was overwhelmingly utilitarian, aimed more at controlling than managing, and focused on the structural manipulation of channels and floodplains in an effort to control flooding, water supplies, and navigational resources. After enormous capital *Costs of control* expenditures, untold environmental damage, and disappointing results, especially in the area of flood control, the balance of things has now shifted somewhat to a more integrated approach that incorporates environmental, cultural, land use, and many other so-called soft agendas.

But a great deal of damage has already been done that is not limited to large rivers and big projects, for local streams have also been abused. Here the responsibility and the jurisdictional circumstances are not as clear as they are with the major rivers. They often involve lists of private property owners as well as drain commissions, forestry operations, state and local highway departments, railroads, and many other parties operating in different ways at different times. To remediate the cumulative environmental impacts and to protect the local stream resources we have remaining has become an issue, often a central one, in nearly every community.

As for large streams, the list of damages is nearly overwhelming, and the responsibility, of course, is much broader. In the United States between 1940 and 1971, the

For the record

U.S. Army Corps of Engineers assisted in projects that modified (e.g., channelized, built levees, diverted, or dammed) more than 10,000 miles of riparian environment. Between the 1940s and 1980, the U.S. Natural Resources Conservation Service was involved in alterations of nearly 11,000 miles of streams. In the Tennessee River Valley, nearly 5000 miles of streams are impounded by reservoirs or have flows regulated by structures such as locks. In the Sacramento River Valley of California, where in 1850 there were 775,000 acres of riparian forest, today only 18,000 acres remain. This huge loss of forest—which is typical of many other California streams as well—was mainly the result of clearing for agriculture, irrigation projects, urban water supply, and flood control.

Reservoir capacity

The Legacy of Dams and Reservoirs. Currently, there are more than 75,000 dams higher than 6 feet in the United States and thousands more in Canada. In a given year, more than 60 percent of the entire streamflow of the United States can be stored in existing reservoirs. In Canada major reservoirs (those behind dams 30 or more meters high) have the capacity to store two years of streamflow from the entire country. Dams on the Colorado River (see Fig. 2.16) can store an astounding four years of discharge when the stream is flowing at a typical rate. About 3 percent of the land area of the United States is covered by reservoirs. In terms of floodplain coverage, between 30 and 40 percent of the area classified as 100-year floodplain is under water in the United States.

Dams and reservoirs are very serious intrusions on riparian and related aquatic environments. A full discussion of their impacts is beyond our scope here, but several highlights are noteworthy:

Dams and sediment

- *Reservoir siltation:* Dams interrupt the normal transport of sediment, trapping huge loads in their reservoirs. Eventually all or part of the reservoir is filled and its utility for water supply, flood control, or recreation is greatly reduced or lost. More than 3000 such dams in the United States have been retired (abandoned), but the original channels and valley floors are lost under massive layers of sediment, and in the main, are beyond restoration to a condition approximating their original states (the case study at the end of Chapter 3 takes up this issue).

Dams and channels

- *Channel erosion:* To make up for the sediment captured in reservoirs, especially bed load sediment, streams scour their channels downstream of dams. Not only is the channel deepened, but the streambed is often reduced to a belt of boulders. After construction of the Glen Canyon Dam on the Colorado, for example, the river cut its bed down by 12 feet or more, destroying most of the benthic (bottom) habitat for some distance downstream (Fig. 14.11*a*).

Dams and discharge

- *Reduced flow variability:* One of the main objectives of dams is to regulate stream-flow by reducing its natural variability (Fig. 14.11*b*). By decreasing the magnitudes and frequencies of peak discharges, particularly flood flows, one of the principal forces acting on the aquatic and riparian habitat is altered. In addition to the change in water supply and distribution, an important source of disturbance is reduced, thereby allowing less tolerant plant and animal species to invade flood-plain ecosystems.

Dams and habitat

- *Habitat fragmentation:* Dams and reservoirs obviously break up habitat corridors and thereby reduce the diversity, productivity, and various functions of riparian and aquatic ecosystems. The most glaring examples are provided by migratory fish such as salmon. On the Columbia River, between 5 and 14 percent of adult salmon are killed at each of the eight dam sites on their way upstream. The survival rate of young salmon migrating downstream through the reservoirs is even lower. Chinook salmon are now listed as a threatened species on the Columbia River.

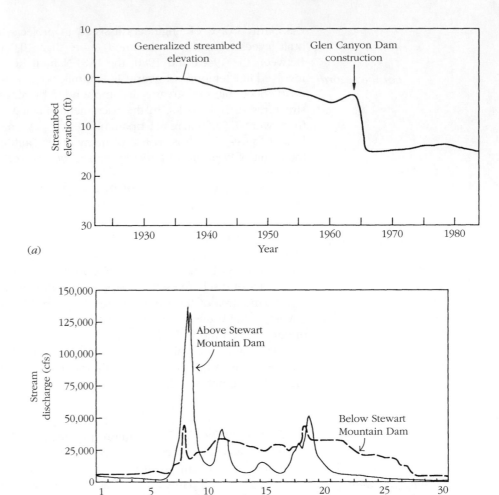

Fig. 14.11 (*a*) Streambed elevation of the Colorado River below the Glen Canyon Dam site before and after dam construction around 1964. (*b*) An example of the influence of dam construction on flow variability, Salt River, Arizona, based on discharge measurements above and below Stewart Dam during January 1993.

Hopeful Signs. The construction of dams in the United States and other developed nations has declined sharply in the past two decades (Fig. 14.12). Relatively

Fewer dams

few potential sites for large dam projects are now under consideration in the United States and Canada. In addition to the several thousand dams in the United States already retired, the next 50 years will see the abandonment of thousands more. These environments, and related ones damaged by levee construction, channelization, and similar activities, will have to be repaired and reincorporated into the larger fluvial/riparian environment (again, see the Chapter 3 case study).

Grassroots movement

A new wave of stream and river restoration programs has emerged with broadly based support from communities, environmental organizations, businesses, educational institutions, and state/provincial and federal agencies (Fig. 14.13). Among these are thousands of local nongovernmental organizations of watershed stewards and streamkeepers. Their programs call for a variety of efforts including repairing damaged habitat, planning open space systems, and managing riparian ecosystems. In one way or another, all these activities are rooted in hydrology, geomorphology, and ecology. Studies in these fields have taught us a lot about the complexities and dynamics of stream corridors.

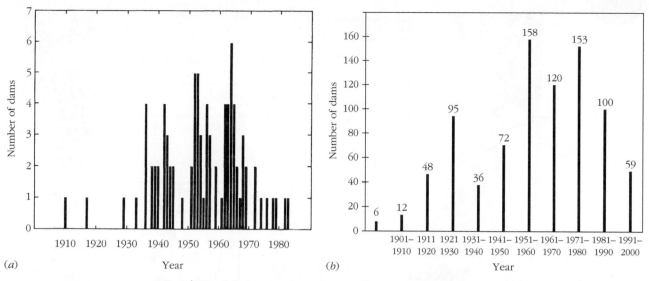

Fig. 14.12 (*a*) Construction of major dams in the United States between 1910 and 1983. A major dam is one with a reservoir capacity of 1 million acre-feet or more. (*b*) Construction of major dams in Canada between 1901 and 2000. A major Canadian dam is one 30 meters high or higher.

Some Governing Principles. Restoring and managing stream corridors effectively takes more than an understanding of the environment based on the traditional measures of site and land use planning, namely, slope, soil, drainage features, topography, vegetation, and habitat. Planners, designers, and managers must embrace certain governing principles based on a process-system perspective, and several are offered here for your consideration:

Flow networks

■ *Streams are parts of flow networks* that gain their water supply from a watershed, concentrate this water in channels, and direct it downstream. It is a one-way flow system; therefore, we must always be cognizant of location in the network and its meaning in terms of upstream/downstream land use relationships. In managing stream systems, the first priority is to protect the headwater areas, and the second is to protect the watershed as a whole against hydrologically abusive land uses.

Hydro corridors

■ *Fluvial systems are more than simple threads of water winding through a network of valleys.* They are actually hydrologic corridors where several water systems—notably, channel, flood, groundwater, wetland, and lake systems—coalesce and interact to shape and change the riparian landscape. Water is decidedly the driving force of stream corridors both on and below the surface, and, therefore, no matter what their objectives may be, planning and management programs must not lose sight of the hydrologic imperative.

Natural fluctuation

■ *Streamflow and related hydrologic systems fluctuate radically over time.* Average flow conditions may have little to do with the forms and features of channels and floodplains that we see on site visits and record in our field inventories. Therefore, in reading cause-and-effect relations into this landscape, we must develop a sense of the magnitude and frequency of significant discharge events based on multiple lines of evidence including discharge records, topographic features, vegetation patterns, old maps and aerial photographs, and local histories, both written and oral (Fig. 14.14).

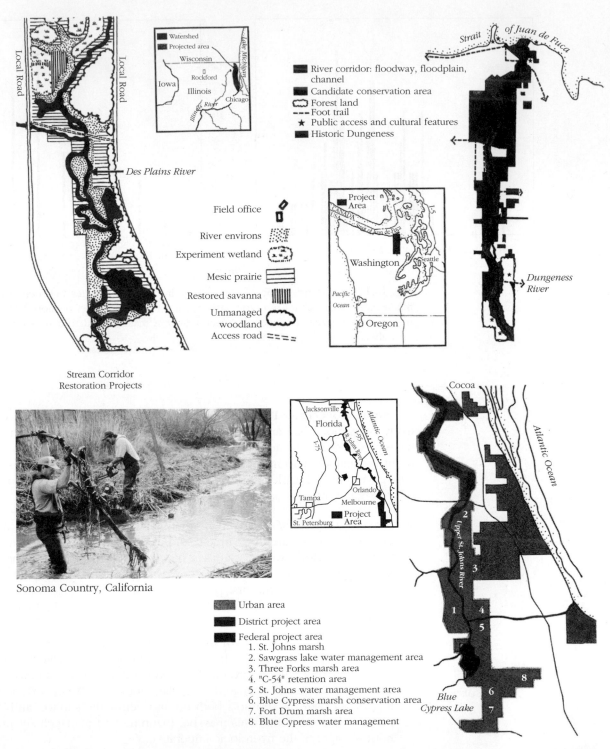

Fig. 14.13 Examples of stream corridor restoration projects carried out under the Rivers, Trails, and Conservation Program of the U.S. National Park Service. The program's objectives include streambank restoration, flood loss reduction, fisheries improvement, and improved realization of recreational resources. The photograph shows volunteer workers restoring a degraded stream in Sonoma County, California.

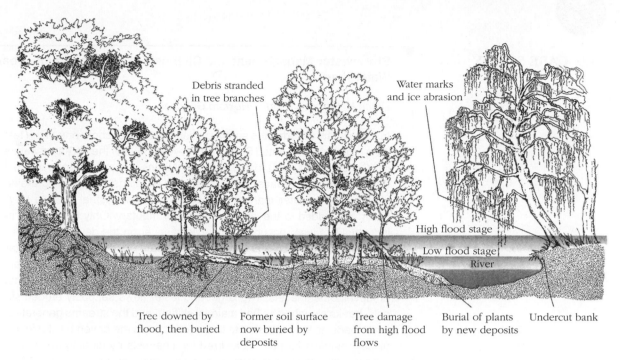

Debris stranded
in tree branches

Water marks
and ice abrasion

High flood stage

Low flood stage

River

Tree downed by
flood, then buried

Former soil surface
now buried by
deposits

Tree damage
from high flood
flows

Burial of plants
by new deposits

Undercut bank

Fig. 14.14 Indicators of past floods, such as water marks, stranded debris, and damaged vegetation, are part of the record of a stream's flow regime.

Geomorphic change

■ *Erosion and deposition are natural processes in stream channels and floodplains.* But it is necessary to distinguish between natural, background levels of geomorphic change and those induced by urban runoff, deforestation, cropland runoff and sedimentation, and structural facilities such as bridges. Undercut banks and point bars are part of a stream's natural dynamics, and most should not be earmarked for stabilization in landscape management programs.

Habitat corridor

■ *The principal biogeographical expression of stream systems is the habitat corridor.* In many respects the stream habitat corridor functions as a biological right-of-way through a watershed, providing linkage and continuity among plant and animal communities. Because they are narrow and continuous over great distances, corridors are prone to breaching from land use systems, which results in fragmentation and reduced biodiversity. The true impact of a breach, however, depends on the relative permeability that a barrier, such as a highway or bridge, poses to plant and animal populations.

Bio-dependence

■ *Many of the species and ecosystems inhabiting stream corridors are not only adjusted to the variable flow regime but are dependent on it* for nutrient supplies, propagule dispersal, seed propagation, and many other life processes and resources. The system's behavior tends to be episodic, and management concepts based on notions of gradual ecological change, as illustrated by the community-succession model, are not tenable for most stream valleys. Further, management programs aimed at reducing variability, such as flood control projects, can seriously alter aquatic and riparian ecosystems. Among many other things, the absence of stress from low- and medium-magnitude flood events allows all sorts of flood-intolerant species to invade these habitats.

14.9 CASE STUDY

Stormwater Management and Channel Restoration in an Urban Watershed

Bruce K. Ferguson and P. Rexford Gonnsen

In the last few decades the southern Piedmont has become one of the fastest urbanizing regions of the United States. The development of the 140-acre watershed draining Coggins Park, an industrial park in Athens, Georgia, began in the 1980s with the construction of roads, utilities, selected buildings, and parking areas. Following standard stormwater management practices for industrial sites, a "dry" detention basin was added to the watershed's major tributary. Only four years later, half the industrial lots were occupied. Impervious cover on the developed parts of the park exceeded 90 percent, and, over the whole watershed, it totaled 30 percent.

The resultant increase in stormwater loading of the three tributary channels draining the park was severe. Stormflow peak rates had more than doubled over predevelopment levels. The larger and faster flows seriously eroded stream channels, banks sloughed in, bed material shifted, and the streams generated huge sediment loads as they adjusted to their new flow regime driven by stormwater. In some places, builders further destabilized the channels by digging trenches to drain old farm ponds and moving streams to make room for construction fill. The detention basin originally constructed as a part of the park's infrastructure proved to be ineffective at either suppressing peak flows or capturing mobile sediment. It neither reduced runoff nor changed the total volume of water released downstream.

Faced with stormflows that eroded away landscape plantings and sediment that turned the stream brown, smothered aquatic vegetation, and even filled a waterfall pool, downstream residents sued, demanding that the problem be corrected. Decisive countermeasures were called for, and a stream and watershed rehabilitation project was conceived to capture mobile sediment, stabilize stream channels, suppress peak storm flows, augment baseflows, establish wetlands, and improve water quality.

To reduce bank erosion and stabilize the channel, numerous check dams were added to the stream (Fig. 14.A). This established a series of intermediate base levels below which the stream could not erode while the pools behind the dams captured sediment. As the pools filled with sediment the channel gradients upstream of dams became gentler, thus reducing the stream's energy to erode and transport sediment. This led to further buildup of the streambed, and as the channel stabilized, wetland and riparian vegetation became established on the deposits of channel sediment. In addition, the sediment promoted streambed infiltration and recharge of the alluvial aquifer, which enhanced stream baseflows during dry periods.

Next, the detention basin was restored and enlarged. It now retains sediment from upstream, and this improvement, coupled with channel stabilization, has resulted in much clearer discharge water. Additional but smaller holding basins were placed in the disturbed streams where existing roads, utilities, and land use allowed, and more check dams were added. The goal was to induce such substantial channel filling that check dams and reservoirs would become buried, resulting in a more uniform channel gradient that would approach equilibrium condition. At this point a stream channel could accommodate the flows delivered to it without excessive channel erosion or sedimentation.

Hydrologic modeling of the overall drainage system indicated that the restoration effort consisting of check dams and reservoirs had indeed reduced the magnitude of peak flows for the 10-year storm to predevelopment levels (Fig. 14.B). However, additional measures were needed to manage the runoff that would be generated by

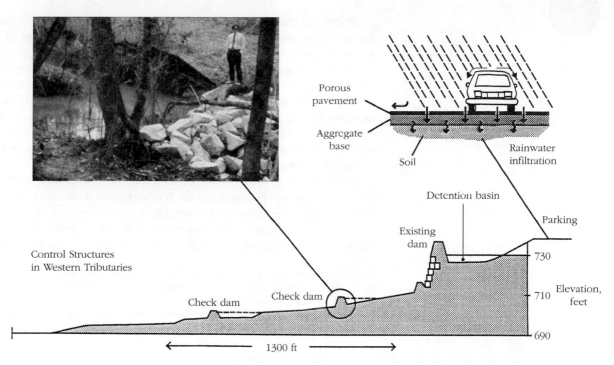

Fig. 14.A Check dams reduced channel erosion by slowing flow and catching sediment. Porous pavement reduced runoff volume from parking and storage areas.

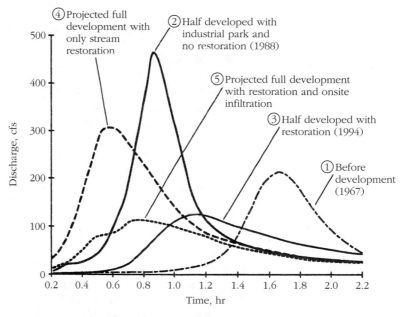

Fig. 14.B The projected performance of different stormwater management strategies given partial and full site development.

future development at full buildout of the industrial park. Analysis showed that onsite reduction of runoff from future building sites was needed. In urban areas, such as Coggins Park, however, open space for surface basins on industrial plots was not available. As an alternative, we turned to porous pavements with aggregate bases as a means of increasing infiltration and reducing runoff.

Infiltration attacks the problem of urban stormwater at its very source. Rainwater can pass through a porous asphalt surface and be stored in the void spaces of stone aggregate while infiltrating the underlying soil (see Fig. 14.A). In addition to reducing the volume of water, infiltration reduces stormwater contaminants such as suspended solids and automotive oils. The porous pavement infiltration systems were designed to work in concert with the check dams and reservoirs to help hold future peak flows to predevelopment levels. Our hydrologic models indicate that in combination these measures will produce acceptable levels of streamflow, bring environmental damage under control, and eliminate serious off-site impacts.

Recently, beaver tracks were seen along the channels of Coggins Park. Beavers may be taking over the business of flow regulation and check dam operation. Eventually, they may eliminate the need for perpetual human vigilance and maintenance of the drainage system. In any case they are a sign of recovery in a small stream system that had been devastated by urbanization and its fugitive runoff.

Bruce K. Ferguson is professor of landscape architecture at the University of Georgia who specializes in stormwater hydrology. P. Rexford Gonnsen is a partner and a landscape planner in Beall Gonnsen & Company, Athens, Georgia.

14.10 SELECTED REFERENCES FOR FURTHER READING

Booth, D. B., and Henshaw, D. C. "Rates of Channel Erosion in Small Urban Streams." In *Land Use and Watersheds*. Washington, DC: American Geophysical Union, 2001.

Collier, Michael, et. al. "Dams and Rivers: A Primer on the Downstream Effects of Dams." *U.S. Geological Survey Circular*. 1126, 1996.

Ferguson, B. K. "Urban Stream Reclamation." *Journal of Soil and Water Conservation*. 46, 5, 1991, pp. 324–328.

Grossman, Elizabeth. *Watershed: The Undamming of America*. New York: Counterpoint, 2002.

Hirsch, R. M., et al. "The Influence of Man on Hydrologic Systems." In *The Geology of North America*. Boulder, CO: Geological Society of America, 1990.

Kaufman, M. M., and Marsh, W. M. "Hydro-Ecological Implications of Edge Cities." *Landscape and Urban Planning*. 36, 1997, pp. 277–290.

Keller, E. A. "Pools, Riffles, and Channelization." *Environmental Geology*. 2, 2, 1978, pp. 119–127.

Leopold, L. B. *A View of a River*. Cambridge, MA: Harvard University Press, 1994.

Leopold, L. B., Wolman, M. G., and Miller, J. P. *Fluvial Processes in Geomorphology*. San Francisco: Freeman, 1964.

McGuane, Thomas. "On Henry's Fork." In *Heart of the Land: Essays on Last Great Places*. New York: Vintage Books, 1994.

McGuckin, C. P., and Brown, R. D. "A Landscape Ecological Model for Wildlife Enhancement of Stormwater Management Practices in Urban Greenways." *Landscape and Urban Planning*. 33, 1995, pp. 227–246.

Palmer, M. A., et al. "The Ecological Consequences of Changing Land Use for Running Waters, with a Case Study of Urbanizing Watersheds in Maryland." *Yale Bulletin of Environmental Science*. 107, 2002, pp. 85–113.

Poff, L. N., et al. "The Natural Flow Regime: A Paradigm for River Conservation and Restoration." *Bioscience*. 47, 11, 1997, pp. 769–784.

Smith, D. S., and Hellmund, C. A. (eds.). *Ecology of Greenways.* Minneapolis: University of Minnesota Press, 1993.

Stanford, J. A., and Ward, J. V. "An Ecosystem Perspective of Alluvial Rivers: Connectivity and the Hyporheic Corridor." *Journal of the North American Benthological Society.* 12, 1, 1993, pp. 48–60.

Swift, B. L. "Status of Riparian Ecosystems in the United States." *Water Resource Bulletin.* 20, 2, 1984, pp. 223–228.

Williams, G. P., and Wolman, M. G. "Downstream Effects of Dams on Alluvial Rivers." *U. S. Geological Professional Paper.* 1286, 1984.

Wolman, M. G. "A Cycle of Sedimentation and Erosion in Urban River Channels." *Géografiska Annaler.* 49A, 1967, pp. 385–395.

Related Websites

Alberta Riparian Habitat Management Society. http://www.cowsandfish.org/
A nonprofit organization educating citizens and communities about management practices in riparian ecosystems. Because of local economics, the focus is on agriculture. Follow links to management suggestions and to learn more about riparian habitat.

Delaware River Keeper Network. http://www.delawareriverkeeper.org/
A nonprofit organization protecting the Delaware River through education, advocacy, and restoration. Following the restoration link, there are several case studies of stream restoration and ideas people can use at their own homes.

National Center for Earth-Surface Dynamics. A National Science Foundation Science and Technology Center. http://www.nced.umn.edu/
Organization that is working with university communities to predict earth-surface dynamics and their effect on the landscape to inform management practices. Stream restoration research, education, and current projects are given.

On the Cutting Edge, Teaching Geomorphology in the 21st Century. "Processes of River Erosion, Transport, and Deposition." 2009. http://serc.carleton.edu/NAGTWorkshops/geomorph/visualizations/erosion_deposition.html
A collaborative project to assist geoscience teachers. Click the photo links to view movies and flash animation of river formation processes.

THE COASTAL LANDSCAPE: SHORELINE SYSTEMS, LANDFORMS, AND MANAGEMENT CONSIDERATIONS

15.1 INTRODUCTION

Emerging crisis

Shore erosion, flooding, and property damage are nagging and costly problems in the coastal zone. In the United States and Canada, which together account for more than 200,000 miles of coastline, these problems have grown significantly in the past three decades, not because the oceans and lakes are behaving much differently than they did in years past, but mainly because of increased development and use of the coast. This has given rise to heavy financial and emotional investment and, for many of the 100 million or more North Americans who live on or near a coast, it has led to a bittersweet relationship with the sea. The costs to individuals and society are rising. In response, efforts to manage and protect coastal lands have reached critical levels at community, state/provincial, and national levels. And with the prospects for rising sea levels, increased storm magnitudes and frequencies, and growing coastal population worldwide in this century, the problem is sure to become exceedingly acute.

Groundwork

In 1972, the U.S. federal government passed the Coastal Zone Management Act, which provides for the formulation of coastal planning and management programs at the state level. Among the responsibilities of the state programs is the classification of coastlines according to their relative stability, including the potential for erosion. To make such a determination, planners need to understand not only the physical makeup of the coast, but also the nature of the forces acting on it. The principal force in the coastal zone is that of wind waves. They are the source of most shore erosion, sediment transport, and deposition, which together shape the beaches, sandbars, cliffs, barrier islands, and related features. These landforms are the settings where the dramatic interplay between land use and the sea is carried out.

Beyond engineering

Finding the proper fit between land use and the sea demands more than muscling the coastline into an engineered fortification to protect land use facilities. It requires a more multidimensional perspective that gives as much weight to ecosystems, geomorphic systems, scenic and recreational resources, history, and culture as it does to protecting real estate. And it must give serious consideration to the remarkable geographic diversity of the coast, not only its differences in carrying capacity and resilience to land use disturbance, but the historical patterns and types of human settlement. In short, we, as a society, must develop more thoughtful and sustainable approaches to our use of the world's seacoast. But in the face of rising sea levels and bigger storms, overcoming the temptation to invoke endless structural defenses will be hard.

15.2 WAVE ACTION, CURRENTS, AND LONGSHORE SEDIMENT TRANSPORT

Most wind waves are generated over deep water where they cannot cause erosion because the motion of the wave does not reach bottom. In shallow water, waves not only touch bottom, but they can also exert considerable force against it. Initially, this force is not very great, but as the wave nears shore, it increases rapidly. Where the shear stress of this force exceeds the resisting strength of the bottom material, displacement of particles occurs. Along coasts comprised of loose sediments, such as sand and silts, large quantities of particles are churned up, especially as the waves break in the shallow water near shore.

Waves deep and shallow

Wave Size and Behavior. The wave form we see on the water surface represents only a fraction of the wave. The rest is under water and has a rotating motion. The water depth at which this rotating motion first touches bottom and begins to move small particles is called the wave base. **Wave base depth** increases with wave size and is roughly proportional to 1.0 to 2.0 times the wave height. Relative to wavelength, wave base falls at a depth between 0.04 and 0.5 wavelength (Fig. 15.1). The

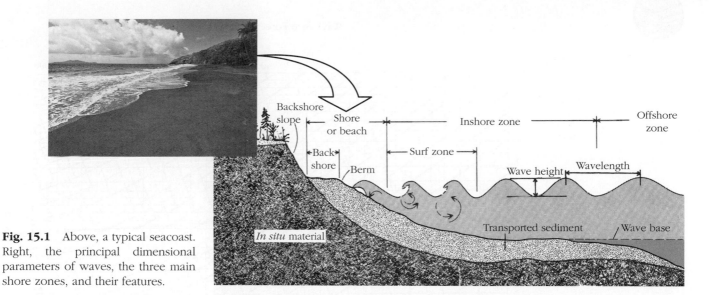

Fig. 15.1 Above, a typical seacoast. Right, the principal dimensional parameters of waves, the three main shore zones, and their features.

shallow-water zone along a coastline is defined by the wave base depth, and because the wave base changes with wave size, the shallow zone actually fluctuates in width with different wave events. In the oceans, however, wave base for large waves is generally considered to average 30 feet; in the Great Lakes it averages around 10 feet.

The ability of waves to erode and transport sediment is a function of the wave size and of the size and availability of sediment. In deep water, **wave size** is the product of three variables: wind velocity, wind duration, and fetch. *Fetch* is the distance *Wave size* of open water over which a wind from a certain direction blows. Fetch varies dramatically depending on the shape and size of a waterbody. It is the main reason why inland lakes cannot generate large waves. The largest waves are produced when wind velocity, duration, and fetch are all large. Forecasts of the size of *deep* water waves can be made using the graph in Figure 15.2. For example, at a wind velocity of 20 mph and a fetch of 100 miles, the wave height reaches 5 feet after about 14 hours duration.

Wave Refraction. Near shore, another variable comes into play: water depth. In shallow water, the lower wave rubs on the bottom, losing energy to friction and to the turbulence generated by the churning motion of the water, causing it to slow down. The rate of deceleration increases toward shore with increasing friction and turbulence, and nearshore wavelength shortens rapidly (and height increases) as waves *The refraction process* "close ranks" before dissipating themselves on the beach. If a wave approaches shore at an angle, and most do, it actually bends or *refracts* as the inner segment slows down much faster than the outer segment. This pattern is easily observed and measured by tracing the curved alignment of the wave axis (crest line) in the shallow-water zone, and it signals a reorientation of the wave's energy so that its force strikes the shore more head on (Fig. 15.3).

Refraction is critical to understanding the distribution of wave energy along the coast. It is a fairly easy parameter to map because it is controlled by just two principal variables: (1) the direction of wave approach and (2) the bottom topography of the shallow-water zone. Therefore, **refraction patterns** for most coasts can be delineated based on deep-water wave direction and offshore bathymetry.

Wave-Size Forecasting Chart

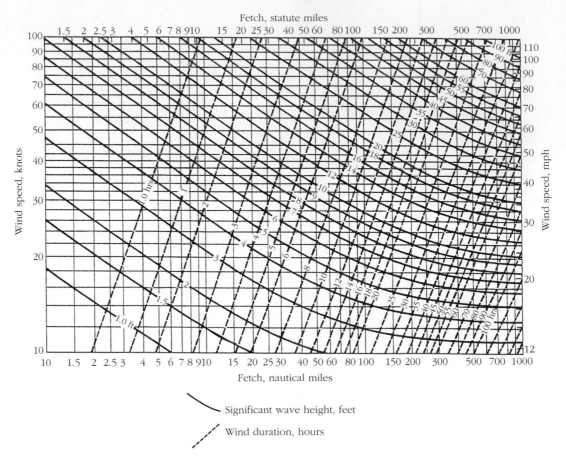

Fig. 15.2 A chart for forecasting the size (height) of deep-water waves based on fetch, wind duration, and wind velocity. The duration curves give the minimum time required to reach maximum height for a given wind velocity and fetch.

Refraction patterns

How refraction translates into wave energy distribution can be illustrated by constructing lines called *orthogonals* (perpendicular lines) across the crest of approaching waves as they appear on an aerial photograph like the one shown in Figure 15.3*a*. Assuming that wave size (and therefore energy) is uniform along the entire wave crest in deep water, we see that any convergence or divergence of the orthogonals in shallow water represents a change in the relative distribution and the orientation of wave energy. Normally, refraction causes wave energy to become focused on headlands and the seaward sides of islands, and diffused in embayments and on the leeward sides of islands. These are typically the sites of erosion and deposition, respectively (Fig. 15.3*b*).

Mobilizing sediment

Sediment Transport. For waves to modify shorelines, sediment must be picked up and moved from one place to another. For this to take place, the force exerted by the motion of the waves must be great enough to dislodge particles. On the outer edge of the shallow-water zone this force is typically slight and particle movement is slight, especially with sand-sized sediment, and is limited to a to-and-fro motion with the passage of each wave. Nearer shore, turbulence rises, producing a lifting motion, especially where waves are breaking, that carries particles into the water column and on a dizzying ride before settling back to bottom. Although the movement of individual particles can be in any direction, the net direction of movement after the passage of many waves is mainly parallel to the shoreline.

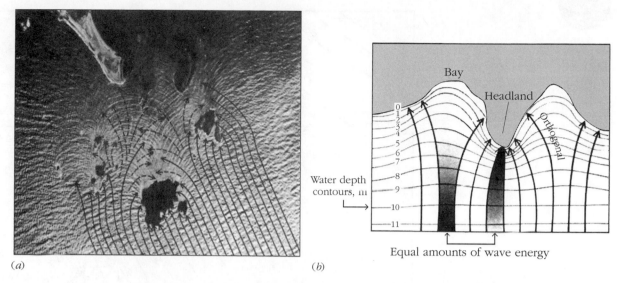

Fig. 15.3 Wave refraction patterns (*a*) as revealed by aerial photography and (*b*) as related to the distribution of energy around a headland and adjacent embayments.

Longshore transport   Two factors account for the parallel, or *longshore transport*, of sediment. One is that most waves approach and intercept the coast at an angle, therefore, the direction of wave force is oblique to the shoreline. Although waves refract into a more direct approach angle near shore, most retain a distinct angle as they cross the shallow-water zone. Thus, a large component of the energy for sediment transport is set up parallel to the shoreline.

The second factor is a current that flows along the coast. This current, called a **longshore current**, moves parallel to the shoreline in the direction of wave movement at rates as great as 2 to 3 feet per second. Longshore currents are driven by
Longshore current   wave energy, increasing in velocity and size (volume of water) with wave size and duration. They work hand in hand with waves in moving sediment. When waves cross the shallow zone and churn up sediment, the longshore currents transport it downshore before it settles back toward the bottom, where it is churned up again and transported another increment. Longshore currents function as coastal conveyor belts (Fig. 15.4). On major ocean coasts, they move hundreds of thousands of cubic yards of sediment a year.

Other types of currents also operate along the shore, and one of the most prominent is the *rip current*. Flowing seaward from the shore across the lines of approaching waves, rip currents are narrow jets of water that intercept the longshore train of
Nearshore circulation   sediments, carrying part of it into deeper water. Together, rip currents, waves, and longshore currents form **nearshore circulation cells**, moving both water and sediments toward, away from, and along the shore (see the inset in Fig. 15.4). Rip currents and nearshore circulation cells are not common to all shores. Most are found on high-energy coasts such as Southern California.

The Longshore Sediment System. The predominant system operating in the coastal zone (and within which nearshore cells are always nested) is the **longshore (or drift) system.** These systems rank only behind rivers in terms of total sediment transport worldwide, and any attempt at understanding how the coastal zone functions,
A river of sediment   virtually no matter where, must include serious consideration of the longshore sediment system. The main workhorse of this system is the longshore currents and related wave processes, but it also includes other forms of sediment movement along the coast, mainly *beach drift* (swash and backwash action of waves on the shore itself).

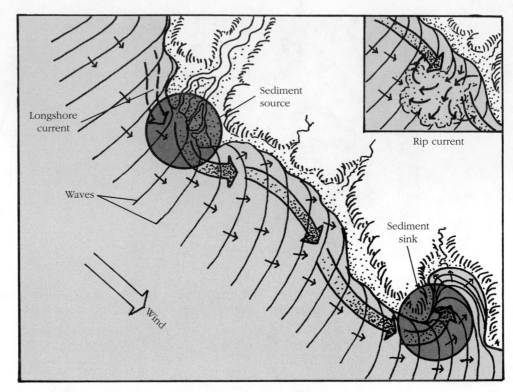

Fig. 15.4 Nearshore circulation showing waves, longshore currents, rip currents, and the transport of sediment from a source (above) to a sink (below).

Longshore systems are basic input/output systems. They are made up of three basic parts: (1) a sediment source, (2) a transport zone, and (3) a sediment sink

System components   (Fig. 15.4). These components, and thus the system itself, can vary enormously in scale, sediment transfer rates, and relationship to coastal forms and features, but when the system is in balance, the rate of sediment input, transport, and output are equal. This, of course, rarely happens in nature, and even less so when humans are involved, and therefore the volume of sediment moving through the system may fluctuate, often radically, from time to time and from place to place.

Sources and Sinks. For most systems, the main source of sediment (or *source area*) consists of streams and rivers. Where the coast is supplied with massive amounts of sediment, such as at the Mississippi Delta, there is often more sediment input to the

Sediment inputs   shore than the longshore systems can carry away. But in most places the longshore systems can accommodate the sediment fed to the coast by streams, and it is carried downshore where we see it in the form of transient features like beaches and sandy islands, among other things.

The second source of sediment in longshore systems is shore erosion. This happens where waves are able to cut into the land behind the shore (the backshore) and release soil and other *in situ* materials into the longshore system. Although the total contribution of sediment from backshore erosion is small compared to rivers, locally it can be considerable and is often a source of great concern because it may involve the loss of valuable coastal real estate.

At the other end of the system, the sediment is deposited in a *sink.* All sorts of

Sediment outputs   environments serve as sinks, but they all share one characteristic: They are places where the longshore system loses a large part of its energy and drops its load. The area on the leeward side of an island is such a place because the island blocks the receipt of wave energy needed to drive the longshore current. One of the most common sinks is at the mouths of bays and harbors because the water is too deep here for the waves and currents to sustain contact with the bottom. As a result, the sediment load is dumped as the longshore conveyor belt passes into deep water.

15.3 NET SEDIMENT TRANSPORT AND SEDIMENT MASS BALANCE

Scale is an important consideration in landscape planning, and the coastal zone is no exception. Longshore systems are almost always the framing model, but within these often large systems are smaller parcels of coastal space that are usually the focus of planning and management problems. Nevertheless, our concern remains one of understanding the dynamics of the coastal environment.

Gross and Net Sediment Transport. On most shorelines, with the passage of storms or the seasons, waves and currents may change direction, in turn, causing reversals in the longshore system. Thus, the same sediments, more or less, may be transported past one point on the shore several times in one year. A measure of this quantity, the total amount of sediment moved past a point on the coast (with no rule against double counting), is called **gross sediment transport**. For obvious reasons, it is not a good indicator of the balance of sediment at a place on the coast because it tells us neither whether the shore is losing or gaining sediment, nor whether the beach is growing or shrinking. Therefore, another measure is used, called net sediment transport.

Total transport

Net sediment transport is the *balance* between the sediment moved one way and that moved the other way along the coast:

Net transport

$$\text{Net } Q_t = Q_p - Q_s$$

where

$\text{Net } Q_t$ = net quantity of sediment per year
Q_p = longshore transport in the primary direction
Q_s = longshore transport in the secondary direction

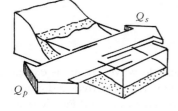

If the longshore system is driven predominantly by waves from one quadrant of the compass, then net sediment transport can be large. By contrast, if the longshore system operates in both directions, then net transport may be small and gross transport may be large. Ultimately, we are concerned with the trend because it can tell us something about the development of the coastline.

Site-Scale Sediment Balance. At a local scale of observation, where only a short segment of beach or small waterbody is concerned, we might consider making a detailed determination of the **sediment balance** (Fig. 15.5). The mass of sediment on the beach for any period of time is equal to all inputs less all outputs.

Local balance

$$\text{Sediment mass balance: } L_i - L_o + O_i - O_o + R_i - W_o = O$$

where

L_i = longshore input
L_o = longshore output
O_i = shoreward inputs from sand bar migration in summer
O_o = offshore outputs from bar migration in fall and winter
R_i = inputs from backshore slope erosion and runoff
W_o = losses (outputs) from wind erosion of beach sand

A value greater than zero means that the reservoir of beach sediment has gained mass; less than zero, that it has lost mass. Generally, these two trends are manifested

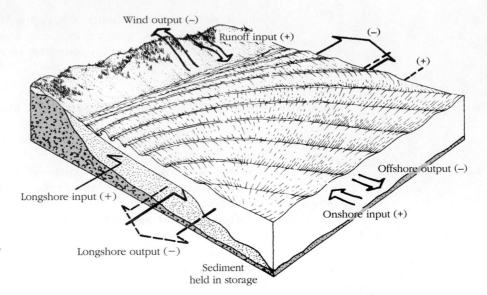

Fig. 15.5 Components in the sediment mass balance of a short segment of beach, a useful model for local planning and management problems.

Application

in larger and smaller beaches, respectively, that is, the beach actually expands and contracts. In most places, detailed data are not available for computation of the sediment mass balance, nor is there adequate time or resources in most planning studies to acquire the necessary data. Therefore, in site planning we must usually resort to an interpretation of local records such as old maps, land survey records, and archival aerial photographs, as well as the testimony of longtime property owners to gain an idea of local trends. Unfortunately, all this will be made more difficult with sea level rise in the decades ahead.

15.4 TRENDS IN SHORELINE CHANGE

Where river sediments are abundant, virtually all available wave energy is expended in transporting the sediment load along the coast. However, where river sediments are scarce, such as in Southern California where reservoirs have deprived Pacific beaches of much nourishment from stream sediments, the body of beach sediment is often small and offers little protection for the *in situ* material under and behind the beach. In other words, if the beach is thin, the bank behind the shore does not get much protection against the onslaught of storm waves.

Retrogradation

Under such circumstances, the backshore slope (which is made up of *in situ* material rather than shore sediment in transit) is susceptible to wave erosion. With the loss of this ground, the shoreline shifts or retreats landward. Such coastlines are characterized by distinctive features, such as sea cliffs, exposed bluffs, and wave-cut banks, and they are referred to as *retrogradational* because over the long run they retreat landward as they give up sediment.

Retreat . . . erosion

Definition of Terms. *Retreat* is defined as the landward displacement of the shoreline. It is usually caused by erosion, but it may also be caused by a rise in water level, the subsidence of coastal land, or any combination of the three. Where retreat is caused by erosion, the amount of *in situ* material actually lost can be computed by multiplying the retreat rate (R) times the backshore slope height (H) times a given length (L) of shoreline (Fig. 15.6):

$$\text{Erosion} = R \bullet H \bullet L$$

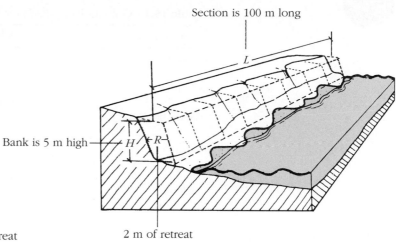

Section is 100 m long

L

Bank is 5 m high — *H* *R*

2 m of retreat

Fig. 15.6 The erosion associated with shoreline retreat is computed from the retreat rate (*R*), the backshore slope height (*H*), and the length of the shoreline (*L*).

Erosion = 5 m × 2 m × 100 m = 1000 m³

Progradation

At the other extreme are coastlines that build seaward, called *progradational* coastlines. Progradation is usually caused by deposition of sediment brought on by decreased wave energy, increased sediment supply, decreased water depth, and/ or construction of sediment trapping structures such as groins and breakwaters. Progradational coastlines are characterized by low-relief terrain in the form of various depositional features such as broad fillets of sand at the heads of bays, spits, and bars near the mouths of bays and behind islands, as well as beach ridges and sand dunes.

Most depositional features are prone to rapid changes in shape and volume; therefore, short-term trends should generally not be used as the basis for formulating land use plans in the coastal zone. In most instances, huge amounts of sediment are needed to build a small area of shore land such as a sand bar or a bayhead beach because the bulk of such features are built under water. It is important to remember this if the rate of growth or decline of the feature is to be related to the net sediment transport rate over some time period. The planning and development implications of depositional coastlines are taken up later.

Regional Coastal Trends. It is instructive to examine regional trends in coastal change based on historical data such as old maps and charts. One such study, conducted for the United States, revealed that as a whole the U.S. coast (including the Great Lakes) is retreating by an average rate of 2.6 feet per year. Further, the rate varies from an average of nearly 6 feet of retreat on the Gulf Coast to less than 1 inch a year on the Pacific Coast. Table 15.1 gives the rate for various regions and coastal (landform) types.

Meaning

Such data are meaningful only in a general way inasmuch as they indicate the direction (+ or −) and the rate of change that, for representative stretches of shoreline, can more or less be expected in the future. Locally, however, they may be meaningless because rates and directions of change vary with the coastal type, local sediment systems, harbor facilities, and the incidence and effect of forceful events such as hurricanes. Generally speaking, predictions of trends in shoreline change, especially long-term rates, should be viewed cautiously, especially in light of the impending sea level and weather changes related to global warming and ongoing engineering manipulation of shorelines that alter longshore sediment transport.

Table 15.1   Coastal Change Rates for the United States Based on Historical Records

Coastal Type	Region	Rate (ft/yr)
Mud flats	Florida	−1.0
	Louisiana–Texas	−7.0
Bedrock	Atlantic	+3.3
	Pacific	−1.7
Sandy beaches	Atlantic	−3.3
	Maine–Massachusetts	−2.3
	Mass–New Jersey	−4.3
	Gulf	−1.3
	Pacific	−1.0
Deltas		−8.3
Barrier islands	Gulf	−2.0
	Louisiana–Texas	−2.6
	Florida–Louisiana	−1.7
	Atlantic	−2.6
	Maine–New York	+1.0
	New York–North Carolina	−5.0
	North Carolina–Florida	−1.3

Source: Dolan et al., 1983.

15.5 SAND DUNE FORMATION AND NOURISHMENT

Put a fresh pile of sand on most beaches and the wind will dry it out and blow it landward. If the pile is consistently replenished, the wind-blown sand may lead to a heap of sand on the backshore or beyond. If the wind drives the heap of sand inland, we have, by definition, created a **sand dune**. An entire complex of mobile heaps of sand constitutes a *dunefield*, one of the prized landscapes of the coastal zone.

Sand Dune Types and Formation.   Sand dunes are a common and integral part of every major coastline from the Arctic to the tropics. Their formation and maintenance depend directly on two key factors: (1) an ample supply of erodible sand and *Ingredients* (2) a source of wind energy to drive the sand landward. The source of sand for the vast majority of coastal dunes is the beach in front of them. Clearly the best sources are broad, sandy beaches that are fed by the longshore system and that have little or no plant cover. A second source of dune sand is wave-cut banks and cliffs of sandy composition. These two sources give rise to two different classes of coastal sand dunes: (1) *low-elevation dunes* that begin near the backshore and rise gradually land *Dune classes* ward (Fig. 15.7*a*) and (2) *perched dunes* that begin high above water level near the brow of a bluff or sea cliff (Fig. 15.7*b*).

Onshore Wind.   The shore is a windier place than locations a little inland from it. The reason has to do with the movement of air at the base of the atmosphere, called the **atmospheric boundary layer**. This layer of air, which measures about 1000 feet deep, drags over the earth's surface and is slowed by the resulting friction. When the boundary layer moves over open water, a comparatively smooth surface, it is less impaired by friction than it is over land; therefore, near-surface wind is generally faster over water. In addition, as the air flows from water onto the land, it also tends to accelerate at the coastline.

Acceleration is caused by the constriction of the boundary layer as the land rises in elevation at the coast. The train of air must be squeezed through a smaller space, and in order to maintain continuity of flow, wind velocity increases. This phenom *Coastal aerodynamics* enon is especially pronounced along high banks and sea cliffs, as is shown in the model in Figure 15.8*a*, where wind energy and erosive power rise dramatically along

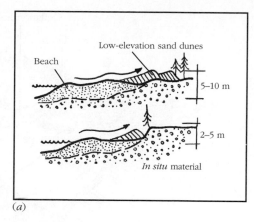
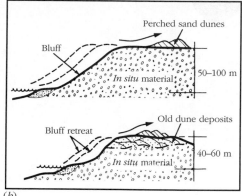

Fig. 15.7 (*a*) Low-elevation sand dunes, the most common coastal dunes, rely on the beach for sand supply. (*b*) Perched dunes rely on the bank, bluff, or cliff in front of them for sand supply. Perched dunes are common to the Oregon and Great Lakes coasts, among other places.

the brow and may give rise to the formation of perched sand dunes. It also accounts for a high stress zone along the brow, where in addition to wind erosion, the landscape here suffers from thermal stress and extreme wind turbulence, which is often reflected in damaged vegetation and facilities (Fig. 15.8*b*).

Blowouts   ***Dune Origins and Forms.*** On most beaches, dune formation begins in the backshore zone with the development of a low ridge of sand paralleling the shore. As the ridge grows, blowouts form at selected spots from which tongues of sand migrate landward. These function as ramps for the transport of sand wafted from the beach. The advancing edge of the dune deposit buries vegetation, soil, and sometimes wetlands as it moves inland. From the blowout to the leading edge of the dune, the form (in planimetric view) is one approximating an elongated *U* and in most places is called a *hairpin dune* (Fig. 15.9).

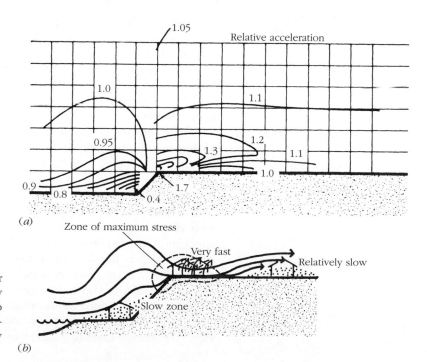

Fig. 15.8 (*a*) Airflow over a high coastal bluff or sea cliff. For wind flowing directly onshore, velocity can accelerate as much as fourfold from the toe to the brow of the slope. (*b*) The corresponding implications for facilities at three sites along the airflow profile.

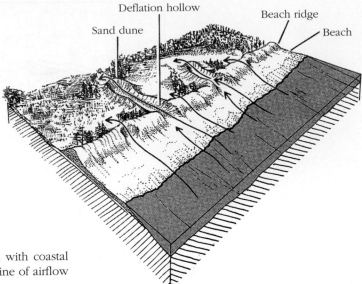

330 THE COASTAL LANDSCAPE

Fig. 15.9   The most common setting and forms associated with coastal sand dunes. The hairpin form of the dune follows the main line of airflow inland from the beach, driving the dune over the landscape.

Migration

As the dune moves inland, it grows not only longer but deeper as sand is added to the main body of the deposit. On some barrier islands and sand spits, dunes may migrate all the way across the landform to the inner shore. On most shores, however, the available wind energy is insufficient at some distance inland to transport enough sand to overcome vegetation. Therefore, the dune ceases to advance, and plants take over the surface.

If the process of dune formation is repeated many times along a segment of coastline, the deposits tend to overlap and merge together forming a complex dune ridge called a *barrier dune*. The barrier dune generally marks the highest landward extent of the dunefield (the total mass of dune features landward of the beach). The barrier dune is usually steep on the leeward (inland) side, where wind deposits are held in place by vegetation, and gentle on the windward side where sand is transported upslope from the blowouts.

Managing Coastal Sand Dunes.   Coastal sand dunes can present difficult land use and management problems. They are extremely attractive to residential, resort, and recreational development by virtue of the open vistas, the juxtaposition of open sand, grass-covered and wooded surfaces, and of course, the access to the shore.

Rationale

They are, however, very fragile environments where vegetation, sand deposits, and slopes may be so delicately balanced that only slight alterations in one can lead to a chain of changes such as renewed wind erosion, blowout formation, and dune reactivation.

Dunes as Systems.   What is often not clearly understood about coastal dunes is that they are themselves small sediment systems with source areas, transportation zones, and sediment sinks. For dune systems to maintain themselves, nourishment must be sustained with windblown sand. This requires maintenance of source

System nourishment

areas as well as the flow of air from source areas to sinks (deposition zones). When development takes place in dunefields, source areas may by reduced with the planting of groundcover on barren surfaces as a part of landscaping and erosion control programs. In addition, transportation zones may be disrupted with the building of roads and structures in the backshore and lower dune areas. In the area of the barrier dune complex, a favorite location for siting homes and resorts, construction inevitably requires alteration of *metastable slopes* held in place by vegetation. (See Section 3.4 for a discussion of metastable slopes.)

Erosion control issue

With respect to reductions in sand dune nourishment, the main problem is the weakening of the dune building and movement processes and the encroachment of vegetation over the dunefield. Successful *erosion control* programs that stabilize source areas (such as backshore slopes) may so limit the supply of sand that the interplay between geomorphic processes and vegetation—which gives the dunefield much of its special character—becomes largely one-sided, and the dunefield becomes overgrown and inactive. In that case, the dunefield is transformed into sand hills, or what are sometimes called *fossil dunes*. Management programs that recognize wind erosion, sand movement, and a changeable landscape as the most desirable components of dune environments—in the same way that wave action and beach dynamics are viewed as attributes of attractive shorelines—may be advisable in many coastal areas.

15.6 APPLICATIONS TO COASTAL ZONE PLANNING AND MANAGEMENT

Risky ventures

Despite calamities from hurricanes, storm surges, erosion, and flooding, population and development are on the rise in coastal areas worldwide. In the United States more than 50 percent of the population now lives in coastal counties, and development pressure there is 40 to 50 percent higher than in noncoastal counties. Many risky and environmentally fragile coastal areas are being taken up by ambitious real estate schemes, and although they may engender serious concern and often outcries from the public, the development proposals are usually not given nearly enough scrutiny by officials, with too little attention to sustainability questions.

Basic needs

Evaluating Land Use Proposals. To evaluate a development proposal, planners must first know the makeup and dynamics of the coastline involved. This usually requires at least a rudimentary understanding of the critical systems, including the direction of water and sediment movement, the types of coastal landforms and lithologies (compositions), and an assessment of recent erosion and deposition trends. It is especially important to map landform indicators of geomorphic trends of coastal change, including where the coast is composed of hard and soft materials and where it is low-lying and where it is elevated. Within this framework, more detailed studies can be carried out for the specific segment where the action is proposed. At the heart of such studies, especially when shoreline structures such as marinas, piers, and seawalls are involved, is the identification of longshore (drift) systems followed by a best approximation of the sediment budget.

Measuring trends

The sediment budget requires an analysis of the net sediment transport, and two types of methods can be employed: shoreline change and wave energy flux. The first and most accurate method is based on measurements of *volumetric changes* in a shoreline. This usually involves either comparing old charts and aerial photographs with newer ones or making detailed measurements of sediment accumulation behind a shoreline barrier such as a large breakwater or jetty. The latter approach necessitates making a topographic survey at the time of construction and another some years later, and then measuring the net change between the two (Fig. 15.10). Where harbor entrances are maintained by sand-bypassing operations such as dredging, the amount of sediment that must be transferred from one side of a barrier to the other each year can be determined by this method. The second method is based on computations of wave energy and established correlations between wave energy flux and sediment transport rates. This method requires wave data for the location in question, and because such data are usually scarce, the wave energy computations are difficult to prepare.

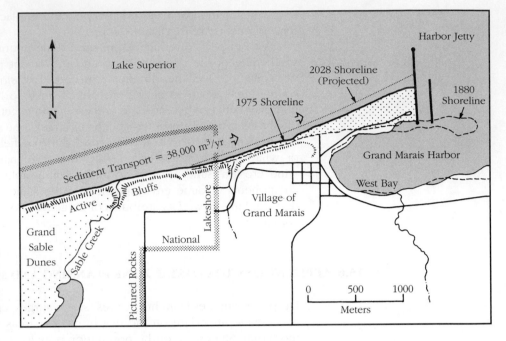

Fig. 15.10 Trends in shoreline change, 1880–1975, with a projection to 2028. Note that, as a result of the harbor jetty (built in the 1960s), the change has been negative on the east and positive on the west. Net sediment transport is eastward.

Key Questions. Upon completion of the coastal inventory and estimated sediment budget, three important questions can be answered:

Please answer

1. What is the nature of the system we are dealing with in terms of the amounts of sediment being moved and the directions of movement? At the most basic level can we identify who is giving up sediment and who is gaining sediment (Fig. 15.11)?

2. What is the relationship among the features, processes, and trends of the coast, such as the growth of spits, the retreat of bluffs, and the nourishment of dunefields?

3. Where is the proposed project located in the system, and what is its relationship to the forms and features of the system?

Answers to these questions can then be used to test the feasibility of development schemes, evaluate community plans, or guide the formulation of coastal zone management plans.

Among the problems commonly faced by private development is how to deal with so-called soft environments such as sand dunes, wetlands, and erodible beaches. Coastal communities, on the other hand, are frequently faced with serious conflict when it comes to the need to protect coastal resources while accommodating economic development schemes involving harbor improvements, marinas, and related facilities.

Site Planning Considerations. Because the coastal zone is so attractive for residential development, it is necessary to examine briefly some of the problems of site planning near shorelines. In evaluating a site for development, as already recommended, landscape planners must first examine the position of the site in the larger sediment system and determine whether it is in a retrograding, prograding, or stable part of the coastline. Next, they need to determine the composition of the site, its topographic configuration, and whether it belongs to a special coastal habitat such as a delta, estuary, or salt marsh. Very risky development sites are low-elevation, soft shorelines, such as barrier islands along the Gulf and Atlantic coasts, which are periodically cut back and overwashed by hurricane waves. On the other hand, sites defended by bedrock are usually safe from rapid erosion, though the bedrock may pose significant problems in building construction, water supply, wastewater disposal, and stormwater drainage.

What to examine

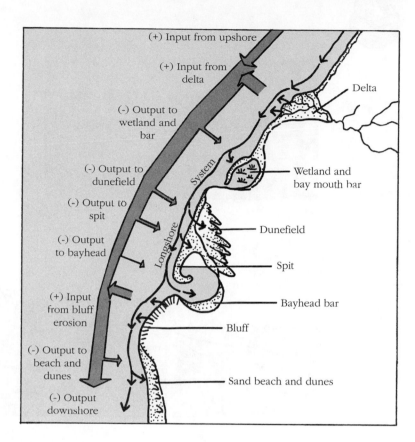

Fig. 15.11 A diverse coastline made up of different shore types and features, each with a definable role in the longshore system. Knowing the basic layout and operation of the system is essential to making prudent planning and design decisions for coastal development at different locations.

Interpreting Change. Shorelines composed of unconsolidated (soft) materials should be treated carefully when development calls for permanent structures because most of these shorelines are prone to alternating seaward/landward oscillations over time. Minor oscillations occur seasonally with winter/summer changes in the sediment mass balance. Over longer time periods, the magnitudes of fluctuations can be expected to increase with the average return period; that is, big changes occur with a lower frequency than the small ones. Studies show that most oscillations are related to climatic fluctuations in the range of 20 to 60 years or more that are marked by periods of increased storminess and shore retreat with intervening periods of quieter weather and beach building. However, the climatic change event in which we are now immersed may render such generalizations suspect in the decades ahead.

Oscillation questions

Geobotanical Clues. But even with records of climate-related fluctuations, determining the proper setback distances for buildings is difficult because the magnitude of shore fluctuations often varies for different segments of a continuous shoreline, and in many areas records lack sufficient detail to help planners figure out local trends. In some instances, however, vegetation and slope conditions can provide insight into trends and fluctuations. Exposed roots, tipped trees, and undercut banks are signs of shore erosion and retreat (Fig. 15.12*a*); whereas stable backshore slopes and numerous young plants on the backshore are signs of progradation. In fact woody plants can be used to measure the rate of progradation.

Tree forms and ages

If tree stands of different *age classes* can be identified on the backshore, they may reveal how far landward the sea or lake has advanced since the stands became established. In Figure 15.12*b*, for example, the presence of 75- to 100-year-old trees beginning at a distance of 100 meters (330 feet) from the shore indicates that the area beyond this point has been free of transgression by waves for at least 75 to 100 years. In contrast, seaward of this point the tree ages tell us that the area has been eroded away at least once in the last 75 to 100 years and has subsequently reformed. Nearest shore, the vegetation reveals that the frequency of destruction and rebuilding is evidently

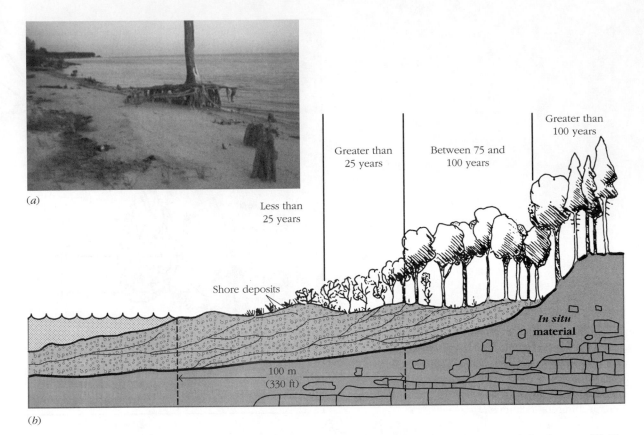

Fig. 15.12 (*a*) Evidence of a retreating shoreline in exposed roots and dead trees. (*b*) The extent of past landward fluctuations in the shore is sometimes revealed by the age patterns of tree stands.

25 years and less. This technique is best suited to areas where one stand gives way sharply to another because the absence of a transition between neighboring stands is a sign that the older stand has been cut back by erosion and that the younger stand subsequently established itself on new deposits.

High Risk–High Stress Sites. Despite their appeal as places to build and live, erodible, low-elevation coastal terrain is generally not suitable for development. Time and time again, nature demonstrates its unsuitability to us, particularly in areas exposed to the open sea. This suggests that such terrain should best be left to open space, special easements, nature preserves, and parks. Where commitments have already been made to development programs, the planner's role is to devise layout schemes in which facilities are not only the least prone to damage but themselves least interfere with coastal processes and seasonal and longer term changes.

The Variable Setback Rule. As a rule, it is advisable to employ a variable approach to setting development setbacks. That is, setback distance should be adjusted to the location in the longshore system, to observable shoreline trends, to storm exposure, and to coastal composition and elevation. A major concern is not *No safe standard* only the threat of wave erosion and shore retreat, but also storm surges and flooding. The ocean front development along the barrier islands of the Atlantic and Gulf coasts is repeatedly damaged or destroyed by hurricane-driven erosion and flooding because, among other things, setbacks are much too small for these low-elevation, soft shorelines (Fig. 15.13).

Unfortunately, the harsh lessons of the hurricanes have not curbed development or improved setbacks as property owners, motivated in part by support from

Fig. 15.13 Damage to the beach environment and residential property from Hurricane Fran, which struck a massive blow to the North Carolina coast in September 1996. The ocean shore lies in the foreground. Safer ground lies across the road behind the houses.

the National Flood Insurance Program, rebuild in the aftermath of hurricanes such as Hugo (1989), Andrew (1992), Fran (1996), Isabel (2003), Katrina (2005), and Ike (2008). Unable to guide land use toward safer and less costly locations and configurations, planning officials are turning to **risk management** programs that include early warning systems, evacuation plans, and disaster relief measures.

Planning and Sea Level Change. Although the coastal development problem seems overwhelming today, all signs are that it is going to grow worse in the decades ahead as the sea plays a new hand of cards dealt to it by climate change. The total force delivered by the sea will rise as a function of increasing storm magnitude and frequency and rising sea level. It will reach farther inland, do much more work in terms of eroding beaches and moving sediment, and cause greater property damage than at anytime in the twentieth century. The economic and legal uncertainties for property owners, developers, and communities will reach new heights as they struggle with temptations and demands to fortify retreating shorelines, rebuild damaged facilities, and fend off the sea.

For coastal planners, a major dilemma will be the configuration of future coastlines and where to establish land use fallback positions. This calls for major mapping efforts in which several sea level rise scenarios are examined. Initially, scenarios depicting rises of 0.5 meter, 1.0 meter, and 1.5 meters are probably appropriate (Fig. 15.14), but as climate change forecasts are modified, additional scenarios will have to be examined, and other factors such as changes in hurricane force will have to be included for selected coastlines. Almost needless to say, any meaningful attack on this problem will require a sizeable GIS army.

The Wind Stress Factor. For development sites situated near the crest of a sea cliff or backshore slope, the threat of wave erosion may not be great; however, the effects of wind and slope instability may be significant. The pattern of onshore airflow over a high coastal slope (where the wind direction is more or less perpendicular to the coast) produces a zone of sharply accelerated velocity on the upper slope (see Fig. 15.8). For long, continuous bluffs and cliffs, the velocity increase from the slope foot to the slope crest is about four times for slopes of 4:1 inclination or greater. Where forest grows up to the edge of the slope crest, the tree canopy forces the wind upward, creating a fast air zone in the canopy crown and a slow air zone beneath it.

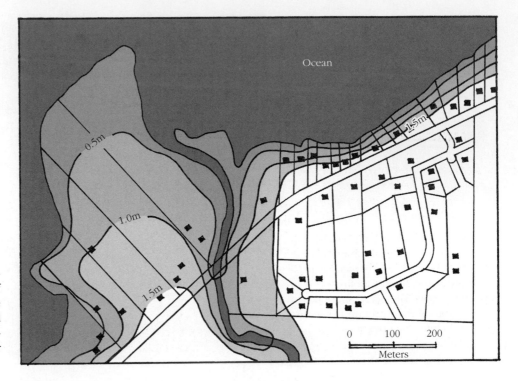

Fig. 15.14 A map showing projected changes in a section of coastline under three sea level rise scenarios: 0.5 meter, 1.0 meter, and 1.5 meters. The limit for each represents the projected high water mark.

However, if tree cover is sparse or absent, fast air may cling to the ground as it crosses the brow of the slope and then separate from the surface a short distance inland.

Since the force of wind increases approximately with the cube of velocity, the brows of bluffs and cliffs are in reality greater risks for site planning and facility design than raw wind data would lead us to believe. Sites situated here are subject to severe wind stress during storms, and in cold climates, to high rates of heat loss and deep penetration of ground frost. Downwind from the brow, sand and snow accumulate where fast air flow separates from the ground. As Figure 15.8*b* hints, this zone is often the site of sand dune and snowbank formation. If the brow of the slope is forested, sand and snow accumulation often take place close to the brow itself.

Great views, but...

Stay back

Clearly, each zone presents limitations to development, and the nearer the development is to the crestslope, where the coastal vistas are best, generally the more serious the limitations are. Therefore, we can safely say the placement of structures on or close to the brow of the slope, even if it is secured by bedrock, is not advisable as a general practice. Also, development is feasible on stable sites only where proper site analysis and appropriate engineering and architectural solutions have been worked out. This calls for an assessment of airflow patterns (vegetative indicators are often helpful in the absence of flow data), slope analysis to determine the stability against mass movement and erosion, soil assessment for placement and design of stormwater and wastewater systems. Above all, the design of structures and landscaping on sites with exposures to strong winds should honor aerodynamics for best performance. The house in Figure 15.15, which represents a fairly typical scene in coastal North America, was evidently sited and designed with little or no such consideration.

15.7 SITE MANAGEMENT CONSIDERATIONS

Besides the various guidelines for planning and design already described, a number of additional considerations have to be made in managing coastal sites. Most of these relate to the backshore area and how to maintain the stability of banks and bluffs. The backshore zone represents a buffer between *in transit* (shore) material and *in*

Fig. 15.15 Houses at the crest of a low bluff on the Pacific Coast that protrude into the fast air zone. An important clue for designers on aerodynamic conditions in such settings is provided by the configuration of tree canopies.

situ (terrestrial) material (see Fig. 15.1). The primary objective in site management is the maintenance of the backshore slope against disturbance by land use and erosion by various processes.

Water Management. Stormwater from impervious surfaces, such as shoreline roads and residential complexes, can cut deep gullies into banks, weakening them and leaving them more susceptible to wave erosion. Another drainage problem is soil saturation from septic system drainfields, downspouts, lawn irrigation, and leaking water and sewer lines. This water often concentrates on the lower slope where it reduces bank resistance to failure (slumps and slides) and wave erosion (Fig. 15.16).

Bank weakening In one dramatic incident of coastal slope failure at Palos Verdes Hills in Southern California, water contributions (losses) to the ground from 150 residences were estimated at 32,000 gallons per day. Elimination of lawn irrigation and septic drainfield infiltration, coupled with prudent stormwater management would have greatly reduced, perhaps averted, slumping and saved millions of dollars in damages to homes and infrastructure.

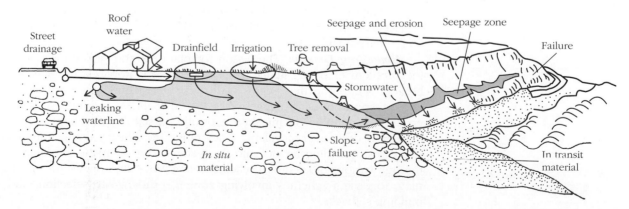

Fig. 15.16 A schematic diagram illustrating the common bank stability problems related to land use alterations of drainage and vegetation.

Bank Vegetation. Vegetation is a source of bank stability that must be managed with care. Not only do root systems provide resistance against erosion and slope failure, but plants draw large amounts of water from the soil, thereby reducing saturation and seepage. Common problems with residential development are (1) the removal of trees to improve vistas, (2) the introduction of exotic species that displace native species and are often less effective as bank stabilizers, and (3) applications of excessive irrigation water to maintain the introduced plants.

Common problems

Another destabilizing factor is the development of footpaths and the placement of structures such as stairways, viewing decks, and beach houses on or near banks. Footpaths not only destroy vegetation but human traffic weakens soil, pushes large amounts of material downslope, and opens the slope to erosion by runoff and wind. Structures on slopes also alter and weaken vegetation, and their footings are often points of concentrated erosion related to turbulence in runoff and wave swash.

The Engineering Options. Finally, we must address the pressing question of structural controls such as seawalls, breakwaters, and groins. Huge expenditures are made in the United States and Canada on engineered structures in an effort to improve navigation, reduce shore erosion, and protect coastal real estate. Although structural controls are a necessity for most navigational, bridge, and harbor facilities, for residential properties, they are not only questionable in halting long-term erosion, but are often a source of serious damage to the coastal environment, especially shore and nearshore habitats and recreational and aesthetic resources.

Serious concerns

Seawalls, for example, which are commonly used to defend individual properties, often increase the scouring effects of waves, resulting in deeper water immediately near shore. Eventually the wall fails, but before it does it may cause increased erosion on adjacent unprotected properties as waves converge on the projecting seawall and refract to the right and left. Groins, jetties, and breakwaters, which are designed to trap sediment and build wider beaches or protect harbor openings, often interrupt the longshore flow of sediment, thereby depriving areas downshore of their sediment supply and leading in turn to beach erosion and shoreline retreat (Fig. 15.10). This problem is so serious around many harbor entrances that sand must be mechanically transferred from one side of the entrance to the other, a process called *sand bypassing* (Fig. 15.17). Where, however, sand bypassing is necessary to protect navigation facilities, its expense must inevitably be added to the costs of managing the coastal zone.

Mitigating measures

Increasingly, *artificial nourishment* of the longshore system is being used to help maintain beaches. This process involves feeding a beach with a supplemental supply of sand by trucking, pumping, or barging it from another source. Miami Beach is often cited as a successful example of a beach nourishment program. Here the shore was fed with 25 million tons of sand that have maintained a beach for more than two decades. Beach nourishment is favored by many communities because it is far less disruptive to the shore and backshore than seawalls and bulkheads and, depending on the location, less expensive. But it is not without problems. Large-scale projects can be costly, and offshore nourishment can result in disturbance of the littoral environment.

An alternative

Emerging Lesson. The lesson for planners and policy makers is to approach the problem of beach maintenance and related problems not site by site but at a broader scale, one that begins with the larger sediment system, its sources, sinks, and transport zones, as suggested in Figure 15.11. At the local scale, property owners and communities facing shore erosion and stability problems need to understand that many options are available. They range from heavy engineering involving large structures extending far out into the water and a complete hardening of the shore, through various biotechnical schemes using wood and plants and site-adaptive planning, to design schemes involving zoning, prudent site selection, and appropriate building setbacks.

Look to options

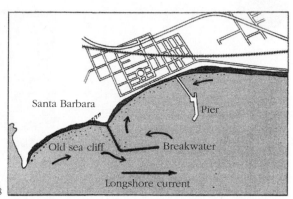

(*a*) Construction 1928

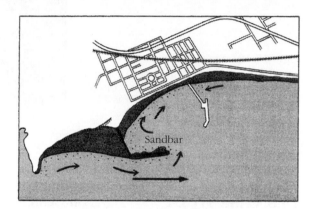

(*b*) 1948

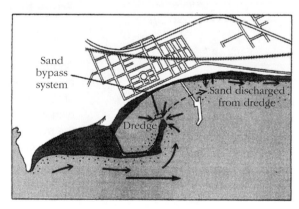

Fig. 15.17 Sand accumulation caused by the harbor facilities at Santa Barbara, California, and the sand bypass system constructed to transport the sand and clear the harbor entrance.

(*c*) 1965

Thoughtful decisions

In all cases, the decision on a management approach must consider not only the land use facilities and capital costs, but give serious consideration to community values, ecological factors, user recreational and safety issues, landscape scenic quality, maintenance costs, and the facility life cycle issue. When people feel the need for action, the tendency is to jump to engineered treatments while ignoring or dismissing other treatments. Not only are engineered options generally overapplied, but the materials and design schemes usually employed in these treatments are neither user friendly (and often downright dangerous) nor animal friendly. Furthermore, the life cycles of all such facilities are not endless, and replacement costs often include the expense of the removal and disposal of old structures. Finally, most are also aesthetically offensive in the coastal landscape, for among other things, they are blocky and angular and constructed of materials with textures and colors alien to the shore (Fig. 15.18).

Fig. 15.18 Seawalls are not only unsightly, but often hazardous to pedestrians and damaging to shoreline ecology and geomorphology.

15.8 CASE STUDY

In Search of Green Alternatives to Conventional Engineered Shore Protection: The Rootwall Concept

Elliott Menashe

Shoreline homesites afford stunning views and are coveted properties, but they can also be unstable and hazardous places to live. Finding a balance between land use pressures and shoreline stability is an ongoing challenge. Not surprisingly, land owners often feel compelled to modify the shoreline to protect their property. One of the most profound and damaging coastal modifications is the widespread construction of engineered structures designed to protect shorelines from erosion. Known variously as seawalls, bulkheads, and revetments, they are widely employed in an effort to stop erosion and to hold banks in place. However, it is becoming evident that building permanent, wall-like structures is a measure both doomed to failure and damaging to coastal ecosystems, to say nothing of the costs involved and the recreational and aesthetic amenities compromised (Fig. 15.A).

Not surprisingly, various so-called soft-shore, or nonstructural, alternatives are being explored. The rootwall is one such promising alternative that is being explored in the Puget Sound region of the U.S. Pacific Northwest. The rootwall employs large tree and root masses, called large woody debris (LWD), as primary structural components to provide immediate protection from wave attack and slope erosion. The concept is based on the observed role of large trees, both living and dead, in helping stabilize shorelines in heavily forested coastal zones.

Trees and their root masses not only add strength to backshore slopes, but adapt to changing conditions as the bank and shore retreat and change from around them (Fig. 15.B). Moreover, when a tree topples onto the beach, its root and trunk mass become natural armor for the shore, dissipating wave energy, capturing sediment, and providing habitat for new plants such as dune grass and sand cherry. Indeed an integral element of the rootwall is the incorporation of desirable native vegetation into the structure's design to encourage the establishment of plant

Fig. 15.A A failing seawall on a beach cleared of large woody debris and trees. Notice the accentuated erosion of the bank on the right.

communities that have been disrupted by past human activities such as the logging of old-growth forests and land development. The purpose of the rootwall is not to prevent all erosion or coastal change. Rather, it is designed to reduce the severity and scale of erosional processes related directly or indirectly to human activities. In addition, the rootwall helps re-establish the ecological character of the shoreline.

Along forested shorelines large woody material is an important structural and biological component of the nearshore environment (Fig. 15.B). Accumulations of large pieces of woody debris, notably whole trees, can support a wide range of biotic functions, including the maintenance of crucial linkages between terrestrial

Fig. 15.B A wooded backshore where trees are playing a vital role in stabilizing the bank. The trees adapt to the bank's slow retreat by bending and reorienting their root masses.

and marine areas. LWD that lodges behind backshore berms provides the framework from which complex microhabitat features and stable shore forms are created, reducing beach sediment suspension and removal by waves, prograding beaches, and maintaining stable rates of sediment yield to coastal systems. Smaller woody debris and cut logs without rootmasses, on the other hand, are too buoyant and

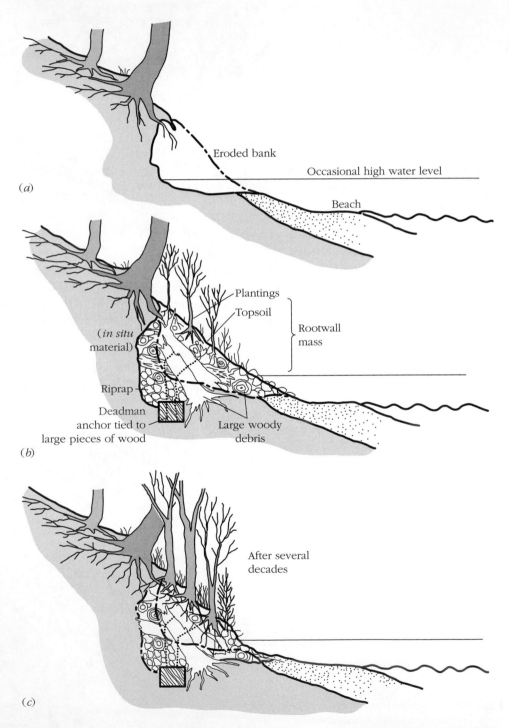

Fig. 15.C   Application of the rootwall concept requires a combination of large woody debris anchored in place, boulder riprap, boulder armoring, fill soil, and woody plantings: (*a*) the problem, (*b*) the design, and (*c*) the result.

mobile to resist tidal action and storm surges. These materials rarely remain in place long enough to provide the stabilizing benefits of LWD, and during storms they can become waterborne and function as battering rams.

The implications of deliberately creating LWD structures that mimic the beneficial natural assemblages found on shorelines is obvious. Additional benefits are the improvement of nearshore and distant-view aesthetics, the low cost of woody debris (much of it can be salvaged from inland site-clearing operations), and the development of high-quality coastal wildlife and fishery habitat features. The transfer of LWD from site-clearing operations (burnpile) to marine shorelines provides the added benefit of re-establishing the link between the forests and the marine environment which has been interrupted for nearly 100 years. Drift logs in inland waterways, a constant hazard to navigation, provide an additional source of LWD for rootwall structures, converting a liability into an asset. Because available LWD is now much smaller than that commonly found on beaches before the 1850s, the structural elements of the rootwall would have to be anchored in beach substrates.

Versions of rootwalls are currently being proposed for selected Puget Sound shorelines as an alternative to conventional bulkheads (Fig. 15.C). They appear to be best suited to low- to moderate-energy beaches and locations sporadically subject to wave attack. Undoubtedly it will take a number of years of field experiments to come up with a successful rootwall design. However, based on the success of engineered logjams in river systems as a means of erosion control and habitat restoration, we are encouraged by the concept. In addition, rootwalls are not a stand-alone device. They can be used in conjunction with other soft shore-protection measures, such as beach nourishment. In addition, plantings of native trees and shrubs within and behind the LWD structure should be a crucial element of this biostructurally engineered system. Finally, design and construction of a rootwall structure would be site specific and require close collaboration among several professionals, including landscape architects, coastal geomorphologists, geotechnical engineers, marine construction contractors, and marine vegetation restorationists.

Elliott Menashe is owner and principal of Greenbelt Consulting, Clinton, Washington. He specializes in vegetation management and the restoration of coastal areas.

15.9 SELECTED REFERENCES FOR FURTHER READING

Allen, James R. "Beach Erosion as a Function of Variations in the Sediment Budget, Sandy Hook New Jersey, U.S.A." *Earth Surface Processes and Landforms.* 6, 1981, pp. 139–150.

Coastal Engineering Research Center. *Shore Protection Manual.* Washington, DC: U.S. Government Printing Office, U.S. Army Coastal Engineering Research Center, 1973.

Davies, J. L. *Geographical Variation in Coastal Development.* London: Longman, 1977.

Dolan, R., et al. "Erosion of the U.S. Shorelines." In *CRC Handbook of Coastal Processes and Erosion.* (P. D. Komar, ed.). Boca Raton, FL: CRC Press, 1983.

Healy, R. C., and Zinn, J. A. "Environmental and Development Conflicts in the Coastal Zone." *Journal of the American Planning Association.* 51, 3, 1985, pp. 299–311.

Inman, D. L., and Brush, B. M. "The Coastal Challenge." *Science.* 181, 1973, pp. 20–32.

Jarrett, J. T. "Sediment Budget Analysis Wrightsville Beach to Kure Beach, North Carolina." Coastal Engineering Research Center Reprint 78-3, U.S. Army Corps of Engineers, 1978, pp. 986–1005.

Kuhn, G. G., and Shepard, F. P. "Beach Processes and Sea Cliff Erosion in San Diego County, California." In *CRC Handbook of Coastal Processes and Erosion* (P. D. Komar, ed.). Boca Raton, FL: CRC Press, 1983.

Marsh, W. M. "Nourishment of Perched Sand Dunes and the Issue of Erosion Control in the Great Lakes." *Journal of Environmental Geology and Water Science.* 16, 2, 1990, pp. 155–164.

Marsh, W. M., and Mewett, A. M. *A Framework Plan for Coastal Zone Management, East Central Vancouver Island.* Courtenay, BC: Regional District of Comox-Strathcona, 2002.

National Research Council. *Beach Nourishment and Protection.* Washington, DC: National Academy Press, 1996.

National Research Council. *Mitigating Shore Erosion Along Sheltered Coasts.* Washington, DC: National Academy Press, 2007.

Peterson, C. H., and Bishop, M. J. "Assessing the Environmental Impacts of Beach Renourishment." *Bioscience.* 55, 2005, pp. 887–896.

Pilkey, O. H., and Dixon, K. L. *The Corps and the Shore.* Washington, DC: Island Press, 1996.

Zenkovich, V. *Processes of Coastal Development* (D. Fry, trans.). Edinburgh, UK: Oliver and Boyd, 1967.

Related Websites

Government of U.S. Virgin Islands, Department of Planning and Natural Resources. "The Virgin Islands Coastal Zone Management Program." 2006. http://www.viczmp.com/. 2006
A program by the government to manage and protect the coastal zone of the Virgin Islands. Find management plans for the area as well as issues of pollution.

Lake Huron Centre for Coastal Conservation. http://lakehuron.ca/index.php?page=home
A nongovernmental organization working on conserving the coastal ecosystem of Lake Huron. As a main focus, the site gives a lot of information about coastal processes. This includes beach and dune formation and the importance of it, as well as vegetation management on bluffs.

National Oceanic and Atmospheric Administration. "Alternatives for Coastal Development." N.d. http://www.csc.noaa.gov/alternatives/
The costs of coastal development financially, socially, and environmentally. The site goes on to compare three alternatives: a conventional design, a conservation design, and a new urbanist design.

National Oceanic and Atmospheric Administration. "Beach Nourishment: A Guide for Local Government Officials." N.d. www.csc.noaa.gov/beachnourishment/index.htm
An NOAA project teaching government officials about beach nourishment. The coastal geology link leads to information about sediment transport and implications for planning. The introduction also gives information about coastal processes.

16

SOLAR CLIMATE NEAR THE GROUND: LANDSCAPE AND THE BUILT ENVIRONMENT

16.1 INTRODUCTION

Solar landscape

The prudent landscape planner should be as sensitive to solar radiation as a part of environmental analysis as he/she is to water supply, stormwater, slope, or watersheds. Solar radiation is a key consideration in all types of residential design, first because it is a primary source of light and energy. Houses and landscaping designed to take advantage of solar heating opportunities in winter and of solar screening in summer are not only less expensive facilities to operate but are often more comfortable and pleasing as living environments. Beyond the home and yard, solar radiation is an essential consideration in many aspects of landscape planning and management, especially in the varied arena of microclimate as it relates to community design, riparian corridors, and landscape ecology, and it is sure to grow more so as the climate changes in the decades ahead.

Analysis of solar radiation begins with an examination of sun angle and its variations with topography and the seasons. This information is essential to understanding the distribution of solar energy in the landscape. However, the translation of solar variables into meaningful information for planning decisions can be a difficult exercise and unfortunately one that few of us are able to accomplish effectively.

Approached properly, the problem often requires the computation of radiation balance and heat budgets, taking into consideration radiation gains and losses, surface reflection, ground materials, and energy flows in the form of ground heat, sensible heat, and latent heat. But there is much to be learned and applied without full-blown quantitative analysis. Our discussion begins with rudimentary sun angle concepts and then goes on to solar heating of the landscape and its implications for local environments, including microclimates and landscape ecology.

16.2 SUN ANGLE AND INCIDENT RADIATION

Solar intensity

The angle formed between sunlight approaching the earth's surface and the surface itself is called the **sun angle**. To envision this angle, think of straight rays of light striking a flat surface such as an airfield. A more direct angle, one that is closer to 90 degrees, causes a greater concentration of solar radiation on the surface. Conversely, smaller angles have weaker solar intensities.

Incident radiation

Sun Angle and Incident Radiation. Understandably, sun angle is an important factor in heating the earth's surface. Compare, for example, the areas bombarded by the beam in the diagrams of Figure 16.1*a* and *b*. The beam in diagram *a*, which is the same strength as the beam in diagram *b*, spreads over more surface area because it strikes the surface at an angle. Because the beam in *a* spreads over more area, its density, or **incidence**, is lower. We can show this with a simple computation that involves dividing the quantity of energy in the beam (S_i) by the area (A) that it strikes:

$$S_I = \frac{S_i}{A}$$

where[1]

S_I = solar radiation incident on the surface
S_i = quantity of energy in the beam
A = surface area intercepted by the beam

---

[1] Energy flux (flow) calculations normally use calories (cal) or watts (W) as the energy units and are written cal/cm².min and W/m².

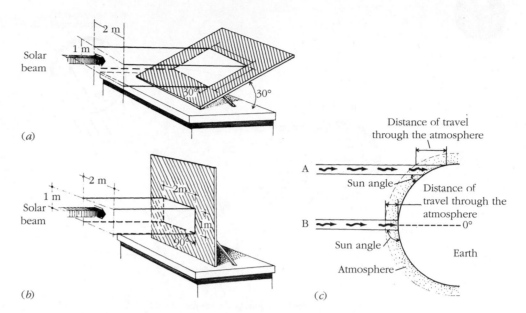

Fig. 16.1 Variations in radiation intensity on an inclined surface (*a*), and a vertical surface (*b*); diagram (*c*) shows the relationship between sun angle and the curvature of the earth including the relative thickness of the atmosphere to incoming solar radiation at high and low latitudes.

Because the earth is curved, most solar radiation enters the atmosphere and hits the surface at an angle. As one goes farther poleward, the angle becomes smaller and the beam of radiation becomes more diffuse and thus less intense. Figure 16.1*c* shows this by contrasting a beam at the equator (B) with one near the Arctic Circle (A). Beam A not only spreads over more surface area but also passes through a greater distance of atmosphere, thereby giving the atmosphere a greater opportunity to reflect and scatter radiation before it reaches the ground.

16.3 VARIATIONS IN SUN ANGLE WITH SEASONS AND TOPOGRAPHY

Understanding sun angles more completely requires taking additional factors into account. Foremost is the seasonal change in sun angle as a function of the earth's axial tilt. Because of the fixed inclination of the earth's axis (at 23.5 degrees off vertical), the earth appears to tip toward and away from the sun as it goes around in its orbit (compare the diagrams on the left and right sides of the orbit in Fig. 16.2). This produces seasonal changes in sun angle for all locations on the earth.

Seasonal sun angles

Seasonal Variations in Sun Angle. We usually think in terms of four key sun angles in the year corresponding to the seasons (or their onset) in the midlatitudes (Fig. 16.2). For the Northern Hemisphere, the highest and lowest angles each year occur on June 20 to 22 and on December 20 to 22, respectively. These dates are called the *summer* and *winter solstices*. (For the Southern Hemisphere they are the reverse, of course.) In fall and spring, the sun angles are intermediate, and there are two dates on which they are exactly intermediate, March 20 to 22 and September 20 to 22, called the *equinoxes*.

From summer solstice to winter solstice, the sun angles for any location in the middle latitudes (defined, for convenience, as the zone between 23.5 degrees and 66.5 degrees latitude) vary by 47 degrees (that is, twice the 23.5 degree axial tilt). Sun angle readings are normally given as the high noon position of the sun represented

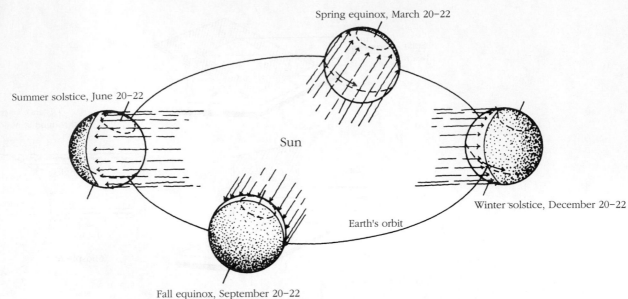

Spring equinox, March 20–22

Summer solstice, June 20–22

Sun

Earth's orbit

Winter solstice, December 20–22

Fall equinox, September 20–22

Fig. 16.2   Seasonal changes in sun angle and the orbital path of the earth about the sun. The 23.5-degree angle of tilt is constant throughout the earth's revolution.

by the angle formed between one's outstretched arm pointed at the sun and the horizon of the landscape.[2] Figure 16.3 illustrates the principal sun angles in the year for 50 degrees north latitude.

Procedure

Calculating Sun Angle.   Computing the sun angle for any latitude and date involves three basic steps. First, the *declination of the sun* must be known. This is the latitude on the earth where the sun angle is vertical (90 degrees) on a given date. It can be read from the graph in Figure 16.4*a*. Second, the zenith angle must be determined by counting the number of degrees that separate the latitude of the location (site) in question from the declination (latitude). The scale in Figure 16.4*b* may be helpful in making this count. *Zenith angle* is the angle formed between the zenith

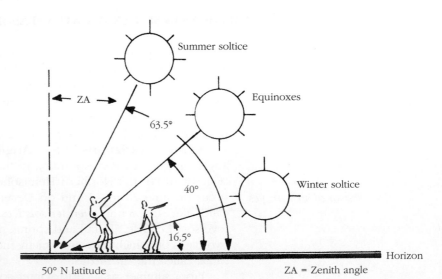

Summer soltice

Equinoxes

ZA

63.5°

40°

Winter soltice

16.5°

50° N latitude

Horizon

ZA = Zenith angle

Fig. 16.3   The annual changes in sun angle for 50 degrees north latitude, a range of 47 degrees between winter and summer solstices.

[2] This is actually a slight approximation because solar radiation is refracted (bent) somewhat as it passes through the atmosphere.

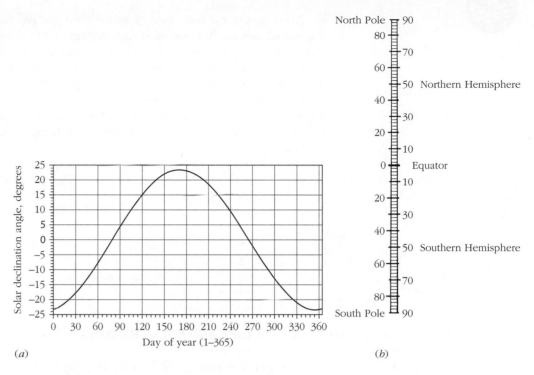

(a) (b)

Fig. 16.4 (*a*) Sun declination chart. Read the chart by finding the date in question on the bottom line; then follow the nearest vertical line upward to the curved graph line. At that point, take the nearest horizontal line to the side of the chart and read the appropriate number. Be sure to read Northern or Southern Hemisphere. (*b*) Zenith angle chart. Find the latitude in question; then find the declination of the sun and count the number of degrees between the two. This will give you the zenith angle. (Source of declination chart: U.S. Department of Commerce, National Oceanic & Atmospheric Administration, NOAA Research, Earth System Research Laboratory, Global Monitoring Division.)

(a vertical line perpendicular to the ground) and the position of the sun in the sky. In Figure 16.3, it is the angle (ZA) between the sun and the broken line at the time of the summer solstice. The last step is to subtract the zenith angle from 90 degrees. This gives us the **sun angle** for the latitude and date in question. The following example shows the steps to be followed in computing a sun angle:

Basic steps

- Location   = 50 degrees north latitude (given)
- Date   = June 15 (given)
- Declination of sun = 23 degrees (Fig. 16.4*a*)
- Zenith angle, *ZA*   = 27 degrees (Fig. 16.4*b*)
- Sun angle, *SA*   = 90 degrees minus *ZA*, that is, 90 degrees − 27 degrees
- *SA*   = 63 degrees

Sun Angle on the Ground. Once the sun angle of a location is known, we can move to the local scale and examine the influence of the landscape, that is, how the sun angle varies with hills, valleys, buildings, and other landscape forms. Hillslopes and roofs that face the sun are brighter and warmer than those that face away from the sun. In addition, the angle changes from dusk to dawn so that slopes with eastward orientations are favored by the morning sun and those with westward orientations are favored by the afternoon sun.

Role of slope

To compute the influence of an inclined surface on local sun angle (let us call it **ground sun angle**), we must first determine (1) the sun angle on flat ground for that latitude, (2) the direction in which the slope faces, and (3) the angle of the slope (that is, its inclination in degrees). If the slope faces the noon sun, the sun angle on the slope is equal to the flat ground angle plus the angle of the slope. If the product is greater than 90 degrees, then subtract it from 180 degrees to get the appropriate angle. For slopes that face away from the sun, the sun angle is equal to the flat surface sun angle minus the angle of the slope. If the product is negative, the slope is in shadow.

Ground sun angle

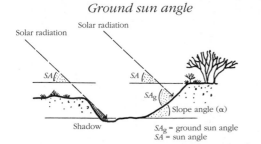

$$\text{Ground sun angle} = SA \pm \alpha$$

where

Ground sun angle (SA_g) = sun angle on slope face
SA = sun angle on flat ground
α = angle of slope in degrees

16.4 RADIATION BALANCE AND SOLAR HEATING

To determine the amount of solar heating on a surface, it is first necessary to understand that, at ground level, solar radiation can be disposed of in only two ways: reflection and absorption. The reflective capacity of a surface, called **albedo**, is expressed as the percentage of incoming solar (shortwave) radiation rejected by the surface:

Albedo

$$A = \frac{S_o}{S_i} \times 100$$

where

A = albedo
S_i = incoming shortwave (solar)
S_o = outgoing shortwave (solar)

All earth materials reflect a portion of the solar radiation that strikes them, but the values vary widely, as Table 16.1 reveals.

Solar Gain. The solar energy absorbed by a surface, which we call **solar gain**, is equal to incoming shortwave (S_i) less the amount reflected (S_o):

Solar gain

$$\text{Solar gain} = S_i - S_o$$

Radiation and temperature

This quantity represents the energy added to the absorbing material in the form of heat. It in turn produces a rise in the material's temperature. The actual amount of temperature rise for a given amount of energy added varies according to the material's composition. In other words, if two different materials—say, water and soil—absorb the same amount of radiation, they do not yield the same temperature. This is explained mainly by differences in a property called *volumetric heat capacity* (or *specific heat*). Water has high volumetric heat capacity and is thus slow to produce a temperature rise with the addition of energy compared to that for dry sand. It follows that moisture differences in soil are often reasonably good indicators of variations in ground temperature over small areas. (See Table 18.1.)

Table 16.1 Albedos for Various Surfaces

Material	Albedo (%)
Soil	
Dune sand, dry	35–75
Dune sand, wet	20–30
Dark (e.g., topsoil)	5–15
Gray, moist	10–20
Clay, dry	20–35
Sandy, dry	25–35
Vegetation	
Broadleaf forest	10–20
Coniferous forest	5–15
Green meadow	10–20
Tundra	15–20
Chaparral	15–20
Brown grassland	25–30
Tundra	15–30
Crops (e.g., corn, wheat)	15–25
Synthetic	
Dry concrete	17–27
Blacktop (asphalt)	5–10
Water	
Fresh snow	75–95
Old snow	40–70
Sea ice	30–40
Liquid water	30–40
30° latitude, summer	6
30° latitude, winter	9
60° latitude, summer	7
60° latitude, winter	21

Source: From W. D. Sellers, *Physical Climatology* (Chicago: University of Chicago, 1974). Used by permission.

Landscape Applications. Given that solar radiation usually strikes surfaces in the landscape at an angle, albedo must be combined with the concept of incident radiation flux (flow over the receiving area) to determine the solar heating for a surface. This can be done computationally using just three variables: (1) the ground sun angle (based on latitude, date, and surface inclination or slope), (2) the intensity of solar radiation, and (3) the albedo of the surface:

Solar heating

$$SH = S_i(1 - A)\, \sin SA_g$$

where

SH = solar heating in cal/cm^2 • min (or watts/m^2)

S_i = incoming solar radiation in cal/cm^2 • min (or watts/m^2)

A = albedo ($1 - A$ gives the percentage absorbed)

SA_g = ground sun angle in degrees

It is instructive to see how important slope and albedo are in the solar heating of a varied landscape. For example, given the surfaces (located at 45 degrees north

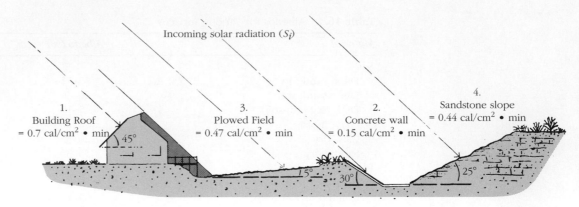

Fig. 16.5 Variation in solar heating related to slope and surface materials at four locations.

Applications latitude) represented by the landscape section (profile) in Figure 16.5, the rates of solar heating at noon on the equinox are as follows:

1. *Building roof:*
   - Slope = 45°
   - Orientation = south
   - Albedo = 10%
   - SA_g = 90°
   - S_i = 0.78 cal/cm² • min

   Solution:

   $$SH = 0.78\,(1 - 0.10)\,\sin 90°$$
   $$= 0.78\,(0.9)\,1.0$$
   $$= 0.70\ \text{cal/cm}^2 \bullet \text{min}$$

2. *Concrete wall:*
   - Slope = 30°
   - Orientation = north
   - Albedo = 27%
   - SA_g = 15°
   - S_i = 0.78 cal/cm² • min

   Solution:

   $$SH = 0.78\,(1 - 0.27)\,\sin 15°$$
   $$= 0.78\,(0.73)\,0.26$$
   $$= 0.15\ \text{cal/cm}^2 \bullet \text{min}$$

3. *Plowed field:*
   - Slope = 5°
   - Orientation = south
   - Albedo = 22%
   - SA_g = 50°
   - S_i = 0.78 cal/cm² • min

   Solution:

   $$SH = 0.78\,(1 - 0.22)\,\sin 50°$$
   $$= 0.78\,(0.78)\,0.77$$
   $$= 0.47\ \text{cal/cm}^2 \bullet \text{min}$$

4. *Sandstone slope:*
   - Slope = 25°
   - Orientation = south
   - Albedo = 40%
   - SA_g = 70°
   - S_i = 0.78 cal/cm² • min

   Solution:

   $$SH = 0.78\,(1 - 0.40)\,\sin 70°$$
   $$= 0.78\,(0.60)\,0.94$$
   $$= 0.44\ \text{cal/cm}^2 \bullet \text{min}$$

The concrete wall gains the least energy, about one-third that of the field and slope and about one-fifth that of the roof. Note that this pattern of solar heating in the landscape could be changed dramatically with some simple design alterations.

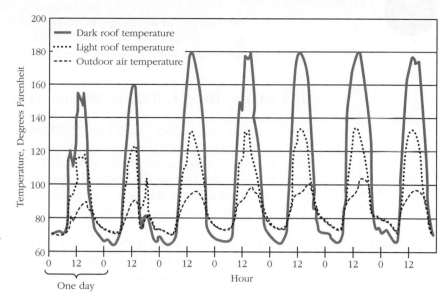

Fig. 16.6 Variations in the temperature of light-colored and dark-colored roofs based on field measurements. The lower line represents air temperature.

Design implications

The color of surface material, for example, can have a remarkable influence on roof temperatures. This is documented in the graph in Figure 16.6, which shows that in summer dark-colored roofs are 40–50°F warmer at midday than light-colored ones. This is attributed to differences in albedo. Of course, several other design options are available, including changing the orientation of the roof, taking advantage of shade trees, and building a green roof. With the latter, not only can albedo be modified by the plant foliage, but much of the heat generated by solar radiation can be dissipated in evapotranspiration as latent heat rather than sensible heat.

On ground-level climate

Working in combination, these landscape conditions (buildings, plowed fields, exposed rock, etc.) can have a profound influence on ground-level climate. Under calm, sunny conditions, the layer of air over surfaces such as these will begin to develop a spatial pattern of temperatures roughly corresponding to the pattern of solar energy absorbed. Depending on the scale of the resultant thermal polygons, this may then induce differential air movement with the warm air rising or sliding upslope and the cool air draining downslope (Fig. 16.7). On some days this pattern may carry over well past the period of peak solar radiation and into the evening, as long as regional weather systems do not obliterate it. Although such information can

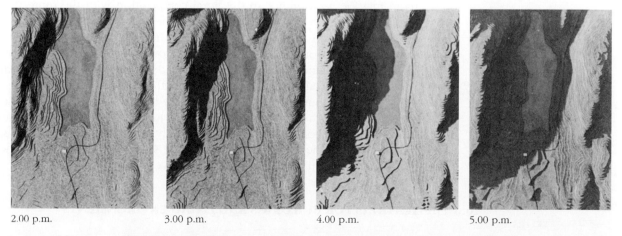

2.00 p.m.   3.00 p.m.   4.00 p.m.   5.00 p.m.

Fig. 16.7 Afternoon patterns of sunlight and shadow at Jordan Pond Valley, Acadia National Park, Maine, on August 1, based on a simulation model. The shadow slope begins cooling by midafternoon, followed by cool air drainage in the late afternoon and evening.

be valuable in landscape planning and design, remember that it represents only one set of conditions (though an important one) in the complex set of factors operating in the surface climate system, and it should be interpreted accordingly.

16.5 IMPLICATIONS FOR LAND USE, VEGETATION, AND SOIL

To gain an idea of the impact of land use and land cover on the gain of solar energy by the landscape, we can compare the differences in slope (both angle and orientation) and surface materials before and after development.

Pre- and post-considerations

Impact on the Solar Landscape. This assessment requires first mapping the pre-development slopes of various angles, orientations, and compositions; measuring their areas; and then computing their total solar gain over some time period. These figures are summed for the entire project area and compared to the parallel figure based on the same computation for the postdevelopment landscape. Although many additional factors, such as means and extremes in air temperature, wind, precipitation, and humidity, have to be taken into account to determine the climatic significance of a local change in solar gain, such a comparison provides one important measure of the relative impact of different land uses and development schemes on thermal conditions at ground level.

Application to the Natural Landscape. The concept of variations in incident radiation related to slope angle and orientation can, of course, be extended to the natural landscape. Indeed, in some places variations in landscape form and content may control the balance of local landscape systems such as the one defined by the interplay among ground heat, soil moisture, vegetation, runoff, soil erosion, and landforms. In semiarid mountainous areas, such as parts of Colorado, New Mexico, California, and

Semiarid mountains

interior British Columbia, the more direct sun angles on south-facing slopes result in greater surface heating and in turn higher evaporation and plant transpiration rates than on north-facing slopes. The resulting difference in soil moisture is often great enough to cause marked differences in vegetation on north- and south-facing slopes. The photograph in Figure 16.8a shows one such example from Colorado, where warmer and drier south-facing slopes lack forest covers.

Arid mountain slopes

In even drier areas, where only herbs and shrubs can survive, the plant cover on south-facing slopes is often measurably lighter than that on north-facing slopes. Less ground cover means higher rates of runoff and soil erosion on south-facing slopes, leading in turn to more and deeper gullies. In time (and all other things being equal), the south-facing slope may be reduced to a lower angle than its north-facing neighbor with larger alluvial fans (sediment from gully erosion) along the footslope. In addition, there may be a difference in the floral composition with the more drought-tolerant species making up a higher percentage of the plant cover on south-facing slopes.

Humid cliff faces

Differences in plant species related to the influence of slope on incident radiation can also be found in humid regions, though the examples are rarely as obvious as those in dry landscapes. In the midlatitudes, combinations of heat and light may ensure the survival of certain plants on extreme slopes. For instance, north-facing cliffs along the south shore of Lake Superior harbor species of ferns and mosses that are separated by hundreds of miles from the main bodies of their populations in arctic and subarctic regions (Fig. 16.8b). Apparently, the low light intensities and cool temperatures along these cliffs have favored the survival of these plants since the end of the last continental glaciation about 10,000 years ago.

16.6 IMPLICATIONS FOR BUILDINGS AND LIVING ENVIRONMENTS

It is common knowledge that the placement and size of buildings and trees in cities can also affect the reception of solar radiation, sometimes more acutely than hillslopes. This fact is gaining increased attention in urban planning and design, and well

(*a*)

(*b*)

Fig. 16.8 (*a*) Differences in vegetation on north- and south-facing mountain slopes. The north-facing slopes sustain forest cover, whereas south-facing slopes are limited mainly to grasses. (*b*) The shadow zone along the north-facing cliffs on Lake Superior helps create a cool microclimate conducive to certain arctic and subarctic plants.

it should with rising concern over urban heating and cooling costs, carbon dioxide loading of the atmosphere, and public health and safety. Some common solar features of cities include shadow corridors and solar windows (or gaps) (Fig. 16.9). **Solar windows** are narrow spaces between tall buildings through which the solar beam passes

Relevance   to ground level. Depending on the orientation and spacing of the buildings, the shaft of light may illuminate a patch of ground for only a short time each day, making it difficult to design pleasing plaza, playground, and garden/park spaces.

The Shadow Factor. Shadow corridors are elongated zones, bordered by a continuous ridge of tall buildings that block the sun. In the most extreme situations, direct (beam) solar radiation is never received in such zones. The only skylight comes from diffused solar radiation and radiation reflected from nearby buildings.

The length of a shadow cast by a building or tree is a function of the height of the object and the sun angle. Computations can be made using the formula:

Shadow length
$$S_l = \frac{h}{\tan SA}$$

where

S_l = shadow length
h = height of the object
SA = sun angle

The tangents can be obtained from Table 16.2.

Table 16.2 Tangents (tan) for Angles 5°–85°

5° = 0.087	45° = 1.0
10° = 0.268	50° = 1.19
15° = 0.268	55° = 1.43
20° = 0.364	60° = 1.73
25° = 0.466	65° = 2.14
30° = 0.577	70° = 2.75
35° = 0.700	75° = 3.73
40° = 0.839	80° = 5.67
	85° = 11.43

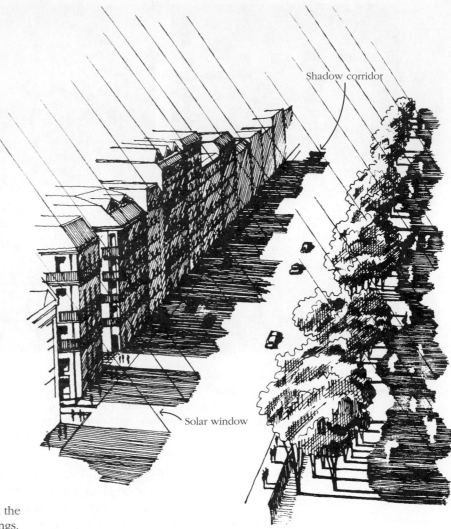

Fig. 16.9 The pattern of solar radiation in the urban environment as altered by tall buildings.

This formula is traditionally used in site planning in areas of excessive heat and intensive solar radiation because it is necessary to provide for shade in pedestrian areas, in parking lots, on building faces, on plazas, and the like. (also see Section 17.6 and Fig. 17.12). The need for shade is generally greatest in the hours between 11 a.m. and 4 p.m., when high solar intensities are coupled with high air and ground temperatures (Fig. 16.10).

Tree shade   The efficiency of trees in providing shade is related to several factors including tree placement, foliage density, and canopy size and structure. The greatest shade efficiency is provided by forests with multiple-storied structures such as the one shown in Figure 17.10 where there is an 80 percent reduction in solar radiation from the top to the base of the canopy. In residential areas where only a single story of yard and street trees is possible and canopy size and density are the chief concerns, species such as red oak and sugar maple (in the North) and live oak (in the South) are particularly effective. Properly placed, a healthy, mature shade tree can reduce beam (direct) solar radiation by 50 percent or more and in turn yield significant savings for air conditioning.

Health and Safety Concerns.   Although shade can be a distinct advantage for local pedestrians and residents of cities prone to hot summers and frequent heat

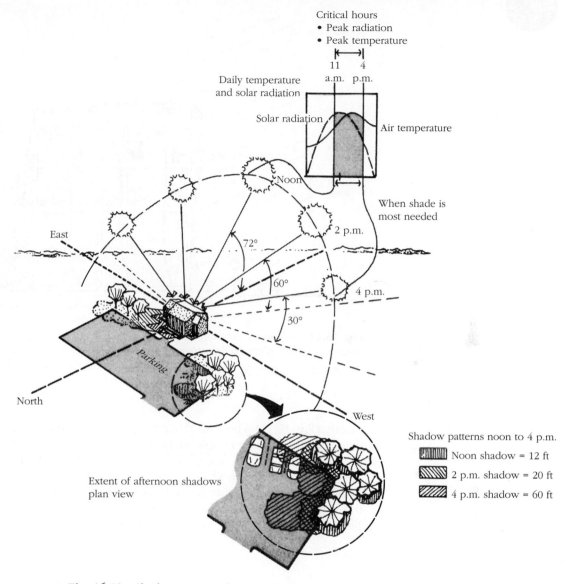

Fig. 16.10 Shadow patterns between noon and 4 p.m. associated with a building and trees near a parking lot. Shade is the most critical between 11 a.m. and midafternoon, when air and ground temperatures are highest.

waves, in many northern cities, such as Minneapolis, Detroit, Calgary, and Montreal, shadow corridors in winter encourage the buildup of ice and snow, making foot travel hazardous (Fig. 16.11). Moreover, the solar gain is very poor in these zones for living units with northerly exposures, resulting in cooler room temperatures and higher heating costs. This is especially significant in light of findings in Great Britain and the United States concerning illness and death among the elderly brought on by hypothermia.

Urban cold pockets

Accidental hypothermia, a disorder characterized by low body temperature (near 90°F), slowed heartbeat, lowered blood pressure, and slurred speech, can be brought on in persons over age 70 by room temperatures as modest as 65°F, inadequate clothing, and prolonged periods of physical inactivity. Solar exposure may make a difference of several degrees in room temperatures, especially during cold spells, and in turn can tip the balance between hypothermia and a normal state of health in the

Hypothermia

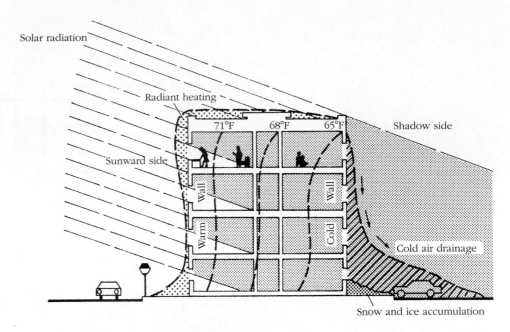

Fig. 16.11 Schematic diagram illustrating the effects of sun angle on winter living conditions in and around a northern apartment building. The north wall is cold on both the interior and exterior of the building. Snow and ice buildup are also favored on the north walk and street.

elderly (Fig. 16.11). The United States National Institutes of Health estimate that more than 2 million elderly people in the United States are vulnerable to accidental hypothermia. Undoubtedly, many of these persons inhabit buildings whose orientation, design, and neighborhood exclude or greatly restrict access to direct solar radiation in living spaces. On the other hand, these same conditions may be an advantage during the summer because they are not prone to excessive heating and thus may be less risky places for persons prone to heat stress.

16.7 CASE STUDY

Solar Considerations in Midlatitude Residential Landscape Design

Carl D. Johnson and Brian Larson

One of the primary objectives in residential and urban design is the mitigation of climatic extremes in spaces occupied by humans. In architecture the focus is on the internal climate of buildings, which is achieved mainly through air conditioning, light control, and so on. In landscape architecture the primary concern is with outdoor spaces, and climatic modification is attained through the use of vegetation, siting of buildings, the use of different types of ground materials, and topographic features, either as they exist or as they could be constructed.

In the continental midlatitudes, discomfort from the cold poses a major restriction to the use and enjoyment of outdoor space. Therefore, in the design of modern residential complexes, it is desirable to modify the microclimate to encourage greater use of patio and yard space, as the scenarios in Figure 16.A suggest. Understandably, the level of modification that can be expected through landscape planning is relatively modest, particularly when located in a Minnesota or Quebec winter. On the other hand, small modifications of marginally cold or cool weather,

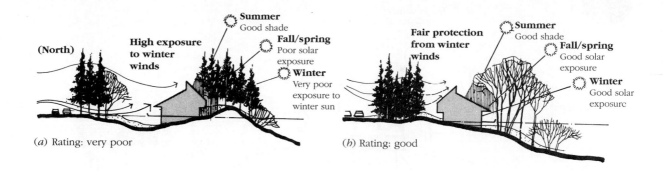

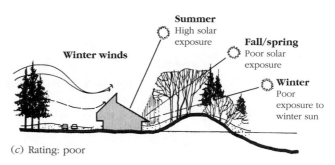

Fig. 16.A Landscape design can make a difference in energy use and microclimate in and around residential structures. Here are three simple scenarios addressing different design schemes based on tree types, wind direction, and roof orientation: (*a*) Conifers on both north and south; exposure to cold wind; shade in all seasons with very low solar gain in winter—*rating: very poor.* (*b*) Conifers on north, deciduous on south; fair protection from cold wind; summer shade with high solar gain in winter—*rating: good.* (*c*) Mixed tree cover on north and south; exposure to cold wind; low solar gain in winter; roof exposed to summer sun—*rating: poor.*

such as that of spring and fall, are indeed possible, and days that would otherwise be uncomfortably cool can in fact be made quite pleasant through sensitive planning and design.

Newport West is a low-density townhouse development located on the northwest edge of Ann Arbor, Michigan. The development site is characterized by hilly terrain with a major swale running through it. The east-facing and south-facing slope of the swale, an old farmfield fringed by trees, was selected for the building site because it offered the greatest opportunity to optimize microclimate conditions and conserve building energy.

The housing units were arranged in a series of clusters to form solar pockets and to provide protective buffers from the cold, windy northern exposures. The sun pockets were designed to provide comfortable outdoor spaces in fall and spring and to afford habitats for exotic plants such as azaleas and rhododendrons. The townhouses were constructed with large south-facing windows to receive solar radiation and to augment interior heating in fall and winter. On the southwest sides of the units, facades were protected from excessive heating by the afternoon summer sun with full-crown deciduous trees.

After construction and landscaping were completed and the units were occupied, a set of temperature readings were taken in early March to determine the effectiveness of the design in modifying microclimate (Fig. 16.B). Ground temperatures varied substantially depending on solar exposure and were 7°C higher at the

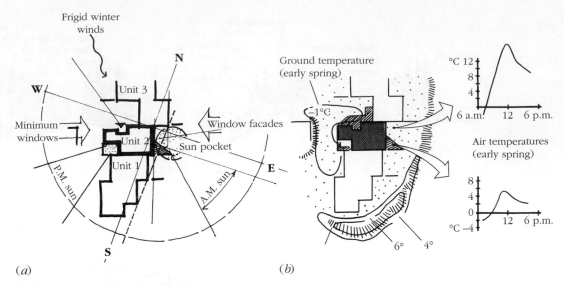

Fig. 16.B A residential design scheme based partially on the climatic considerations on the building fringe and bordering landscape. The results are based on field measurements taken on a March afternoon. (*a*) The layout—microclimate vectors and design; (*b*) the results—air and ground temperatures.

5-centimeter depth on inclines near south-facing walls compared to surfaces near north- and northwest-facing walls. In the patio spaces, air temperatures varied with shade and beam radiation receipt; in the southeast-facing sun pocket, the daily high temperature (at surface level) on one bright day was nearly 10°C higher than in permanently shaded areas nearby. On cloudy days and windy days the difference was negligible, however.

An examination of the climatic records shows that in the U.S. Midwest a total of 10 to 20 days in fall and spring can be classed as calm and sunny with uncomfortably cool ambient air temperatures. This brief study suggests that these sorts of conditions can be improved in near-building spaces through climate-oriented building design and siting. It also suggests that conditions for exotic plants are more favorable in sun pockets, at the same time recognizing that summer heat may be excessive and that protective shading with deciduous trees may be necessary. The study also implies that residential units that offer both warm and cool outdoor exposures are preferable to those with single exposures in the warm/cold climates because they increase the opportunities for the seasonal use of outdoor space. This consideration is growing increasingly important as people are faced with smaller residences situated in community or neighborhood clusters with limited yard space.

Carl D. Johnson was a founding partner of Johnson, Johnson and Roy, Inc., Ann Arbor, Michigan. Brian Larson is the senior principal in Larson, Burns and Smith, a landscape architecture firm in Austin, Texas.

16.8 SELECTED REFERENCES FOR FURTHER READING

American Institute of Architects Research Corporation. *Solar Dwelling Design Concepts.* Washington, DC: U.S. Department of Housing and Urban Development, 1976.

Buffo, John, et al. "Direct Solar Radiation on Various Slopes from 0 to 60 Degrees North Latitude." U.S.D.A. Forest Service Research Paper PNW-142. Portland, OR: U.S. Department of Agriculture, Forest Service, Pacific Northwest Research Station. 1972.

City of Davis (California). *A Strategy for Energy Conservation.* Davis, CA: Energy Conservation Ordinance Project, 1974.

Griggs, E. I., et al. "Guide for Estimating Differences in Building Heating and Cooling Energy Due to Changes in Solar Reflectance of a Low-Sloped Roof." ORNL Report 6527. Oak Ridge, TN: Oak Ridge National Laboratory, 1989.

Heisler, G. M. "Effects of Individual Trees on Solar Radiation Climate of Small Buildings." *Urban Ecology.* 9, 1989, pp. 337–359.

Huang, J., et al. "The Potential of Vegetation in Reducing Summer Cooling Loads in Residential Buildings." *Journal of Climate and Applied Meteorology.* 26, 1987, pp. 1103–1116.

Kuman, Lalit, et al. "Modeling Topographic Variation in Solar Radiation in a GIS Environment." *International Journal of Geographic Information Science.* 11, 5, 1997.

Land Design/Research, Inc. *Energy Conserving Site Design Case Study, Burke Center, Virginia.* Washington, DC: U.S. Department of Energy, 1979.

Marsh, William M., and Dozier, Jeff. "The Radiation Balance." In *Landscape: An Introduction to Physical Geography.* Reading, MA: Addison-Wesley, 1981, pp. 21–35.

National Institute on Aging. "A Winter Hazard for the Old: Accidental Hypothermia." NIH Pub. 78–1464. Washington, DC: U.S. National Institutes of Health, Department of Health, Education and Welfare, 1981.

Sellers, William D. *Physical Climatology.* Chicago: University of Chicago, 1974.

Tuller, S. E. "Microclimatic Variations in a Downtown Urban Environment." *Géografiska Annaler.* 54A, 1973, pp. 123–135.

Related Websites

London Metropolitan University. Low Energy Architecture Research Unit. http://www.learn.londonmet.ac.uk/packages/clear/index.html
A project to educate academics and laypeople about thermal heating focusing on architecture. Follow the links to information about passive heating and cooling and microclimate.

Environmental Protection Agency. "Heat Island Effect." 2009. http://www.epa.gov/hiri/
EPA's site with information about the heat island effect. The site also includes suggestions for mitigation at the community level and what is already going on in your local area.

Portland Bureau of Environmental Services. City of Portland. "Ecoroofs." 2009. Ecoroofs. http://www.portlandonline.com/BES/index.cfm?c=44422
The City of Portland's site about their green roof program, established to help mitigate urban heat and pollution. The city now offers incentives to buildings with ecoroofs. Program information is given as well as facts about ecoroofs.

US Forest Service. Climate Change Resource Center. "Urban Forests and Climate Change. 2008. http://www.fs.fed.us/ccrc/topics/urban-forests/
Government site looking at how trees impact the urban climate and how urban forestry impacts climate change. Read basic information about urban forestry as well as case studies. The site offers several articles for further readings under the "Energy conservation" subtitle.

17

MICROCLIMATE, CLIMATE CHANGE, AND THE URBAN LANDSCAPE

17.1 INTRODUCTION

Long before the specter of global climate change hit the headlines, another form of human-induced climate change was identified and documented in urban regions. It was found that the wholesale displacement of landscapes from rural forms into manufactured forms can cause significant changes in atmospheric conditions near the ground. In extreme situations, such as in the heavily built-up areas of large cities, these changes extend hundreds of meters into the atmosphere and are of such magnitudes that they produce a distinct climatic variant, the **urban climate**. Generally speaking, the urban climate is warmer, less well lighted, less windy, foggier, more polluted, and often rainier than the regional climate in which it is nested.

Urban variants

In this manufactured landscape, micro- or less-than-microclimatic variations can also be considerable. Air quality may be exceptionally poor along transportation corridors and in industrial sectors. Residential sectors may be warmer than average in summer, and in the inner city some areas may get so hot during heat waves that people perish from heat stress. On the other hand, during warm weather, parks may be several degrees cooler than the surrounding urban landscape.

Implications

These variations can be important considerations in urban planning and design. The documentation of the desirable climatic effects of vegetated areas, for example, helps provide a rationale for the inclusion of parks and greenbelts in urban master plans. Transportation planning in American and Canadian cities invariably includes air quality guidelines and goals, including options for reducing carbon dioxide loading. Proposals for industrial development must include forecasts on gaseous and particulate emissions and the patterns of exhaust plumes under different atmospheric conditions. But that is not the case with urban thermal conditions, and it is not the case in the developing world, where most urban growth will take place in this century. This is an important omission because it appears that global and regional warming will force urban temperatures even higher.

Megacities

Within a few decades, several urban centers will exceed 50 million, and one, Shanghai, will approach 100 million or more. Growth will be accompanied by an infusion of automobiles, rising air pollution, massive areas of hard surface, and unprecedented levels of fuel consumption, all of which will combine to drive urban climate to greater and greater extremes. In some sectors summer heat levels will be especially extreme, particularly when combined with the forcing effects of global warming. This chapter summarizes what we know about urban climate and offers recommendations for climate considerations in urban planning and design.

17.2 THE URBAN HEAT ISLAND

Urbanization transforms the landscape into a complex environment characterized by forms, materials, and activities that are vastly different from those in the rural landscape. Not surprisingly, the flow of energy to and from the urban landscape is also different.

Radiation and heat flows

Radiation and Heat. As a whole, the receipt of solar radiation is substantially lower, whereas the generation of sensible heat (the heat of dry air) at ground level is greater in cities compared to the neighboring countryside. Furthermore, the rate of heat loss from the envelope of air over the city through outflows of infrared radiation and the flushing of heated surface air is lower. On balance, the increase in sensible heat output, coupled with lower rates of heat loss, is more than enough to offset the reduced input of solar radiation, resulting in generally higher air temperatures in urban areas throughout most of the year and much higher temperatures over time periods ranging from a few days to a week or more. The spatial pattern of these temperatures is often concentric around the city center, producing a **heat island** in the

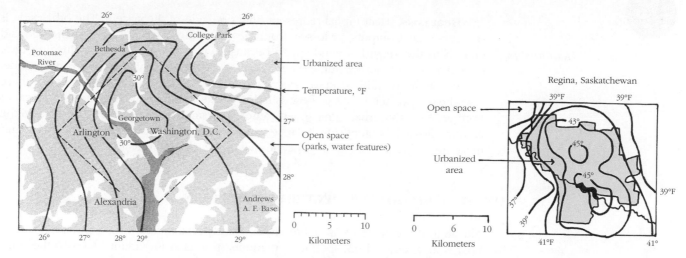

Fig. 17.1 Two versions of the urban heat island over cities of different sizes: (*a*) Washington, D.C., metropolitan area mean winter heat island based on the period 1946–1965. (*b*) Regina, Saskatchewan, urban area on one October evening at 10:00 p.m.

Heat island   landscape (Fig. 17.1). The geographic extent and intensity of the urban heat island varies with city size (based on population) and with regional weather conditions. In general, large cities under calm, sunny weather produce the strongest heat islands.

The overall structure of the urban atmosphere can be envisioned as a large dome centered over the urban mass. This envelope of air is called the **urban boundary**
Urban boundary layer   **layer**. Because of a heavy particulate content, it is highly efficient in back-scattering solar radiation, often effecting a reduction as high as 50 percent in the lower 5000 feet above a city. A growing number of research findings also reveal that increased cloudiness and precipitation are associated with the urban atmosphere and that these trends carry downwind to neighboring areas in the urban region. Thus, it appears that the domelike structure of the boundary layer is pronounced only during relatively calm atmospheric conditions, whereas during a steady airflow across the region, the dome is tipped downwind and develops a plume that diffuses over a regional airshed.

Surface Heating. At ground level, the causes underlying the formation of the urban heat island, and indeed the urban climate in general, are many and complex. First, the materials of the urban landscape possess different reflective and thermal characteristics than those of the rural landscape. For many cities, especially those in
Causes   dry regions, the albedos of street and roof materials are lower than the landscapes they displaced, resulting in higher levels of solar absorption. For cities in forested regions, the urban surface is less shaded than the tree-covered one it supplanted. The *volumetric heat capacities* (or specific heats) of street and building materials are lower than those of materials such as moist soil in the rural landscape (see Table 18.1 in Chapter 18). In general, street and roof materials register heat capacities about half that of moist soil. This means that urban surfaces generally reach a higher temperature with the absorption of a given quantity of radiation and, in turn, heat the overlying air faster.

Second, the **Bowen Ratio**, which is a measure of the heat released from a surface in sensible form relative to latent form, is much higher in cities owing to the limited areas of open water, vegetation, and exposed soil. With a paucity of vapor sources,
Sensible and latent heat   *latent heat flux* from the surface is relatively low; conversely, *sensible heat flux* is relatively high, giving rise to higher air temperatures. Added to this is the heat released from artificial sources (automobiles, buildings, etc.). In the midlatitudes, these sources typically contribute more energy to the interior of a large city such as New York City in midwinter than the solar source does.

Heat flushing

Heat Loss. Heat inputs represent only one side of the system. The other side is, of course, heat outputs, or losses, from the urban atmosphere. The principal consideration in this regard is wind speed because wind is responsible for flushing away the envelope of heated surface air. Because of the aerodynamic roughness added by tall buildings, cities tend to have much lower wind speeds at ground level; therefore, heated air tends not to be flushed away as readily as it is in rural landscapes. Furthermore, the urban atmosphere tends to retain more heat because of a higher carbon dioxide content. On balance, then, the urban landscape yields and retains more heat, thereby accounting for the overall heat island effect.

17.3 CLIMATIC VARIATIONS WITHIN THE URBAN REGION

Sector variations

Although the climate of an entire city is an important issue for air pollution control boards and regional transportation planners, the most important issues to the urban planner, landscape architect, and architect are small-scale climatic variations within the urban area. Perhaps the easiest variation to visualize is that associated with solar radiation around tall buildings, but other parameters, including temperature, wind, fog, and pollution, also show considerable variation within the urban landscape (Fig. 17.2). With the exception of air pollution, the nature and significance of these variations are generally not well documented. But it is widely agreed that the extremes in these atmospheric conditions within the urban landscape affect the health, safety, and

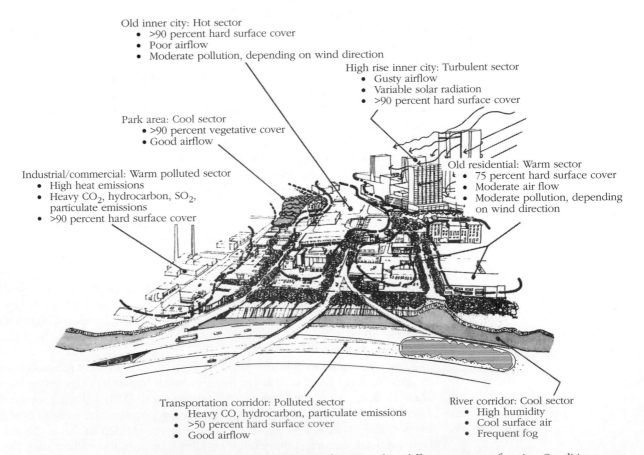

Fig. 17.2 Climatic conditions near the ground in different sectors of a city. Conditions vary with surface cover, solar radiation, airflow, and air pollution, among other things.

comfort of a significant number of people in most cities. Here are brief summaries of six key attributes of the urban climate.

Solar Variations. Beam radiation is intercepted by buildings, and, depending on sun angle and building height, a shadow of some size is created. Where buildings are closely spaced, a **shadow corridor** may form (see Fig. 16.9 in Chapter 16). For a single building site, incoming solar radiation varies with season, wall orientation, and

Solar exposure

time of day, and in cold climates, there may be distinct differences in heating and comfort from the sunny to the shadow sides of buildings, such as the one shown in Figure 16.11. Similarly, building sides with high solar exposure may be subject to extreme heating in summer, especially during heat waves when high ambient air temperatures combine with solar heating.

In the 1995 heat wave that killed 700 people in Chicago, more than half the victims lived on the top floor of their buildings, where, of course, solar heating was the most extreme. On the other hand, pockets sheltered from beam radiation by tall buildings may perpetually be in shade and lighted only by diffuse solar radiation. These pockets can be cool and damp, especially where airflow is poor. Coupled with the refuse that often collects in back lanes and alleyways, the resultant environmental conditions can be very unsavory.

Temperature Variations. Most cities are geographically diverse in surface materials, physical forms, and activities, and we would expect settings as different as people parks and industrial parks to develop markedly different temperature regimes. Studies show, however, that this is so only where thermal variations are not masked by strong regional weather systems or extreme local influences on climate. The latter is exemplified by a pocket park of vegetation in the midst of an inner city; whatever modification in temperature is achieved by the park is masked by the thermal umbrella of the surrounding mass of buildings.

Park effects

The thermal modification of the urban heat island by a large park or greenbelt can be significant, however. One set of readings made in San Francisco shows how extreme the cooling related to a large park can be (Fig. 17.3). The temperature difference between Golden Gate Park and downtown near Union Square was more than 15°F. Other investigations have shown older residential areas with mature trees to be cooler than new residential areas and other urban surfaces. In Washington, D.C., it is not uncommon for the corridor of parks and water features along the Potomac River Valley to be cooler during summer days and evenings than the heavily built-up areas on either side of it.

Island edge

On the perimeter of the city, the urban heat island may decline sharply where the urban landscape quickly gives way to the rural landscape. Pictured in a temperature profile such as the one shown in Figure 17.4, this sort of border is characterized by a *cliff* in the graph line. From a planimetric perspective, the border configuration appears to be very irregular in detail with cool inliers represented by parks and river corridors and warm outliers represented by large shopping centers and industrial parks.

Wind and Convective Mixing. The general influence of a city on airflow is to reduce wind speed at levels near the ground. This phenomenon can be illustrated

Profile displacement

by comparing the profiles of wind speed over urban and rural surfaces. The elevated topography of the urban environment displaces the **wind velocity profile** upward, leaving a thicker layer of slow-moving air near the ground (Fig. 17.5). At a more detailed level of observation, however, large variations in wind speeds can be found in relatively small areas. Much of this variation is related to the size, spacing, and arrangement of buildings.

Airflows around buildings

Three example are noteworthy. First, in the case of an individual building, the structure represents an obstacle to airflow and, to satisfy the continuity of flow principle, wind must speed up as it crosses the building. In a two-dimensional model, the highest speeds are reached on the windward brow of the building and across the roof. Air is also deflected from the brow down the face of the building (labeled A in

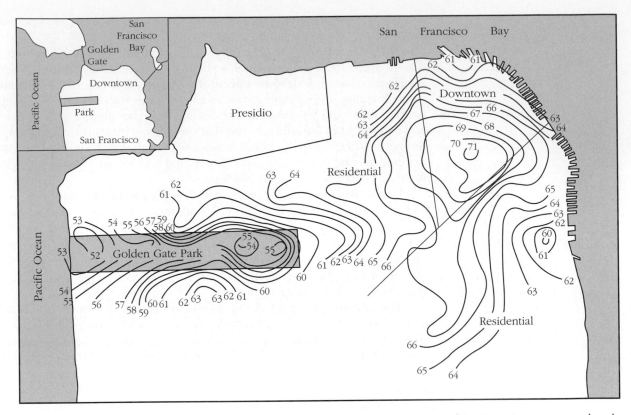

Fig. 17.3 The influence of a large park on the temperature of San Francisco contrasts sharply with the heat island over the heavily built-up downtown area.

Fig. 17.6); on the leeward side, speeds decline and streamlines of wind spread out, with some descending toward the ground.

Where two tall buildings of similar heights are spaced close to each other, the streamlines of fast wind do not descend to the ground but are kept aloft by the roof of the second building. This gives rise to a small pocket of calm air between the buildings where mixing with the larger atmosphere of the city is limited (labeled B in

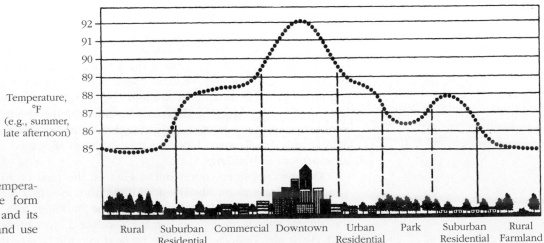

Fig. 17.4 A schematic temperature profile depicting the form of the urban heat island and its relationship to different land use zones.

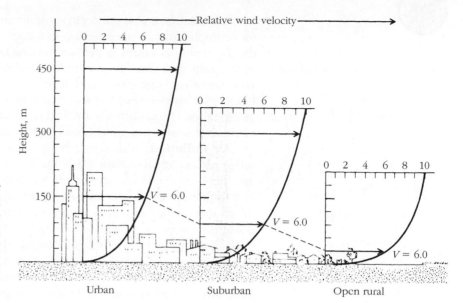

Fig. 17.5 Profile of wind velocity over urban and rural landscapes. Fast air is forced upward by the building mass as revealed by the elevation of the 6.0 velocity line compared to the rural landscape. Although ground-level wind velocities are markedly lower in cities, turbulence tends to be higher because of tall buildings.

Fig. 17.6). Depending on local conditions, the air in such pockets, as noted earlier, may be measurably different from the surrounding atmosphere.

The third example involves the alignment of buildings and streets. Streets bordered by a continuous mass of tall buildings have the topographic character of canyons, and if aligned in the direction of strong winds, tend to channel and constrict airflow. This produces heightened wind velocities at street level, especially during gusts, and increased turbulence along the canyon walls (labeled C in Fig. 17.6).

Fog and Precipitation. The incidence of fog in cities may be twice that of surrounding country landscape. Most of this is usually attributed to the abundance of condensation nuclei (particulates) from urban air pollution. Local concentrations of fog are more common in selected areas, especially under calm atmospheric conditions coupled with strong nighttime cooling at the surface. Of the several contributing factors besides air pollution, one is related to the availability of water vapor near the ground. In low-lying coastal areas and in river valleys, the concentration of vapor may be appreciably higher than elsewhere in the city. In addition, cold air drainage into low-lying areas promotes fog development. Conversely, heated buildings and hard surfaces may locally reduce fog development because they tend to limit the normal rate of fall in nighttime air temperatures.

Contributing factors

Fig. 17.6 Airflow over and around buildings. The highest velocities are reached on the windward brow and across the roof of the tallest building. A strong flow of air is also deflected down the building face (A), but a calm zone develops in the space between the buildings (B). Accelerated flow (C) is associated with the canyon between large buildings.

Precipitation

Precipitation is also greater in some urban areas. In the Detroit metropolitan region, for example, annual precipitation in some areas is as much as 8 inches greater than the mean annual values in the surrounding nonurban areas. Other urban centers such as St. Louis and Chicago may induce greater precipitation downwind from the city, whereas other large cities have little apparent influence on precipitation rates even though the incidence of fog and clouds may be greater. The chief cause of increased precipitation is undoubtedly the higher incidence of particulates in the urban atmosphere. Instability (rising air) caused by the urban heat island may also be a factor.

Air Pollution. Although a body of heavily polluted air may blanket an entire urban region under certain atmospheric conditions, pollution levels are on the average higher in the inner city than in surrounding suburban areas (Fig. 17.7). In addition, pollution levels on many days vary sharply from one quadrant or sector of an

Spatial patterns

urban area to another. Two factors account for this: (1) the site-specific nature of many pollution sources such as power plants, highway corridors, and industrial plants; (2) short-term changes in the mixing and flushing capacity of the urban boundary layer. During windy and unstable weather, pollutants are mixed into the larger mass of air over the city and flushed away, thereby limiting heavy concentrations, if any, to relatively small zones downwind of discharge points.

During calm and stable atmospheric conditions, however, pollutants tend to build up over source areas, and if these conditions are prolonged, the concentrations coalesce to form a composite mass over the urban region. This buildup is especially pronounced with *particulates*, considered the most harmful pollutant to human

Urban dust dome

health, which often form a visible **dust dome** over cities (Fig. 17.8). The largest particles are discharged from sources such as power plants, foundries, and construction projects, and they tend to settle close to their sources. Near factory and power plant exhaust stacks, the fallout pattern can be traced in the distribution of dirt blanketing the neighboring landscape.

Intermediate-sized particles usually remain suspended for several days and spread over a much greater volume of air than large particles. They and smaller particles con-

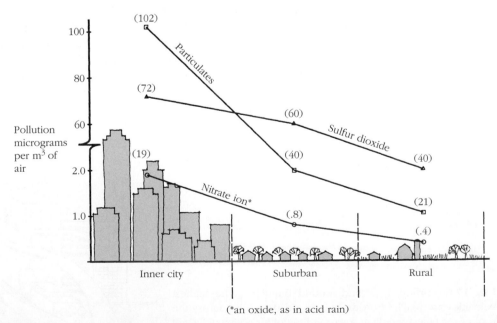

(*an oxide, as in acid rain)

Fig. 17.7   Representative changes in air quality from the inner city to suburbia and the rural landscape beyond.

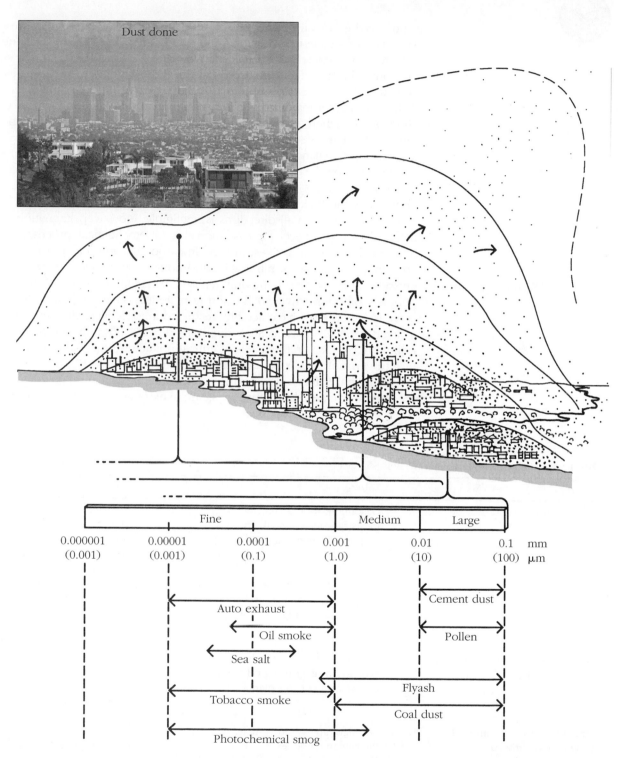

Fig. 17.8 A schematic portrayal of a dust dome over a large city during relatively calm atmospheric conditions. Larger particulates tend to be concentrated close to their sources. The inset photograph shows the hazy conditions associated with the dust dome over New York City on a winter day.

Diffusion

tribute to the formation of huge dust domes that may envelop an entire urban region. The class of smallest particles, which are less than 1.0 micrometer (one-millionth of a meter) in diameter, may remain suspended in the atmosphere for several weeks or months because the slightest motion of the air moves these particles upward and laterally. Most gaseous pollutants also have long residence times in the atmosphere. If they are lighter than air, such as methane and CFCs, they may float upward in the atmosphere, where they remain until they are altered chemically and/or recycled through the earth's surface environments.

Mass balance

The mass balance of pollutants for a given volume of atmosphere can be estimated based on the total rate of pollutant emission and the rate of removal by airflow. Removal includes both lateral and vertical components; therefore, it is easy to imagine how heavy the buildup of pollution can become during a prolonged thermal inversion when airflow in all directions is negligible. Moreover, stagnation of polluted air increases the prospects for oxidation and photochemical processes involving sulfur dioxide, nitrogen oxides, and hydrocarbons, leading to the formation of **smog** containing sulfuric acid, nitric acid, and noxious gases such as ozone. Under severe

Secondary impacts

episodes of air pollution, the only realistic management option (other than regulating people's outdoor activities) is short-term reduction in emissions. Officials in cities such as Los Angeles have actually restricted automobile traffic and industrial activity to avert a health disaster. Such decisions depend not only on the gross level of air pollution, but also on the levels of critical pollutants, especially hydrocarbons, oxides of nitrogen, sulfur dioxide, and airborne particles (Table 17.1).

Regional Effects. The plumes of polluted air generated from metropolitan areas are known to extend tens, hundreds, and in extreme cases, thousands of miles beyond

Table 17.1 Major Air Pollutants and Their Sources

Pollutant	Source	Effects
Carbon monoxide and dioxide	Gasoline-powered vehicles Industry using oil, coal, and gas Building heating using oil and gas Power plants using coal, oil, and gas	CO_2 promotes heat retention and atmospheric warming CO enters human bloodstream rapidly, causing nervous system dysfunction and death at high concentrations
Sulfur oxides (sulfur dioxide and sulfur trioxide)	Industry using coal and oil Heating using coal and oil Power plants using coal, oil, and gas	Irritate human respiratory tract and complicate cardiovascular disease Damage plants, especially crops Promote weathering of building skin materials
Nitrogen oxides (nitric oxide and nitrogen dioxide)	Gasoline-powered vehicles Building heating using oil and gas Industry and power plants	Irritate human eyes, nose, and upper respiratory tract Damage plants Trigger development of photochemical smog
Hydrocarbons (compounds of hydrogen and carbon)	Petroleum-powered vehicles Petroleum refineries General burning	Toxic to humans at high concentrations Promote photochemical smog
Oxidants (secondary pollutants in smog including ozone)	Vehicle exhausts Industry Photochemical smog	Irritate eyes Damage plants Promote cancer
Particulates (liquid or solid particles generally smaller than 100 micrometers)	Vehicle exhausts Industry Building heating General burning Spore- and pollen-bearing vegetation	Some are toxic to humans Some pollens and spores cause allergic reactions in humans Promote precipitation formation

Acid rain

their source areas. The effects on regional climate are not well documented, but they are known to be more pronounced in certain regions and seasons and are characterized by increased cloudiness, precipitation, and turbulent weather. Another regional effect has been added in the last few decades, **acid rain** in southeastern Canada and northeastern United States. Much of this is caused by the formation of sulfuric acid from the combination of atmospheric moisture and sulfur trioxide in polluted air. Because of the regional flow of weather systems across the Midwest, the acidic moisture is carried from industrial areas and precipitated in Ontario, Quebec, and New England where the biota of thousands of lakes and ponds have been damaged by the increase in water acidity.

17.4 AIR POLLUTION MANAGEMENT IN THE URBAN REGION

Air quality policy

Under the **U.S. Clean Air Act** (originally enacted in 1970 and last amended in 1992), the U.S. Environmental Protection Agency (EPA) was directed to set air quality standards to protect human health with an adequate margin of safety. The resultant standards, called the **National Ambient Air Quality Standards (NAAQS)**, defined two levels of air quality standards: (1) primary standards designed to protect the health of people; (2) secondary standards designed to protect against environmental degradation. The Clean Air Act amendments also set air quality standards for certain land use areas. Three classes of areas were defined for the purpose of preventing air quality deterioration beyond certain base performance levels.

- *Class I:* Areas such as national parks and wilderness areas have the highest standards (that is, where least deterioration is allowed).

- *Class II:* These are broadly defined as intermediate areas and have intermediate standards.

- *Class III:* Industrial urban areas have the lowest standards.

In practice, all areas are defined as Class II unless they are moved into Class I or Class III by the EPA.

The Automobile Dilemma. In the 1970s and 1980s, pollution control in the United States focused on the least expensive measures and largest sources of pollution. Two major areas of activity, large industrial facilities and motor vehicles, were pressed with most of the responsibility for cleanup. Emission standards were implemented *Emissions standards* for automobiles that included exhaust system devices called catalytic converters on all new vehicles. In addition, lead was banned from use as a gasoline additive in new automobiles. Fuel economy standards, called **CAFE standards** (corporate average fuel economy standards) were also mandated.

As a result of these actions, pollution rates per automobile dropped significantly and lead levels in air and water declined over most of the United States. The Great Lakes, for example, saw lead concentrations drop dramatically between 1970 and 1980. On the other hand, the number of automobiles in use increased significantly between 1970 and 1988, and the total miles traveled by Americans annually doubled (from about 1000 billion to over 2000 billion), both of which offset much of the gains made by reduced emissions and better mileage (Fig. 17.9).

In the late 1980s, further amendments were added to the Clean Air Act. These amendments addressed three major areas:

- Reductions in sulfur dioxide emissions from power plants with a target of 10 million tons of reduction per year over the 1990s.

Current targets
- Reductions in hazardous air pollutants (mainly synthetic compounds such as benzene, and PCBs).

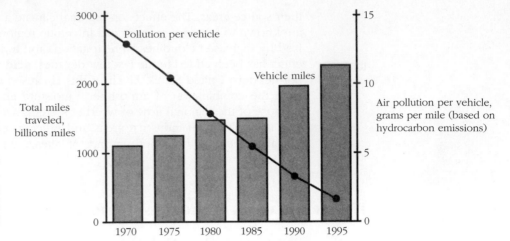

Fig. 17.9 Trends in automobile miles traveled and pollution emissions per automobile in the United States, 1970–1995. The improved emissions performance was and still is offset by increased travel.

■ Improved air quality in selected areas, mostly urban areas, where ambient air quality standards are consistently being violated.

The California Effort. California has stepped ahead of the U. S. EPA and set more stringent standards in several areas. The California Air Resources Board (CARB) has defined a number of vehicle emission categories with the overall aim of reducing hydrocarbon and nitrogen oxide emissions beyond federal standards. Sixteen other states have also elected to follow CARB standards. California is also pushing to enact greenhouse gas regulations (despite EPA resistance).

The LA experience   In Southern California, where regional air pollution is extremely critical, the Los Angeles Air Management District has implemented a market incentives program. Under this program, polluters agree to a pollution allowance, that is, a fixed quantity of allowable emissions. The total allowance for the district is designed to meet air quality standards set by the EPA. The novelty of the program is that it allows polluters—excluding the more than 8 million vehicles on the road daily—to buy and sell allocations. This rewards those who are able to achieve lower emission rates by allowing them to sell allocations for profit to less efficient operations. The overall balance, given both more efficient and less efficient operations, is designed to lower pollution levels over the district. Automobile pollution, on the other hand, is being addressed by the strict implementation of state emissions (CARB) standards.

Traffic loads   The Los Angeles situation is an extreme example of the urban air pollution problem faced by virtually every metropolitan center in the United States. The principal source of the problem is increased automobile traffic brought on by population growth, increased use of cars, and longer travel distances. Expressways in and around large cities carry an astonishing daily load of vehicles, typically hundreds of thousands per route. Many cities have sought various means of reducing the traffic load, including the use of ride sharing and mass transit systems. At the same time, however, the states and the federal government continue to expand expressways and build new links, thereby encouraging more automobile travel over greater distances as well as continued urban sprawl.

Commuting   **Land Use Implications.** In fact, highway-induced sprawl has given rise to a new urban form, the **edge city**, at key interchanges around and between urban regions. Served exclusively by automobile and truck traffic, edge cities foster even greater travel, especially among commuters. Daily one-way travel distances between home and workplace of 50 miles or 75 miles are not uncommon today. The 1990 U.S. census revealed that the trend toward long commuting distances continues to rise; today nearly 25 percent of the workers in the United States actually drive to another county to work compared to 15 percent in 1960.

On balance, the United States and Canada have developed urban transportation systems that are very expensive for the environment, the citizen, and local units of government. The systems rely overwhelmingly on cars and trucks to move people and materials, and they encourage long daily travel distances. They not only promote *Excessive travel* massive fuel consumption and air pollution, but necessitate a huge and expensive highway infrastructure, sprawling land use patterns, and highly inefficient communities. Indeed, in many respects the traditional form and function of the North American community in regions overwashed by expressway systems are being replaced by more transient forms of settlement and gathering places.

The incentive to reverse sprawl and increase urban density is gaining momentum in some quarters. Increasing density can improve air quality by decreasing automo- *Change?* bile use and emissions. However, there is a limit to this strategy. Studies show that increasing urban density, even significantly (by, say, twofold), is limited to emission reductions of 30 percent or less. To improve urban air quality (including carbon dioxide emissions) much beyond this level will require switching fuels and transport modes.

17.5 CLIMATE CONSIDERATIONS IN URBAN PLANNING

Most planners recognize climate as an important element in urban planning and design, but few have been able to incorporate variables such as wind, heat, and solar radiation effectively into the information base for decision making. Several factors are responsible, and one of the most important is the level of scientific understanding of microclimate in the built environment.

The Problem. Although architects and engineers understand many of the influences of climate on a building—such as wind stress, solar exposure, and corrosion of skin materials—comparatively little is known about the influence of buildings, *Need to know* especially building masses, on urban climate at the sector and local scales. In contrast to urban hydrology, for example, technical planning is able to provide fewer models and less accurate forecasts to guide the urban planner in setting the heights of buildings, the balance between hard surfaces and vegetative surfaces, the widths of streets, and the like.

A second factor is the general lack of *planning regulations* pertaining to climate. Although it is widely recognized that urban climate affects the health and well-being *Local policy* of people, resulting in, among other things, greatly increased medical costs, few ordinances/bylaws have been enacted establishing climate performance standards for environments outside buildings. As noted, one exception is in the area of air quality, and a second one may be emerging related to global warming.

National regulations on industrial and automotive emissions seek to improve living conditions in cities. In local transportation projects involving federal funds, a trans- portation master plan is required that takes air quality into consideration. In locales *Air quality* that are subject to severe episodes of air pollution, such as Los Angeles County, local agencies are responsible for regulating the outdoor activity of schoolchildren and, under emergency conditions, for reducing automotive and industrial activity.

Signs of Change. Beyond examples related to air quality, however, planning agencies have traditionally paid little attention to climatic parameters such as surface heating, airflow, fog, and radiation. Prospects for change in this state of affairs are *Convention* not good where a direct relationship to human safety or to capital costs is not appar- ent, that is, where direct savings to individuals, companies, agencies, or institutions are not clearly evident. A case in point was the emergence of solar energy–oriented communities in the 1980s where ordinances on so-called solar rights were legislated because access to the sun could be given some economic value.

New concern

Facing the Heat. But today, with cities heating up with population growth and increased car traffic superimposed on global warming, energy consumption and carbon dioxide outputs are rising, providing new justification for giving serious attention to urban climate as a planning and design consideration (see Case Study 17.7). Indeed, the rising specter of climate change has cast new light on urban climate in planning circles. Climate is now an agenda item in many large cities, and while most struggle to get a handle on the problem, cities seem committed to doing something to reduce energy use and greenhouse gas outputs. In fact, in the first decade of this century, cities generally did more to address the issue than did the federal governments in the United States and Canada.

Uncertain forecasts

Climate is invariably addressed in environmental impact statements, but it usually consists of descriptions of existing conditions with some "forecasts" about potential changes given a proposed action. Only cases involving air quality changes have been a source of serious concern and may be the basis for recommending against or altering a proposed action. As for changes in the physical components of climate, few guidelines or performance standards have been established to aid planners in formulating plans and reviews. As a result, in most cases no one is quite sure how much importance should be ascribed to a suspected change in some aspect of physical climate despite ample documentation of an urban climate and its effects on energy use and human health.

Emerging policy

But this is changing, and the driver is climate change. The concern is that the effects of global warming will be most acutely felt in cities because it will drive up base temperatures, superelevating the urban heat island effect. Accordingly, policies are being directed toward reducing traffic by closing streets, increasing density, and building mass transit and toward reducing building energy demands by endorsing LEEDS (Leadership in Energy and Environmental Design) standards, adding shade trees to residential areas, replacing paving on closed streets with vegetation, and adding green roofs and gardens to the tops of large buildings.

17.6 CLIMATIC CRITERIA FOR URBAN PLANNING AND DESIGN

The challenge

Throughout the world cities are growing, and urban leaders are being pressed to accommodate new facilities, technologies, and people. This need demands that the urban planner and designer be on constant watch for opportunities to improve pedestrian movement, commuter traffic, land use, air quality, wastewater disposal, energy consumption, and so on. The overriding challenge is to achieve a proper balance between the economic functions that are necessary to the city's existence and an environment that allows for the health and well-being of the populace. For years the debate hovered around the costs of improving air quality through such measures as stricter automobile emission standards on the one hand and economic growth on the other. Apparently, the economic growth argument is now edging toward improving the quality of air and slowing the deterioration of urban climate because it is increasingly clear that both are necessary to building a healthy business climate.

Improving Street Climate. Cities should be healthy places for both buildings and people. Sullied, dim, sullen atmospheres drive up maintenance and health costs and reduce the life cycles of both structures and inhabitants. Though little can be done in designing cities to combat regional atmospheric conditions, such as those governed by the movement of air masses, measures can be taken to minimize thermal extremes and high levels of air pollution associated with microclimates of various scales within the city. Basically, only four types of climatic controls or changes are possible through urban planning and design, given the goal of improving living conditions in cities prone to excessive heat and poor air quality:

Considerations

1. *Reduce summer solar radiation* by shading critical surfaces. These include pedestrian walks, waiting areas, and busy streets as well as homes, office buildings, and even factories. If trees are used, not only is shade provided, but latent heat flux is increased with transpiration, which lowers air temperatures. The trees thus function as sinks for carbon dioxide. (In winter, on the other hand, the objective in northern cities is to provide shelter against the cold on streets and to improve the solar efficiency of buildings and public spaces.)

2. *Reduce the abundance of concrete and asphalt* and other hard surfaces, and increase the amount of vegetation and open water. This control reduces the so-called urban desert effect by raising volumetric heat capacities and increasing rates of latent heat flux, thereby lowering ground-level air temperatures. Green roofs, rooftop gardens, street gardens, recovered (daylighted) stream corridors, and street trees should be considered.

3. *Increase airflow at ground level* to flush heated and polluted air away from streets, buildings, and the city in general. Consider using more than real estate economics and zoning in siting land uses and building forms. Consider also seasonal airflow patterns, that is, the implications of upwind and downwind locations in land use and facility siting.

4. *Reduce air pollution* by decreasing emission rates, improving flushing rates, and increasing pollution sinks (such as carbon dioxide sequestering by trees and soil). This includes a host of factors, such as reducing automobile traffic by converting selected streets to neighborhood parks and gardens, improving and expanding mass transit, improving automobile fuel efficiency, as well as reducing building energy use.

Scale Considerations. In applying climatic factors to urban design, planners must first consider the scale of the problem. For problems of citywide scope, the location, structure, and layout of streets, building masses, and industrial parks must be weighed against airflow patterns, sources of pollution (such as existing traffic corridors and industrial parks), and the ratio of open to developed space. In inner cities, the maintenance of ground-level airflow in summer is very important; therefore, inner city street corridors should be wide, aligned with prevailing winds, and kept free of major obstacles.

Location, direction, and flow

Where vacant land is available in and around built-up areas, it should be converted to vegetation (rather than left barren), and, as far as possible, vegetation should be expanded into inner city areas along streets, in pocket parks, and on rooftops. Where pollution is a problem, seasonal patterns of airflow and weather should be considered in locating industry, power plants, and the like. In the midlatitudes, winter weather often produces plume patterns that direct polluted air toward the ground. Where this phenomenon is known to occur, polluting activities should be situated so as to minimize their impact on residential areas.

At the scale of individual blocks of buildings, attention must be given to orientation with respect to airflow and solar radiation and to building sizes and forms. The relative heights of buildings in masses must be taken into account. For example, to minimize the nuisance and danger of wind gusts to pedestrians, studies have shown that the taller of two adjacent buildings should not exceed the shorter by more than twofold. Generally speaking, the taller the buildings are, the more attention should be given to the effects of their spacing, orientation, form, skin materials, and other factors on microclimate.

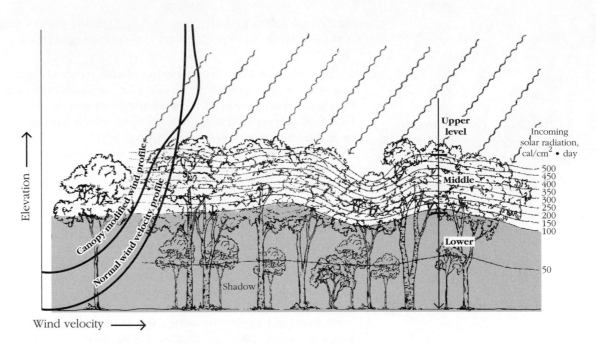

Fig. 17.10   Vertical zonation of climate, the canopy of a large forest, and the related levels of solar radiation. The canopy reduces incoming solar radiation by more than 80 percent, leaving the forest floor mostly in shadow. The middle level, where most tropical forest animals live, also holds promise for human habitation in cities.

Climate aloft

Vertical Zonation. In considering the vertical dimension of urban climate, it is important to ask what level (elevation) is most appropriate for different human activities. Clearly, ground level in the inner city has several distinct disadvantages including heavy air pollution and competition with automotive traffic. Similarly, high elevations pose the hazard of high-speed winds that can damage structures and impair human safety. At the middle level, however, in the four- to ten-story range, air is generally cleaner than it is at ground level, but substantially less windy than that at higher elevations. On balance, this level provides a somewhat healthier and safer climate.

Forest model

Therefore, where heat and pollution at ground level are problems, rooftop spaces and balconies in this zone seem to offer promise for expanded human use if, as New Yorkers and others are finding, appropriate landscaping can be introduced to ensure shade and safety. This concept is similar to the model of bioclimatic zonation in large, tropical forests where the middle level of the canopy is the optimum climatic zone for many creatures, the top being too windy and hot and the floor too humid and shaded (Fig. 17.10).

Roof water

Rooftop storage of stormwater may also be desirable in modifying the urban climate. Upon evaporation, large quantities of sensible heat are taken up and released with the water vapor, thereby cooling the roof surface and the air over it. In Texas, for example, as much as 80 inches of water can be evaporated from open water surfaces in the average year—twice the average annual precipitation for most of the state. Therefore, it is easily possible to dissipate the total annual quantity of stormwater if the water can be held on rooftops.

Streets for people

Heat Syndrome in Humans. At the street scale, consideration must be given to the potential for thermal stress on people in waiting and walking spaces. This is especially critical in cities that record official temperatures above 90°F (32°C) on many days per month in summer because in the urban thermal climate such readings translate

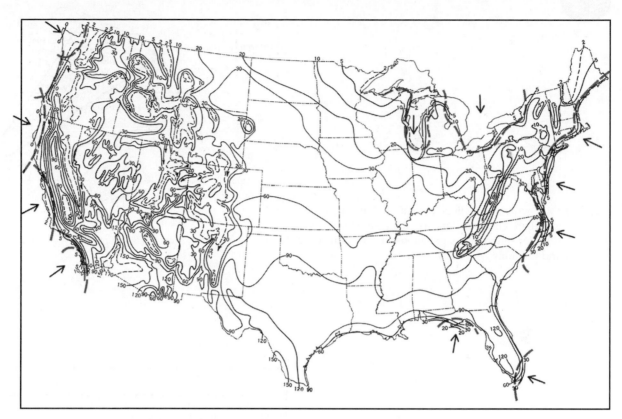

Fig. 17.11 The distribution of the average number of days per year with air temperature reaching 90°F or more, a serious consideration to public health in large urban areas. The arrows identify relatively cool coastal locations.

into temperatures above 100°F (38°C). Interior locations in cities of the American South and Southwest are especially prone to such conditions and will likely grow more extreme with global warming in the decades ahead (Fig. 17.11). Phoenix, San Antonio, and Orlando are currently subject to more than a 100 such days a year on average.

Health risk   In addition to high temperatures, intensive solar radiation, poor airflow, high humidity, and physical exertion also contribute to heat stress or heat syndrome. **Heat syndrome** refers to several clinically recognizable disorders in the human thermoregulatory system that reduce the body's ability to shed heat or cause a salt imbalance because of excessive sweating. In the extreme, the result is a lethal heat stroke. Thus, along pedestrian corridors with high solar exposures, poor air circulation, and long walking distances, the potential is great for heat syndrome among walkers. To avoid this, shaded rest stops with good ventilation should be provided at appropriate locations. The distribution and location of stops should be based on origin and destination patterns for different walkers, especially the elderly and disabled.

Comfort measures   The chart in Figure 17.12 defines three comfort zones based on temperature, humidity, wind, and solar radiation. Notice that the condition most conducive to heat stress combines high temperature with high humidity (above 75 percent), full sun around midday, and calm wind conditions. The effect of humidity is illustrated further in Figure 17.13, which gives the apparent or effective temperature for different combinations of temperature and relative humidity. Notice that at a temperature of 95°F and a relative humidity of 65 percent, the effect on the body is equivalent to a 110°F temperature, ranking in the danger category.

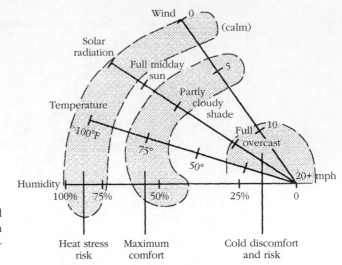

Fig. 17.12 A generalized climatic comfort chart that can be used to evaluate urban environments for their suitability for human occupation under different combinations of radiation, wind, temperature, and humidity.

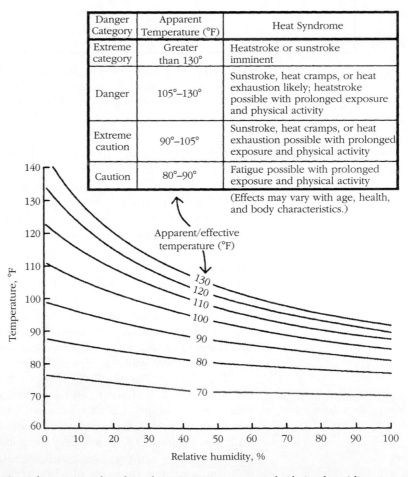

Danger Category	Apparent Temperature (°F)	Heat Syndrome
Extreme category	Greater than 130°	Heatstroke or sunstroke imminent
Danger	105°–130°	Sunstroke, heat cramps, or heat exhaustion likely; heatstroke possible with prolonged exposure and physical activity
Extreme caution	90°–105°	Sunstroke, heat cramps, or heat exhaustion possible with prolonged exposure and physical activity
Caution	80°–90°	Fatigue possible with prolonged exposure and physical activity

(Effects may vary with age, health, and body characteristics.)

Fig. 17.13 A heat stress chart based on air temperature and relative humidity. At a temperature of 95°F and a relative humidity of 65 percent, the effective temperature (what the body feels) is 105°F.

17.7 CASE STUDY

Modifying Urban Climate and Reducing Energy Use Through Landscape Design

W. M. Marsh

Almost everywhere cities are growing, and, as they expand, the size and intensity of the urban heat island increase. While this is taking place, regional climate is heating up with global warming. Peak temperatures in Los Angeles have increased by 5°F or more since 1940, primarily in response to growth of the urban region. Researchers estimate that the added cost of air conditioning for the 5° increase is at least $150,000 per hour. Similar trends in other large U.S. cities are producing increases in energy costs for air conditioning of $50,000 to $100,000 per hour. A rough estimate of the total air-conditioning costs related to the urban heat island in the United States is $1 million per hour, or $1 billion per year.

To reduce the urban heat island significantly would require major changes in the way we design, operate, and live in cities. To begin with, we can reduce air-conditioning demands by making just two modifications in the urban landscape: (1) more light-colored (that is, higher albedo) building materials and (2) more shade trees, that is, less direct solar radiation on houses and streets. Studies in California show that a mature tree cover can reduce the daytime temperatures in residential neighborhoods by up to 9°F, and tree shade on residential buildings can lower air-conditioning costs by 40 percent or more, depending on housing type. A change from dark to light roof and walls can reduce home air-conditioning costs for some houses in some cities by as much as 22 percent. Extended over an entire city, the use of lighter materials could reduce heat island temperatures by as much as 5°F. Because less fossil fuel would be burned to generate power for cooling, the indirect benefits of these energy-reducing measures would also include lower carbon dioxide inputs to the atmosphere and healthier living conditions with cleaner, cooler urban air.

Based on computer simulations, researchers found a difference in the effects of increased albedo between older, poorly insulated houses and newer houses with better insulation. In older houses, an increase in albedo from 30 to 40 percent produced an energy savings for air conditioning of 11 to 22 percent. In a well insulated house, the same albedo change produced an energy savings of 8 to 13 percent (Fig. 17.A). Although few data are available for summer/winter comparisons, it is

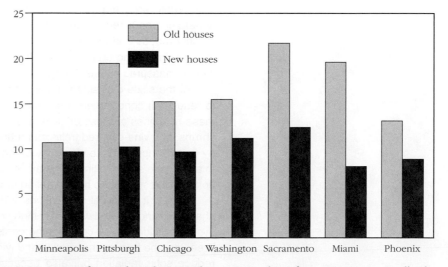

Fig. 17.A Energy savings for residential air conditioning resulting from an increase in albedo from 30 to 40 percent (Source: Taha, 1988).

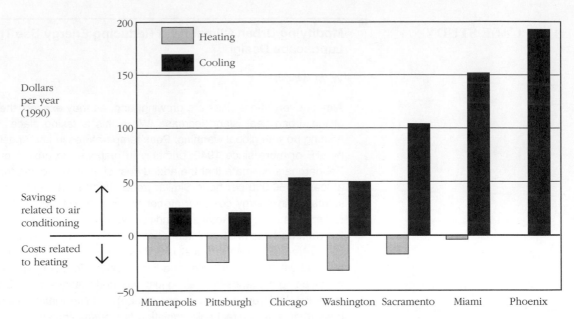

Fig. 17.B Energy savings (black bars) and added costs (gray bars) related to shading typical houses with trees. The added costs do not take into account foliage changes and wind buffering (Source: Huang et al., 1990).

apparent that the real advantage in changing urban landscapes to higher albedo materials is found in southern cities. For northern cities, some consideration must also be given to the heating advantage of lower roof albedos in winter.

To some extent, the same holds true for tree shade. Ideally, we should design for more shade in summer and less shade in winter, as is illustrated in Case Study 16.7. Therefore, the choice of tree types and their placement along streets and around residential buildings are very important. The general advantages and disadvantages of summer and winter tree shade in terms of energy use in northern and southern cities are shown in Figure 17.B. Again, the data are based on computer simulations. As landscape architects know all too well, community officials are often reluctant to add sizable street trees to infrastructure projects. With these findings, however, the debate can be taken beyond the usual case for the role of trees in creating livable neighborhoods and into the realm of dollar savings, to say nothing of the implications for improving urban climate. Indeed, the cost of trees and their maintenance can be measured against energy savings over the life cycles of the trees.

At the scale of residential lots, research reveals that the least expensive way to reduce air conditioning costs is to shade the air-conditioning unit itself because these machines are less efficient at high temperatures. The shade from trees, tall shrubs, or a vine-covered trellis can reduce temperatures around the air conditioner by 6–7°F, improving the efficiency of the air conditioner by as much as 10 percent. In addition to shade, siting of the air conditioner unit to take advantage of the cooler, north-facing side of the building is an obvious first step.

Finally, it is critical in locations with heavy air-conditioning needs to provide shade on windows and walls with high solar exposure. Here, with an efficient shade system, residential air-conditioning costs can be reduced by 40 percent or more. Trees are the best source of wall and window shade not only because they sharply reduce solar radiation but also because they cool the air around them through transcription. Trees should be planted so that, near maturity, the limbs reach within 5 feet of the east or west walls and 3 feet of the south wall. This configuration helps create a cool envelope between the trees and the house, thereby reducing building skin temperatures and lowering heat penetration. ∎

17.8 SELECTED REFERENCES FOR FURTHER READING

American Society of Landscape Architects Foundation. *Landscape Planning for Energy Conservation.* Reston, VA: Environmental Design Graphics, 1977.

Chandler, T. J. *The Climate of London.* London: Hutchinson and Co., 1965.

Duckworth, F. S., and Sandberg, J. S. "The Effects of Cities upon Horizontal and Vertical Temperature Gradients." *Bulletin American Meteorological Society.* 35, 1954, pp. 198–207.

Ellis, F. P. "Mortality from Heat Illness and Heat-aggravated Illness in the United States." *Environmental Research.* 5, 1, 1972, pp. 1–58.

Federer, C. A. "Trees Modify the Urban Microclimate." *Journal of Arboculture.* 2, 1976, pp. 121–127.

Huang, Y. J., et al. The Wind-Shielding and Shading Effects of Trees on Residential Heat and Cooling Requirements. *ASHRAE Transaction.* Atlanta, GA: Society of ASHRAE, 1990.

Landsberg, H. E. "The Climate of Towns." In *Man's Role in Changing the Face of the Earth.* Chicago: University of Chicago Press, 1956, pp. 584–606.

Marsh, William M., and Dozier, Jeff. "The Influence of Urbanization on the Energy Balance." In *Landscape: Introduction to Physical Geography.* Reading, MA: Addison-Wesley, 1981.

Oke, T. R. *Boundary Layer Climates.* New York: Halsted Press, 1978.

Oke, T. R. "Towards a Prescription for the Greater Use of Climatic Principles in Settlement Planning." *Energy and Buildings.* 7, 1984, pp. 1–10.

Olgyay, Victor, *Design with Climate*, 4th ed. Princeton, NJ: Princeton University Press, 1973.

Quayle, R., and Doehring, F. "Heat Stress." *Weatherwise.* 34, June 1981, pp. 120–124.

Taha, N. G. *Site-Specific Heat Island Simulations: Model Development and Application to Microclimate Conditions.* LBL Report No. 24009. Berkeley, CA: Lawrence Berkeley Laboratory, University of California, 1988.

Thurow, C. *Improving Street Climate Through Urban Design.* Chicago: American Planning Association, Planning Advisory Service Report 376, 1983.

Related Websites

BC Climate Action Toolkit. 2009. http://www.toolkit.bc.ca/
   A collaborative effort to provide resources to cities, communities, and individuals about climate change. The sections about transportation and land use planning give ideas of successful policies and designs.

Environmental Protection Agency. Air Pollution. 2009. http://www.epa.gov/ebtpages/airairpourbanairpollution.html
   All the basics on urban air pollution. The quality planning and standards give information on the U.S. Clean Air Act and pollution standards. Subtopics include sources of air pollution and the effects.

International Association for Landscape Ecology. "Climate Change–Landscape Change: Where Is Landscape Ecology in the Climate Change Debate?" 2009. http://www .landscape-ecology.org/climate_change.html

A look at climate change and landscape ecology by the International Association of Landscape Ecology. The site provides further reading material and general information about landscape ecology.

Smart Growth America. "Energy and Climate." http://www.smartgrowthamerica.org/ climate.html.

A coalition dedicated to making American cities more livable. The climate section explains the connection between energy, climate, and smart growth and how to combat it.

U.S. Government, interagency. Air Now. 2007. http://www.airnow.gov/index .cfm?action=static.background

An inter agency and international group delivering information about smog and ozone. Maps provide local and national forecasts and links to publications give further information about these two issues.

GROUND FROST, PERMAFROST, LAND USE, AND ENVIRONMENT

18.1 INTRODUCTION

Practically everywhere in the landscape we can see the direct or indirect influences of ground heat. Farmers remind us that the germination of many seeds depends on ground temperature. They also remind us that the evaporation of soil moisture and irrigation demands are strongly influenced by soil heat. In cities, the urban heat island intensifies with the heating of ground under dry, unshaded streetscapes, driving up temperatures both day and night. Permafrost, which occupies 25 to 30 percent of the land area of this planet, is a form of ground frost that can place severe stress on most modern land uses. In North America, the largest areas of permafrost lie in Canada and Alaska, arguably the continent's last frontier environment, where we struggle to impose modern land uses and infrastructures and bicker over its impacts on the fragile tundra landscape.

Looming crisis   Now another major concern has entered the scene: global warming. The most acute effects of early global warming are apparently being manifested in the arctic and subarctic regions where the largest annual temperature increases on the planet are being recorded. The results are glaring. Sea ice is declining dramatically, and, with it, coastal lands are losing the protection of shore ice and permafrost (Fig. 18.1). Inland, the loss and reduction of permafrost will trigger landscape change, affecting slope stability, streamflow, and ecology, and will damage pipeline, road, and community infrastructure.

Though less serious than permafrost, seasonal ground frost can be an important consideration in planning and engineering facilities in the midlatitudes. In particular, water pipes and sewer lines must be laid below the frost line, and building foundations and roadbeds must be designed to minimize disruption and damage from frost. Despite modern road designs, frost is a major source of pavement damage, a fact vividly reflected in the contrasting condition of paved roads and the associated costs of highway maintenance in northern and southern states.

Our agenda   This chapter first examines the nature of heat in the ground, how it changes with daily and seasonal climatic conditions, and the thermal properties of soil that govern ground heat behavior. Because ground heat originates at the surface, the landscape itself has a pronounced influence on it. Accordingly, most of the remainder of the chapter explores the thermal influences of vegetation, snow cover, land use, and related factors that are central to landscape planning and design interests.

Fig. 18.1 An effect of global warming? Erosion of an Arctic shoreline in Alaska weakened by the loss of permafrost.

18.2 DAILY AND SEASONAL VARIATIONS IN SOIL HEAT

Air-soil system

The thermal condition of the soil is principally a product of the atmospheric climate over it. As weather and climate change, so ground heat changes, particularly in response to solar heating and air temperature fluctuations. Differences in temperature between the soil and the base of the atmosphere set up vertical energy exchanges. When the ground surface is relatively warm and the underlying soil cool, as it would be on a day in early summer, then heat flows into the soil. When the soil is warmer than the atmosphere, heat flows out of the soil into the overlying air. Hardly ever are the soil and atmosphere at the same temperature, because the atmosphere is subject to such rapid, large temperature changes, whereas the soil is slow to change temperature, especially at depth.

Heat input

The Diurnal Cycle. We can examine soil heat flow in various time frames, beginning with a *diurnal period* or 24-hour day/night cycle. On a summer day, for example, the ground surface may heat to a temperature of more than 100°F (38°C), while just a few feet below the soil temperature may be only 70°F (21°C). The heat flow under these conditions is downward, following the temperature gradient. But because soil is not a good thermal conductor, the total penetration is only a few feet into the ground before the sun sets and the supply of surface heat is lost. This is verified by the graph line in Figure 18.2 representing the 50-centimeter (1.5-foot) soil depth. Notice that below this depth, at 100 centimeters (3 feet), there is no diurnal temperature change.

Heat output

Once the ground has taken on a load of heat, and the night air over it has cooled, the heat flow reverses, and the ground conducts heat upward into the atmosphere. This continues throughout the night until all or most of the previous day's load of ground heat is released. This variation in soil heat and daily directions of flow goes on to a lesser or greater extent on most days of the year at all terrestrial locations experiencing light and dark phases. Depending on the intensity of surface heating and the ability of the soil to conduct heat, there is a maximum depth, called **diurnal damping depth**, that marks the limit of daily temperature change for every soil. Many factors in the landscape affect soil heating and the diurnal damping depth, in turn affecting ground level climate, including plant cover, topsoil, the intensity and duration of solar radiation, albedo, snow cover, soil moisture, and land use.

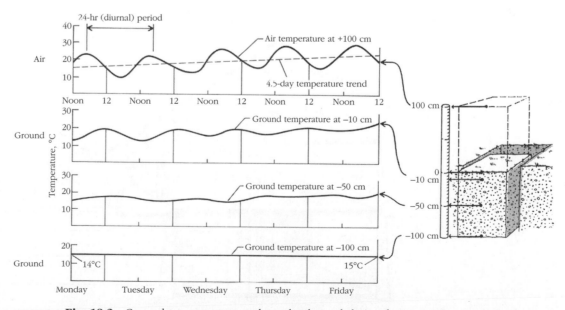

Fig. 18.2 Ground temperatures at three depths and their relation to air temperature over a 4.5-day period. The diurnal damping depth appears to lie close to 50 centimeters; beyond that depth, the daily variation in surface temperature is not apparent.

Seasonal heat flows

The Seasonal Cycle. The ground temperature also varies with the seasons. If we examine the average surface temperatures for summer, fall, winter, and spring, it is apparent that from winter to summer the soil should be heating up and that from summer to winter it should be cooling down. The depth of the seasonal change is much greater than that of the diurnal change; so the *seasonal damping depth* is much greater, on the order of 10 feet in the midlatitudes. However, owing to the time it takes heat to reach this depth, the soil does not reach its maximum temperature until a month or more after the surface has reached its maximum. Thus ground heat and surface heat are always out of phase with each other.

Lag times

The heat seasons in the soil lag behind the heat seasons on the surface and in the atmosphere. Therefore, the soil is always cooler in summer and warmer in winter than the surface. These facts have some important implications for building design and energy conservation. In regions with warm summers and cold winters, such as southern Canada and the U.S. Midwest, subterranean structures have a distinct thermal advantage over above-ground structures. Basements, for example, are cooler in summer and less expensive to heat in winter than a comparable structure above ground (Fig. 18.3).

18.3 CONTROLS ON SOIL HEAT AND GROUND FROST

The rate at which heat flows into and out of the soil depends on two main factors:

1. The *thermal gradient* (or temperature differential) between the ground surface and the soil at some depth.
2. The soil's *thermal conductivity*, which is determined by its composition and water content.

Thermal conductivity

In the first column of Table 18.1, the thermal conductivities are given for nine different earth materials. Notice that sand and clay conduct heat better than organic material and that conductivity increases with soil moisture content. Organic matter is a very poor heat conductor, about 10 times slower than sand and clay, whether wet or dry. As a

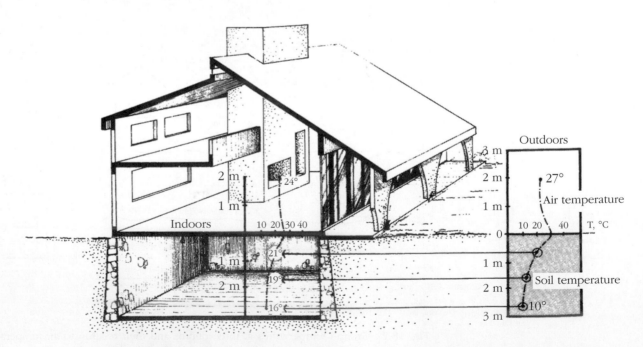

Fig. 18.3 Summer ground temperatures at depths of 1, 2, and 3 meters, compared to those at the same depths in the basement of a house.

Table 18.1 Thermal Properties of Some Common Earth Materials

Substance	Thermal Conductivity[a]	Volumetric Heat Capacity[b]
Air		
Still (at 10°C)	0.025	0.0012
Turbulent	3,500–35,000	0.0012
Water		
Still (at 4°C)	0.60	4.18
Stirred	350.00 (approx.)	4.18
Ice (at –10°C)	2.24	1.93
Snow (fresh)	0.08	0.21
Sand (quartz)		
Dry	0.25	0.9
15% moisture	2.0	1.7
40% moisture	2.4	2.7
Clay (nonorganic)		
Dry	0.25	1.1
15% moisture	1.3	1.6
40% moisture	1.8	3.0
Organic soil		
Dry	0.02	0.2
15% moisture	0.04	0.5
0% moisture	0.21	2.1
Asphalt	0.8–1.1	1.5
Concrete	0.9–1.3	1.6

[a] Heat flux through a column 1 m² in W/m when the temperature gradient is 1°K/m.
[b] Millions of joules needed to raise 1 m³ of a substance 1°K.

result, organic matter often serves as an effective thermal insulator, which helps explain why permafrost is particularly prominent and lasting in areas of muck and peat soils.

Thermal diffusivity   The rate at which a given temperature, such as the freeze line (0°C), actually moves into the soil is somewhat different than is suggested by the conductivity value. This rate, called *thermal diffusivity*, is a product of the volumetric heat capacity (given in the third column of Table 18.1) and the conductivity of the soil. Diffusivity is highest at moisture contents between 8 and 20 percent. Thus, in saturated soils, frost penetration is usually not as great as it is in damp soils, owing to the higher heat capacity of soggy soil.

Frost and Landscape. The landscape, particularly vegetation, snow cover, and land use, also plays a major part in ground frost penetration. Among other things, snow cover and vegetation dampen surface wind velocities, thereby tending to reduce soil heat loss. Land use, on the other hand, has a variable effect. For instance, a building *Land use/cover influence*   reduces heat flow from the soil, whereas a barren highway usually accelerates it by facilitating wind chill and soil heat outflow (Fig. 18.4a). The combined effects of land use, vegetation, and snow cover can be dramatic. Where forest cover has been cleared for agriculture or urban development and snow has been removed by plowing and/or wind, soil heat loss and frost penetration are maximized. The graphs in Fig. 18.4b illustrate the influence of snow cover for a grass-covered site in Minnesota. The key factor is the wind chill effect of fast moving cold air directly on or very close to the soil surface.

For most parts of North America, the depth of frost penetration is not well documented and is usually estimated from climatic records. The standard map of frost depth in the coterminous United States is based on representative winter tempera-*Regional frost data*   tures and does not take into account factors such as snow cover, vegetation, or buildings (Fig. 18.5). As a result, field measurements of frost penetration will often show appreciable deviation from this map for any winter. A snow-covered swamp in Maine may rarely if ever receive ground frost, whereas a nearby airfield, windswept and barren, may receive 6 feet or more penetration. Local building codes setting frost depths for foundations and piping are usually based on gross approximations.

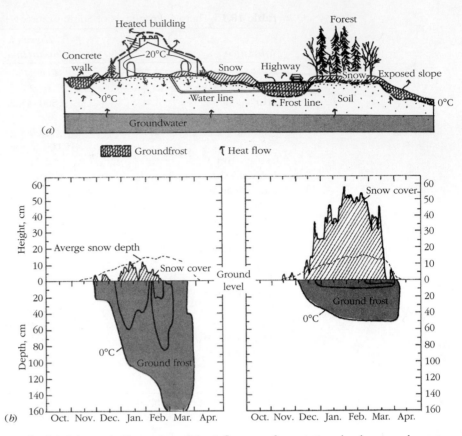

Fig. 18.4 (*a*) Schematic illustration of the influence of vegetation, land use, and snow cover on ground frost distribution and depth. (*b*) Measured frost penetration related to snow cover. In the first graph, snow cover was light and frost penetration great; the second illustrates the opposite condition.

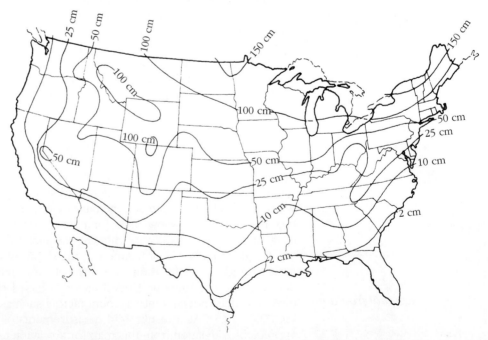

Fig. 18.5 Expected ground frost penetration by the end of winter. Departures from these values may be considerable, depending on the year and local snow, soil, and land use conditions.

Snow Cover and Wind Exposure. When snow cover *is* taken into account, the following formula and graph can be used to estimate the depth of frost penetration in the northern United States and southern Canada. The formula combines snow depth and heating degree days to give degrees temperature per inch (or centimeter), denoted as *T*:

Snow cover effect

$$T = \frac{\sum HDD_{\text{OCT-MAR}}}{\sum (S \bullet n)_{\text{NOV-MAR}}}$$

where

T = degrees temperature per inch (or centimeter) (To find the frost depth, this figure is read into the vertical scale of the graph.)

HDD = heating degree days, summed October through March

S = average monthly snow depth in inches

n = number of days with snow cover of 0.5 inch or more ($S \bullet n$ is computed month by month and summed November through March.)

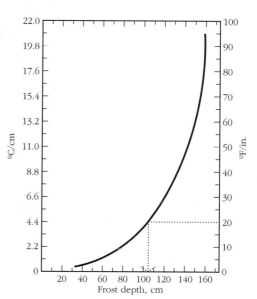

Taking the result of a computation using this formula, we find that the maximum depth of the freezing (0°C or 32°F) isotherm for winter is read from the graph in the manner illustrated for 20°F per inch (4.4°C/cm). For this example, the depth of the freezing isotherm is about 105 cm (41 inches).

In addition, the orientation and exposure of the ground can be critical. On barren slopes a southward exposure may make an appreciable difference in radiation receipts (see Fig. 16.8*a* in Chapter 16), whereas north-facing slopes may not only receive less radiation but also lose ground heat more rapidly because of exposure to northerly *Slope wind chill* winds (Fig. 18.6). A related consideration on slopes is the upslope increase in wind velocity, which induces accelerated rates of soil heat loss and frost penetration at higher elevations. The crests of high windy slopes may develop the deepest frost layers, especially if they are north facing. This can be a critical consideration along coastal slopes (also see Fig. 15.8).

Mapping Ground Frost Susceptibility. On balance, then, we must take many factors into account when we attempt to forecast the pattern of ground temperature fluctuations and frost penetration, especially in areas of varied terrain. Unfortunately, mathematical models that integrate many variables are difficult to manipulate, require data from the field to be set up, and do not lend themselves to mapping programs. In environmental inventories for impact studies, master planning, or constraint studies, we must instead often turn to simpler and less expensive methods in identifying areas susceptible to heavy frost penetration.

One of these is the Geographic Information System (GIS) map overlay method, in *A GIS application* which individual criteria, such as those given in Table 18.2 (based on the relative influences of topography, vegetation, snow cover, exposure, land use, and soils) are mapped and then superimposed on one another. The resultant combinations are designated high,

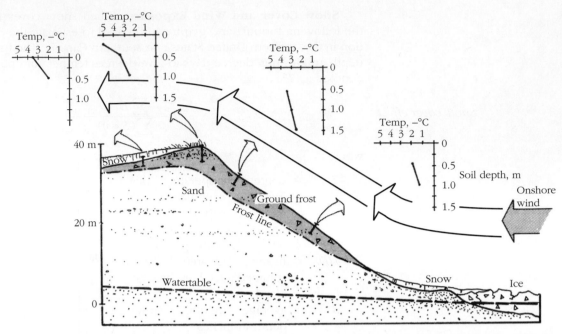

Fig. 18.6 Frost penetration on a sandy slope exposed to northerly wind blowing off Lake Superior. Wind velocity increases upslope, and heat loss increases, with the highest wind velocity producing the greatest frost penetration in the upper slope (also see Fig. 15.8).

medium, or low susceptibility. Each of the maps is usually broken down into categories that are numerically coded prior to overlaying; for example, soils might be classed as moist organic (1), moist mineral (2), and well drained mineral (3), three being the most susceptible to seasonal frost penetration. The results of this method do not indicate how deep frost penetration should be; rather, they indicate only the relative penetration and may be used to isolate areas where more detailed analysis can be carried out (Table 18.2).

Table 18.2 Ground Frost Susceptibility

	Low	*Medium*	*High*
Soil type	Organic	Wet mineral (clay, loams, sand)	Well drained mineral (loams, sand)
Soil moisture	Saturated	Moist (near field capacity)	Damp (less than field capacity)
Vegetation	Heavy forest	Grass, low shrubs	Barren
Wind exposure	Low exposure to cold, fast wind (usually south- and southwest-facing slopes)	Intermediate exposure (such as east-, northeast-, and west-facing slopes)	High exposure to cold, fast wind (usually north- and northwest-facing slopes)
Snow cover	>50 cm (November–March)	10–50 cm (November–March)	<10 cm (intermittent cover throughout winter)
Solar exposure	South-facing slope >20%	Flat ground or locally irregular terrain	North-facing, shaded
Land use/cover	Parks, wetlands, stream corridors, woodlots	Pasture, planted fields, golf courses	Airfields, highways, parking lots, barren fields

18.4 PERMAFROST

Definition

From the southern part of the United States, the depth of ground frost penetration increases northward to a point in Canada where the inflow of summer heat is inadequate to melt completely the deeper winter frost. The layer of frozen ground that remains is **permafrost**. Under persistently cold ground conditions, permafrost can penetrate hundreds of feet downward until its advance is offset by heat flow from the earth's crust.

Patches and pockets

Discontinuous Permafrost. In varied terrain, permafrost first appears in isolated pockets on north-facing slopes where solar heating is weakest or under thick insulating layers of organic soil. In North America, such pockets of permafrost are reported as far south as 50° north latitude in the Canadian Shield and even farther south at high elevations in the Rocky Mountains, but their occurrence is very sporadic and usually hidden under several meters of soil material. These pockets mark the southern fringe of a broad permafrost region called the *discontinuous zone* that stretches across North America from Newfoundland in eastern Canada to the Bering Strait in western Alaska (Fig.18.7).

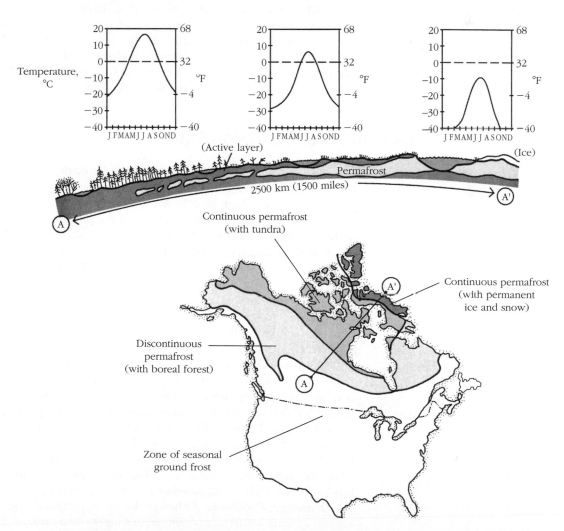

Fig. 18.7 The general distribution of permafrost in North America and the representative climatic condition and landscape type associated with each zone.

90–100 percent coverage

Continuous Permafrost. Northward, the patches of permafrost grow much broader and thicker and are overlain by a layer of soil called the **active layer**, which freezes and thaws seasonally. Near the Arctic Circle, the discontinuous zone gives way to the *continuous zone* where permafrost extends uninterrupted over vast areas of land and in the extreme reaches depths of a 1000 feet or more. The thickness of the active layer also changes northward with shorter and cooler summers. In the discontinuous zone it is generally 6 to 12 feet thick, but poleward declines to a very thin layer or disappears altogether in the upper reaches of the continuous zone (Fig. 18.7). Although permafrost is decidedly a terrestrial phenomenon, it is also known to extend under the shallow waters of the Arctic Ocean, an apparent relic condition of the Ice Age when sea levels were significantly lower.

Thaw/freeze-up

Seasonal Change. Nowhere is the seasonal flux of soil heat more apparent than in permafrost regions. In summer, the active layer develops with the penetration of heat from the surface (Fig. 18.8*a*). As the ground thaws, the soil becomes saturated with meltwater, shallow ponds develop (called *thaw-lakes*), and local runoff systems are activated. But with the onset of cold weather in fall, the heat flow reverses in the upper active layer, surface water freezes, local runoff ceases, and frost begins to penetrate the soil from the surface. Because the lower active layer is still thawed at this time, heat flows both upward and downward from this relatively warm zone sandwiched within the frozen ground (Fig. 18.8*b*).

Ground temperature

The temperature profile assumes a spoon shape at this time. But in the ensuing months of winter, when the active layer freezes out completely and surface temperatures fall far below 0°C, the thermal gradient is fully reversed from that of summer, and the heat flow is upward. Paradoxically, the permafrost layer is the primary source of heat for the landscape during winter (Fig. 18.8*c*).

18.5 LAND USE AND FROZEN GROUND

Modern land uses have proven to be problematic in most permafrost regions. In the past half century many military installations, pipelines, railroads, highways, and communities have been built in Alaska, northern Canada, and Russia, and these have provided ample evidence of the nature of the problem. Almost always the problem is tied to changes in the energy balance at ground level and below.

Thermal alteration

Soil Heat and Landscape Change. The energy flow (both heat and solar radiation) at ground level is altered, often dramatically, first with the clearing of vegetation and surface grading (Fig. 18.9*a*). When a foreign material such as concrete or asphalt is placed on the ground, the energy balance of the active layer is altered, making it grow colder or warmer. In soils where ice comprises a large part of the soil bulk, thawing can reduce the permafrost (because of a volume reduction of about 9 percent as water changes from ice to liquid) and cause the ground surface to subside.

Impacts

This process leaves sinkholelike depressions in the surface known as **thermokarst**. As the depressions form, they collect water, thus causing further alteration of vegetation and soils that in turn may advance thawing and thermokarst development. When heated buildings, utility lines, or oil lines are set on the ground without adequate insulation to check the flow of heat into the active layer, ground subsidence can be more dramatic and very damaging to both the structures and the environment (Fig. 18.9*b*) In Russia, it is speculated that pipeline oil leaks from permafrost-related damage and poor line maintenance are responsible for spilling millions of barrels of oil into the tundra, boreal forest, and Siberian streams every year.

Drainage Problems. Other problems experienced by land use in permafrost regions include inadequate drainage in summer and unstable surface materials subject to settling and shifting. Along the Trans-Alaska Pipeline, for example, service roads were built on elevated gravel beds to minimize their thermal influence on the underlying

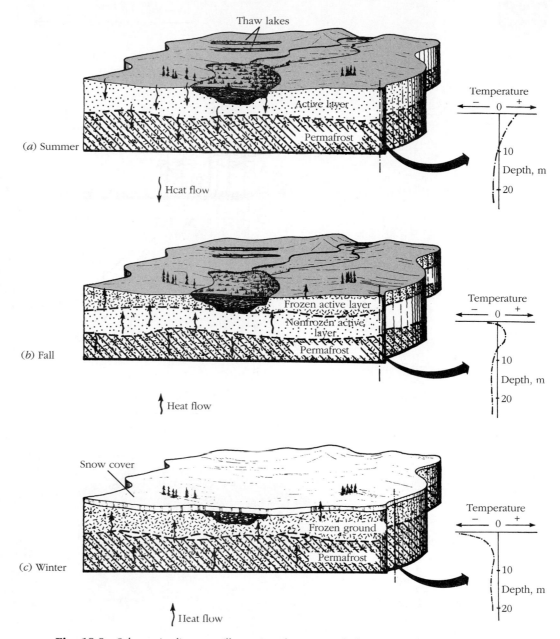

Fig. 18.8 Schematic diagrams illustrating the seasonal changes in the freezing and thawing of the upper permafrost; (*a*) summer; (*b*) fall; (*c*) winter. Notice the changes in the direction of heat flow.

Ponding and runoff

permafrost, but the roadbeds often interrupted summer drainage, raising water levels in ponds and wetlands that may flood roadways. Drainage problems, coupled with ground subsidence, water supply, and waste disposal problems, have placed severe limitations on urban development in permafrost regions. In North America, the northern limit of urban development roughly coincides with the 0°C mean annual temperature line, which falls a few hundred miles south of the permafrost zone (Fig. 18.10).

Waterlines and Roads. Frozen ground is also a problem outside permafrost areas. In northern Europe, southern Canada, and the northern United States, as well as in many other areas of the world, ground frost causes highway buckling, damage

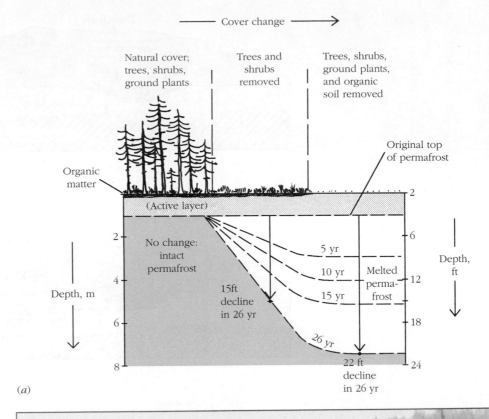

Fig. 18.9 (*a*) The measured change in permafrost depth in response to land clearing, illustrating the profound influence that the vegetation and topsoil have on ground heat flow. (*b*) Subsiding caused by melting of the permafrost from the heat generated by a furnace in this house.

Ground frost problems to building foundations, and water pipe freezing and breakage. In the United States and Canada, building codes generally recommend that foundations be set below frost depth (usually given as 4 to 5 feet in the northern tier of states and in southern Canada) to minimize frost heaving and damage.

But despite design codes and special engineering, water mains frequently freeze in extremely cold years in Minnesota, Wisconsin, Michigan, and Ontario, costing cities millions of dollars in repairs. The reason for the freeze-up is that most water lines are

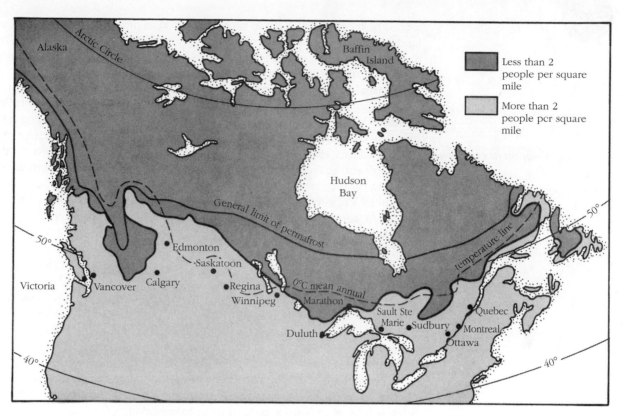

Fig. 18.10 The southern limit of permafrost, the 0°C mean annual temperature line, and the northern fringe of urban development in Canada.

located under streets which are plowed clean of the thermally protective cover of snow. In highway construction, frost heaving caused by the growth of ice lenses in the road-bed, called *ice segregation*, is a serious problem. To prevent it, gravel-based roadbeds are required because gravel does not transmit capillary water upward from the underlying soil fast enough to allow ice lenses to form under the cold concrete or asphalt.

18.6 PLANNING AND DESIGN APPLICATIONS

With the exception of engineering design standards, such as those previously mentioned, little formal attention is given to ground frost in community planning outside permafrost regions. Within permafrost regions the picture is quite the opposite, though permafrost is by no means universally recognized and addressed in planning methodology and practice, even in the most hostile settings.

Action and time lags

The Impact Dilemma. Although modern engineering makes it possible to construct and successfully operate many types of infrastructure systems in permafrost environments, the design and management methods and techniques needed to mitigate the environmental impacts of these systems are not yet fully developed. At least part of the problem is inadequate knowledge of the full range of impacts because of the sizable time lags, often decades, between an action and/or disturbance and the resultant environmental impacts. In other words, some impacts from projects dating from the 1970s, such as the Trans-Alaska Pipeline, are still unfolding.

This problem will likely grow more complex in the decades ahead as development advances and the effects of global warming unfold. Infrastructure designs with a record of success may have to be re-evaluated as the stability thresholds in permafrost landscapes change with warming. Moreover, warming will make it even more difficult to measure the full effects of land use–driven change, and lag times between cause and effect may be even longer.

Planning Applications. Permafrost is a serious land use planning consideration in and around communities like Fairbanks, Alaska, and Yellowknife, Northwest Territory in Canada. Development proposals involving road building and residential infrastructure, for example, are reviewed with an eye to potential permafrost problems. The first level of evaluation involves checking the location of the proposed development against the distribution of soils known to have permafrost problems. Drawing on the results of permafrost research in the Fairbanks area, the U.S. Natural Resources Conservation Service (NRCS) has classified the soils of the Goldstream, Saulich, Ester, and Lemeta series as those with greatest susceptibility to permafrost (Fig. 18.11).

Development concerns

For projects that involve these soils, the review may be taken to a second level of evaluation, which includes an examination of the types of activities and facilities actually proposed. In some cases the project may be compatible because it is judged to be neither prone to damage from the environment nor itself a significant threat to the environment. In other cases, modifications may be recommended, such as changes in building sites or the use of special engineering technology for footings and utility lines (Fig. 18.12). In still other cases, the project may be viewed as incompatible with the environment and not recommended for approval.

Review process

The recommendation is then passed on to the staff of the planning agency in charge, where it is combined with recommendations from other technical fields and interest groups to form a general recommendation on the proposal. This statement is then studied and discussed by a decision-making body such as a planning commission, county board, or city council. A vote is taken, and the proposal is approved, denied, approved with specified reservations that ask the applicant to agree to certain changes, or sent back to the applicant or the reviewer for further work and documentation.

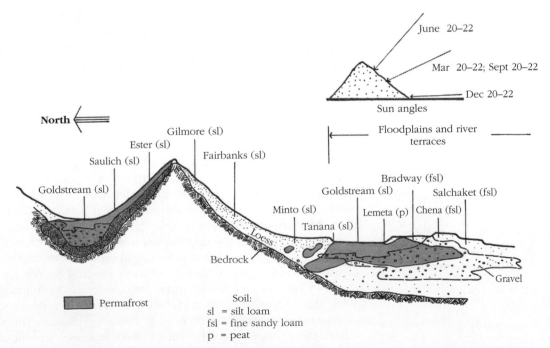

Fig. 18.11 Schematic diagram showing the locations of soils and their susceptibility to permafrost in the Fairbanks, Alaska, region.

Fig. 18.12 Design modifications in residential development in response to permafrost (Norman Well, Northwest Territory, Canada). Homes, water lines, and sewage lines are built above ground to minimize thermal disturbance of the permafrost.

One of the questions increasingly addressed is how the proposed project might be affected by the component of permafrost decay related to climatic warming.

18.7 CLIMATE CHANGE IN A PERMAFROST LANDSCAPE

Warming trend For reasons not fully understood by climatologists, the arctic and subarctic regions of North America and Eurasia are feeling the strongest effects of early global warming. Mean annual temperature change maps show two large zones of warming in the 1.5–3.0°C range, one over Siberia and one over northcentral Canada and Alaska. Environment Canada, the federal agency responsible for national weather and climate monitoring, reports that in the twentieth century the mean annual temperature of the MacKenzie Region in northwestern Canada increased by 1.7°C. By 1990 the effects of longer and warmer summers were being documented on several fronts, most notably in declines in sea ice coverage and thickness and in the retreat of permafrost in coastal areas and interior lowlands.

Forecasts Barring a major change in global politics, economic development, and population growth, global warming is expected to continue in the twenty-first century. With a doubling of global carbon dioxide, the mean annual air temperature over northern North America will rise by 4–5°C over twentieth-century levels. The result will be a partial or complete depletion of permafrost over large areas. In the discontinuous permafrost zone, where ground temperature is within 1–2°C of melting, permafrost will disappear altogether.

Expected impacts **Effects on Landscape and Infrastructure.** Because permafrost is a major player in arctic and subarctic landscapes, there is little doubt that its decline will lead some to dramatic geographic changes. Where they can be expected depends on the magnitude of warming coupled with topographic, soil, hydrologic, and vegetative conditions. The map in Figure 18.13 identifies the areas in Canada where low, moderate, and high landscape impacts are expected according to Environment Canada, and Figure 18.14 offers an interpretation of what some of those impacts might be.

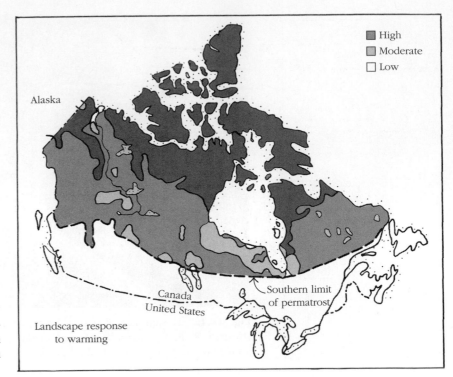

Fig. 18.13 The regional distribution of land-scape response to climate warming in northern Canada. The heaviest impacts are expected in the zone of continuous permafrost.

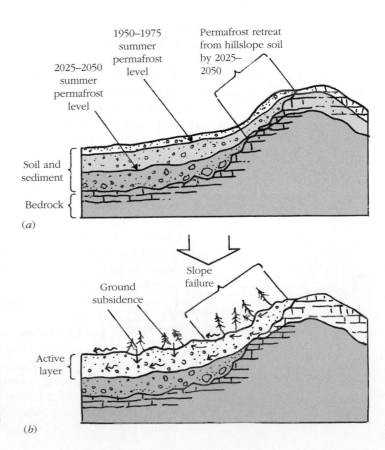

Fig. 18.14 Some of the landscape impacts expected from a major decline in permafrost: (*a*) permafrost retreat; (*b*) possible results. Notice that the hillslope is left without a permafrost foundation and thus fails, and the ground settles differentially, affecting drainage and vegetation.

As permafrost recedes from hillslopes, the ground will lose stability and likely fail due to slumping, sliding, and erosion. The ground will also undergo differential settlement, causing changes in surface drainage and vegetation patterns. Where facilities, such as pipelines, roads, and building foundations, have been designed to withstand the permafrost conditions of 1970, 1980, or 1990, the decay of the permafrost may undermine them and render them obsolete or environmentally dangerous. The following case study describes one such facility.

18.8 CASE STUDY

Permafrost and the Trans-Alaska Pipeline

Peter J. Williams

Remarkable as it now seems, the earliest proposals in the Alaska Oil Pipeline Project were to construct the pipeline in a conventional manner, burying it in the ground for virtually the entire distance of the 800 miles from Prudhoe Bay to Valdez. What was not understood was that the warm pipeline might thaw the underlying permafrost, melting its interstitial ice, leading to settlement or subsidence. At one planning stage, serious consideration was given to using a chilled pipeline, that is, cooling the oil so that the pipe could be buried without thawing the permafrost. This proposal was rejected because of the effect of low temperatures on the oil; it would be too viscous to flow satisfactorily. The interaction of a chilled pipeline with the soil, which was to present problems for subsequently proposed gas pipelines, was apparently not foreseen.

Probably three-quarters of the proposed Trans-Alaska Pipeline route overlay permafrost, and at least half of the permafrost was estimated to contain ice, which on melting would cause settlement. The effect of the warm pipeline was such that up to 30 feet of soil could be expected to thaw in the first year. The amount would vary, of course, depending on the preexisting ground temperatures and the type of soil.

Near Prudhoe Bay the permafrost is some 2000 feet thick. Southward it becomes generally thinner, although the thickness is highly variable. Thawing around a buried, warm pipeline would progress downward, although at a decreasing rate, for many years. The degree of consequent settlement, of subsidence, would depend on the amount of "excess" ice in the thawed layer, but quite often there would be several meters of displacement. Obviously, such effects, left unchecked, would cause great disruption of the pipeline.

As soil surveys proceeded much more ground ice was discovered than was initially predicted. This information, coupled with findings as to the amount of thaw that would occur, led gradually to the decision to build more and more of the pipeline above ground, elevated on pile supports, rather than buried in the ground as first envisaged (Fig. 18.A). The air passing beneath an elevated line would dissipate most of the heat from the pipe and greatly reduce the thawing of the permafrost. Furthermore, the pile supports, or vertical support members (VSMs), could be designed to permit lateral movements of the pipe as it expanded or contracted with temperature changes.

Raising the pipe above ground on the VSMs did not ensure that there would be no thawing of the ground. The disturbance inflicted on the ground during the course of construction was sufficient to initiate temperature changes in the soil that could result in thawing to a significant depth, in all except the coldest, most northern areas. Climatic change, too, might in places initiate a continuing thawing.

Fig. 18.A Air view of service road and Alaska Pipeline under construction, 1975.

Fig. 18.B Where it crosses areas of permafrost, the pipeline is elevated on vertical support members with thermal devices to resist heat flow and keep footings frozen in the ground.

Ultimately, the solution of the problem lay in the so-called thermal VSMs (Fig. 18.B), which are equipped with devices known as heat pipes. These are sealed 2-inch-diameter tubes within the VSMs. They extend below the surface and contain anhydrous ammonia refrigerant. In the winter months this evaporates from the *lower* end of the tube and condenses at the top, where there are metallic heat exchanger fins. The evaporation process occurs because, during the winter months, the ground is *warmer* than the air outside. The evaporation process itself cools the lower end of the pipe and the surrounding ground, which is the point of the device. So, by cooling the permafrost in winter, its temperature is sufficiently lowered to prevent thaw during the summer. As the mean ground temperature falls, the heat pipes thus prevent the warming that would otherwise occur following disturbance of the ground surface. About 380 miles of pipe were built above ground, and about 80 percent of this length relies on thermal VSMs.

Peter J. Williams is professor of geography, director of the Geotechnical Science Laboratories at Carleton University, Ottawa, and a specialist in problems of development in cold environments. [Source: *Pipelines and Permafrost: Physical Geography and Development in the Circumpolar North* (Longman, 1979). Used by permission of the author.]

18.9 SELECTED REFERENCES FOR FURTHER READING

Baker, Donald G. "Snow Cover and Winter Soil Temperatures at St. Paul, Minnesota." *Water Resources Research Center Bulletin.* University of Minnesota, 37, 1971.

Brown, R. J. E. "Influence of Climate and Terrain on Ground Temperatures in the Continuous Permafrost Zone of Manitoba and Keewatin District, Canada." *Third Conference of Permafrost Proceedings.*, Edmonton, vol. 1, 1978, pp. 16–21.

Bush, E., et al. *Climate Change Impacts on Permafrost Engineering Design.* Ottawa: Environment Canada, 1998.

Brownson, J. M. J. *In Cold Margins: Sustainable Development in Northern Bioregions.* Missoula, MT: Northern Rim Press, 1995.

Ferrians, O. J., et al. "Permafrost and Related Engineering Problems in Alaska." *U.S. Geological Survey Professional Paper,* 678, 1969.

French, H. M. *The Periglacial Environment.* New York: Longman, 1976.

Péwé, Troy L. "Effect of Permafrost on Cultivated Fields, Fairbanks Area, Alaska." In *Mineral Resources of Alaska, Geological Survey Bulletin.* 989, 1951–1953, pp. 315–351.

Smith, M. W. "Microclimatic Influences on Ground Temperatures and Permafrost Distribution in the Mackenzie Delta, Northwest Territories." *Canadian Journal of Earth Science.* 122, 8, 1975, pp. 1421–1438.

U.S. Natural Resources Conservation Service. *Soil Survey: Fairbanks Area, Alaska.* Washington, DC: U.S. Government Printing Office, 1963.

Walker, D. A., et al. "Cumulative Impacts of Oil Fields on Northern Alaskan Landscapes." *Science.* 238, 1987, pp. 757–760.

Washburn, A. L. *Periglacial Processes and Environments.* New York: St. Martin's Press, 1973.

Williams, Peter J. *Pipelines and Permafrost: Physical Geography and Development in the Circumpolar North.* New York: Longman, 1979.

Related Websites

International Permafrost Association. 2009. http://ipa.arcticportal.org/
> *A group educating and bringing together groups working on permafrost issues. "What Is Permafrost" gives general information about permafrost and explains how global climate change is affecting it. The site also includes several publications for further reading.*

Natural Resources Canada, Geological Survey of Canada. "Permafrost." 2007. http://cgc.rncan.gc.ca/permafrost/index_e.php
> *In addition to general permafrost information, the site informs and shows pictures of development issues in permafrost zones. It provides a case study of a northern Canadian community-scale project.*

National Snow and Ice Data Center. "All About Frozen Ground." N.d. http://nsidc.org/frozenground/index.html
> *Insight into how different types of frozen formations affect land. The "People and Frozen Ground" section covers building, transportation, and natural resource use in northern climates.*

National Oceanic and Atmospheric Administration (NOAA). "A Near-Realtime Arctic Change Indicator Website." http://www.arctic.noaa.gov/detect/index.shtml
> *A division of NOAA, this governmental group researches Arctic climate change and its effect on ecosystems and development. The "Permafrost" section shows great pictures of permafrost and thermal karsts.*

19

VEGETATION, LAND USE, AND ENVIRONMENTAL ASSESSMENT

19.1 INTRODUCTION

Perhaps no component of the landscape is more directly related to land use and environmental change than vegetation. Besides being the most visible part of most landscapes, it is also a sensitive gauge of conditions and trends in parts of the landscape that are otherwise not apparent without the aid of detailed observation and measurement. The loss of vigor in tree species near highways, for example, may be an indication of impaired drainage or heavy air pollution, thereby drawing attention to environmental impact problems that might otherwise be overlooked. In agricultural regions, changes in shrub and tree species in swales and floodplains may be a response to heavy sedimentation, pointing up the need for erosion control. In urban regions, the invasion of floodplains by alien plant species in response to decreased peak streamflows is pointing up the need to change our approach to stormwater and flood management.

Vegetation also plays a functional role in the landscape because it is an important control on runoff, soil erosion, slope stability, microclimate, and noise. In site design, plants are used not only for environmental control, but also to improve aesthetics, to frame spaces, to influence pedestrian behavior, and to control boundaries. Whereas other landscaping methods and materials can be used for the same purposes, few are as versatile and inexpensive as vegetation. Not surprisingly, much of the work of the landscape architect involves designing planting plans.

Although they are not widely recognized, vegetation has some negative aspects. The most serious are probably the noxious plants that inhabit disturbed areas in and among urban, suburban, and agricultural lands. These are mainly weed plants, such as poison ivy and ragweed, that are poisonous (usually in the form of allergic reactions) to many people. In addition, many introduced plants displace native species and degrade ecosystems. Two salient examples are kudzu, an Asian vine, that has smothered large areas of landscape in the American South, and purple loose-strife, a tall European herb, that is displacing rushes, reeds, and other native plants from wetlands throughout eastern United States, southeastern Canada, the Midwest, and the Pacific Northwest.

19.2 DESCRIPTION AND CLASSIFICATION OF VEGETATION

The need   Most planning projects involving the alteration of the environment call for a description of vegetation types and distributions. This may be combined with land use inventories to produce land cover maps or treated as an independent task that includes wetland mapping, species inventories, and related activities. In either case, the objective is to document the distribution and makeup of the vegetative cover, and this task requires the use of an appropriate plant or vegetation classification scheme. Detailed descriptions of vegetation are virtually a universal requirement of environmental impact statements, and the conscientious investigator typically provides inventories and descriptions that draw on several classification schemes.

Floristic   **Classification Systems.** Three types of classification systems are used for plants and vegetation. The *floristic* (or Linnaean) system is the most widely used. It classifies individual plants according to species, genera, families, and so on, using the universally recognized system of botanical names. Plants are normally referenced by their species and genus names given in Latin or in latinized forms such as *Pinus resinosa* (red pine).

Form and structure   *Form and structure* (or physiognomic) schemes classify vegetation or large assemblages of plants according to overall form, with special attention to dominant plants (the largest and/or most abundant). The traditional classification for vegetation that begins with forest, savanna, grassland, and desert is a form and structure scheme.

Ecological *Ecological* schemes classify plants according to their habitat or some critical parameter of the environment such as soil moisture or seasonal air temperatures. Various ecological-type systems have been devised, but the only one commonly used is that associated with ecosystems. Large-scale or macro ecosystems are named for their habitat types or, more broadly, for their physiographic associations, for example, wetlands, floodplains, sand dunes, and stream channels. Owing to our interest in habitat conservation planning as a part of species protection programs, the need to address plants, vegetation, and ecosystems in terms of habitat relations is stronger today than ever.

The problem **Selecting a System.** Unless dictated by environmental regulations, the type of description and classification used in a project should be governed by the nature of the problem and the character of the landscape. Increasingly, we are faced with settings where several waves of previous land uses have used, reused, and rearranged the landscape, leaving a hodgepodge of vegetation. This typically includes remnant patches of managed forest such as woodlots, belts of planted vegetation such as hedgerows and windbreaks, patches of alien plants, and artifacts of primeval ecosystems such as isolated wetland pockets (see Fig. 19.3*a*). These features need to be examined not only as biological phenomena made up of species, populations, communities, and so on, but as part of the site's resource base having annual productivity in organic matter; utility in landscape design; and value in managing runoff, soil erosion, and other environmental systems. Therefore, some mix of floristic, form and structure, and ecological schemes is called for in most planning problems.

Five-level system The scheme given in Table 19.1 draws on all three classification systems. It is organized into five levels, each addressing a different classification element. Level I is based on overall structure, level II on dominant plant types, level III on plant size and density, level IV on site and habitat, and level V on significant species. Level V is included to provide for rare, endangered, protected, and highly valued species—a standard requirement today for virtually any planning project—as well as plants of value in landscape management or design for a proposed or existing land use.

19.3 TRENDS IN VEGETATION CHANGE

The first and most dramatic event in the history of vegetation change was the nineteenth-century clearing of the North American interior. Nowhere in the world has such a large area been cleared and settled as rapidly as southern Canada and the eastern half of the United States. Between 1800 and 1900, more than 500,000 square miles of virgin forest were cut in the United States alone (Fig. 19.1).

Farms and lumber **Deforestation.** Most forests were cut to make way for farms and to provide lumber for new and expanding towns and cities. Although much of the land cleared for lumber reverted back to woodland, much of that cleared for farms remained in agriculture, from which a landscape best described as agricultural parkland emerged. In this landscape, only in nonarable sites such as swamps, steep slopes, and deep stream valleys did the original vegetation escape destruction. But even these sites have been dramatically reduced and altered.

In the United States (less Alaska), more than 50,000 square miles of wetland have been destroyed since settlement, and since 1950, agriculture has accounted for 70 to 90 percent of the losses. In addition, the remaining patches of vegetation are often quite different floristically from the original plant cover because they may be too small to support viable populations and because they have been subject to disturbances such as wood harvesting, fires, sedimentation, and flooding that have eliminated certain tree species and ground plants and encouraged alien species.

Decline of Farmland. A second trend has been the decline of the agricultural landscape, first in the nineteenth century with the failure of farms in marginal areas

Table 19.1 Multilevel Vegetation Classification

Level I (vegetative structure)		Level II (dominant plant types)	Level III (size and density)	Level IV (site and habitat or associated use)	Level V (special plant species)
Forest (trees with average height greater than 15 ft with at least 60% canopy cover)		E.g., oak, hickory, willow, cottonwood, elm, basswood, maple, beech, ash	Tree size (diameter at breast height); density (number of average stems per acre)	E.g., upland (i.e., well drained terrain), floodplain, slope face, woodlot, greenbelt, parkland, residential land	Rare and endangered species; often ground plants associated with certain forest types
Woodland (trees with average height greater than 15 ft with 20–60% canopy cover)		E.g., pine, spruce, balsam fir, hemlock, douglas fir, cedar	Size range (difference between largest and smallest stems)	E.g., upland (i.e., well drained terrain), floodplain, slope face, woodlot, greenbelt, parkland, residential land	Rare and endangered species; often ground plants associated with certain forest types
Orchard or plantation (same as woodland or forest but with regular spacing)		E.g., apple, peach, cherry, spruce, pine	Tree size; density	E.g., active farmland, abandoned farmland	Species with potential in landscaping for proposed development
Brush (trees and shrubs generally less than 15 ft high with high density of stems, but variable canopy cover)		E.g., sumac, willow, lilac, hawthorn, tag alder, pin cherry, scrub oak, juniper	Density	E.g., vacant farmland, landfill, disturbed terrain (e.g., former construction site)	Species of significance to landscaping for proposed development
Fencerows (trees and shrubs of mixed forms along borders such as roads, fields, yards, playgrounds)		Any trees or shrubs	Tree size; density	E.g., active farmland, road right-of-way, yards, playgrounds	Species of value as animal habitat and utility in screening
Wetland (generally low, dense plant covers in wet areas)		E.g., cattail, tag alder, cedar, cranberry, reeds	Percent cover	E.g., floodplain, bog, tidal marsh, reservoir backwater, river delta	Species and plant communities of special importance ecologically and hydrologically; rare and endangered species
Grassland (herbs, with grasses dominant)		E.g., blue stem grass, bunch grass, dune grass	Percent cover	E.g., prairie, tundra, pasture, vacant farmland	Species and communities of special ecological significance; rare and endangered species
Field (tilled or recently tilled farmland)		E.g., corn, soybeans, wheat; also weeds	Field size	E.g., sloping or flat, ditched and drained, muckland, irrigated	Special and unique crops; exceptional levels of productivity in standard crops

Source: Adapted from W. M. Marsh, *Environmental Analysis for Land Use and Site Planning*. Copyright © 1978, McGraw–Hill, New York. Used with the permission of McGraw–Hill Book Company.

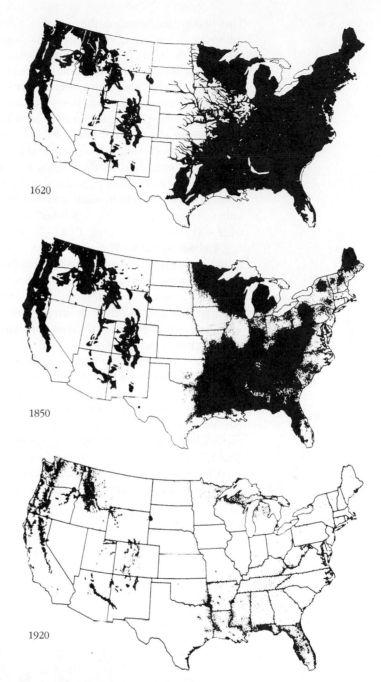

1620

1850

1920

Fig. 19.1   Reduction in virgin forest in the United States from the 1600s to the early 1900s. By 1920 vast areas of forest had been transformed into agricultural parkland. Since 1950 or so, farmland has declined and woodland has become reestablished over large areas beyond urban centers.

and second in the twentieth century with the shift of rural population to the cities. With the abandonment of small farms, much cultivated land has reverted back to some sort of woodland cover, though rarely with a composition and vigor comparable to primeval forest. This is explained partly by the degraded condition of the land after farming.

Farming impacts

With few exceptions (one being the farmers of German heritage in Pennsylvania), American farmers in the nineteenth and early twentieth centuries were very hard on the land. Most known soil conservation and crop management practices were ignored, and, after a generation or two, fields carved from woodlands were often abandoned.

The plants that invaded these lands were typically second-rate, initially little more than weedy species from the field margins, some native and some inadvertently introduced by the farmers. (One estimate places species of foreign origin at nearly 20 percent of the total number of plant species in the northeastern United States and eastern Canada.) On the other hand, these plants helped stabilize the ground and rebuild the soil. Runoff and soil erosion declined, and animal habitat was reestablished, though it favored different species than the forest of 50 or 100 years before.

Also on the positive side of the landscape ledger were the contributions made by conservation programs sponsored by the U.S. federal government as part of the New Deal legislation in the 1930s. Prominent among these was the CCC (Civilian *CCC reforestation* Conservation Corps), an organization that enlisted 3 million unemployed men to work on a host of projects in rural and wilderness lands, including landscape restoration and reforestation. The CCC worked in most states and was responsible for planting over 400 million trees (Fig. 19.2).

Urban Sprawl. The next trend was massive urban sprawl after World War II. Initially, much of this growth was absorbed by abandoned farmland, but as the development rate accelerated and land values increased, active farmland was also absorbed *Wholesale clearing* (Fig. 19.3). Nearly everywhere that conventional urban sprawl has taken place, it has resulted in wholesale destruction of most existing vegetation including the fencerows, woodlots, and orchards of active farmland as well as the second-growth woodland of abandoned farmland. In addition, large tracts of habitat such as riparian woodland corridors have been fragmented and reduced in area.

Only in the past few decades has suburban development begun to take a less all-consuming approach toward the landscape. This newer approach has been stimu- *Changing times* lated by three changes: (1) the enactment and enforcement of environmental regulations, particularly wetland and protected species ordinances; (2) public pressure, especially from nongovernmental environmental organizations such as streamkeepers and sportsmen's associations; and (3) the realization by experienced developers that features like woodlots, wetlands, and old hedgerows actually have economic value as real estate, most notably in residential development.

Fig. 19.2 Between 1933 and 1942, the Civilian Conservation Corps helped stabilize a declining American rural landscape by planting millions of trees.

(*a*)

(*b*)

Fig. 19.3 The loss of farmland to suburban development in the Valley Stream area, Long Island, New York, between (*a*) 1933 and (*b*) 1959.

Suburban landscape

As landscapes go, most of suburbia is new and the vegetative cover is still developing, meaning that it is undergoing comparatively rapid change as it adjusts to its new habitat. In most areas, street trees are one or more decades from maturity, hedgerows are still being planted, and property owners are still in the process of making adjustments in yard plants by replacing exotic species with poor survival records with hardier species. One measure of the level of maturity of suburban vegetation is the diversity and abundance of wildlife such as songbirds, squirrels, opossums, and raccoons.

Wildlife

Generally, animal habitat improves with the density, diversity, and productivity of the plant cover. The species diversity of the suburban plant cover, however, usually remains far below that of the rural cover it replaced because of the severe restrictions property owners place on the plant species allowed to grow in yards and the narrow range of habitat types owing to the absence of topographic and hydrologic diversity in most built-up areas.

Urban Landscape Decay. A fourth trend in vegetation is the deterioration of the managed urban and suburban landscape with urban decay. As inner cities, industrial areas, and old residential neighborhoods decline, the built landscape in many cities is falling into ruin, being abandoned, and taken over by weedy species and resilient survivors of old yards and parkways. Vast areas of abandoned neighborhoods in Detroit, for example, are being overgrown by weed trees such as box elder, old shrubbery such as junipers, and noxious plants such as ragweed, as vacated houses and streets literally decay (Fig. 19.4). At the other extreme are efforts by cities—including some of the same cities with decaying neighborhoods—to revive their business districts by enriching streetscapes with trees, pocket parks, and greenways along streams and waterfronts (Fig. 19.5).

Greenish brownfields

Managed forests

Beyond the urban scene are the forests managed for lumber and other forest products. This vegetation, too, has been substantially changed not only in the harvesting of original forests but also in the formation of new forest types. Forests managed for a sustained yield of timber based on selective cutting practices, for example, not only favor certain species but eliminate old-growth forests from the mix of forest

Fig. 19.4 Vast areas of abandoned residential neighborhoods in cities such as Detroit are being taken over by weed trees, shrubs, herbs, and fugitive yard plants.

Fig. 19.5 The introduction of vegetation to otherwise barren commercial districts has become a common practice in design programs aimed at revitalizing inner cities.

Fig. 19.6 Planted forests such as this one, usually made up of a single tree species, have replaced diverse, species-rich forests over vast areas of the United States and Canada.

habitats. In addition many native forest species, including shrubs and herbs, have been displaced by introduced species, particularly around settlements and farming areas. And research has demonstrated that nutrients (particularly nitrogen and phosphorus) from atmospheric pollution and agricultural runoff, coupled with reduced magnitudes and frequencies of flood flows on engineered streams, has selectively favored certain shrub and tree species, often introduced ones, resulting in widespread floristic changes even in protected forests. Lastly, sizable areas of planted forests have replaced "natural" managed forests and/or farmland in many areas. These forests are essentially monocultures that lack the diversity of species, structures, and habitats of the forests they replaced (Fig. 19.6).

19.4 THE CONCEPT OF SENSITIVE ENVIRONMENTS

Minority environments

The concept of sensitive environments in planning has grown in part from a reaction to the wholesale mistreatment of what we might call *minority environments*, such as wetlands, small stream valleys, and coastal sand dunes, whose value cannot be measured accurately by standard economic criteria, that is, how much the land and its ecosystems are worth in the real estate market. The concept has also emerged from the improved understanding of the role of such environments in the maintenance and quality of the larger landscape. It is common knowledge in today's society, for example, that wetlands are not only important biologically but also play a role in groundwater recharge and that small stream valleys are important riparian habitats as well as sources of water for lakes and wetlands.

Community values

Accordingly, many communities are incorporating special provisions for sensitive environments into their master plans based on economic rationale (because it costs more to build in and manage such environments), social rationale (because these environments are valued by people for their scenic and general aesthetic value), and scientific rationale (because they are often necessary to the maintenance of larger environmental systems). Wetlands, stream corridors, and shorelands are examples of landscapes that are widely recognized as sensitive environments.

Quiet victim

Wetlands. Prior to the 1950s, wetlands were generally viewed as third-rate landscapes of little or no economic value. They were indiscriminately altered and destroyed on a wholesale basis to provide cropland, improve agricultural production, facilitate navigation, control pests, and furnish land for urban development. Wetlands are now widely recognized for their roles as hydrologic, ecologic, recreational, forestry, and agricultural resources. The definition and mapping of wetlands (taken up in detail in Chapter 21) is generally based on three sets of criteria: vegetation, soils, and hydrology. Although hydrologic processes are usually the controlling force in the origin and formation of wetlands, the definition of wetlands based on hydrologic criteria has proven difficult for a variety of reasons. However, vegetation, especially indicator species, has provided a more useful set of criteria.

Hydrologic dustbins

Stream Valleys. Stream corridors are among North America's most maligned environments. For much of the twentieth century, streams served as hydrologic dustbins for farmlands and settlements, taking on heavy loads of wastewater from sewers and field drains. By the second half of the century, they were often viewed as liabilities to development and subjected to channelization, piping, and damming. This problem and society's effort to recover these prized environments are discussed in Chapter 14. Suffice it to report here that today thousands of grassroots organizations, with the help of governmental agencies, are working to restore local streams and the corridors of woodlands and marshes that stretch along them, all in an effort to resurrect their character as landscapes and their roles as habitats for a rich assortments of plant and animal species.

Delicate balance

Shores and Dunes. Coastal areas have long been attractive places to visit, but in the past 50 years development for residential and commercial purposes has increased substantially in these areas. This development has been facilitated by the growth of highway and road systems, the increased availability of land with the decline of agriculture, and the rising popularity of water-oriented living. Today the coastal zone is gaining population faster than any major geographic setting in the world. Development has led to the widespread alteration of coastal environments, especially the "softer" ones such as barrier beaches, beach ridges, sand dunes, coastal marshes, and backshore slopes where prized plant and animal communities are often found (Fig. 19.7). Because of the delicate balance that typically exists between these communities and the hydrologic and geomorphic environment, the imposition of roads, houses, and navigational facilities often leads to the loss of entire communities and, in turn, to the decline of an environment valued by many for ecological and aesthetic reasons. Furthermore, aggressive development has placed facilities in vulnerable situations that call for protective structures that damage ecosystems, reduce access, and degrade scenic quality (see Figs. 15.18 and 15.A).

19.5 VEGETATION AS A TOOL IN LANDSCAPE PLANNING AND DESIGN

Trees and real estate

Economic Value. The place of vegetation as a planning and design tool has improved significantly in the past several decades. Among other things, the costs of land clearing and landscaping alone have motivated some developers to incorporate more existing (predevelopment) vegetation into site plans. In residential areas, a

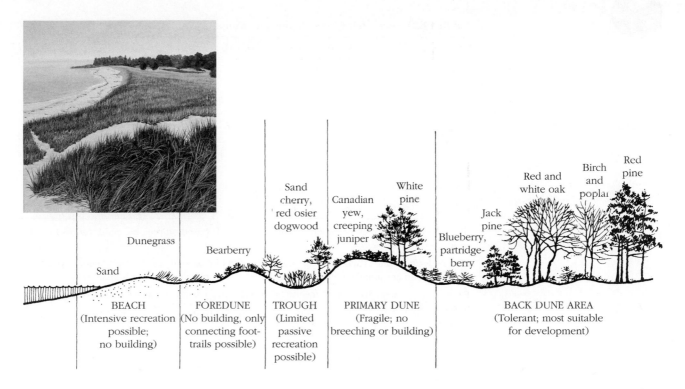

White
pine

Red
pine

Red and
white oak

Birch
and
poplar

Sand
cherry,
red osier
dogwood

Canadian
yew,
creeping
juniper

Jack
pine

Dunegrass

Bearberry

Blueberry,
partridge-
berry

Sand

BEACH
(Intensive recreation
possible;
no building)

FOREDUNE
(No building, only
connecting foot-
trails possible)

TROUGH
(Limited
passive
recreation
possible)

PRIMARY DUNE
(Fragile; no
breeching or building)

BACK DUNE AREA
(Tolerant; most suitable
for development)

Fig. 19.7 A profile across a belt of sandy shoreline showing the relationship between plants and landforms, including an assessment of the relative resistance of microenvironments to disturbance.

mature shade tree may have an estimated value of $10,000 to $25,000 or more, and the composite assemblage of plantings may improve the real estate value of an average residential lot by $50,000 or more.

Value added   A major residential developer in the Midwest claims that a site-adaptive approach that incorporates existing vegetation such as woodlots, hedgerows, and wetlands into design schemes yields 30 percent higher sales prices for lots. Nevertheless, a great many developers insist on stripping away vegetation and topsoil to make way for infrastructure installation and house construction. The cost of rebuilding the landscape is usually passed on to the homeowner. In the end much more is lost than money, however.

Landscape Design. Beyond its direct economic value, vegetation is also recognized for its role in landscape design. Indeed, it can be fairly argued that vegetation is the principal design tool of landscape architecture. In site planning, for example, *Function and aesthetics* vegetation is regularly used to screen certain land use activities and features, to abate noise, to modify microclimate, to improve habitat, and to stabilize slopes. As a visual barrier, hedgerows and border trees can help separate conflicting land uses such as residential, commercial, industrial, and institutional. In this capacity, the density and permanency of the foliage are critical because they control the transmission of light. In parks, estates, campuses, and similar areas, vegetation is used to frame outdoor spaces, to orient pedestrian circulation, and to provide the landscape with much of its essential aesthetic character. The latter purpose includes combinations of form, color, and texture integrated with buildings, landforms, and nuances of seasonal life cycle changes.

Noise Abatement. Vegetation can also be used to help control noise. Under barrier-free conditions, the level (magnitude) of sound from a point source decreases at

Table 19.2 Locomotive Noise Reduction With and Without a 200-Foot-Deep Forest

Frequency (Hz)	Noise at 250 ft Without Forest (dB)	Noise at 250 ft With Forest (dB)
31.5	39	39
63	57	56
125	63	61
250	68	65
500	73	69
1000	74	68
2000	72	64
4000	68	56
8000	61	41
	79 dBA	73 dBA

dB = decibel; unit of measurement of sound magnitude based on pressure produced in air from a sound source

dBA = decibel scale adjusted for the sensitivity of the human ear; a correction factor applied to dB units that takes into account the pattern of sound frequencies perceived by the human ear

Frequency = the pitch of sound measured in cycles per second; higher pitches have higher frequencies (more cycles per second)

Hertz (Hz) = cycles per second

a rate of 6 decibels (dB) with each doubling of travel distance. (From a linear source such as a highway, the decay rate is nearer to 3 decibels per doubling distance.) Placed in the path of sound, vegetation absorbs and diverts sound energy, and it is somewhat more effective for sound in the high-frequency bands (those above 1000–2000 hertz). In forests, the litter layer (decaying leaves and woody materials) appears to be very effective in sound absorption. Table 19.2 gives an example of the reduction in locomotive noise associated with a 250-foot-wide belt of forest. For the most effective use of vegetation in noise reduction, it is best to combine plantings with a topographic barrier such as a berm or an embankment.

Microclimate and Energy. The influence of vegetation on ground-level climate can be very pronounced. A plant cover effectively displaces the lower boundary of the atmosphere upward from the ground onto the foliage. A microclimate of some depth is thus formed between the foliage (for example, under the tree canopy) and *Cooling effect* the ground, where solar radiation, wind, and surface temperatures are lower than those over a nonvegetated surface (see Fig. 17.10). Heat exchange between the landscape and atmosphere is also influenced by vegetation. The *Bowen Ratio* (sensible heat to latent heat flux) is lower because of transpiration, producing lower air temperatures over vegetated than over nonvegetated surfaces, an important consideration in urban climate modification.

As a barrier to airflow, vegetation tends to force wind upward, thereby increasing the depth of the zone of relatively calm air over the ground, forming a *boundary sublayer*. The thickness (or roughness length) of the sublayer increases with the height and density of the vegetation. Under a mature fir forest it is usually around 2.5 meters *Airflow* (8 feet) deep, whereas in grass it is a hundred times less at 2.5 centimeters (1 inch) deep (Fig. 19.8). A related effect can be found on the downwind side of a vegetative barrier (shelter belt) where a calm zone forms under the descending streamlines of wind. The breadth of this sheltered zone also varies with the height and density of the barrier; however, significant wind reduction can generally be expected over a distance of 10 to 15 times the height of the barrier (Fig. 19.9). In snowfall areas, this

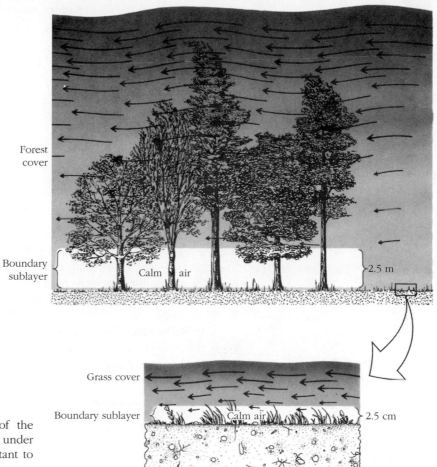

Fig. 19.8 The difference in the thickness of the boundary sublayer, the zone of relative calm air under coniferous trees and grass. This layer is important to the formation of ground-level microclimates.

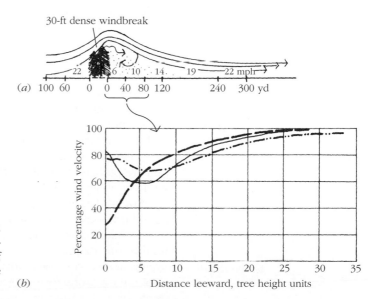

Fig. 19.9 (*a*) Pattern of wind velocity across a barrier. (*b*) Curves representing the percentage change in wind velocity downwind from a tree barrier. Greatest depression of wind velocity can be expected within 10 tree lengths of the windbreak.

zone is subject to the formation of snow drifts, the length of which can be estimated using this formula:

Snow drifts

$$L = \frac{36 + 5h}{K}$$

where

L = snow drift length in feet
h = barrier height in feet
K = barrier density factor (50 percent density is equal to 1.0, and 70 percent is equal to 1.28.)

Shade and energy

Case Study 17.7 addresses several ways in which vegetation can be used to reduce residential energy consumption. Chief among these is the use of shade trees to decrease solar heating and reduce air conditioning costs. Studies in California reveal that a mature tree cover can reduce daytime temperatures in residential neighborhoods by as much as 9°F and, depending on housing type, air conditioning costs by 40 percent or more.

Air Pollution. The influence of vegetation in reducing contaminants in polluted air is not well documented for urban areas, but the existing evidence suggests that it is relatively small. Plants are known to absorb certain gaseous pollutants, for example, carbon dioxide, ozone, and sulfur dioxide. But the absorption is apparently limited to the air immediately around the leaf and thus has only minuscule effects on these pollutants in the larger urban atmosphere. Heavy herb covers and dense stands of shrub and tree-sized vegetation with full covers of foliage act as *sinks* (catchments) for airborne particulates, but their net effectiveness is questionable because a sizable percentage of particulates initially caught appears to reenter the atmosphere within hours or days.

However, vegetation does appear to be effective in mitigating air pollution by means of trapping large particles in air moving laterally within several meters of the ground. In areas of heavy air pollution, a more pressing question may be that of the impact of pollutants on the health and survival of vegetation. Ozone and sulfur dioxide are the pollutants causing greatest damage to woody plants. Other pollutants such as fluorides, dust, and chlorine are also known to cause damage, but the damage is usually localized around the point of emission.

Erosion Control. We have long recognized the value of plants in erosion control and soil stabilization. Even a light herb cover can reduce wind erosion to negligible levels. Indeed, a thin cover of lichens over silty soils is credited with bringing erosion control to much of the North American Dust Bowl of the 1930s. In the universal soil loss equation (examined in Chapter 12), plant cover is one of the principal factors with both canopy cover and ground cover used in estimating soil loss rates. In combating gullying, the most insidious form of erosion, plant-based *biotechnical* approaches are very effective. *Contour wattling* on side slopes and *brush dams* in channels, for example, are particularly noteworthy as a means of slowing runoff and capturing sediment while establishing soil-stabilizing seedling masses (see Fig. 12.13).

Ground cover

Social Value. People have long recognized the desirability of vegetation in neighborhoods, towns, and cities. This appeal is related not only to the perceived role of plants in climate control, noise abatement, and the like, but also to sociocultural norms that place value on living plants and the habitats they create. Residential preference surveys bear this out when people identify parks and green spaces as important reasons for choosing one neighborhood or community over another. Real estate data also support this because wooded and landscaped lots consistently bring higher prices than those without vegetation or with unkempt vegetation. In senior citizen communities, people appear to enjoy better health where plants are abundant,

Aesthetic value

especially plants requiring their attention. Indeed, the case can be made that vegetation is the environmental glue that helps transform settlements, neighborhoods, and institutions into communities.

19.6 APPROACHES TO VEGETATION ANALYSIS

In most landscapes the distribution of plants can be highly variable, even at the local scale. The reasons for the variation are often complexly tied to existing environmental (mainly physiographic) conditions as well as to past events such as fires, floods, and land use change, and to the geographic availability of species that inhabit the area. Which of these myriad variables exerts the greatest control on the composition and distribution of the plant cover is usually a difficult question to answer.

Spatial Correlations. In searching for explanations for the distributions of plants, three basic types of studies or approaches are used. One approach is to map the distribution of plant types and selected environmental features and then examine the two distributions to see what correlations can be ascertained. For example, a comparison of topography and tree species may show that certain species consistently appear *Geographic coincidence* in stream valleys (see Fig. 14.8). What such a relationship means is not revealed by the correlation, but it may provide clues about what questions should be raised for analysis. The floors of stream valleys are usually wetter, subject to more flooding, and comprised of more diverse soils than upland settings. Plants must spend much of the year under conditions of saturated soil and/or standing water. Moisture tolerance may thus prove to be a good candidate for the detailed analysis of plants that are found in stream valleys, especially if the other settings that support different vegetation in the study area are appreciably drier (Fig. 19.10).

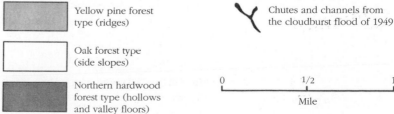

Fig. 19.10 The distribution of forest types in the central Appalachians related to soil moisture conditions associated with landforms and a major runoff event from a massive cloudburst.

Yellow pine forest type (ridges)

Oak forest type (side slopes)

Northern hardwood forest type (hollows and valley floors)

Chutes and channels from the cloudburst flood of 1949

0   1/2   1
Mile

Habitat controls

Controlling Factors. A second approach begins with an examination of the environment in an effort to identify those features and processes that may influence plant distributions. The purpose is to identify factors known to exert control on certain plants and, by virtue of their distribution, hypothesize which plants should and should not be found there. The analytical part of this approach involves testing the proposed or expected distributions by making measurements in the field and then investigating the detailed relations between the affected plant(s) and the environmental (or habitat) variable (Fig. 19.11). An example of this approach is provided by Pitcher's thistle, a protected plant that grows on sand dunes along the Great Lakes. To complete its life cycle, the plant requires a habitat of shifting sand intermediate in the dunefield between areas of wind erosion and heavy sand deposition. Therefore, a landform is the key to finding Pitcher's thistle: You must first find active coastal sand dunes and next find the sites of intermediate geomorphic activity within the sand dune system.

Geobotanical pointers

Plants as Indicators. The third approach uses key plants as indicators of environmental conditions and events. Based on existing knowledge of the habits and tolerance levels of selected plants, the controlling forces in the environment can often be identified according to which plants are found in an area. At the simplest level, this entails determining the presence or absence of certain types of vegetation over an area. In the example previously cited, the presence or absence of Pitcher's thistle at different locations within a dunefield can be used as an indicator of geomorphic activity, specifically sand erosion and deposition patterns. On a more general level, consider a region where forests are the predominant natural vegetation but there are sites where forest cover is absent. This condition could indicate that (1) levels of *stress* (particularly solar radiation, water, heat, or nutrients) are too great for trees; (2) the *resources* of the site, for example, soil cover and water supply, are limited; and/or (3) a recent *disturbance* such as a tornado or some land use activity destroyed the forest at the site and it has not grown back yet. Table 19.3 lists a number of site conditions in different bioclimatic regions that can be interpreted from vegetation.

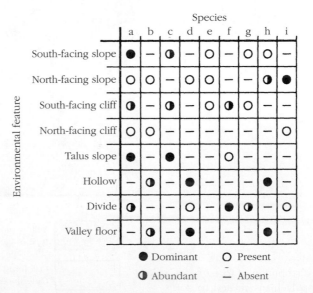

Fig. 19.11   A simple matrix showing the expected correlation between selected plant species and selected processes and features of the environment.

Table 19.3 Vegetation Indicators of Site Conditions

Climatic Region	Absence of Plant Cover	Sparse Herb and Shrub Cover	Thick Herb and Shrub Cover	Brush and Small Trees	Blade and Reed Plants	Highly Localized Tree Cover
Humid (Eastern North America, Pacific Northwest, South)	• Bedrock at or very near surface • Active dunes • Recent human use, cultivation, etc. • Recent fire • Recent loss of water cover	• Bedrock near surface • Recent or sterile soils • Dunes, fill • Recently disturbed (fallow, fire, flood) • Active slopes/ erosion	• Recently logged or burned • Too wet for trees • Managed grazing • Organic soil • Old field regrowth	• Landslide/fire, flashflood scars • Old field or woodlot regrowth • Shale/clay substrate • Organic soil • Moisture deficiency	• Organic soil • Standing water • High groundwater table • Springs, seepage zones	• Wet depression, organic soil • Steep slopes in agricultural areas • Flood-prone areas
Semiarid (High Plains, Central California)	• Caliche or salt pan (playa) at or very near surface • Desert pavement	• Localized water sources • Wind erosion • Overgrazing	• Overgrazing • Free from burning • Too dry for trees	• Channels with available moisture • Aquiferous substrate	• Same as above	• Aquiferous substrate • Seepage zone or spring • Stream valley (galleria) forest • Plantation
Arid (Southwest, Great Basin, Southern California)	• Rock surface • Unstable ground such as dunes or rockslides • Too dry			• Protected pockets • Favorable (moist) slopes • Logged/ burned		
Arctic and Alpine (N. Canada, Alaska, Rockies)	• Rock surface • Active slopes • Persistent ice, snow, or ground frost • Ponded water during growing season	• Above treeline • Persistent ice, snow, or ground frost • Active slopes • Periglacial processes active	• Above treeline • Ice, snow, and wind pruning • Mildly active slopes • Wet depressions	• Wind/ice pruning • Avalanche, landslide scars, fire • Recent logging near tree line, permafrost near surface	• Same as above	• Protected pockets

Beyond this sort of exercise, the particular types of plants, their densities, and physiological conditions (health) can be examined to learn about the detailed nature of the environment and its forces.

Source: From W. M. Marsh, *Environmental Analysis for Land Use and Site Planning* (New York: McGraw–Hill, 1978). Used by permission.

19.7 SAMPLING VEGETATION

Why sample?

Whether the objective is to inventory and describe vegetation for the environmental impact and assessment studies, analyze vegetation for scientific purposes, or use vegetation as an indicator of site conditions and past events, vegetation usually has to be sampled in some way. Sampling is a means of selective observation that enables us to estimate various aspects of a plant population or vegetation community based on measurements of only a small portion of it. Sampling is attractive because it saves time and money; however, the proper use of sampling techniques can be difficult. Among the techniques used in vegetation studies are quadrat sampling, stratified sampling, transect sampling, systematic sampling, and windshield-survey sampling.

The Sample Area. In any sampling problem, the first task is to define the relevant population. In the case of vegetation, this is usually accomplished by defining the geographic area occupied by the vegetation under study. This area, which may be a development site, an environmental zone, or a subdivision of a larger area, is outlined on a large-scale map. The vegetation within it can then be sampled, using either the quadrat or transect method.

Selecting subareas

Subdividing the population or study area into subareas or sets prior to drawing the sample is referred to as **stratified sampling**. The strata are coherent subdivisions based on observable characteristics or prior knowledge of the area. A typical subdivision for many areas involves three strata: floodplains, valley slopes (walls), and uplands. Within these strata, quadrats are chosen on a random basis or transects are laid out.

Stratified sampling is widely used today with remote sensing imagery to define regional vegetation patterns. Indeed, aerial photographs are almost indispensable in modern vegetation studies, and more technically sophisticated remote sensing systems, such as line scanners, are showing promise for discriminating major types of vegetation. A form of stratified sampling is also used in soil mapping for land use projects where the strata are defined and sampling points assigned according to development and use zones.

Sampling Techniques. In quadrat sampling the first step involves subdividing the whole area into grid squares. For small study areas, whole squares may be used as a sample quadrat, whereas for large areas, it may be necessary to use some fraction of a square as the quadrat depending on the type of vegetation. Generally speaking, a

Quadrat size

1- to 2-square-meter quadrat is used for grasslands and marshes, 4 to 6 square meters for low shrubs, 15 to 30 square meters for brushland, and 30 to 100 square meters for woodland and forest.

In **random transect sampling** the study area is divided into a number of strips, called transects. The width of the transects should vary with the type of vegetation; 5 to 10 meters, for example, would be appropriate for most forests. Of the transects selected for sampling, the entire transect may be sampled, or individual quadrats may

Options

be selected for sampling within the transect.

Systematic sampling also uses quadrats. This type of sampling is used where there is little or no prior knowledge of the population or area under consideration. A grid is drawn over the area, and a sample is taken at each intersect in the grid. Quadrats can be used as the sample unit, and it is generally recommended that, together, the quadrats cover a minimum of 20 percent of the study area.

The **windshield survey** is the quickest and least expensive sampling technique. Though more commonly employed in land use surveys than in vegetation studies, it can be helpful in gaining an overview of vegetation types. Be aware, however, that roadside vegetation may not be representative because it may be planted, cut back in road construction or maintenance, or atypical of the area owing to the establishment of second-growth trees and weedy plants in the road right-of-way.

19.8 VEGETATION AND ENVIRONMENTAL ASSESSMENT

Finally, vegetation plays a key role in environmental assessment and impact analysis. In fact, as mentioned earlier, few components of the landscape lend themselves to the identification of environmental stress and change as well as vegetation does. At least five parameters or measures of impact related to vegetation can be highlighted in the context of evaluating a proposed action.

Gauging loss

Measures of Impact. First is the sheer loss of cover, measured, for example, by the area of forest lost to development. This is arguably the most visible change that can be made in the character of a landscape, and it often evokes a powerful emotional response in the public. Second, the loss of valued species, communities, and habitats is a critical measure of environmental impact as mandated by law at various levels of government. Third is the economic loss represented by the loss of merchantable vegetation such as timber; the loss of other economic opportunities such as maple syrup production, hunting, and fishing; and the decline of real estate value on the site and in neighboring lands.

System impacts

The fourth measure is based on the role of vegetation in larger environmental systems. Vegetation is often integrally linked to other parts of the landscape system such as microclimate, ecosystems, slope stability, soils, and hydrology, and alteration or loss of plant cover can spell serious disruption to these systems. Fifth, it is important to remember that natural vegetation is adjusted to a certain set of environmental conditions, and changes in these conditions, even subtle ones, are often reflected in changes in the vigor, reproduction capacity, and makeup of plant communities. Therefore, plants serve as valuable "thermometers" of environmental performance, giving us warnings when things are not working well and when landscapes are not sustainable.

19.9 CASE STUDY

Vegetation and Wildlife Habitat in Residential Planning, Central Texas

Jon Rodiek and Tom Woodfin

In southcentral Texas, as in many other parts of North America, urbanization has promoted habitat fragmentation. Fragmentation is characterized by the compartmentalization of otherwise large, spatially continuous areas of habitat. It has two negative impacts on wildlife: (1) reduction in total habitat area, which primarily affects population size; and (2) the subdivision of the remaining area into disjunct patches, often separated by barriers, which primarily affect dispersal and migration routes. According to recent studies, temperate ecological communities appear to be more resistant to the effects of fragmentation than are tropical communities. Among the reasons cited are that temperate species tend to occur in higher densities, are more widely distributed, and have better dispersal powers.

On the other hand, it can be argued that the effects of fragmentation only *seem* less severe in temperate zones because most of the habitat damage there was done long before the original landscapes and wildlife could be documented. Such is likely the case in the region of College Station, Texas, where much of the clearing in the floodplains and woodlands took place around the 1860–1900 period (Fig. 19.A). In the past 50 years, however, the trend toward fragmentation has actually been reversed in many rural areas with the regeneration of small woodland and riparian patches following the abandonment of farming. Near cities this recovery process is interrupted by conventional suburban sprawl.

In one 2800-acre drainage area, which was analyzed for cover change over a 47-year period (1940–1987), losses were measured in the following cover types: pastureland, old fields, young stands of upland hardwoods, cultivated fields, shrub-grasslands, hardwood-shrub lands, and mature bottomland hardwoods. Increases were measured in grasslands, mature upland hardwoods, savannahs, and riparian

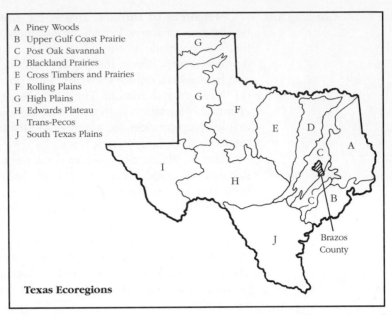

A Piney Woods
B Upper Gulf Coast Prairie
C Post Oak Savannah
D Blackland Prairies
E Cross Timbers and Prairies
F Rolling Plains
G High Plains
H Edwards Plateau
I Trans-Pecos
J South Texas Plains

Texas Ecoregions

Fig. 19.A The ecoregions of Texas. Brazos County, where the study site is located, lies within the post oak savannah region.

woodlands. Drainage lines changed little, judging from a slight decrease in the total lengths of first-, second-, and third-order streams. Roads and houses increased significantly, however. Roads increased from 7.96 to 22.6 miles, and the number of houses and other buildings increased from 30 to 124.

On balance, the trend in land cover change can be characterized by a three-phase sequence: floodplain/riparian woodland to agrarian land to mixed suburban/urban/agrarian land. This trend is, of course, typical of land use changes in and near urban regions across the United States and Canada, and it has serious implications for wildlife. As agrarian ecotones (or their remnants) are converted to residential tract development, they become segmented and reduced in size. The challenge for environmental planners and landscape architects is to devise more creative styles of land conversion that increase habitat areas and geographic continuity.

In this particular case, an experiment was set up to design a residential landscape that incorporates basic biogeographic principles for wildlife habitat improvement. The effort was organized around two objectives: (1) to lay out two single-family residences on 15 acres of land and, if the layout proved successful, (2) to extend the site design concept over a larger, remaining residential tract. The ultimate goal would be to apply this approach to an entire watershed habitat reserve, a network made up of subregion habitat reserves tied together with a large number of site-scale habitat reserves.

In the first phase of the project, plant cover associations and related land uses were analyzed at three essential scales: watershed, subregion, and sites. Oak mottes, riparian woodlands, managed grasslands, and savannah were found to be common at all three scales. Mesquite grassland habitats were common only to the subregion and site level (Fig. 19.B).

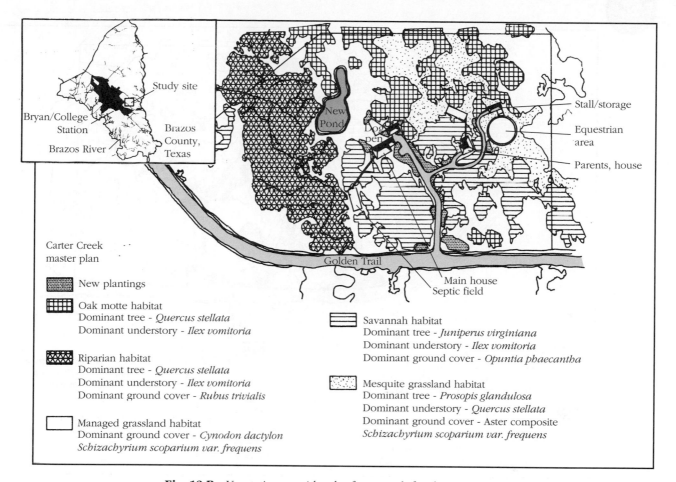

Carter Creek master plan

New plantings

Oak motte habitat
Dominant tree - *Quercus stellata*
Dominant understory - *Ilex vomitoria*

Riparian habitat
Dominant tree - *Quercus stellata*
Dominant understory - *Ilex vomitoria*
Dominant ground cover - *Rubus trivialis*

Managed grassland habitat
Dominant ground cover - *Cynodon dactylon*
Schizachyrium scoparium var. frequens

Savannah habitat
Dominant tree - *Juniperus virginiana*
Dominant understory - *Ilex vomitoria*
Dominant ground cover - *Opuntia phaecantha*

Mesquite grassland habitat
Dominant tree - *Prosopis glandulosa*
Dominant understory - *Quercus stellata*
Dominant ground cover - Aster composite
Schizachyrium scoparium var. frequens

Fig. 19.B Vegetation provides the framework for the site master plan. Grassland with mixed woodland dominates the site.

The initial planning task involved formulating a framework plan that attempted to integrate a residential complex and wildlife habitat. This plan first calls for the siting of facilities, including two residences, drives, storage facilities, and an equestrian area, in a configuration that least impacts existing woodland habitat. Vegetation was then added to provide screening, browse, and cover in selected areas, mainly along drives, open areas, and the riparian corridor. To further diversify the habitat, a pond was added adjacent to the riparian habitat area. Finally, feed areas were added to the site in the form of winter and summer pasture (Fig. 19.C).

The ultimate goal of this project is to mitigate the impacts of fragmentation of wildlife habitat brought on by urbanization. The strategy employed is based on the idea of establishing a skeletal corps of wildlife habitat along the riparian woodland and bottomland-hardwood landscapes. These habitats are the most critical to the resident wildlife in the region. Although wooded landscapes are increasing in acreage, they currently represent only 6 percent of the subregion total. Protective zoning and enhancement of these landscapes, especially edges and linking segments, are seen as the most effective means of improving the balance between wildlife and residential development in this area.

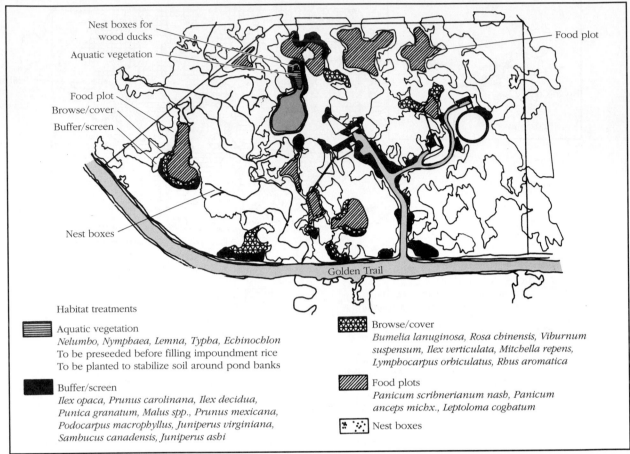

Fig. 19.C   The proposed vegetation changes designed to enrich and diversify habitat.

Jon Rodiek and Tom Woodfin are landscape architects at Texas A&M University who specialize in wildlife habitat planning as a part of landscape design.

19.10 SELECTED REFERENCES FOR FURTHER READING

Carpenter, Philip L., et al. *Plants in the Landscape*. San Francisco: Freeman, 1975.

Davis, Donald D. "The Role of Trees in Reducing Air Pollution." In *The Role of Trees in the South's Urban Environment* (Symposium Proceedings). Athens: University of Georgia, 1970.

Gleason, H. A., and Cronquist, Arthur. *The Natural Geography of Plants*. New York: Columbia University Press, 1964.

Grey, Gene W., and Deneckie, F. J. *Urban Forestry*. New York: Wiley, 1978.

Hack, J. T., and Goodlett, J. C. "Geomorphology and Forest Ecology of a Mountain Region in the Central Appalachians." *U.S. Geological Survey Professional Paper* 347, 1960.

International Union of Forestry Organizations. *Trees and Forests for Human Settlements.* Toronto: University of Toronto Centre for Urban Forestry Studies, 1976.

McBride, J. R. "Evaluation of Vegetation in Environmental Planning." *Landscape Planning.* 4, 1977, pp. 291–312.

Maher, Neil M. *Nature's New Deal: The Civilian Conservation Corps and the Roots of the American Environmental Movement.* New York: Oxford University Press, 2008.

Mooney, P. F. *Plants: Their Role in Modifying the Environment; A Selected and Annotated Bibliography.* Mississauga, ON: Landscape Ontario Horticultural Trades Foundation, 1981.

Schmid, J. A. *Urban Vegetation: A Review and Chicago Case Study.* Department of Geography Research Paper 161. Chicago: University of Chicago, 1975.

Thurow, Charles, et al. *Performance Controls for Sensitive Lands.* Reports 307 and 308. Washington, DC: American Society of Planning Officials, 1975.

U.S. Forest Service. *Better Trees for Metropolitan Landscapes.* USDA Forest Service General Technical Report NE-22. Washington, DC: U.S. Government Printing Office, 1976.

U.S. Forest Service. *National Forest Landscape Management.* Agricultural Handbook No. 478. Washington, DC: U.S. Government Printing Office, 1974.

Related Websites

Buell-Small Succession Study. 2004. http://www.ecostudies.org/bss/index.html
A research group looking at how old-field vegetation in succession from farmland to rich habitat occurs. Read about vegetation succession and explore their research results.

Elwha River Watershed Information Resource. http://www.elwhainfo.org/elwha-river-watershed
A collaborative effort to educate the public about the Elwha Watershed. This group emphasizes revegetation as a method of river restoration. The site also goes into the method of revegetation following dam removal.

Nature Conservancy. "Climate Change." 2009. http://www.nature.org/initiatives/climatechange/
A conservation group working to protect the environment. The site talks about vegetation and landscape change as an effect of climate change, including habitat biodiversity. It also suggests ways people can change their behavior and calculate their own carbon footprint.

City of Minneapolis. "Minneapolis Urban Forest." 2009. http://www.ci.minneapolis.mn.us/sustainability/urbantreecanopy.asp
The benefits environmentally and economically of urban trees; an example of city policy to protect vegetation.

LANDSCAPE ECOLOGY, LAND USE, AND HABITAT CONSERVATION PLANNING

20.1 INTRODUCTION

Landscape fragmentation

Landscape fragmentation is the inevitable by-product of settlement and land use. Scholars tell us that many of the revolutions in Chinese history were actually aimed at reversing excessive farmland fragmentation and reforming rural land use. The North American landscape is juvenile compared to China's, but the rate at which it was cleared and settled is unprecedented in the world. In the continental heartland, it took only a matter of decades to fragment habitat systems such as riparian networks, wetlands, and forests as the landscape was filled with farms, railroads, highways, and settlements. Until now we have had little concern with landscape fragmentation beyond the traditional problems related to access, land ownership, water rights, and so on. But recently, with recognition of the effect of fragmentation on biodiversity, landscape ecology has gained a place in the environmental planning agendas of North America and Europe.

The application of spatial (geographic) analysis to problems of habitat planning and management in rural and suburban landscapes is called **landscape ecology**. It is a direct response to the decline in biodiversity and biological productivity as a result of habitat and ecosystem fragmentation, reduction, simplification, and contamination. From the air, these landscapes have a patchy geographic aspect that some writers in the field have termed mosaics. *Mosaics* are made up of pieces of landscape systems: drainage nets, soils, woodlands, wetlands, and the like, all representing the disaggregated remnants of various types of habitats and ecosystems.

Mosaics

The study of landscape ecology focuses on three traits of landscape mosaics: structure, function, and change (or form, process, and change). The basic aim is to discover the relationships between landscape form and function in order to design landscapes that support richer and more productive mixes of plant and animal species. As a landscape planning problem, the objective is to reduce fragmentation and weld fractured landscapes back together into more functional patterns with greater ecological resilience and sustainability.

20.2 THE BIOGEOGRAPHICAL FOUNDATION

The scientific underpinnings of landscape ecology come from biogeography, particularly island biogeography. Studies in **island biogeography** have revealed that two geographic factors, *habitat area* and *distance* between island habitats, have a strong influence on species diversity.

Key parameters

Controls on Species Diversity. The larger the area is of an island or its terrestrial counterpart (the patch), the larger the number of species it can support. This principal was demonstrated in the original island biogeography studies based on freshwater bird species inhabiting islands of Southeast Asia (Fig. 20.1). It was also demonstrated that the greater the distance is between neighboring islands, the lower the genetic mixing and the lower the biodiversity will be.

Whether the principles of island biogeography are strictly applicable to fragmented terrestrial landscapes is debatable, but the concepts unquestionably apply. According to the area factor, island remnants of habitat, despite the fact that they look like the original habitat, are incapable of performing biologically as the former habitat once did. As a general rule, a 90 percent reduction in habitat area results in a 50 percent reduction in biodiversity. In addition, as more islands are taken down, the remaining forest islands are left farther and farther apart. This, of course, reduces mixing among isolated members of island populations (Fig. 20.2).

90/50 percent rule

In tropical forest habitats, the loss of species may actually exceed 50 percent, given a 90 percent habitat reduction. The reason is that the ranges of many tropical

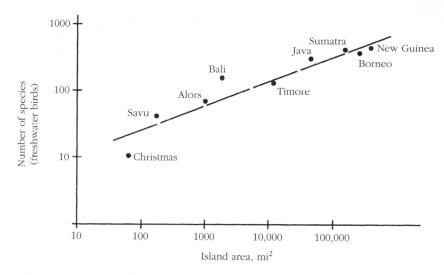

Fig. 20.1 The relationship between island size and number of species of freshwater birds from early biogeography studies in Southeast Asia. Species numbers increase with island size.

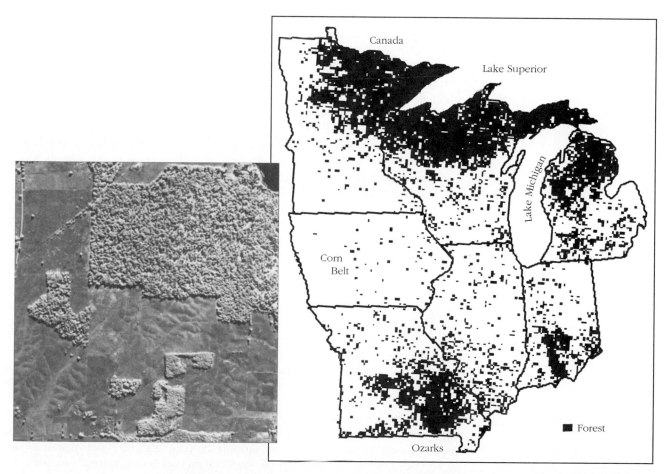

Fig. 20.2 A map of forest fragmentation in the Midwest. Michigan, Wisconsin, and Indiana were nearly fully forested only 200 years ago. The inset photograph shows the fragmentation at a finer level of resolution with variously sized "islands" of forest in a "sea" of farmland.

species are highly localized (i.e., endemic), often occupying not more than a few acres or a few square miles. This contrasts with midlatitude species, which, though fewer in total number, are often distributed in much larger ranges. Thus, in the tropics, the eradication of large areas of forest is more likely to destroy the entire range of many endemic species, resulting in greater than 50 percent species loss as forest loss approaches 90 percent.

Delayed response

Time Lag Factor. In addition, studies of the short-term relationship between species decline and forest loss sometimes reveal lighter than expected losses. The reason, as other studies have shown, is that reduction in biodiversity does not occur suddenly with habitat reduction. Rather, species counts decline gradually over many years or many decades. Thus, short-term postclearing evaluations of biodiversity changes, based on indicator organisms such as birds, must be extended over the long run to get a reliable picture of the impact of habitat loss on species numbers in general. Finally, we must recognize that our current understanding of these problems is very limited. Much remains to be learned about biogeographical relations as a whole, including differences between tropical and temperate species in terms of population densities, range sizes, and fragmentation effects.

20.3 HABITAT, LAND USE, AND BIODIVERSITY

Habitat and niche

Habitat is the local environment of an organism. It is sometimes referred to as the environmental address of an organism, but more frequently it is thought of as a unit of space and its environmental features, principally microclimate, soil, topography, water, available nutrition, and other organisms (Fig. 20.3). Although different organisms may occupy the same or very similar habitats—for example, different birds in the same wetland—interaction with the habitat is different for each species.

The term **niche** describes an organism's way of life or what it does in its habitat. Insects occupying a tree canopy habitat, for instance, may survive by acquiring food in different ways: Some eat leaves, some suck nectar, and others chew bark. Niche defines an organism's functional relationship with its habitat; through its niche, each organism "sees" its habitat as unique.

Habitat replacement

Habitat Versatility. Because of the special relationship each organism has with its habitat, trading one habitat for another is virtually impossible. Unlike humans, who have a wide range of habitat versatility, most organisms displaced from a habitat cannot simply take up life in another. In addition, when habitats are destroyed, they cannot be recreated in a manner suitable to sustain most of the organisms that once occupied them. This is especially true for complex habitats, such as many forests and wetlands, where niches are formed by many and often subtle interrelationships with other organisms, such as insects and microflora. Herein lies the fundamental dilemma of preserving in zoos or seed banks organisms whose habitats have been destroyed. If their habitats are gone and cannot be resurrected, then there is no place to sustain the organism, and the zoo becomes an artificial life support system.

Wetland experience

Wetland construction is the most widely practiced habitat recreation enterprise in North America today. It is promoted by wetland mitigation programs that allow or require developers to create wetlands to replace wetland acreage destroyed in development projects. Although scientific evaluation of constructed wetlands is incomplete, studies to date reveal that constructed wetlands are poor replicas of the habitats they replaced because they do not support the diversity and types of species that their prototypes supported. Measured by other criteria such as hydrologic function, they may be more successful.

Land Use Impacts. What do land uses do to habitat? The effects are highly variable, depending on the type and density of land use. The least disruptive land use

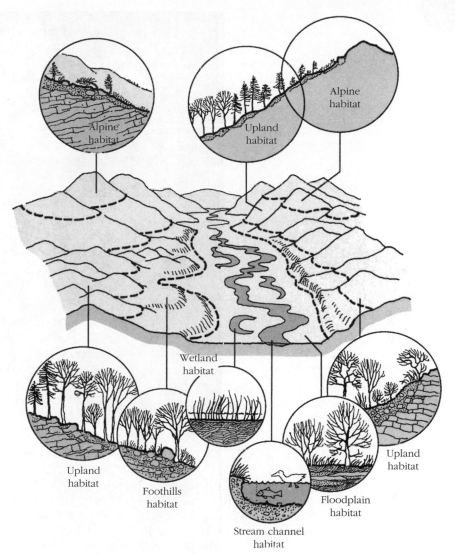

Fig. 20.3 A schematic diagram illustrating the different habitats associated with a stream valley in mountainous terrain, ranging from alpine at high elevation to aquatic and floodplains on valley floors.

is decidedly hunting-gathering, as was once practiced by nomadic North American native tribes, because land clearing and agriculture are not involved and human populations are very light. In fact, it is not inappropriate to consider hunting-gathering societies as part of the natural landscape because they, like most other mammal populations, do relatively little to manipulate the environment.

When agriculture replaces hunting and gathering, the impact on habitat increases, but initially it may not be significant because farms are usually small and *Land use and habitat* widely separated. But as the landscape fills in and agriculture is organized into large-scale systems, the balance between cleared land and natural habitat shifts in favor of cleared land. The actual ratio between the two depends on the suitability of the land for agriculture and the regional pressure to use more land. As Figure 20.4 illustrates, the conversion of natural landscapes into landscapes dominated by farmland can be described more or less as a stagewise process.

Land Clearing and Development. In the early stages, land clearing follows a reasonably predictable pattern corresponding to the physiography of local watersheds.

Fig. 20.4 Infilling of the landscape by agriculture, resulting in the reduction of open space and fragmentation of habitat corridors. Land use eventually spills over slopes and into stream valleys and wetlands.

Toward fragmentation

In areas of low to modest relief such as the U.S. Midwest, farmland first takes up belts of arable land lying on the interfluves between corridors of less arable land such as floodplains and wetlands stretching along valley floors. Next, formal roads and railways are driven through the landscape along surveyed routes that often have little relationship to local patterns of drainage, soils, and vegetation. As the landscape is cracked open, as it were, and population and agriculture increase, the pressure to clear more land often pushes farming and settlement beyond the reasonable limits of cropland into marginal lands, such as slopes and floodplains. Wetlands are drained, dams are built, streams are channelized, woodlands are removed, and the remaining habitat corridors are broken up or fragmented into smaller segments. The remaining uncleared landscape is left in small, widely separated patches (Fig. 20.4).

Landscape degradation

Habitat Loss. In the agriculture sector of the landscape, habitat potential is severely limited. Trees and shrubs, the habitats of many birds and insects, are reduced to field margins or *edge habitats*. Natural ground cover is largely eliminated, and topsoil is reduced significantly by the runoff and wind erosion associated with crop farming. Plowing greatly disrupts or destroys root, insect, and microorganism habitat, and pesticide applications further limit the soil's ability to serve as habitat. Humus declines dramatically, and earthworm, beetle, ant, and other invertebrate populations decline precipitously or disappear. Wildflowers, such as trillium, wild geranium, mayapples, and trout lilies, are eliminated from cropland or are forced into woodland fragments along and between fields—geographic refugia. In addition, many of the original plant and animal species are replaced by introduced species.

Monoculture patches

Simplification. Under intensive cultivation, the landscape becomes *biologically simplified* and often impoverished. Farmfields are reduced to biological **monocultures** dominated by several crop species with a few hardy weed species and greatly reduced and simplified surface and soil biodiversities and/or greatly imbalanced remnant populations. The remnant island patches of habitat are too small and widely spaced to support more than a fraction of the original species. As agriculture intensifies with commercialization and/or expansion into marginal lands, the remnant habitat is chopped up into even smaller fragments that are subject to severe damage from surrounding cropland and farm animals (Fig. 20.5).

Favored species

Species Change. Whereas many species are reduced or eliminated by land clearing, agriculture, and settlement, a number of species are favored by such changes. Among the plants, **weed species** like bull thistle, sumac, ragweed, and poison ivy have increased their geographic coverage and populations with the spread of development. These plants have responded to (1) edge habitats such as fence lines and hedgerows along farm fields where the original species cannot sustain themselves and (2) disturbed areas such as road margins, playgrounds, and construction sites. As we noted in the previous chapter, in extreme cases, plants such as kudzu have overrun and smothered local landscapes (Fig. 20.6).

Opportunism

Many mammals and birds have also increased, taking up edge habitats and disturbed areas. The fragmentation of forested landscapes in the U.S. Midwest and southern Ontario, for example, has favored **opportunistic species** such as cowbirds, blue jays, and crows. After being introduced into North America,

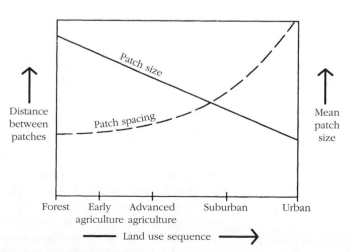

Fig. 20.5 Schematic graph illustrating the changes in patch size and spacing with land use change from wilderness to urban.

Fig. 20.6 A landscape smothered by kudzu, a vine introduced to the American South for the purpose of arresting soil erosion in the 1930s.

Habitat suburbia

European house sparrows and starlings have been very successful in agricultural areas as well as in cities and suburbs. Coyotes, rabbits, opossums, raccoons, and white-tailed deer have shown a remarkable capacity to adapt to new and changed habitats. Such versatility has enabled them to reinhabit suburban areas from which they were displaced during the clearing and construction phases of the development process. The trend to reoccupy is explained in part by the observation that, as the suburban landscape ages and tree and shrub cover increase, the landscape becomes more diverse and richer as habitat. In addition, the geographical linkage within and among suburban areas is often improved as habitat corridors are created with the establishment of parks and wooded areas around schools, in vacant lots, and along streets.

20.4 ENDANGERED, THREATENED, AND PROTECTED SPECIES

Protected classes

In 1973 the United States Congress passed the **Endangered Species Act**. This controversial law established two classes of protected species. **Endangered species** are defined as those in imminent danger of extinction in all or a significant portion of their ranges. **Threatened species** are those with rapidly declining populations that are likely to become endangered in the foreseeable future. Although the act applies to all lands and marine environments (private, state, and federal), it has focused particularly on federal construction and land use projects (e.g., military bases, interstate highways, and flood control facilities) and on private projects using federal money or requiring federal permits. The act also makes it illegal to capture, kill, possess, buy, sell, transport, import, or export threatened or endangered species.

Ecosystem protection

The Larger Issue. Although the importance of individual protected species should not be trivialized, it is not the central problem in species protection. The central problem is the much broader issue of biodiversity and habitat protection. The greatest loss of species is caused by habitat destruction, not by the exploitation of individual organisms; therefore, the focus should be on the protection of entire ecosystems. In protecting ecosystems, habitat must automatically be incorporated because habitat, represented by soil, topography, water features, and so on, is the very foundation of ecosystems. The passage and enforcement of wetland laws represent a move toward habitat protection and the conservation of biodiversity.

On the other hand, the protection of individual organisms is not without broader ecological benefits. To protect individual species, their ecosystems must also be protected, and in turn a larger network of organisms and habitats also receive protection. The least prudent approach to species protection is the creation of zoo-like preserves where special organisms are given showcase status without the appropriate habitat, area, and ecosystem arrangements for long-term viability.

Vulnerable species

Extinction-Prone Species. Some species are decidedly more prone to extinction than others. At the top of the list are **endemic species** with ranges so small that the species can be swept to extinction after only several years of land clearing for agriculture or lumber. Endemic species that are relics of once large populations and ranges are also vulnerable. Among other things, these species often have very narrow habitat requirements and are therefore subject to decline from subtle imbalances in the environment. A celebrated example is the Chinese giant panda, which depends on only a few species of bamboo for virtually its entire food supply. Should these bamboos (which are already severely limited by land use) be destroyed by development or natural change, the pandas are likely to starve. Only about 1000 pandas remain in the wild in China.

Breeding population

Species with small populations are also prone to extinction because the loss of relatively few individuals may pull the population below the critical threshold of a breeding population. This is especially significant for animals, such as whales, accustomed to living and breeding in groups. If the group becomes too small, breeding may cease even though breeding adults are available. Small populations, such as those of cheetahs, also make some species vulnerable if they reproduce slowly. These species average one or fewer offspring a year, and the young require a number of years to reach breeding age. The recovery time is dangerously long for such populations because they are subject to other perturbations during recovery.

Risk of isolation

Susceptible Island Species. Another extinction-prone class is species that have evolved in an isolated environment that has protected them from competition and predation. Many natural island species, such as the giant tortoises on the Galápagos Islands, declined when goats were introduced to the islands because goats consumed the tortoise's food sources. In other cases, island species were decimated by predators introduced by humans or by humans themselves. Among the predators commonly associated with humans are pigs, dogs, chickens, and rats. Many bird populations in New Zealand and Hawaii, for example, evolved without serious natural predation, making them vulnerable to the new and vigorous predators introduced with early settlers. Introduced snakes have been particularly effective in decimating bird populations on a number of tropical islands. Witness that two-thirds of the birds in the Hawaiian Islands became extinct over the two hundred years following the initial settlement of the islands by Polynesians. The World Conservation Union estimates that since 1600 nearly 40 percent of the animal extinctions in the world have been caused by introduced species.

20.5 PATTERNS AND MEASURES OF LANDSCAPE FRAGMENTATION

Ecosystem networks

Before the intrusion of agriculture, settlements, and associated transport systems, the biological landscape was covered by an interlocking network of ecosystems. These ecosystems were of different sizes, shapes, and compositions, and, despite the seemingly random geographic patterns displayed by some, their distributions were actually quite structured. The basic structural framework for ecosystems was provided by physiography, and since the scale, pattern, and makeup of physiographic systems were different in different parts of the continents, the landscape ecology was accordingly different as well.

The physiographic framework

Physiography and Fragmentation. In the eastern United States early settlers remarked on the patchy character of the New England forest cover, whereas settlers in the Midwest marveled at the vast uniformity of the tall grass prairie of central Illinois. Neither response is surprising in light of the physiographic character of the two locations. The physiography of Illinois favored large and continuous ecosystem areas that conformed to the broad pattern of stream corridors and the interfluves between them. In New England, where the Northern Appalachian terrain was diverse and irregular with numerous rock outcrops and various glacial deposits interspersed with wetlands, physiography favored small and discontinuous ecosystems, more patches, and irregular corridors.

Breaking the frame

Into these landscapes were superimposed the land use systems like those we described earlier. Some land uses such as early farming honored the physiographic framework, but modern ones such as highway systems tend to treat the physiographic framework indiscriminately, especially in nonmountainous regions. Each land use system added a layer to the landscape, and each fragmented the ecological patterns of the original physiographic landscape. Habitat patches became smaller, and corridors became narrower and more segmented. In some instances new corridors emerged, such as utility rights-of-way and systems of field hedgerows. But on balance, the landscape's ecological linkage declined as its economic (land use) linkage increased. Fewer original species could be supported, and, for those remaining, populations were smaller, less productive, and less viable biologically.

Riparian corridors

Corridor Systems. At least five types of corridor systems can be found in most rural and/or partially developed landscapes. The most fundamental corridor system is the **riparian network**. This system has a hierarchical structure that follows networks formed by stream systems (Fig. 20.7*a*). The geographic pattern of riparian systems varies widely, but the fundamental structure is governed by the principle of *stream orders* with first-order corridors (no branches), second-order corridors (formed by at least two first-order branches), third-order corridors, and so on down the network (see Fig. 9.1). The size of the corridors in a riparian network increases with the corridor rank, and critical to their ecological function is the fact that corridors of all orders are linked together in comprehensive, self-sustaining water systems.

Interfluve corridors

The upland counterpart to the riparian corridor system is the **interfluve corridor system** (Fig. 20.7*b*). Named for the fingers of upland terrain that lie between individual riparian corridors in a drainage network, these corridors follow drainage divides. They support distinctly different ecosystems than the adjacent stream lowlands and are less functionally integrated because they are not tied together by a flow system of drainage channels. Nevertheless, they also tend to be organized in a hierarchical fashion with main interfluves and connecting branches of progressively lower orders (see Fig. 19.10).

Linear corridors

Linear corridors are perhaps the least complicated corridor systems. The most glaring examples of linear corridors are rights-of-way for utilities and roads, but there are also striking natural examples, such as seashores and lakeshores. Shoreline corridors may be circular, as around an inland lake, or elongated, as along the seashore. When shoreline corridors are combined with riparian corridors, it is possible to define **corridor loops** or cells, as illustrated in Figure 20.7*c*.

Grid corridors

Grid corridor systems are common by-products of the land survey system in the American landscape. The rectilinear pattern of land use has given rise to straight hedgerows, tree corridors, and similar border habitats (Fig. 20.7*d*). In agricultural and residential areas, even drainage lines are often adjusted to the rectilinear grid. The resultant pattern is a grid network with cells of different sizes and compositions. Most cells are farmfields, but some are woodlots, wetlands, lakes, and ponds.

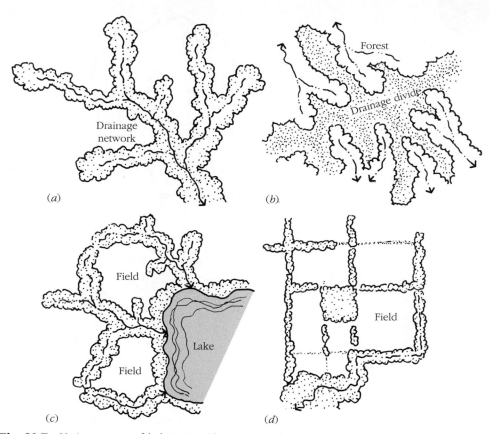

Fig. 20.7 Various types of habitat corridor systems that are common to the North American landscape: (*a*) riparian system; (*b*) interfluve system, which follows the divides between damage nets; (*c*) loop system; (*d*) grid system, which follows field margins, woodlots, and drainage ditches.

Fragmentation

Fragmentation of any corridor leads to **segmented** or **disjointed corridor systems**. Most riparian corridors are segmented as a result of road crossings, wetland eradication, farmland clearing, urban development, and many other uses. In many coastal areas, linear corridors are often so disjointed that the corridor is reduced to a series of patches or of remnant pieces protected from development by rugged terrain, wetland, or parks.

Spatial Parameters. Landscape ecologists and others have attempted to quantify some of the spatial or geographic attributes of patches, corridors, and other landscape features in an effort to provide systematic measures of a landscape's ecological potential and problems. Although the effort is laudable and brings a more analytical

Analysis

perspective to landscape ecology, much remains unknown about the biogeographical or ecological significance of these measures, that is, about their correlations with the makeup of plant and animal populations, productivity, animal migrations, and other biological attributes of landscape. Nevertheless, these measures are valuable in descriptive landscape ecology, and we review several here.

Descriptive measures

The density of landscape fragmentation is referred to as *patchiness*, which is simply the total number of patches of all sizes and types per unit area of land. A related measure is the *index of connectivity*, which is the ratio between the number of actual connections among patches in a study area and the maximum possible number of connections. The interaction or flow of organisms among patches appears to be influenced by the size of patches and the distance separating them. This is described

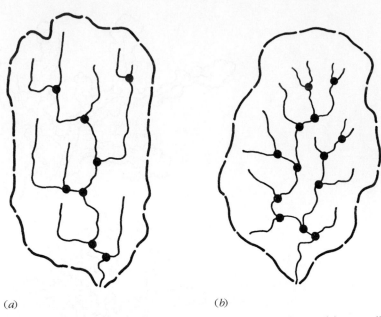

(a)   (b)

Fig. 20.8   The difference in the number of intersections, or nodes, in (*a*) a parallel drainage network (basin order = 3, nodes = 8) and (*b*) a dendritic drainage network (basin order = 3, nodes = 13). Nodes increase with drainage density.

by a model from geography, borrowed from physics, called the *gravity model* in which the relative attractiveness of two places (patches) to each other is a function of their sizes (masses) and the distance between them. In other words, the bigger the patches are (and therefore the larger their resource bases) and the closer they are, the more attractive they are for species.

Networks and nodes   The lines of linkage among patches and/or corridors define *networks,* and the intersections within networks are *nodes.* We have already examined various types of corridor networks, and it is easy to see how widely network connectivity can vary, as defined by the occurrence of nodes. For example, a riparian corridor system aligned with a parallel pattern of stream channels may have relatively few nodes (low linkage) for the total length of corridors in the system. However, a corridor system defined by a dendritic drainage pattern may have many nodes (high linkage), especially if a few intersecting land use corridors are added (Fig. 20.8).

20.6 HABITAT CONSERVATION PLANNING

Conservative programs   There are several approaches to minimizing habitat reduction and species loss. **Conservation programs** are one of the traditional approaches. These programs focus on open space: parks, wildlife preserves, wilderness areas, forest reserves, and landscape quality. Although parks and preserves, which are the most popular measures, have drawbacks for biodiversity preservation, they are nonetheless an important line of defense in the effort to slow habitat and species losses. Both Canada and the United States have taken leading roles in establishing national conservation programs and today have the largest areas of national parks and reserves in the world. But more conservation programs are needed, particularly in less developed countries where biodiversity is threatened by, among other things, tropical forest loss. Because most of the areas in need of conservation programs already contain local human populations, accommodations have to be made for these people.

Joint efforts

Integrated Approaches. *Integrated conservation-development* projects have been proposed as a way of reducing the land use pressure on protected areas such as parks and preserves while assisting local populations. The objective is to provide people with sustainable, income-generating opportunities. Such opportunities could include (1) purchasing additional land or negotiating for the use of land for farming and (2) setting up buffers around protected lands where certain economic activities are encouraged. Other approaches call for *integrated landscape management*, which involves coordinating government agencies, businesses, community leaders, landowners, and others in a region to ensure that biodiversity objectives are included in the overall planning and management process.

Area protection

In the United States, a program called **habitat conservation planning** emerged in 1982 as part of the Endangered Species Act. The objective of this program is to preserve specific areas for protected species within larger use areas such as lumbering. The selection of the habitats set aside for preservation involves both landowner and agents of the government, and the plan that is worked out is binding for 50 years. For landowners, this represents a no-surprise approach that enables them to develop a land use plan and use the remaining land without unexpected government interventions because of new species or habitat findings. Not surprisingly, the latter has become a point of contention, as government planners attempt to modify plans in light of new scientific findings about species ranges, migratory patterns, and so on.

Open Space Approach. Greenbelts, or *greenway systems,* are one of the oldest and most popular approaches to habitat conservation. In Europe and North America they have long been used in urban planning as a part of park and recreation planning. Although all sorts of land has been used for greenbelts in urban areas, stream corridors have been the favored settings for them because they frequently offer the only large belts of land available. Coincidentally, stream corridors are also the only remaining areas of rural type habitat in many urban regions (see Fig. 14.13). Today urban planners still justify greenbelts as multiuse areas for flood management, recreation, nature preservation, and other uses. Increasingly, however, habitat conservation and biodiversity are heading the list of rationales for creating greenbelts.

Stream corridors

Opportunity

New opportunities to create greenbelts for improved biodiversity are emerging in rural areas with the decline in family farming and the shift in rural land use patterns. As small farms fall out of production, fields are abandoned, especially where the land tends to be marginal, such as in floodplains and hilly terrain. The fields fill in with woody vegetation, topsoil begins to rebuild, and habitat generally improves. This process, however, does not automatically build habitat corridors because plowed fields, communities, and roads often remain as barriers. Thus, planning action that includes land acquisition and cooperation from private landowners is usually required to facilitate the linkage process (Fig. 20.9).

Fortunately, the concept of networks and spatially integrated systems is very much a part of the traditional land use infrastructure in North American rural areas. Small roads link to larger roads, villages and towns are linked by railroads and highways, and small streams are linked to larger streams to form drainage networks. Thus habitat restoration programs advocating development of corridor systems by linking habitat fragments together are not foreign to our way of thinking about landscape organization.

Counterforces

Barriers. Despite the increased attention to biodiversity and habitat conservation and restoration, strong counterforces are at work in the landscape. Chief among these is the emergence of major ecological barriers in the North American landscape. Consider the system of interstate highways and similar large roads, especially where development has added several tiers of land use along the highway. The resultant corridor is an imposing barrier, an ecological wasteland, that not only limits seasonal migrations and interbreeding among divided populations but interferes with longer-term

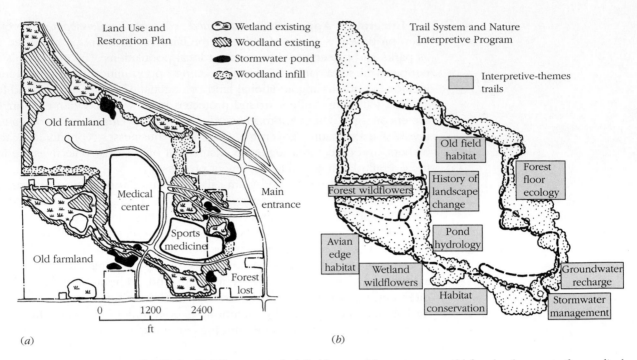

Fig. 20.9 Building a greenbelt/habitat corridor system on old farmland as part of a medical center campus. (*a*) The land use and restoration plan. (*b*) The resultant corridor with trail system and nature interpretative program.

Stand aside

migrations as well. A restriction on regional migration may be a serious impediment to the maintenance of certain populations because they will need to shift their ranges in response to the climate changes expected in the next century. Some experts expect the ranges of many midlatitude plants and animals to be displaced northward in response to global warming trends, and there is concern that the massive regional barriers imposed by farmland, urban areas, and transportation corridors in the U.S. Midwest, for example, will greatly inhibit this biogeographical trend.

Bioplanning Guidelines. We end this chapter with some general guidelines for local and regional projects involving habitat conservation planning and design.

Systems first
- Focus on **habitat systems** and associated ecosystems rather than on selected species, for without the proper habitat types and species mixes, most target species cannot be sustained.

Key locations
- Focus on **critical locations** because the most promising places selected for preservation and management should be those that meet organism preferences. High preference nodes, such as lakes and wetlands in riparian corridors, are the anchor sites in a habitat system.

The foundation
- Habitat systems should be founded on a firm **physiographic base**. Without a supporting physiographic system such as a fluvial corridor, habitat systems lack the resilience necessary for long-term sustainability. Accordingly, corridors made up of habitat fragments representing widely different types of physiography are less secure than corridors based on a single physiographic system or a set of interdependent physiographic systems.

Linkage
- Be mindful of **scale and connectivity**. In general, plan for fewer larger areas rather than for many smaller areas, but, where parcelization is natural and unavoidable, the linkage between parcels should not be forced. In addition, be cognizant that critical scales or spatial thresholds may exist for certain species and ecosystems,

and these may be tied to specific physiographic features, such as prairie potholes, seepage zones, cliff faces, and sinkholes.

Geographic scale

- **Species richness** decreases and range size increases over broad reaches of latitude poleward. This rule holds true for birds, reptiles, amphibians, and trees as well as for other groups of organisms. Therefore, the scale factor in habitat conservation planning may vary appreciably from south to north in countries as large as the United States and Canada.

Evaluation

- The changes in biodiversity, population densities, and other measures targeted in **habitat conservation planning** may show unexpected and/or disappointing results because (a) they often lag habitat improvements by years or even decades and (b) the scales and configurations of planning efforts may not be adequately understood for certain habitats and species. Thus, projects may inadvertently miss the mark.

20.7 CASE STUDY

Marsh Restoration for Bird Habitat in the Fraser River Delta, British Columbia

Patrick Mooney

Iona Island is located at the mouth of the Fraser River, in the Vancouver region of British Columbia, on Canada's west coast (Fig. 20.A). The primordial delta landscape here provided the native peoples with abundant food from the land, river, and sea. In the late 1880s, the diking and draining of the delta for farming were begun. By 1950 much of the native vegetation had been removed, the marshes drained, and tidal flooding halted. In about 1955 the first bridges spanned the north arm of the Fraser River, linking the delta to Vancouver and opening it to urbanization. Nevertheless, the delta remains a migratory bird habitat of international significance and supports the largest wintering population of raptors in Canada.

In 1987, much of Iona Island, situated in the northwestern corner of the delta was designated as a regional park. A constructed pond in the park had been recently filled, destroying wildlife habitat. In addition, the proposed expansion of the nearby Vancouver International Airport (YVR) would soon eliminate important habitat of the regionally rare yellow-headed blackbird (*Xanthocephalus xanthocephalus*). In response, a wetland restoration with the goals of maximizing general avian diversity and providing yellow-headed blackbird habitat was implemented in 1992.

Restoration of wildlife habitat requires a two-step modeling process. Using site analysis, literature review, expert advice, fieldwork, and laboratory testing, the restorationist seeks to understand first the habitat requirements of the intended species and then how the site conditions necessary to support the intended species can be developed.

The yellow-headed blackbird habitat model showed that the bird nests in open marshes, in hardstem bulrush (*Scirpus acutus*), adjacent to open water. Transects through nearby McDonald Slough revealed that hardstem bulrush grows at about a 2-foot depth and will not grow in much deeper water (Fig. 20.B). Cattails (*Typha spp.*) tolerate drier conditions and thus grow at higher elevations in the marsh. In the restoration, the pond bottom was contoured, and the marsh plants were placed to provide these depths while limiting the expansion of marsh vegetation into open water areas. Soil fertility, pH, and nutrient levels between the slough and the restoration site were laboratory tested to ensure that they matched. Existing

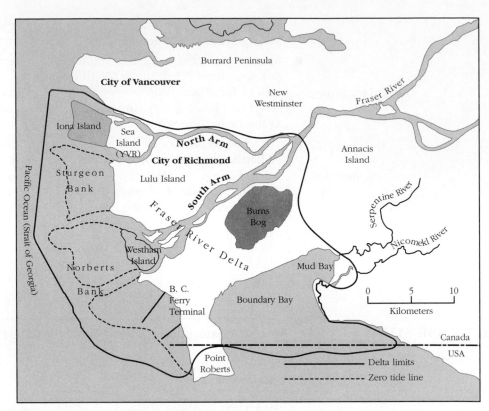

Fig. 20.A The Fraser River Delta stretches from the City of Vancouver to the American border. It originally contained extensive areas of wetland. The project site is located on Iona Island.

yellow-headed blackbird habitat at YVR provided emergent plants for the marsh and native shrubs and trees for the areas around the marsh.

The site was monitored for three years after completion. All the emergent plants colonized in their intended locations. In the spring of 1993, yellow-headed blackbirds moved into the restored pond. Two years later, 11 yellow-headed blackbird nests and 54 eggs were recorded. In the five years after the restoration, general avian diversity increased dramatically (Fig. 20.C).

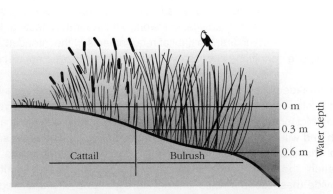

Fig. 20.B Hardstem bulrush is limited to water depths in the range of 1 to 2 feet.

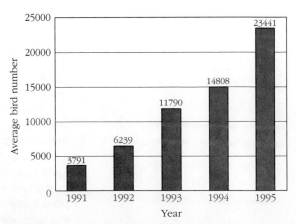

Fig. 20.C Iona Island bird survey totals, 1991–1995.

The project site was contaminated, however, with two exotic invasive plants: Scotch broom (*Cyticus scoparius*), and purple loosestrife (*Lythrum salicaria*). In subsequent years, manual cutting of the broom and the use of biological controls on the loosestrife have controlled these plants.

The lesson of this project is that significant biodiversity can be restored even in disturbed urban habitats. It demonstrates the benefits of ecological modeling in restorations as well as the dangers of not eliminating exotic invasive plants from a restoration site. Restoration projects should be monitored so that midcourse corrections can be made and so that the successes and failures of the restoration can be used to improve future projects.

Patrick Mooney is a landscape architect at the University of British Columbia who specializes in West Coast habitat restoration. He is a fellow in the Canadian Society of Landscape Architects.

20.8 SELECTED REFERENCES FOR FURTHER READING

Beatley, Timothy. *Habitat Conservation Planning: Endangered Species and Urban Growth.* Austin: University of Texas Press, 1994.

Berger, J. J. (ed). *Environmental Restoration: Science and Strategies for Restoring the Earth.* Washington, DC: Island Press, 1990.

Butler, R. W., and Campbell, R. W. "Birds of the Fraser River Delta: Populations, Ecology, and International Significance." *Occasional Paper 65.* Canada Wildlife Service, 1987.

Collinge, S. K. "Ecological Consequences of Habitat Fragmentation: Implications for Landscape Architecture and Planning." *Landscape and Urban Planning.* 36, 1996, pp. 59–77.

Cronon, William. *Changes in the Land: Indians, Colonists, and the Ecology of New England.* New York: Hill and Wang, 1983.

Edwards, P. J., et al. (eds.). *Large-Scale Ecology and Conservation Biology.* Oxford: Blackwell Scientific, 1994.

Falk, D. A., et al. *Restoring Diversity: Strategies for Reintroduction of Endangered Species.* Washington, DC: Island Press, 1996.

Forman, R. T. T., and Godron, M. *Landscape Ecology.* New York: Wiley, 1986.

MacArthur, R. H., and Wilson, E. O. *The Theory of Island Biogeography.* Princeton, NJ: Princeton University Press, 1967.

Orians, G. H. *Blackbirds of the Americas.* Seattle: University of Washington Press, 1985.

Robinson, G. R., et al. "Diverse and Contrasting Effects of Habitat Fragmentation." *Science.* 257, 1992, pp. 524–526.

Smith, D. S., and Hellmund, P. C. (eds.). *Ecology of Greenways.* Minneapolis: University of Minnesota Press, 1993.

Vos, C. C., and Ophdam, P. *Landscape Ecology of a Stressed Environment.* London: Chapman and Hall, 1993.

Related Websites

Greenbelt Land Trust. 2008. http://www.greenbeltlandtrust.org/
A nongovernmental organization working to conserve connecting pieces of land to create habitat corridors. Learn about current projects and their education and other environmental initiatives.

International Association for Landscape Ecology. U.S. Regional Association. 2009. http://www.usiale.org/

An international group of professionals interested in landscape ecology. Find general information and links to related resources.

National Oceanic and Atmospheric Administration (NOAA) Fisheries. Restoration Center. http://www.nmfs.noaa.gov/habitat/restoration/

The goals of this fisheries restoration group. The site also gives project information and links to publications for further reading.

U.S. Environmental Protection Agency. "Landscape Ecology." 2008. http://www.epa .gov/esd/land-sci/intro.htm

The EPA's site about landscape ecology. Includes an introduction to the subject, history, and links to current and past projects.

WETLANDS, HABITAT, AND LAND USE PLANNING

21.1 INTRODUCTION

Much of environmental planning has to do with edges—the lines and ribbons in the landscape where one environment gives way to another. No edge is more important than that between land and water. More than a billion people live on this edge, and the most productive ecosystems on earth are found there. Among these ecosystems are wetlands, the organically rich shallow-water environments along streams, lakes, and seas.

Why wetlands?
Long regarded as fringe environments of marginal utility for land use, wetlands have been the object of severe misuse for centuries. In the United States (less Alaska), it is estimated that since about 1800 as much as 50 percent of the original area of wetlands has been destroyed. In the past few decades, however, wetlands have found a solid place in the environmental agendas of the United States and a few other countries. Laws have emerged calling for their protection against the pressures of land development. The environmental rationale behind these laws is basically twofold: (1) Wetlands are important habitats necessary to the survival of a host of aquatic and terrestrial species. (2) Wetlands are integral parts of the hydrologic system necessary for the maintenance of water supplies and water quality.

In this chapter we examine wetlands as landscape systems, in particular as hydrologic systems that foster ecosystems. Attention is given to wetland types based on their physiographic settings, followed by a brief description of the U.S. comprehensive wetland classification system, wetland mapping, and approaches to wetland mitigation and management.

21.2 WETLAND MANAGEMENT RATIONALE

Practical considerations
Besides environmental quality, there are a number of purely practical reasons for respecting wetlands in land use planning. First, the places in the landscape where wetlands form are characterized by drainage conditions that are extremely limiting to most land uses. The sources of these conditions usually extend over an area larger than the wetland itself, and in most instances the conditions do not simply disappear by scraping the wetlands away. Therefore, attempts to build in wetland sites may significantly increase overall development costs because of the need for special allowances for site drainage, flood protection, and facility maintenance.

Second, most wetlands are usually underlain by organic soils that are unstable for most forms of development. To use such soils often requires special and often elaborate engineering schemes, or the soils must be removed by excavation and replaced with stable fill material. In either case, costs are increased significantly. Third, wetlands are landscape amenities and, like lakes and streams, can improve land values and design opportunities for the insightful developer. For many land uses, wetlands clearly enhance property values if they are properly integrated into land use planning schemes.

Definitions. Wetlands cover such a wide spectrum of physical conditions and ecological characteristics that it is difficult to arrive at a succinct definition of them. Scientists generally agree, however, that all wetlands have three characteristics, and these serve as a general definition:

■ The presence of water on the surface, usually relatively shallow water, all or part of the year.

Defining traits
■ The presence of distinctive soils, often with high organic contents, that are clearly different from upland soils.

■ The presence of vegetation composed of species adapted to wet soils, surface water, and/or flooding.

The regulatory agencies responsible for environmental policies have formulated various definitions of wetlands. The U.S. Fish and Wildlife Service uses the following definition, which planners have widely accepted:

Regulatory definitions

> Wetlands are lands transitional between terrestrial and aquatic systems where the water table is usually at or near the surface or the land is covered by shallow water. Wetlands must have one or more of the following three attributes: (1) at least periodically, the land supports predominantly hydrophytes, (2) the substrate is predominantly undrained hydric soil, and (3) the substrate is nonsoil and is saturated with water or covered by shallow water at some time during the growing season of each year.

In Canada, the following definition is used in the Canadian Wetland Registry:

> Wetland is defined as land having the water table at, near, or above the land surface or which is saturated for a long enough period to promote wetland or aquatic processes as indicated by hydric soils, hydrophylic vegetation, and various kinds of biological activity which are adapted to the wet environment.

21.3 WETLAND HYDROLOGY

Water, water . . .

Water is the most fundamental component of wetlands. Although vegetation is usually the most visible component of the wetland environment and is conventionally used to define wetlands for regulatory purposes, water is decidedly the driving force behind the origin and maintenance of wetlands. Indeed, the traditional practice of wetland eradication almost always involved draining the problem area and limiting the influx of the normal water supply.

The Hydrologic System. Each wetland can be described as a hydrologic system with inflows and outflows of water. Most wetlands exhibit a pattern or regime to the inflows and outflows that is manifested in rises and falls in the internal water level. These fluctuations have a wide variety of periods depending on wetland setting, water sources, and climatic situation. Some are rhythmical, as in the daily flux

Variable regimes

of water in tidal marshes; others are seasonal, as in stream valleys that flood in the spring; and still others are sporadic, as in isolated locales where water comes any time with the runoff from rainfall events. In all cases the hydrologic regime is essential to understanding and managing wetlands. Among other things, plant productivity and animal life cycles are adjusted to it.

Water Sources. There are four possible sources of water for the wetland system: (1) direct *precipitation*, (2) *runoff* from surrounding lands including inflowing streams, (3) *groundwater* inflow, and (4) ocean *tide water* (Fig. 21.1). All wetlands receive

Inputs and outputs

precipitation, and virtually all receive runoff of some sort in the form of streamflow, stormwater, and/or overland flow. Depending on the soil and geologic conditions controlling subsurface water, however, only certain wetlands receive groundwater, and, of course, only coastal wetlands receive ocean water or lake water. On the water loss side, wetlands lose water to evapotranspiration, seepage into the ground, stream discharge, and tidal outflow. Taken together, the inputs and outputs of water define the wetland's water balance (Fig. 21.1).

Water Storage. To maintain most wetlands, the water balance cannot fluctuate so radically that there are long periods without substantial inputs of external water. To buffer against such deteriorating events, wetlands usually hold a large reserve of storage water in the form of soil moisture and groundwater. Organic soils, which

Drought buffer

form the substratum of most wetlands, have a very high moisture-holding capacity. At full saturation, organic matter such as muck and peat can hold more than 6 inches

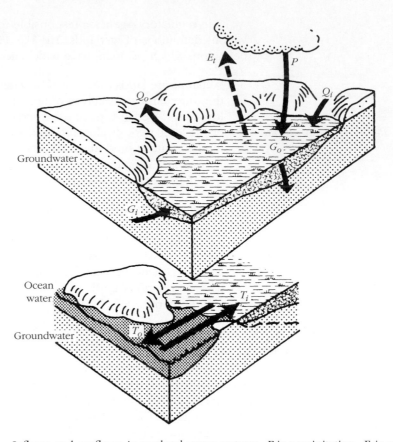

Fig. 21.1 Inflows and outflows in wetland water systems. P is precipitation, E_t is evapotranspiration, Q_i is stream inflow, Q_o is stream outflow, T_i is tidal inflow, T_o is tidal outflow, G_i is groundwater inflow, and G_o is groundwater outflow.

of water for each foot of soil. In addition, these materials have a high moisture transfer capacity whereby moisture is readily conducted from depth to the root zone by capillary action. These two hydrologic conditions account for the survival of many terrestrial wetlands—particularly those not attached to a major source of water such as a stream, lake, or aquifer—during prolonged periods of summer drought.

21.4 THE WETLAND ECOSYSTEM

Overview

Wetlands are attractive habitats for many plants and animals. The diversity of species is often higher in wetlands than in nearby upland landscapes, and where species diversity is not great, the population of individual species may be great. *Biomass* (the total weight of living matter per square meter of surface) is usually greater in wetlands than in nearby deepwater habitat or adjacent upland areas, and the productivity of the wetland vegetation, measured by the amount of new organic material produced each year, is typically greater than that of other habitats.

Energy flow

Ecosystem Processes. Ecological character and function are clearly the ranking attributes of wetlands in today's planning and management agendas. Among other things, wetlands are often cited as model ecosystems. An **ecosystem** is a biological energy system made up of food chains along which energy is passed from one group of organisms to another. The ecosystem's basic source of energy is the solar radiation, heat, and other essential resources taken up by plants in photosynthesis and converted into organic energy in the form of organic compounds (sugars and carbohydrates). Within the plant itself, this energy is moved along two paths. Some is used in respiration (plant maintenance processes), and some in growing new tissue (leaves, seeds, etc.). The total amount of new tissue manufactured in a year is termed *primary productivity*, and it is the source of energy upon which *all* other organisms in the ecosystem depend, either directly or indirectly, beginning with the herbivores and ending with the specialized predators (Fig. 21.2).

Productivity rates

Productivity. Wetland **productivity**, which is measured in grams of organic matter generated per square meter of surface per year, may typically be twice that of nearby upland vegetation. A salt marsh along the Gulf of Mexico, for example, may average close to 2000 grams (per square meter) per year compared to 1000 to 1500 grams per year in a neighboring subtropical forest. Many factors influence wetland productivity. Obviously, climatic conditions are important; northern bogs average only 500 to 600 grams per year because of growing season limitations. Hydrology is also important because it apparently influences the supply of nutrients. Wetlands with the highest productivity are those with a moderate but continuous flow of water;

Nutrient input

stagnant wetlands, by contrast, generally have low productivities. Wastewater can also affect productivity. Wetlands receiving enriched waters from stormwater and sewage treatment facilities usually show accelerated productivities, whereas those receiving pollutants such as heavy metals and petroleum residues not only show a decline in productivity but a decline in species diversity as well.

Organic Mass Balance. If wetlands are capable of producing large amounts of organic matter, they must also have the means of disposing of it. The balance between productivity and disposal or loss determines the **organic mass balance** of the wet-

The system

land. Changes in the organic mass balance are reflected in changes in the reserve of organic matter in the wetland substratum, that is, in the muck and peat deposits. When the mass balance is negative, loss exceeds productivity, and this reserve declines.

The processes responsible for the loss of organic matter in wetlands are decomposition by microorganisms, consumption by herbivores, erosion by surface waters,

Export

and leaching (chemical disintegration) to groundwater. The export of organic matter by erosion and leaching varies greatly depending on the local setting and hydrology.

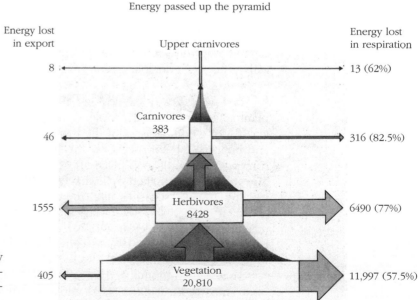

Fig. 21.2 Model of an ecosystem, called an energy pyramid, showing the four basic levels of organization, Silver Springs, Florida. All units are in kilocalories per square meter per year.

For tidal wetlands such as mangrove swamps and salt marshes, export may be as great as 40 to 50 percent of annual productivity. In wetlands that tend to be hydrologically closed, however, export is often negligible, and decomposition is virtually the sole means of loss of organic matter.

Decomposition   Among the controls on decomposition, water depth is the most critical because it, along with the mixing motion of the water, governs the availability of oxygen to many of the decomposing organisms. Under fully flooded conditions, little oxygen is available in the organic deposits and decomposition is slow, especially in stagnant water. If water is drained away to the point where the organic deposits are exposed to the atmosphere, decomposition rates rise dramatically. This is a principal reason for the decline in wetlands when they are artificially drained.

Vegetation Form and Composition. In a general way, the ecology of a wetland is also reflected in the character of its plant cover. Traditionally, wetlands are described according to the structure (or form) and floristic composition of the plant cover. Various terms, such as swamp, marsh, and bog, have been used over the years, and, although their meanings tend to vary somewhat from place to place, these terms *Marshes*   are still meaningful and still widely used. **Marshes** are dominated by herbaceous vegetation, typically bladeleaf plants such as cattails, reeds, and rushes (Fig. 21.3). Although these plants may reach a height of 6 feet or more, marshes often have the look of a grassland or meadow; indeed, some marshes are called wet meadows. Soils are typically rich with relatively high alkaline (pH) levels, making marshes attractive to agriculture in many areas.

Swamps are dominated by trees and shrubs (Fig. 21.3). There are many varieties of swamps in the United States and Canada. At the climatic extremes, for example, are cypress swamps in the American South (see Fig. 14.10) and northern conifer swamps *Swamps*   in the U.S. North and Canada. Northern conifer swamps may be dominated by various tree covers: spruce, tamarack, cedar, or balsam fir, which may occur in various associations with other trees and shrubs. Owing to the short growing season and persistently wet (or flooded) soils, the trees of the northern conifer swamps are often stunted, and at full maturity they may reach heights of only 10 to 20 feet.

Bogs are northern wetlands containing a wide diversity of vegetation. They are characterized by deep organic deposits, typically peat, and tend to be acidic. Bogs often form in ponds or small lakes where the vegetation is organized in concentric bands ranging from trees in the outer band to emergent and floating vegetation near the middle. Although bogs tend to fill in and become grown over in the long term (Fig. 21.3), many show a capacity to expand and contract with rises and falls in water level in response to changes in groundwater, streamflow, and obstructions such as beaver dams.

21.5 WETLAND TYPES AND SETTINGS

Wetlands can be classified in a variety of ways—for example, on the basis of vegetative cover (as previously described), hydrologic regime, or geographic (or physiographic) setting. The most basic control in shaping the wetland system is its *physiographic* setting, which includes the topographic situation, the proximal landscape (surrounding land use and vegetation), soils, and subsurface conditions (deposits and bedrock). Some combination of these factors produces a state of impeded (slow) drainage and/or abundant water supply that gives rise to wetland habitat.

Physiographic Setting. Learning to recognize the physiographic conditions that produce wetlands is the first step in understanding their function and maintenance because setting is critical to wetland hydrology. In many instances, the relationship between setting and hydrologic function is readily apparent from casual field observation, such as in cases of shallow waters along the shore of a pond or estuary. In other

Fig. 21.3 Photographs showing examples of major North American wetland types: marshes, swamps, and bogs: (*a*) coast salt marsh; (*b*) inland freshwater marsh; (*c*) deciduous swamp; (*d*) conifer swamp; (*e*) shrub swamp; (*f*) northern bog; (*g*) aquatic bed; (*h*) open stream.

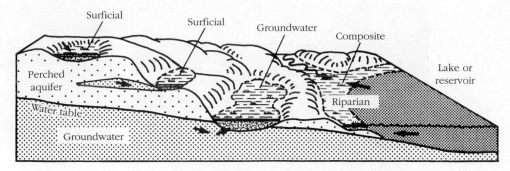

Fig. 21.4 A schematic diagram showing some types of physiographic and hydrologic conditions associated with surficial, groundwater, riparian, and composite wetland sites.

settings, however, it is not so apparent because sources of water or the conditions responsible for regulating water loss may be hidden underground or tied to sporadic hydrologic events. We can define four general classes of wetlands based on physiographic setting and hydrologic conditions: (1) surficial, (2) groundwater, (3) riparian, and (4) composite.

Localized systems

Surficial Wetlands. Sites that collect and hold surface water from direct precipitation and local runoff in the form of overland flow, ephemeral channel flow, and interflow often give rise to surficial wetlands (Fig. 21.4). These sites include shallow swales, closed depressions, and disturbed places where drainage has been blocked by deposits, tree throws, construction, or farming activity. Wetland formation begins with the establishment of a plant cover over the wet spot, which is followed by two changes that help to stabilize the site as a wetland environment: (1) the accumulation of organic debris and mineral sediments leading to the formation of a wetland soil and (2) the sealing of the wetland floor with fine sediments. These changes improve the wetland's overall water balance by increasing its water-retention capacity; nevertheless, surficial wetlands typically suffer from radical seasonal variations in water supply.

Local seepage

Surficial wetlands also include those supported by *perched groundwater*. These are lenses of groundwater that lie near the surface above the main groundwater body (Fig. 21.4). Where perched lenses intercept the surface, water seeps out, saturating the overlying soil and giving rise to wetland conditions. This often accounts for the occurrence of wetlands on hillslopes and at the heads of swales and small stream channels (see Fig. 7.6). In certain northern bogs where sphagnum moss is the dominant plant cover, the wetland may actually expand upward and outward beyond the limits of the seepage zone. This is attributed to the rise of water by capillary flow in the organic mass.

Below the watertable

Groundwater Wetlands. These sites are usually found at lower elevations in the landscape such as on the floors of stream valleys, sinkholes, and glacial kettles. These sites lie at or below the water table of the main groundwater body, and, as low-pressure points in the groundwater system, they receive groundwater discharge. Because the supporting aquifers are often large, the water supply to the wetland is substantial and, unlike surficial wetlands, is not subject to radical fluctuations with variations in precipitation and alterations in the surrounding land cover and surface drainage patterns (Fig. 21.4). Nevertheless, some groundwater wetlands, such as northern bogs, are subject to water level fluctuations in the range of several feet on a seasonal and longer-term basis as the water table rises and falls around them.

Riparian Wetlands. Riparian sites are found in and around major water features such as lakes, large streams, and estuaries (Fig. 21.4). These wetlands usually show a strong gradation in habitat with water depth from deepwater aquatic on the wet side to upland mesic on the terrestrial side. As the principal source of water, the controlling

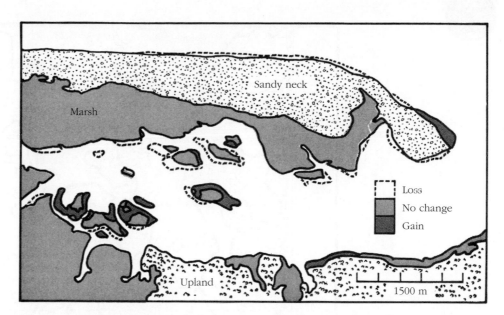

Fig. 21.5 Wetland changes in an estuarine marsh on the Atlantic Coast, 1859–1957. This illustrates a characteristic of wetlands that is not widely appreciated: Wetlands can retreat as well as grow under natural conditions.

On the water's edge water feature governs the wetland's hydrologic regime. This is a two-sided coin, however, because the water feature is also the source of destructive processes such as storm waves, floodflows, and ice movements that can cut wetlands back and in some instances obliterate them entirely (Fig. 21.5).

Composite Wetlands. Composite sites are those supported by two or more major sources of water. Most large, enduring wetlands fall into this class. For example, cypress swamps in river floodplains are dependent on both floodwaters and groundwater. Coastal marshes are often supported by a combination of tidal water, *Multiple systems* stream discharge, and groundwater. Because each source has a different regime, the principal supply of water to the wetland may change from season to season. For example, in the Great Lakes, stream discharge peaks in spring, whereas the lakes themselves reach their highest levels in late summer. Therefore, as the streamflows feeding coastal marshes subside in late summer, the Great Lakes often augment the wetland's shrinking water budget.

Understandably, the composite class of wetlands is often the most complex of the wetland systems and therefore the most difficult to manage because alterations may *Management implications* have compound effects on its various sources of water, on mechanisms for internal water transfer, and on water discharge. This is often illustrated by the construction of roads, canals, navigational facilities, and flood control structures where the impact on the wetland may initially appear to be minor but leads to gradual change that builds into substantial impact in the long run. One of the most controversial management dramas of this sort is currently being played out in the Florida Everglades.

21.6 COMPREHENSIVE WETLAND CLASSIFICATION SYSTEM

U.S. regulatory agencies Among the many governmental agencies concerned with wetlands, the U.S. Fish and Wildlife Service, the U.S. Environmental Protection Agency, and the U.S. Army Corps of Engineers are at the heart of the regulatory review process. In Canada, where there is no federal wetland policy comparable to that in the United States, wetland policy and regulation are handled mainly by provincial agencies. These agencies are responsible

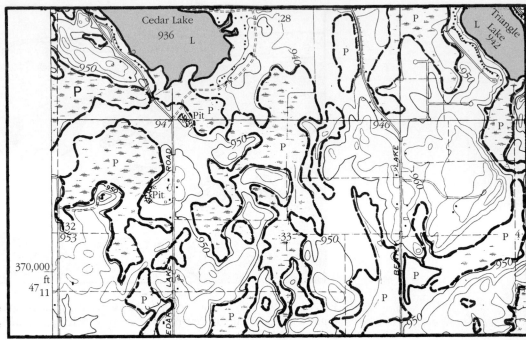

P: palustrine
L: lacustrine

Fig. 21.6 A portion of a map from the National Wetlands Survey. The base map is a standard U.S. Geological Survey topographic contour map. The code refers to wetland classification.

mainly for the review of proposals for permitting projects in stream and coastal habitats where wetlands may be involved.

Classification System. To facilitate the review process, the regulatory agencies in the United States have adopted a wetland classification system that is comprehensive in scope and part of the U.S. National Wetlands Survey (Fig. 21.6). In this system, both wetlands and deepwater habitats are addressed. Deepwater habitats are permanently flooded lands that lie below the deepwater boundary of wetlands. In saltwater settings, deepwater habitat begins at the extreme low water mark of low spring tide. In other waters, the boundary line is at 2 meters (6.6 feet) below the low water mark. This is taken as the maximum depth of growth of emergent aquatic plants. Wetland begins landward of the deepwater line.

Deepwater limits

System Types. The classification scheme is organized into three basic levels beginning with wetland systems (Table 21.1). Five **wetland systems** are recognized: marine, estuarine, riverine, lacustrine, and palustrine. The *marine system* consists of the deepwater habitat of the open ocean and the adjacent marine wetlands of the intertidal areas along the mainland coast and islands. The *estuarine system* is associated with coastal embayments and drowned river mouths and includes salt marshes, brackish tidal marshes, mangrove swamps, as well as deepwater bays. The *riverine system* is limited to freshwater stream channels, and the *lacustrine system* is limited to standing waterbodies, mainly lakes, ponds, and reservoirs. Both the riverine system and the lacustrine system include deepwater habitat. By contrast, the fifth system, the *palustrine system*, includes only wetland habitat. The palustrine system is the major system because it encompasses the vast majority of North America's wetlands, namely, inland marshes, swamps, and bogs.

Wetland systems

Wetland Classes. Beyond the system level, *subsystems* are defined for all but the palustrine system (middle column, Table 21.1). W*etland classes* are defined at the third level, and they represent either basic habitat types or wetland types based on vegetation. Among the palustrine wetlands, the three major classes are (1) emergent wetland, (2) scrub-shrub wetland, and (3) forested wetland. *Emergent wetlands* are dominated

Palustrine classes

Table 21.1 Three-Level Classification of Wetlands and Deepwater Habitats

Systems	Subsystems	Classes
Marine	Subtidal	Rock bottom Unconsolidated bottom Aquatic bed Reef
	Intertidal	Aquatic bed Reef Rocky shore Unconsolidated shore
Estuarine	Subtidal	Rock bottom Unconsolidated bottom Aquatic bed Reef
	Intertidal	Aquatic bed Reef Streambed Rocky shore Unconsolidated shore Emergent wetland Scrub-shrub wetland Forested wetland
Riverine	Tidal	Rock bottom Unconsolidated bottom Aquatic bed Streambed Rocky shore Unconsolidated shore Emergent wetland
	Lower Perennial	Rock bottom Unconsolidated bottom Aquatic bed Rocky shore Unconsolidated shore Emergent wetland
	Upper Perennial	Rock bottom Unconsolidated bottom Aquatic bed Rocky shore Unconsolidated shore
	Intermittent	Streambed
Lacustrine	Limnetic	Rock bottom Unconsolidated bottom Aquatic bed
	Littoral	Rock bottom Unconsolidated bottom Aquatic bed Rocky shore Unconsolidated shore Emergent wetland
Palustrine[a]		Rock bottom Unconsolidated bottom Aquatic bed Unconsolidated shore Moss-lichen wetland Emergent wetland Scrub-shrub wetland Forested wetland

The whole table is grouped under the vertical heading: WETLANDS AND DEEPWATER HABITATS

[a] The Palustrine system does not include deepwater habitats.

by herbaceous vegetation including grasses, cattails, rushes, and sedges. *Scrub-shrub wetlands* are dominated by short, woody vegetation (shrubs and trees less than 20 feet high), and *forested wetlands* are dominated by trees taller than 20 feet.

21.7 WETLAND MAPPING

Wetland mapping has become one of the most important inventory activities in environmental planning. No matter the nature of the project, it is essential to identify not only wetlands but their geographic limits as well. As with virtually all efforts to map landscape features, three problems must be faced at the outset: (1) the criteria or indicators to be used as the basis for boundary delineation, (2) the sources of data, and (3) the mapping resolution, that is, the level of geographic detail required.

Prerequisites

Mapping Criteria. Three criteria are consistently named in regulatory policy for wetland mapping purposes: vegetation, soils, and hydrology. Depending on state and local guidelines, all three, two, or just one criterion may be required to define a wetland.

Vegetation. Vegetation is the most commonly used criterion in wetland mapping. Many wetland edges, such as the one shown in Figure 21.7, can be identified by a relatively abrupt change in the *vegetative structure*, especially where the lowland/upland topographic transition is sharp. This border may also be marked by a third vegetative form, a belt of shrub-size plants fringing the wetland, which is also evident in Figure 21.7. Whereas structure is useful in wetland mapping, *floristic composition* of the plant cover is considered more reliable for boundary demarcation. Certain plants can be used as wetland indicator species, and their dominance may be taken as sound evidence of wetland habitat, although there can be much debate over this. Indicator species vary from region to region, and in most locales a preferred list of indicator species can be obtained from environmental agencies, a university botanist, or environmental organizations (also see website Appendix F).

Structure and composition

Fig. 21.7 Wetland border marked by the abrupt change in vegetative structure.

Hydric soil

Soil. The term *hydric* is widely used to describe wetland soils. These are mainly organic soils, such as muck or peat, but may also include saturated mineral soils. Soils are generally used as a secondary indicator after vegetation. Care should be taken in using published soil maps to corroborate wetland borders initially mapped on the basis of vegetation patterns because many of the boundaries that appear on soil maps are based in part on vegetation. Thus, the discovery of a strong correlation between hydric soils on published soil maps and wetland vegetation may constitute circular reasoning. Field checks of soils are usually advisable in wetland mapping.

Often inexact

Hydrology. The hydrologic criterion is defined by flooding. If an area is periodically or frequently flooded and lies in a flood zone such as a designated floodplain, these conditions are taken as evidence of wetland. Whereas flood-prone areas such as floodplains are readily identifiable, the boundaries of such areas are often inexact because they are normally defined by hydrological calculations rather than by field measurement of floodwater coverage. Therefore, the delineation of wetlands in terms of fixing a boundary line should usually be based more on vegetation and soils than on flood zones. (Also see Section 10.4.)

Topographic maps

Data Sources. The sources of data and information for wetland mapping fall into two classes: (1) published sources in the form of topographic maps, aerial photographs, and soil maps; (2) field surveys. Large-scale *topographic contour maps*, published in the United States by the U.S. Geological Survey and in Canada by the National Mapping Branch and various provincial agencies, mark wetland areas larger than 10 acres or so. These areas are mapped from aerial photographs on the basis of visible surface water, vegetation patterns, topographic trends, and proximity to water features such as lakes and streams. *Soil maps,* such as those published by the U.S.

Soil maps

Natural Resources Conservation Service, do not necessarily show wetlands, but they do depict hydric soils. As noted, however, the original placement of the soil borders that appear on the NRCS maps was often guided by vegetation and topographic lines. Therefore, these borders cannot always be taken as reliable independent indicators at the local scale. Once again, field verification is usually necessary when using soils for wetland delineation.

Aerial photographs

Aerial photographs are especially helpful in wetland mapping. The U.S. Fish and Wildlife Service has used small-scale (high-altitude) infrared photographs to build crude wetland maps for much of the United States. But for project planning purposes a much higher level of mapping resolution is needed that requires large-scale aerial photographs enhanced by stereoscopic perspective. Stereoscopic models afford an exaggerated view of the vegetation structure and forms, and textural differences in forest cover and bladeleaf vegetation, for example, help in distinguishing wetland areas. In addition, infrared imagery provides enhanced scenes in which water, different vegetation types, and exposed soil contrast more sharply with each other than they do on standard aerial photographs.

Ground truthing

Field Verification. In cases calling for the development or alteration of sites containing wetland, *field verification* and the refinement of wetland borders are usually necessary: walking the border and taking notes of indicator plants, soils, and evidence of high water. The evidence of high water includes water marks on trees, debris stranded on low branches and foliage, stranded driftwood on the ground, and shallow and exposed tree roots (see Fig. 14.14). As the border is identified, it should be flagged so that it can be fixed by field survey and plotted on site maps.

21.8 MANAGEMENT AND MITIGATION CONSIDERATIONS

Public values

Wetland mitigation begins not on the development site or at the water's edge, but with programs dealing with public attitudes and information on wetlands. As with many environmental problems, the first steps toward solutions involve improved public awareness followed by understanding of the value of wetlands in the environment. In other words, society has to know enough about something and feel strongly enough about it to make a place for it in the great agenda of environmental problems. That mindset has already been established with wetlands in a general way and has led to enforcement of wetland protection laws such as Section 404 of the U.S. Clean Water Act, as well as the enactment of new laws at the state and local levels.

Managing land use

Setting the Agenda. How to balance society's need to use land and the need to protect wetlands, however, requires the attention of planners and scientists working at the level of individual land use sites. This involves two levels of activity: (1) management of existing land uses and (2) planning future land uses. The principal challenge with existing land uses arises from those activities that continue to damage and destroy wetlands. Chief among these is agriculture, which in the United States has been responsible for 70 to 90 percent of the wetland loss since the 1950s.

Exempt uses

Limiting future wetland loss to agriculture is important, but the mitigation of damage from existing operations is also needed. Unfortunately, few agricultural activities are subject to wetland regulation, and until national policy is extended to include agriculture as well as forestry and mining activities, U.S. wetland protection law will remain somewhat hollow. This is not the case, however, with new urban development. Residential, industrial, and commercial land use plans in most areas are carefully scrutinized for wetland conflicts by federal, state, and local agencies.

Types of mitigation

Wetland Mitigation. The search for land use compatibility with wetlands may be approached along two lines: mitigation and management planning. **Mitigation** involves taking certain actions to counterbalance wetland losses or damage. Three types of mitigation are usually practiced: (1) restoration and enhancement of damaged environments, (2) creation of new wetlands, and (3) wetland preservation through the control of potentially damaging actions. The third type usually includes various means of controlling construction activity, stormwater runoff, soil erosion, sedimentation, and building encroachment, but it may also involve various land protection arrangements such as deed restrictions and environmental easements.

Replacement

With respect to wetland restoration, enhancement, and creation, there are two approaches: replacement mitigation (also called on-site mitigation) and mitigation banking. **Replacement mitigation**, in its simplest form, involves building new wetlands to offset acreage lost or damaged as a result of development. Normally, a replacement ratio, such as 2 acres of replacement for 1 acre of loss or damage, is scheduled as a part of the mitigation plan. Replacement may be on the project site or in a neighboring location, but it is always tied to an individual project. **Mitigation**

Banking

banking, on the other hand, is a system of compensation credits in which surplus wetland mitigation acreage (in the form of replacement, enhancement, restoration, and/or preservation) is banked against future needs. This system is not project specific, and, if structured as part of a habitat conservation plan within a rational geographic framework such as a local or regional watershed, it is a promising approach to compensatory wetland mitigation.

Management Planning. Management planning is central to wetland preservation but not widely practiced. Almost invariably management for preservation must address the larger system of which the wetland is a part. This begins with an inventory

Systems approach

of the principal sources of water and the controls on the waterflow system, followed by an evaluation of the vulnerability of the flow system to alterations from land use that could affect the wetland.

Example

In the case of the wetland in Figure 21.8, for example, the principal sources of water happened to be vegetated swales heading up in the surrounding upland. According to the state regulations governing the site, a buffer zone of fixed width was to be designated around the entire wetland to protect it against land use encroachment. Because of the character of the topography and drainage patterns around the wetland, however, such a buffer definition did not offer adequate protection for the sources of

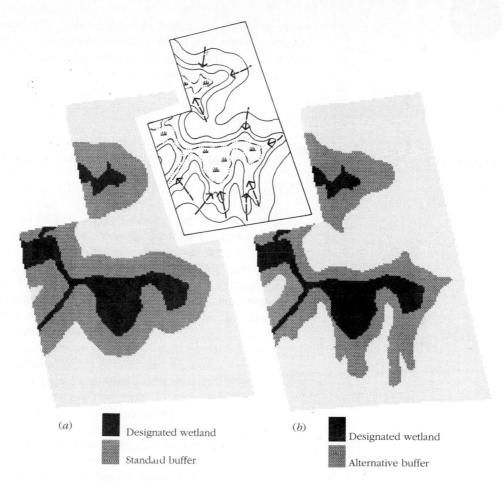

(a) ■ Designated wetland
▓ Standard buffer

(b) ■ Designated wetland
▓ Alternative buffer

Fig. 21.8 Maps showing (*a*) a standard 100-foot wetland buffer and (*b*) a buffer adjusted to the configuration of the drainage patterns around the wetland. The inset shows the topography and associated drainage pattern around the wetland.

runoff. Therefore, an alternative buffer concept was devised based on the configuration of the swale drainage system. The total area of buffer remained about the same, but the performance of the buffer in terms of the inflow of water and the long-term maintenance of the wetland improved with the functional buffer concept.

The main point of wetland management planning is to approach wetlands in much the same way as we would an inland lake, that is, to treat the contributing watershed as the main vehicle for preserving the waterbody. This includes understanding not only the extent and nature of the water sources, but also the role of water storage and the controls on water release in the overall maintenance of the wetland. As with

Watershed attention lake watersheds, the sources of water have to be evaluated in terms of the directness of their linkage to the wetland, their relative contributions, and whether they tend to fall into the manageable or nonmanageable class. Nonmanageable sources are those beyond the reasonable reach of management, such as precipitation and deep sources of groundwater. For manageable sources that make significant contributions to the wetlands, care should be taken to ensure that the availability of water is maintained and that the delivery system is not impaired by structures, grading, or stormwater diversion.

The use of wetlands in stormwater management is widespread today, and, while generally encouraged, the practice is seriously questioned for wetland habitats prone to damage from changes in water regime and water quality. Marshes polluted by many years of urban and industrial runoff, for example, have been discovered to decline sharply in species diversity, eventually reaching a state best described as a monoculture. The levels of permissible pollutant loading in wetlands are clearly a matter of much uncertainty, which at this writing awaits further research.

21.9 CASE STUDY

In Search of Better Wetland Regulation and Management

Gary F. Marx

It is a paradox that, although public support is nearly unanimous for programs that protect the environment and preserve natural areas, wetlands are the only critical habitats actually protected by national law. Despite this national mandate, developers and property owners of all stripes complain publicly that stringent wetland regulations rob them of their constitutional rights to the use of their property. If one believes their rhetoric, wetlands apparently receive comprehensive and unbending protection from development. A closer inspection of the nuts and bolts of these programs suggests that in reality things are quite different. Though some wetland areas receive protection under wetland laws, many acres of wetlands are exempt from protection, others are degraded by nearby development, and still others are lost to piecemeal encroachment by individual property owners.

First, it is commonly believed that the law prevents development in wetlands. Actually, the law just requires a permit before regulated activities can take place. Though the restrictions are stringent, wetlands are routinely filled in under permit for the construction of roads and driveways, ponds, and large commercial developments where the requirements for permit issuance have been met (Fig. 21.A)

Second, many activities are exempt from wetlands regulation. The most significant of these activities is farming. No permits are required for farming and farming-related activities in wetlands, including many projects to drain, plow, and plant wetlands as part of normal farming practice. The result is that thousands of wetland acres annually are converted to farmland and their wetland value as habitat is lost. Forestry activities, including logging, are also exempt from wetland regulation. Though the cutting of trees does not completely alter the hydrology of these

Fig. 21.A Front yard fill for additional lawn destroyed a patch of wetland and reduced the runoff buffer on the lakeshore.

wetlands, habitat quality is lost or greatly reduced when the vegetation is removed. In addition, the heavy equipment used in most logging operations causes considerable disturbance to surface drainage and soils, leaving them vulnerable to serious erosion problems long after the loggers have left the scene.

This brings us to the problem of making quality distinctions among wetlands. Current wetland regulations do not distinguish between high- and low-quality wetlands. If enacted, such provisions could focus protection efforts on high-quality ecosystems, while allowing more liberal treatment of those of lower quality. But such an allowance would be nearly impossible to administer for several reasons. First, defining the quality of wetlands is technically very difficult. Second, any scheme developed for this purpose is open to attack from "experts" hired by the property owner. The result would be something resembling a full employment act for lawyers and wetland experts and requiring heavily increased costs and time to administer the program. This explains the opposition of environmental groups to include a quality factor in the wetland regulatory scheme. Consequently, all wetlands are treated as equal in the eyes of the law, requiring that as much effort is expended protecting marginal and submarginal wetlands as in protecting high-quality systems.

Next is the problem of excluding wetland watersheds from protection. In some states, wetland regulations do not extend beyond the immediate boundaries of the wetland itself. This means that development can occur right up to the very edge of the wetland, including important runoff source areas. Wetland watersheds may be cleared, graded, stormsewered, and developed without regard to effects on the hydrology and biota of adjacent wetlands. Frequently, large-scale grading associated with site preparation produces massive soil erosion, resulting in heavy sedimentation of wetlands. Wetland regulations do not consider this to be a regulated situation even though wetland quality and functional capabilities are lost. The diagrams in Figure 21.B illustrate three approaches to development in wetland watersheds. The top one is clearly inappropriate, whereas the lower one is clearly preferred.

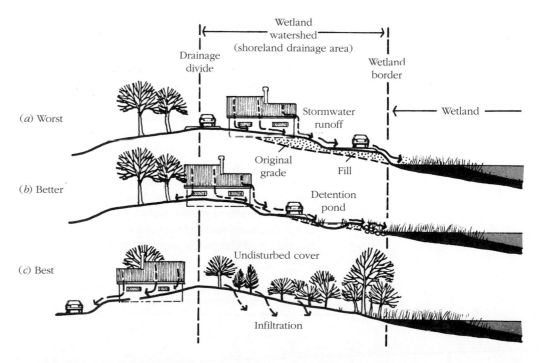

Fig. 21.B Wetland watershed considerations. Development on the back side of the drainage divide (option *c*) would protect the wetland from development runoff.

Another problem is the difficulty in dealing with the secondary impacts of proposed development activities, especially residential developments. For example, where a proposed subdivision is designed around existing wetland areas, only small areas of wetland fill may be required for road crossings and utilities. The issuance of wetland fill permits for such projects is routine but often fails to address the larger impacts that occur when individual homes are built in the development. Many of the lots in these subdivisions extend into the regulated wetland with a building site provided in the upland portion of the lot. Many owners assume the developer handled all the necessary permits as part of the original planning process and believe they can do as they please within their own lot, including draining and filling of the wetland.

In one particularly troublesome case, an application was filed to place fill in less than one-tenth of an acre of wetland for access to a development area of 30 lots on a peninsula in a freshwater lake. A belt of wetland fringed the entire peninsula, and the lots were platted so that each lot extended from interior upland across the wetland to the lake shore. A permit was issued for the road crossing, with the stipulation that lots would be deed restricted against the placement of fill material and the cutting of trees larger than 4-inch diameter.

The first purchaser of a lot in the subdivision cut every tree smaller than 12 inches in diameter, placed stone riprap shore protection along his entire shoreline, and graded the site level between the remaining few trees to facilitate lawn development. The purchaser of the next lot also removed all but the largest trees and graded the site down to the water's edge. Though enforcement actions will be brought against these owners, the damage is already done, and for the most part, it is irreversible. Twenty-eight lots remain to be sold in this subdivision, and if each owner is similar to the first two, the enforcement effort to keep up with the situation—given the huge caseload of field officers—will be overwhelming.

This raises an issue that is at the heart of many compliance problems with private property owners: education. Though people hear about wetlands and are aware that they should be protected, most do not recognize the variety of wetlands that are protected and are further resistant to regulation when it affects activities on their own property. "What effect can my little infringement possibly have? It's the developers who do the real damage." But, of course, they fail to see the big picture and the cumulative impacts of numerous small wetland encroachments in a drainage basin. Such impacts can have devastating effects on the fish and wildlife of such watersheds, as well as significant losses of water quality, groundwater recharge, flood damage amelioration, and other wetland values that can be replaced only at great expense if they can be replaced at all.

Though some of these situations can be addressed by more aggressive enforcement and attention to detail, others require changes in the law to reduce the loss of wetland area that continues to occur under the current laws. Although the current political climate makes such changes to the law unlikely, it is important to the long-term vitality of our natural resources to hold the line on any changes in the law that would further weaken the government's ability to protect the wetlands that remain.

Gary F. Marx is district supervisor in charge of wetland permitting and regulation for the Michigan Department of Environmental Quality.

21.10 SELECTED REFERENCES FOR FURTHER READING

Cowardin, L. M., et al. *Classification of Wetlands and Deepwater Habitats of the United States.* Washington, DC: U.S. Government Printing Office, 1979.

Environmental Law Institute. *Wetland Mitigation Banking.* Washington, DC: Environmental Law Institute, 1993.

Frayer, W. E., et al. *Status and Trends of Wetlands and Deepwater Habitats in the Coterminous United States, 1950s to 1970s.* St. Petersburg, FL: Fish and Wildlife Service, U.S. Department of the Interior, 1983.

Mitsch, W. J., and Gosselink, J. G. *Wetlands.* New York: Van Nostrand, 1986.

National Wetlands Policy Forum. *Protecting America's Wetlands: An Action Agenda.* Washington, DC: Conservation Foundation, 1988.

Redfield, A. C. "Development of a New England Salt Marsh." *Ecological Monographs.* 42, 2, 1972.

Salvesen, David. *Wetlands: Mitigating and Regulating Development Impacts.* Washington, DC: Urban Land Institute, 1990.

Sather, J. H., and Smith, R. D. *An Overview of Major Wetland Functions and Values.* Washington, DC: Fish and Wildlife Service, U.S. Department of the Interior, 1984.

Tiner, R. W., Jr. *Wetlands of the United States: Current Status and Recent Trends.* Washington, DC: U.S. Government Printing Office, 1984.

U.S. Fish and Wildlife Service. *National Wetlands Priority Conservation Plan.* Washington, DC: U.S. Department of the Interior, 1989.

Related Websites

Ducks Unlimited. http://www.ducks.org/
A large wetland and waterfowl conservation organization. Learn about their programs, methods of conservation, and wetlands as habitats. Also find things going on in your state.

Environmental Protection Agency. Wetlands Mitigation Banking Factsheet. 2009. http://www.epa.gov/wetlands/facts/fact16.html
Governmental website giving information about wetland mitigation banking. Read about the status, history, and benefits of the program. Also find further reading sources.

Montana Audubon. Streams and Wetlands. 2008. http://www.mtaudubon.org/issues/wetlands/index.html
A large national group working for the preservation of bird habitat. Read their position on wetlands and their services. The site also includes information on wetland land use planning at the community level.

U.S. Fish and Wildlife Service. "National Wetlands Inventory." 2009. http://www.fws.gov/wetlands/index.html
Wetland mapping systems and status of wetlands across the country. This federal site also talks about related topics such as wetlands and climate change.

FRAMING THE LAND USE PLAN: A SYSTEMS APPROACH

22.1 INTRODUCTION

Early progress

When Ian L. McHarg first published his classic statement, *Design with Nature,* in 1969, and the U.S. Congress passed the National Environmental Policy Act a year later, we in academia and the professions were hopeful that a new era of landscape planning and design had emerged, one that would see the rise of environmentally responsive land use development. Happily some progress did take place in the ensuing decades. Today almost everywhere we give consideration to floodplains, wetlands, air quality, and stormwater management. Species protection is given national attention in both the United States and Canada, and many jurisdictions have enacted policies concerning streams, shorelands, watersheds, groundwater, and open space. Unfortunately, however, the character of development, as well as how it relates to the landscapes it occupies, has not changed much, and on some fronts has declined over the past 40 years. All the while, our knowledge base on the North American landscape and our technical and economic power as individuals, communities, and nations have grown substantially.

Character of development

Modern land use in North America tends to occupy the landscape rather than live within it as other life-forms must do. But if we quiz citizens about their habitat preferences, invariability we find the desire to live with the landscape, even to embrace it and celebrate it. Why, then, can we not design and build communities, neighborhoods, and homes to satisfy that desire? We have the knowledge to do it and a clientele with open arms. Perhaps the clientele does not quite offer open arms because the desire to "live with nature" carries a variety of meanings in North America. But even if we assume that society's notion of living with nature is broadly similar to ours, we must agree that barriers, often substantial ones, stand in the way of designing and building with nature. The proof is in the human landscape around us (Fig. 22.1).

An idea of nature

Common barriers

Some of these barriers are mentioned in the opening chapters of the book. High on the list is the absence of a basic understanding of landscape among community officials, developers, and their agents followed by the assumption that one landscape is pretty much like another, especially in how they function. Granted, recognition is usually accorded to extremes in the landscape such as between seashore and floodplain, but less salient phenomena—especially those at the site scale, such as hillslopes, swales, seasonal streams, rock outcrops, and wet pockets—are usually written

Fig. 22.1 Common scenes of residential development in the North American landscape. Although residents express the desire to live with nature, little of natural landscape character remains after most modern development is in place.

Form and function gap

off as so much noise with little or no meaning in a functional sense (Fig. 22.2). This sort of thinking has, in many quarters, given tacit endorsement to shoddy site planning and design practices that ignore most form and function details and that replace thoughtful attempts at understanding sites as part of the larger working landscape—as parts of systems—with various shorthand approaches involving simple checklists, templates borrowed from other projects, makeover engineering schemes, and so on.

The systems thread

This chapter presents an approach to building landscape plans that draws on systems concepts. It follows, more or less, the main thread of the book, arguing that landscape planning and design schemes must extend into the arena of landscape dynamics and grapple with the systems and processes that shape sites and their settings, for unless we address systems, there is little chance of achieving sustainability in the landscape. The approach outlined is not intended as a methodology but more as a conceptual model aimed at providing a rationale for framing the plan and providing perspective. We begin with systems.

22.2 GETTING A HANDLE ON SYSTEMS

Objective

Using systems as a beginning point in landscape planning might, on first thought, appear a bit daunting, implying the need for all sorts of scientific knowledge and detailed field investigations. But for planning purposes, we are not after an analytical understanding of systems. Our objective, rather, is more contextual: first to identify the kind of landscape system we are dealing with and second to use that information to frame our thinking in the early stages of the planning process. Of course, analytical insight into landscape systems is not to be ignored if it is available. But the process of designing a land use plan is usually not an analytical one, as preparers of environmental impact statements have come to learn over the years. Rather, the process, we argue, involves responding to the system (or systems) that drives, or shapes, the landscape, what we call the formative system. Table 22.1 lists a number of major formative systems, headed by the principal landscape system, watersheds and drainage nets.

Identifying Formative Systems. Formative systems consist of the systems and related conditions that govern the character and operation of a site and its contextual space. The opening challenge to planners and designers is to identify which among the candidate landscape systems is the formative one or ones at a given

Fig. 22.2 A landscape rich in natural detail that has been bullied by heavy-handed site engineering in which features important to landscape character and function, such as swales and wooded hillslopes, have been eradicated.

Table 22.1 Major Formative Landscape Systems and Their Components

System	System Component Contributing/Input	Transport/storage	Receiving/Output
Watershed and drainage net	Headwaters, uplands, wetlands, lakes, springs	Stream channels, floodplains, swales, wetlands	Deltas, bays, estuaries, wetlands, ponds, reservoirs
Groundwater	Recharge zones: valley floors, basins, floodplains, wetlands, lakes	Transmission zones, aquifers	Discharge zone: stream channels, springs, lakes, wetlands, wells
Longshore (drift)	Source areas: deltas, shores, banks, bluffs, cliffs	Transport zone (current and beach drift)	Sediment sinks: bays, spits, bars, beaches, barrier islands
Wind	Source areas: beaches, denuded soil, floodplains, deflation hollows	Transport zone (e.g., wind corridors)	Sand dunes, loess deposits, beach ridges
Wetland	Watershed, aquifer, precipitation, flood flows	Stream channels, interflow, groundwater transmission	Stream discharge, springs, groundwater seepage, evapo-transpiration
Terrestrial ecosystem	Climate (light and heat), soil (water and nutrients)	Food chains, biomass, topsoil	Organic matter, heat, water vapor, nutrients

location—the one or ones that deliver the forces that most shape a site's essential character. The route to the answer lies in first finding the *geographic context* of a site, or more precisely, defining the site's physiographic character.

Every site has a place, and every place is the product of the systems and processes that operate there. All places have **physiographic character**, and it is that character that leads us to the identity of the formative systems. In Chapter 2 we discussed the broadly regional physiography of North America, but here we are interested more in local physiography and what it means in terms of defining formative systems. For example, if the site's physiographic setting is the floor of a stream valley, then the formative system is *Physiography and systems* the stream, its drainage net, and the watershed that feeds them. If the physiographic setting is in the coastal zone, then the longshore (drift) system, driven by wind, waves, and currents, is the likely candidate. To most coastal sites, drainage nets and watersheds are probably of little or no consequence compared to longshore systems, unless, of course, the site lies on or near a delta, river mouth, or major stormwater outfall.

A surprising amount of insight can be gained from this simple step. Among other things, it tells us what to put on the list of planning considerations, that is, what to look for, examine, and address. Many communities miss this point because they prepare standardized inventory lists that usually ask about things such as wetlands and floodplains, no matter where the site is located, while completely ignoring other systems that operate in and around their jurisdiction.

The Significance of Location. The next step is determining *where* you are in the *Location in the system* system. All landscape systems function as *open systems*, meaning they receive inputs of matter and/or energy and release outputs of matter and/or energy. Within the system, work may be performed, and energy and matter may be stored. In a watershed, work is performed when water and sediment are moved downstream. Storage takes place when water is taken up in aquifers as groundwater or when sediment is deposited in floodplains and locked in place by vegetation. Generally speaking every site in *System components* a watershed belongs to one of the following: (a) a zone that gives up (or contributes) water and sediment, (b) a zone that collects water and sediment and/or conveys it downslope or stores it, and (c) a zone that releases water and sediment at the output end of the system (Fig. 22.3).

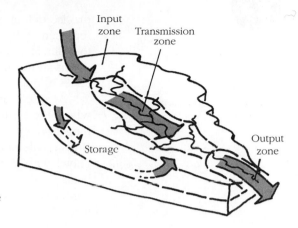

Fig. 22.3 The basic components and functions of an open system using the watershed as an example.

Design implications

Consider the significance of location for the sites at (*a*), (*b*), and (*c*) in Figure 22.4. A site located in a zone that gives up water [a headwaters setting at location (*a*), for example] must be given serious consideration with regard to actions that alter system performance and cause impacts downstream. To maintain the system's long-term performance, that is, to achieve sustainability in the watershed system, a land use plan should be designed to mimic the natural performance of the site (or predevelopment performance, whichever is used as a performance target). That is, if the site discharged no overland flow into surface channels before development, then it should not release surface runoff (stormwater) into natural channels after development.

Other locations

The same objective holds for a site located in the transmission (or conveyance) zone [(*b*) in Fig. 22.4]. The continuity of flow—that is, allowances for inputs and outputs—should be the same after development as before. At the output end of the system, a critical concern should be the performance of the entire system upstream because a change in the rate and amount of water delivery, for example, may have serious effects on flood magnitude and frequency, water supply, and other consequences. In a functional sense, sites at the output end of a system should be thought of as extending way beyond the site's formal boundaries, all the way to the head of the system where the inputs begin. The whole system, then, is the planning arena for such sites.

Other systems

The significance of location in the system is easily demonstrated for watersheds, which are the systems of greatest concern in most areas, but the concept is equally applicable to other landscape systems. Groundwater is also a three-part system, as are longshore (drift) systems, wind systems, wetland systems, and others (see Table 22.1). The geographic configurations of some are more easily defined than others, of course, but at the very least it is important to remember that even though some systems defy

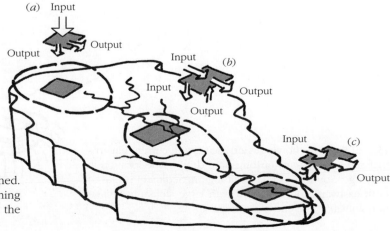

Fig. 22.4 Sites at three different locations in the watershed. Each location carries different implications for site planning and design because each has a different function in the system.

precise definitions, all systems operate according to a definable order that is usually represented by *vectors* (directions) of motion in matter and energy. Vectors enable us to delineate patterns, trends, and linkages in the landscape. *Linkages*, in turn, lead to observations about the connections among different places and ultimately to inferences about cause-and-effect relationships.

Application to a Segment of Coastline. Let us examine a segment of coast where the formative system is a longshore (or drift) cell. Figure 22.5 shows a typical longshore system setup in the Puget Sound–Georgia Strait region of Washington and British Columbia. The coast here is composed of glacial deposits, mostly sand and gravel with a small percentage of cobbles and boulders. The shoreline in this area has a northeast-southwest orientation, which is close to perpendicular to the force of winter storm waves (from the southeast) that drive the coastal currents and that, in turn, do the lion's share of the work in moving sediment down the shore.

The formative system

The resultant longshore system has two arms or cells: one that moves sediment northeastward and the other that moves sediment southwestward. Since no streams supply sediment to the shore here, there is only one source of sediment for the longshore system, namely, the glacial deposits that make up the shore and the banks behind it. Storm waves erode this material and feed it to the longshore system. From the source area (the *in situ* deposits), the sediment is moved downshore, both to the northeast and to the southwest, forming two cells (Fig. 22.5). At the end of each cell, the coastline breaks (turns abruptly) into two bays, and the sediment load is deposited, forming large bars on one end and a long spit on the other.

System form and function

The configuration and operation of these two drift cells are easy enough to identify, and their implications for planning and design are no mystery. In the source area, the shore is giving up sediment and slowly retreating. In the transport zone, the sediment is mainly passing by, en route downshore. From year to year, the beach in this zone may fluctuate, shifting seaward in some years and landward in others, but as a whole the action is overwhelmingly lateral. By contrast in the sinks at the ends of the cells, the shore is growing as sediment fills in the bays. So, on balance, the character and behavior of the shoreline are best explained by examining the form and work of the longshore system. The operative word here is "behavior" because understanding

Understanding performance

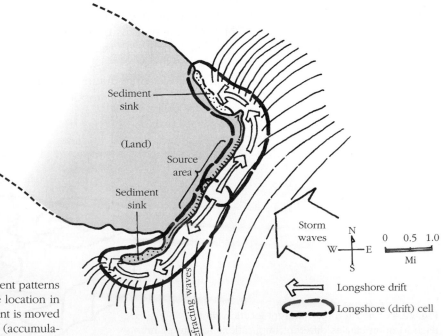

Fig. 22.5 Map showing wind, wave, and current patterns that produce systems of longshore drift at one location in the Puget Sound–Georgia Strait region. Sediment is moved both eastward and westward with large sinks (accumulation zones) at the end of each cell.

the coastline is impossible without a basic appreciation of its dynamics. Unfortunately, landscape dynamics is something rarely entertained in land use planning.

A conventional response

As for land use along this coastline, the entire area, save for two small parks and two military installations, is dedicated to residential development, and, according to existing policy, the only planning regulation that applies to site design is setback distance from the shore. Residential structures must be at least 15 meters (50 feet) back from the high-water mark unless the applicant builds an erosion protection wall, and in that case, the required minimum setback can be reduced to 7.5 meters (25 feet). No mention is made of the longshore system.

Preferred response

If the system *were* considered, the first response would be to vary setback distance according to the site's location in the system. Setbacks should be much larger in sediment source areas to accommodate the inevitable shoreline retreat there. Provision for erosion protection structures and sediment control structures should be eliminated throughout both cells, especially in transport zones, to maintain the continuity of flow. In sink zones, the requirement for erosion protection structures is probably meaningless because such facilities will end up lying idle behind accreting beaches or, in the case of the spit, repeatedly covered and exposed because this narrow neck of sand shifts about over time. Beyond site-scale considerations, however, there is a more meaningful level of planning to consider.

Implications for Community Planning. Recognition of the broader patterns and features of systems early in the planning process could have helped guide planners toward more prudent zoning decisions. From a system's perspective, where, for example, should residential land use and public open space be assigned? What better

Land use allocation

land use to allocate to the sediment source area (where the shore is retreating) than, say, public park with large setbacks and modest, low-cost facilities, such as trails, decking, and parking? Let the shore retreat its few inches a year while the sediment produced feeds the rest of the system.

Further downshore, residential zoning is appropriate, but only with provisions for setbacks large enough to accommodate shoreline fluctuation and rules against clearing and manipulation of the backshore to guard against the destabilization of banks and bluffs. Near the sinks, at the beginning of the accretion zones, residential zoning can be considered, but setback distances should take into account another system—onshore wind and sand dune formation—with development restrictions on landforms such as dunes, fore dunes, and beach ridges. With respect to the spit itself, development there would be highly risky in light of the fact that these long, narrow features are prone to major shoreline shifts and breaches by stormwaves.

22.3 FRAMING SPACE IN A WATERSHED

Landscape order

Landscape is not a random assortment of natural and human artifacts tossed together in different ways at different places. Viewed through the proper lens, landscapes virtually everywhere have remarkable order, and recognizing that order is paramount to building the land use plan. Without this step—that is, figuring out what sorts of land use activities and facilities should go where—we cannot hope to build sustainable landscapes. For sites within watershed systems, which represent the vast majority of the sites we work with, a good beginning point is simply dividing land into two classes: that which (1) gives up water and that which (2) collects water.

Trend surface mapping

Differentiating Terrain. One way of making this determination is to map the pattern of surface runoff based on topography. We describe this procedure briefly in Chapter 9. Start with a topographic contour map. Draw vectors (short arrows) orthogonal to the contour lines. The result is a trend surface map showing two patterns: convergent where water is collecting and divergent where water is dispersing. In relative terms, the lands served by these patterns are, respectively, lowlands and uplands.

Corridors and land units

Another way of differentiating lowlands and uplands in a watershed is by mapping the lowland corridors housing streams and connecting wetlands. If these corridors are mapped as a system and extended to include all water collection features, as is illustrated in Figure 22.6b, then the drainageways housing intermittent and ephemeral streams should also be tied into the corridor network. The parcels of land lying between the corridors (the upland surfaces) define rooms of landscape space (*land units*), which are features central to the planning process (see Fig. 22.6c).

As for the lowland corridors, nature has given them the role of arteries in the watershed system. This is a critical function, and unless it is respected in land use

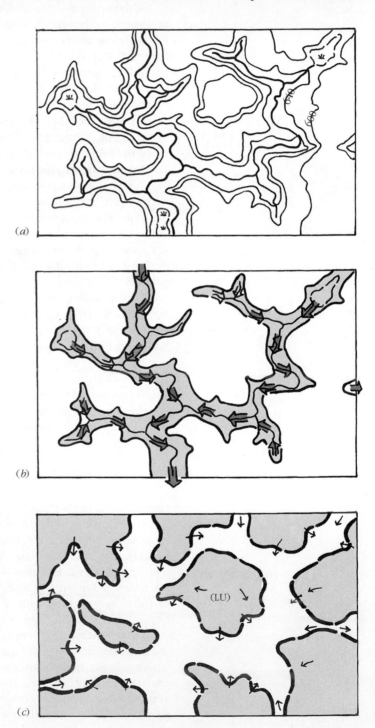

Fig. 22.6 Beginning with a topographic contour map (*a*), the system of lowland corridors is delineated (*b*). The remaining space defines uplands, or land units (*c*).

Corridor uses

planning, we run the risk of interrupting the system and compromising its capacity to serve the landscape as a whole, including the various the systems within it. Therefore, it is entirely appropriate as a first approximation in the planning exercise to translate these two types of lands in terms of suitability for land use. Lowland corridors are generally appropriate as open space and other light-facility land uses such as tree farms, whereas the upland units are more or less appropriate as buildable ground.

Refining Land Use Potential. Additional factors should now be brought into consideration. The first is to refine the definition of buildable land. High on the list is slope, for along the margins of the land units, where the topography breaks away to the lowland corridors, are often found steep slopes that are ill suited for development. Next might be wetlands, but provisions for wetlands should have already been made in mapping the lowland corridors. In that case, we might move on to soils or vegetation, significant habitat, or land use amenities and artifacts. The list and the priorities will vary depending on the geographic area, public policies, and the requirements of the development program, but the outcome is fundamentally the same; that is, to refine the definition of land with development potential by following a coherent line of reasoning within a systems framework. Out of this comes a definition of land parcels (polygons) based on their suitability for different land uses as prescribed by the development program (Fig. 22.7).

Additional criteria

A similar process can be followed for the open space system. Within this area are stream channels, floodplains, wetlands, terraces, forests, historic sites, and so on, which can be mapped and evaluated for different kinds of open space uses. But the palette of land uses is decidedly different from the palette used for the upland areas. Here the considerations are park, nature preserves, flood management areas, and scenic trails and drives, as well as certain economic activities such as tree farming, selected types of agriculture, and hunting reserves.

Refining open space

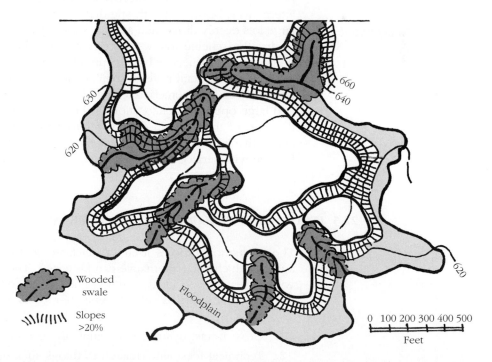

Wooded swale

Slopes >20%

0 100 200 300 400 500
Feet

Fig. 22.7 A land suitability map for the land unit (LU) in Figure 22.6. In this case, two attributes, slopes above 20 percent and upland swales, are added to the lowland corridor (marked as floodplain) as open space. The land remaining is that with the fewest limitations for development.

22.4 DESIGNING CONCEPTUAL ALTERNATIVES

Variables

Attributes

Once a rudimentary understanding of the site and its framing system(s) has been achieved, we are in a position to build a set of conceptual design schemes. The form and character of these alternatives must be based on an understanding of both land and program as well as selected contextual factors such as local policy and the character of existing land use and landscape. Generally speaking, it is advisable to think in terms of three design alternatives, and each should follow a different theme or vision. The ideas leading to a set of themes can come from a variety of sources, but in the end the design alternatives that emerge should have three attributes: (1) They are realistic in the context of the site, region, and program, especially program. (2) They lend themselves to graphic portrayals that do not demand much time and resources to produce. (3) They lend themselves to simple testing (evaluation).

Rationale

Objectives and Application. The objective of this exercise is first *exploratory* in the sense that it requires bringing different, often competing, ideas to the table that will push us to stretch our imaginations and thought processes. The second is *analytical* in the sense that we are forced to think critically and reasonably in selecting one alternative over another. In the end we want to be assured that an appropriate range of design options has been seriously examined, that the options have been scrutinized rigorously and fairly, and that, if none is suitable (that is, solves the problem), then all can be discarded (before much time has been invested) and another set of options advanced. It may also call for a reexamination of the program,

Conceptual themes

Consider a site like the one shown in Figure 22.7 and a program that calls for residential development. Like most sites, this one is part of a functioning watershed with modest topographic relief and a fringing stream valley. Zoning limits development to detached, single-family units with an average net density not to exceed two units per acre and at least 10 percent of the site dedicated to open space. Consider three alternatives, beginning with one that follows a more or less conventional layout with large lots evenly distributed along a winding street system. The theme here is familiar in that the scheme follows a conventional subdivision layout, one that community officials and prospective buyers can relate to. A second alternative might use a cluster concept with several pods of small, tightly spaced lots woven into a large network of open space. The theme is site-adaptive design featuring a compressed infrastructure and large open space allocation. A third alternative might follow a grid layout along the lines of that used in designs referred to as *new urbanism*. The theme with this alternative is community efficiency built around a traditional grid model that facilitates linkage with services as well as bicycling and walking.

22.5 TESTING AND EVALUATING CONCEPT PLANS

Setup

The evaluation process itself is facilitated in the early stages by selecting criteria against which each alternative can be tested. The criteria can be organized into sets with increasing specificity as the evaluation process advances. The testing process, it follows, can be carried out in a series of rounds. First-round criteria might include:

Criteria

1. Gross infrastructure costs (mainly the costs of streets and piping).
2. Technical feasibility (related to things such as the number and structural complexity of stream crossings, slope reinforcements, and pumping stations).
3. Environmental compatibility (including public policies, impact risk, and amenity opportunities such as scenic vistas).

4. Social compatibility (potential for community-building based on opportunities for trails, parks, and access to local services).

5. System compatibility (appropriateness of the proposed action based on location and potential impact in the formative system).

Rank and weight

Weighting Criteria and Assigning Value. Of course, all criteria are not equally important. Therefore, each criterion can be assigned a numerical weight, which can be used as a multiplier in the evaluation process. For example, if a criterion such as environmental compatibility is considered to be very important, it might be given a weight of 3 on a 3-point weighting scale. In the evaluation process, if the design (plan) gets a rank of fair (2 on a 3-point scale) for environmental compatibility, then the score would be 6 for this criterion.

The benefits

Please understand that the numbers are completely arbitrary and that the scores themselves carry no real quantitative value. They are meaningful only in a relative sense. In a system in which the scores for a single criterion can range from 1 to 9, no significance should be ascribed to 1- or 2-point differences. There is, however, real value in the process of deriving an evaluation because (1) it necessitates discussion, debate, and sometimes even research among project team members; (2) it affords clients an opportunity to participate in the evaluation process; (3) it provides a systematic way sorting out alternatives; and (4) it provides rationale for and facilitates communication about the direction of a project.

22.6 ADVANCING THE PLAN

Levels of approximation

Once a design scheme has been selected and rationales have been established for the choice, the process leading to a final design can be launched. There are various ways of approaching this process, and one commonly used in landscape design is what might be called *levels of approximation*. This approach involves advancing the design through stages of refinement, beginning with sketches and followed by more and more refined levels of articulation. It may also involve building graphic models that feature key concepts and principles of the project, including, for example, how to apply system-adaptive design in different terrain settings in the project area.

Conventional application

Among the tasks of particular significance in residential development planning is the delineation and layout of streets, lots, and open space. Too often this task is left to surveyors, and the result is a design scheme such as the one in Figure 22.8*a*, which shows little or no sensitivity to landscape features and even less to landscape systems. Figure 22.8*b* compares this plan to the pattern of wooded swales and steep slopes and earmarks the parcels that conflict with these attributes. If these parcels were deleted from the plan, as they should be, the lot count would be reduced by more than 20 percent. But under conventional development practices this would not happen. Instead these parcels would be mechanically reworked and—with cuts, fills, and pipes—transformed into salable properties, all to the detriment of the drainage system and supporting landscape.

Alternative approach

The plan in Figure 22.9 offers an alternative to the conventional approach. It begins with an open space plan that ties swales and steep slopes together in a network linked to the fringing floodplain. The lots are smaller, less than half the size of those in Figure 22.8, but the lot total is the same, and each lot adjoins the open space system. In addition to honoring the drainage system that feeds the lowland corridors, the open space system should, among other things, also facilitate animal movement and the development of parks and pedestrian trails.

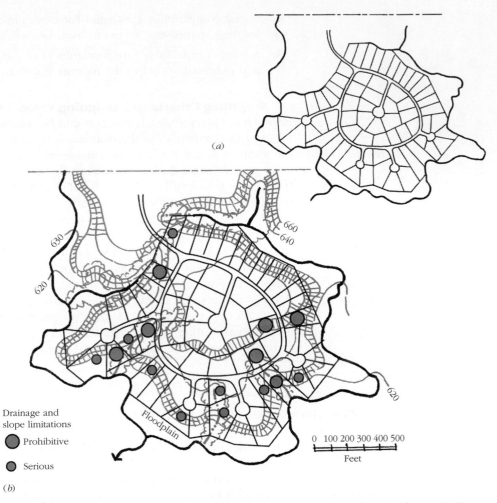

Drainage and
slope limitations

⬤ Prohibitive

⬤ Serious

(b)

Fig. 22.8 (*a*) A conventional residential design scheme based on the site shown in Figure 22.7, in which the entire site (minus an open space fraction) is allocated to lots. (*b*) Evaluating this scheme based on wooded swales and steep slopes. Lots with circles are in conflict.

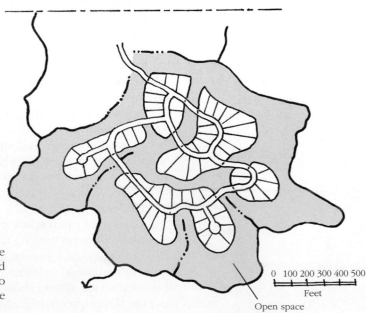

Open space

Fig. 22.9 An alternative layout to the one shown in Figure 22.8 with the same number of lots (65). Lots are clustered on buildable ground with an open space system designed to protect steep slopes and swales while providing open space access for each lot.

But this is not the end of the story, for the system-based approach, which we began at the macroscale with a look at watersheds as formative landscape systems, should not end with the layout of the lots, streets, and open space. It should also be applied at the microscale, first to clusters of lots and then to facility and landscape design within individual lots to ensure the proper fit with systems at the cellular scale and with the larger landscape systems within which they are nested.

22.7 CASE STUDY

The Mountain Watershed Variation

W. M. Marsh

Most of the watersheds we deal with in landscape planning and design are made up of lands of low and intermediate elevation and relief. Most include little or no terrain rugged enough to qualify as mountainous, that is, terrain dominated by large, steep slopes, high and narrow ridges, and deep valleys. The reality of landscape planning in North America, however, must include mountainous terrains because increasingly rugged lands are being pressed into service for development, and, strictly speaking, the version of watershed-based planning outlined in this chapter does not usually apply there.

In particular, the idea of defining buildable land units as the area situated between the lowland corridors is inappropriate in most mountainous watersheds. The reasons are apparent at a glance. Slopes are high and steep, and sites that appear to have development potential are usually inaccessible and high in elevation, among other things. As a whole, only small, selected sites (best described as unique) lend themselves to most land uses, beyond park and wilderness uses, of course.

We are left to work with the lowland corridors in building the land use plan. As with the low-relief watershed, the approach begins by defining the drainage system and identifying where in the system a proposed action is targeted. Given a location in one of the drainage corridors, planners need to acknowledge that virtually all corridor locations in mountain watersheds pose a risk to development. Take a look at the map in Figure 22.A, which shows the upper part of a mountainous watershed in California. Side slopes, which range from 60 to 100 percent, are not only too steep for development but are unstable and prone to landslides. The valley floor, by contrast, is gently sloping; however, it is prone to debris (mud and rock) flows and flooding. Farther down the watershed, the risk of slides and debris flows can be expected to decline, but flooding remains a problem, sometimes even more so than upstream. But here the valley floors are typically larger and more diverse, and they may offer more opportunity for finding sites with lower susceptibility to flooding, slides, and debris flows.

One approach to identifying safer sites in the lowland corridor begins with a survey and evaluation of valley landforms. In virtually every mountain stream corridor is a variety of landforms including backswamps, alluvial fans, and river terraces. Terraces are fragments of former floodplains, remnants of valley floors when streams flowed at somewhat higher elevations. They form when a stream downcuts and establishes a lower channel and floodplain elevation. Downcutting can be caused by various changes—such as an uplift of the land as a part of regional mountain building—but it leads to a new, lowered floodplain, which often includes remnant patches of the old floodplain. These remnants are terraces, and they are usually found along the valley walls, sometimes as shelflike benches several meters or more above the active floodplain and sometimes as isolated parcels of various sizes (Fig. 22.B).

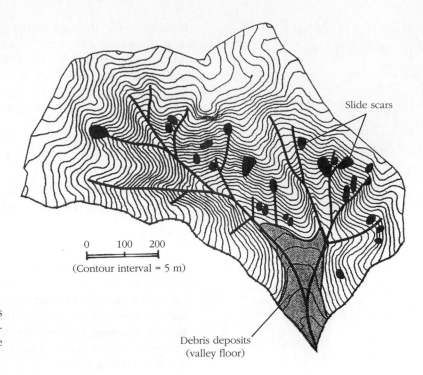

Fig. 22.A The upper reaches of a mountainous watershed in California showing the locations of landslides and the area inundated by debris flows on the valley floor (from Montgomery and Dietrich, 1994).

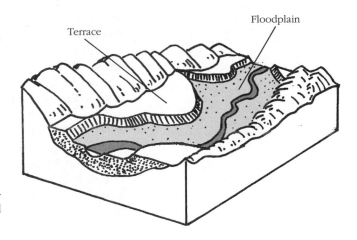

Fig. 22.B A block diagram illustrating river terraces in a mountain lowland corridor. The terraces offer development potential because they are not part of the active floodplain.

In virtually all cases they are fairly flat and well drained and, compared to most other sites in the valley corridor, are reasonably well suited for development.

A second approach is hydrographic and based on the concept of nonbasin drainage area. Lowland corridors are usually joined by many tributary streams, of mostly first, second, and third orders, and the confluences of these streams with the corridor are inherently risky places because they are prone to flooding, debris flows, and alluvial deposits. The land that lies between the lower reaches of neighboring tributaries, however, belongs to neither drainage basin and thus qualifies as nonbasin drainage area. Such areas are typically small with little upslope drainage area and thus are exempt from the usual upslope drainage problems of the basins. In addition, they lack integrated drainage systems and release their runoff by some

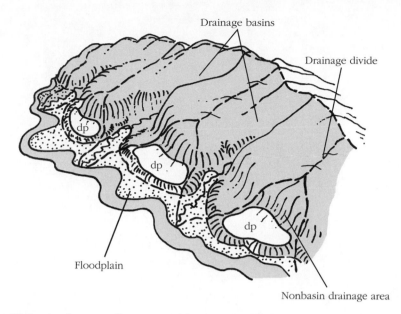

Fig. 22.C A schematic illustration of lands with development potential (dp) based on the concept of nonbasin drainage area, the land between the mouths of adjacent basins.

combination of overland flow and interflow. As the map in Figure 22.C illustrates, the lower limits of the nonbasin area extend into the floodplain of the corridor while the upper limits extend onto the steep mountain slopes. If we discard both the steep slope fraction and the floodplain fraction, the remainder is a reasonably safe parcel with better development potential than surrounding ground above and below.

22.8 SELECTED REFERENCES FOR FURTHER READING

Beatly, T., and Manning, K. *The Ecology of Place Planning: Planning for Environment, Economy, and Community.* Washington, D.C.: Island Press, 1997.

Beer, A. R. *Environmental Planning for Site Development.* London: E.&FN. Spon, 1990.

Bookout, L. W. *Value by Design: Landscape, Site Planning, and Amenities.* Washington, D.C.: Urban Land Institute, 1994.

Capra, Fritof. *The Web of Life: A New Scientific Understanding of Living Systems.* New York: Bantam, 1996.

DeFries, Ruth, et al. (eds). *Ecosystems and Land Use Change.* Washington, D.C.: American Geophysical Union, 2004.

LaGro, James A., Jr. *Site Analysis: Linking Program and Concept in Land Planning and Design.* New York: Wiley, 2001.

Lynch, Kevin. *Site Planning.* Cambridge, Mass.: MIT, 1971.

Waugh, Frank. "Physical Aspects of Country Planning." *The Journal of Land and Public Utility Economics.* 13, 3, 1937.

Woldenberg, M. J. "Horton's Laws Justified in Terms of Allometric Growth and Steady State in Open Systems." *Geological Society of America Bulletin.* 77, 1966.

GLOSSARY

Acid rain   Precipitation with pH levels much below average as a result of the integration of oxides in polluted air with moisture.

Active layer   The surface layer in a permafrost environment, which is characterized by freezing and thawing on a seasonal basis.

Aggradation   Filling in of a stream channel with sediment, usually associated with low discharges and/or heavy sediment loads.

Albedo   The percentage of incident radiation reflected by a material. Usage in earth science is usually limited to shortwave radiation and landscape materials.

Alluvial fan   A fan-shaped deposit of sediment laid down by a stream at the foot of a slope; very common features in dry regions, where streams deposit their sediment load as they lose discharge downstream.

Alluvial plain   The lowland of stream-deposited materials on the floors of stream valleys. The landform commonly referred to as floodplain.

Alluvium   Any material deposited by running water; the soil material of floodplains and alluvial fans.

Angiosperm   A flowering, seed-bearing plant. The angiosperms are presently the principal vascular plants on earth.

Angle of repose   The maximum angle at which a material can be inclined without failing. In civil engineering the term is used in reference to clayey materials.

Aquifer   Any subsurface material that holds a relatively large quantity of groundwater and is able to transmit that water readily.

Atterberg Limits Test   A test used to determine a soil's response to the addition of water based on its changes in the physical state, as from plastic to liquid.

Backscattering   That part of solar radiation directed back into space as a result of diffusion by particles in the atmosphere.

Backshore   The zone behind the shore—between the beach berm and the backshore slope.

Backshore slope   The bank or bluff landward of the shore that is comprised of *in situ* materials.

Backswamps   A low, wet area in the floodplain, often located behind a levee.

Bankfull discharge   The flow of a river when the water surface has reached bank level.

Barrier dune   A large mass or ridge of coastal sand dunes that parallels the shoreline in areas of extensive dune building.

Baseflow   The portion of streamflow contributed by groundwater. It is a steady flow that is slow to change even during rainless periods.

Base level   In a stream system, this represents the lowest elevation, such as sea level, controlling downcutting.

Bay-mouth bar   A ribbon of sand deposited across the mouth of a bay.

Bed shear stress   The force of flowing water dissipated against the streambed; a product of two main variables, channel slope and water depth.

Berm   A low mound that forms along sandy beaches; also used to describe elongated mounds constructed along water features and site borders.

Best management practices   Measures used to prevent or reduce the detrimental environmental impacts caused by development, especially impacts from stormwater runoff.

Bifurcation ratio   The branching ratio in stream networks; the number of streams in one order to that in the next, higher order.

Biological oxygen demand   Oxygen requirements by aquatic microorganisms that deprive fish and other aquatic animals of essential oxygen supplies.

Biomass   The total weight of organic matter within a prescribed surface area, usually 1 square meter.

Bioremediation   The use of communities of plants, animals, and microorganisms to treat wastewater; for example, wetlands are used to remediate stormwater.

BMPs   *See* **Best management practices**

BOD   *See* **Biological oxygen demand.**

Bog   A cool/cold climate wetland characterized by peat deposits, an acidic pH, and often a diverse plant cover.

Boreal forest   Subarctic conifer forests of North America and Eurasia; floristically homogeneous forests dominated by fir, spruce, and tamarack. In Russia, it is called *taiga*.

Boundary layer   The lower layer of the atmosphere; the lower 300 meters of the atmosphere where airflow is influenced by the earth's surface.

Boundary sublayer   The stratum of calm air immediately over the ground that increases in depth with the height and density of the vegetative cover.

Bowen Ratio   The ratio of sensible-heat flux to latent-heat flux between a surface and the atmosphere.

Buffer   The zone around the perimeter of a wetland or lake where land use activities are limited in order to protect the water features.

Buildable land units   Parcels of various size within a designated project area that are suitable for development as defined by a prescribed development program.

CAFE standards   Corporate average fuel economy standards for automobiles as mandated by the U.S. Clean Air Act as part of the National Ambient Air Quality Standards.

Capillarity   The capacity of a soil to transfer water by capillary action; capillarity is greatest in medium-textured soils.

Carrying capacity   The level of development density or use an environment is able to support without suffering undesirable or irreversible degradation.

Chaos theory   A concept of landscape change in which the environment experienced by humans behaves more or less erratically.

Choropleth map   A map comprised of areas of any size or shape representing qualitative phenomena (e.g., soils) or quantitative phenomena (e.g., population); often has a patchwork appearance.

Climate   The representative or general conditions of the atmosphere at a place on earth. It is more than the average conditions of the atmosphere, for climate may also include extreme and infrequent conditions.

Closed forest   A forest structure with multiple levels of growth from the ground up; a forest in which undergrowth closes out the area between the canopy and the ground.

Clustering   A land use development concept in which facilities are grouped closely together with large areas of surrounding open space.

Coastal dune   A sand dune that forms in coastal areas and is fed by sand from the beach.

Coefficient of runoff   A number given to a type of ground surface representing the proportion of a rainfall converted to overland flow. It is a dimensionless number between 0 and 1.0 that varies inversely with the infiltration capacity; impervious surfaces have high coefficients of runoff.

Collection zone   The central, upper part of a small watershed where runoff from the contributing zone concentrates and forms channel flow.

Colluvium   Any material made up of a mixture of runoff and mass wasting (e.g., landslide) deposits.

Concentration time   The time taken for a drop of rain falling on the perimeter of a drainage basin to go through the basin to the outlet. Also called travel time.

Conditional stability   A condition in the landscape in which stability is dependent on one or two essential factors, such as plants holding an oversteepened slope in place; also called metastability.

Conduction   A mechanism of heat transfer involving no external motion or mass transport. Instead, energy is transferred through the collision of vibrating molecules.

Constraint   Any feature or condition of the built or natural environment that poses an obstacle to land use planning.

Conveyance zone   The central route of drainage, usually a channel and valley, in a drainage basin.

dBA   Decibel scale that has been adjusted for sensitivity of the human ear.

Decibel   Unit of measurement for the loudness of sound based on the pressure produced in air by a noise; denoted dB.

Declination of the sun   The location (latitude) on earth where the sun on any day is directly overhead; declinations range from 23.27°S latitude to 23.27°N latitude.

Degradation   Scouring and downcutting of a stream channel, usually associated with high discharges.

Density *See* **Development density**

Depression storage Rainwater and overland flow held in shallow, low spots in the terrain.

Design storm A rainstorm of a given intensity and frequency of recurrence used as the basis for sizing stormwater facilities such as stormsewers.

Detention A strategy used in stormwater management in which runoff is detained on site to be released later at some prescribed rate.

Development density A measure of the intensity of development or land use; defined on the basis of area covered by impervious surface, population density, or building floor area coverage, for example.

Discharge The rate of waterflow in a stream channel; measured as the volume of water passing through a cross-section of a stream per unit of time, commonly expressed as cubic feet (or meters) per second.

Discharge zone An area where groundwater seepage and springs are concentrated.

Disturbance An impact on the environment characterized by physical alteration such as forest clearing.

Disturbance theory An alternative to the community-succession concept in plant ecology in which the principal habitat change agent is external (mainly abiotic) forces such as fire, floods, and land use.

Diurnal damping depth The maximum depth in the soil that experiences temperature change over a 24-hour (diurnal) period.

Drainage basin The area that contributes runoff to a stream, river, or lake. Commonly called a watershed.

Drainage density The number of miles (or km) of stream channels per square mile (or km²) of land.

Drainage divide The border of a drainage basin or watershed where overland separates between adjacent areas.

Drainage network A system of stream channels usually connected in a hierarchical fashion. *See also* **Principle of stream orders.**

Drainfield The network of pipes or tiles through which wastewater is dispersed into the soil.

Earth mound A type of soil-absorption system for residential development in which the drainfield is constructed above ground and covered with a soil medium.

Ecosystem A group of organisms linked together by a flow energy; also a community of organisms and their environment.

Ecotone The transition zone between two groups, or zones, of biota.

Edge city A type of urban center characterized by business and commercial land uses that has developed at selected intersections in the U.S. interstate highway system near large urban centers.

Effective impervious cover Impervious cover that is connected by a ditch or pipe to a drainage network and natural water features.

EIS *See* **Environmental Impact Statement.**

Emergent wetland Wetland dominated by herbaceous vegetation growing in shallow water.

Endangered species According to the U.S. Endangered Species Act, a species in imminent danger of extinction in all or a significant portion of its range.

Energy balance The concept or model that concerns the relationship among energy input, energy storage, work, and energy output of a system such as the atmosphere or oceans.

Environmental assessment A preliminary study or review of a proposed action (project) and the influence it could have on the environment; often conducted to determine the need for more detailed environmental impact analysis.

Environmental impact statement A study required by U.S. federal law for projects (proposed) involving federal funds to determine types and magnitudes of impacts that would be expected in the natural and human environment and the alternative courses of action, including no action.

Environmental inventory Compilation and classification of data and information on the natural and human features in an area proposed for some sort of planning project.

Ephemeral stream A stream without baseflow; one that flows only during or after rainstorms or snowmelt events.

Erodibility   The relative susceptibility of a soil to erosion.

Erodibility factor   A value used in the universal soil loss equation for different soil types representing relative erodibility; called the *K*-factor by the U.S. Soil Conservation Service.

Erosion   The removal of rock debris by an agency such as moving water, wind, or glaciers; generally, the sculpting or wearing down of the land by erosional agents.

Estuarine wetland   Coastal wetlands associated with bays and estuaries of the ocean.

Eutrophication   The increase of biomass of a waterbody leading to infilling of the basin and the eventual disappearance of open water; sometimes referred to as the aging process of a waterbody.

Evapotranspiration   The loss of water from the soil through evaporation and transpiration.

Exchange time   *See* **Residence time.**

Facility   Any part of the built environment, especially structures and mechanical systems.

Facility planning   Planning for facilities such as power-generating stations or sewage treatment plants; usually carried out by engineers.

Feasibility study   A type of technical planning aimed at identifying the most appropriate use of a site.

Fetch   The distance of open water in one direction across a waterbody; one of the main controls of wave size.

Filtration   A term generally applied to the removal of pollutants, such as sediment, with the passage of water through a soil, organic, and/or fabric medium.

Floodway fringe   The zone designated by U.S. federal flood policy as the area in a river valley that would be lightly inundated by the 100-year flood.

Floristic system   The principal botanical classification scheme in use today. Under this scheme the plant kingdom is made up of divisions, each of which is subdivided into smaller and smaller groups arranged according to the apparent evolutionary relationships among plants.

Formation   A structural unit of vegetation that may be considered a subdivision of a biochore; a formation may be made up of several communities. In the traditional terminology, it is called a physiognomic unit; in geology, a major unit of rock.

Fragmentation   The process by which the landscape and its various habitats are broken down into smaller and smaller parcels as a result of land use development.

Frequency   The term used to express how often a specified event is equaled or exceeded.

Frost wedging   A mechanical weathering process in which water freezes in a crack and exerts force on the rock, which may result in the breaking of the rock; a very effective weathering process in alpine and polar environments.

Geographic Information System (GIS)   Computer mapping system designed for ready applications in problems involving overlapping and complex distributional patterns. Two classes of GIS are vector and raster.

Geomorphic system   A physical system comprised of an assemblage of landforms linked together by the flow of water, air, or ice.

Geomorphology   The field of earth science that studies the origin and distribution of landforms, with special emphasis on the nature of erosional processes; traditionally, a field shared by geography and geology.

Global coordinate system   The network of east-west and north-south lines (parallels and meridians) used to measure locations on earth; the system uses degrees, minutes, and seconds as the units of measurement.

Gradient   The inclination or slope of the land; often applied to systems such as streams and highways.

Grafting   The practice of attaching additional channels to a drainage network. In agricultural areas, new channels appear as drainage ditches; in urban areas, as stormsewers.

Gravity water   Subsurface water that responds to the gravitational force; the water that percolates through the soil to become groundwater.

Greenbelt   A tract of trees and associated vegetation in urban and rural areas; may be a park, nature preserve, or part of a transportation corridor.

Green infrastructure   Infastructure that relies on soft or green measures rather than structural or hard measures in stormwater management. Green infastructure includes grasslined swales, infiltration galleries, and porous pavers.

Groin   A wall or barrier built from the beach into the surf zone for the purpose of slowing down longshore transport and holding sand.

Gross sediment transport   The total quantity of sediment transported along a shoreline in some time period, usually a year.

Ground frost   Frost that penetrates the ground in response to freezing surface temperatures.

Ground sun angle   The angle formed between a beam of solar radiation and slope of the surface that it strikes in the landscape.

Groundwater   The mass of gravity water that occupies the subsoil and upper bedrock zone; the water occupying the zone of saturation below the soil-water zone.

Gullying   Soil erosion characterized by the formation of narrow, steep-sided channels etched by rivulets or small streams of water. Gullying can be one of the most serious forms of soil erosion of cropland.

Habitat   The local environment of an organism from which it gains its resources. Habitat is often variable in size, content, and location, changing with the phases in an organism's life cycle.

Habitat conservation planning   A program under the U.S. Endangered Species Act designed to protect habitat areas based on plans involving both government agents and landowners.

Habitat corridor   A belt or zone representing a habitat system (one or several related habitat types), such as a stream valley.

Hardpan   A hardened soil layer characterized by the accumulation of colloids and ions.

Hazard assessment   Study and evaluation of the hazard to land use and people from environmental threats such as floods, tornadoes, and earthquakes.

Heat island   The area or patch of relatively warm air that develops over urbanized areas.

Heat syndrome   Various disorders in the human thermoregulatory system brought on by the body's inability to shed heat or by a chemical imbalance from too much sweating.

Heat transfer   The flow of heat within a substance or the exchange of heat between substances by means of conduction, convection, or radiation.

Hillslope processes   The geomorphic processes that erode and shape slopes; mainly mass movements, such as soil creep and landslides, and runoff processes, such as rainwash and gullying.

Horizon   A layer in the soil that originates from the differentiation of particles and chemicals by moisture movement within the soil column.

Hydraulic gradient   The rate of change in elevation (slope) of a groundwater surface such as the watertable.

Hydraulic radius   The ratio of the cross-sectional area of a stream to its wetted perimeter.

Hydric soil   Soil characterized by wet conditions; saturated most of the year; often organic in composition.

Hydrograph   A streamflow graph that shows the change in discharge over time, usually hours or days. *See also* **Hydrograph method.**

Hydrograph method   A means of forecasting streamflow by constructing a hydrograph that shows the representative response of a drainage basin to a rainstorm; the use of "normalized" hydrograph for flow forecasting in which the size of the individual storm is filtered out. *See also* **Hydrograph.**

Hydrologic cycle   The planet's water system, described by the movement of water from the oceans to the atmosphere to the continents and back to the sea.

Hydrologic equation   The amount of surface runoff (overland flow) from any parcel of ground is proportional to precipitation minus evapotranspiration loss, plus or minus changes in storage water (groundwater and soil water).

Hydrologic versatility   The hydrologic flexibility or diversity of an area in terms of its capacity to process rainwater and runoff.

Hydrometer method   A technique used to measure the clay content in a soil sample that involves dispersing the clay particles in water and drawing off samples at prescribed time intervals.

Hypothermia   A physiological disorder associated with cold conditions and characterized by the decline of body temperature, slowed heartbeat, lowered blood pressure, and other symptoms.

Impervious cover   Any hard surface material, such as asphalt or concrete, that limits infiltration and induces high runoff rates.

Infiltration beds   A general term applies to beds underlain by gravel and/or covered by vegetation designed to infiltrate stormwater. Also called infiltration galleries or biomediation beds.

Infiltration capacity   The rate at which a ground material takes in water through the surface; measured in inches or centimeters per minute or hour.

Inflooding   Flooding caused by overland flow concentrating in a low area.

Infrared film   Photographic film capable of recording near infrared radiation (just beyond the visible to a wavelength of 0.9 micrometer), but not capable of recording thermal infrared wavelengths.

Infrared radiation   Mainly longwave radiation of wavelengths between 3.0–4.0 and 100 micrometers, but also includes near infrared radiation, which occurs at wavelengths between 0.7 and 3.0–4.0 micrometers.

In situ   A term used to indicate that a substance is in place as contrasted with one, such as river sediment, that is in transit.

Interception   The process by which vegetation intercepts rainfall or snow before it reaches the ground.

Interflow   Infiltration water that moves laterally in the soil and seeps into stream channels. In forested areas this water is a major source of stream discharge.

Island biogeography   The study of biodiversity as related to island size and separation distance.

Isopleth map   A map comprised of lines, called isolines, that connect points of equal value.

Lacustrine wetland   Wetland associated with standing waterbodies such as ponds, lakes, and reservoirs.

Land cover   The materials such as vegetation and concrete that cover the ground. *See also* **Land use**.

Landfill   A term applied to various managed waste disposal sites involving ground burial.

Landscape   The composite of natural and human features that characterize the surface of the land at the base of the atmosphere; includes spatial, textural, compositional, and dynamic aspects of the land.

Landscape design   The process of laying out land uses, facilities, water features, vegetation, and related features and displaying the results in maps and drawings.

Landscape ecology   The application of geographic (spatial) analysis to problems of habitat planning and management in natural landscapes. A field of biogeography concerned with habitat fragmentation and biodiversity.

Landscape planning   The decision making, technical, and design processes associated with the determination of land uses and the utilization of terrestrial resources.

Landscape texture   The pattern or fabric of a landscape; the grain of a landscape related to the composite scale and trends of salient forms and features.

Landslide   A type of mass movement characterized by the slippage of a body of material over a rupture plane; often a sudden and rapid movement.

Land use   The human activities that characterize an area, for example, agricultural, industrial, and residential.

Latent heat   The heat released or absorbed when a substance changes phase as from liquid to gas. For water at 0°C, heat is absorbed or released at a rate of 2.5 million joules per kilogram (597 calories per gram) in the liquid/vapor phase change.

Leachate   Fluids that emanate from decomposing waste in a sanitary or chemical landfill.

Leaching   The removal of minerals in solution from a soil; the washing out of ions from one level to another in the soil.

Levee   A mound of sediment that builds up along a river bank as a result of flood deposition.

LID   *See* **Low impact development.**

Life form   The form of individual plants or the form of the individual organs of a plant. In general, the overall structure of the vegetative cover may be thought of as life form as well.

Lineament   Straight features in the landscape marked by slopes, segments of stream channels, soil patterns, or vegetation.

Line scanner A remote sensing device that records signals of reflected radiation in scan lines that sweep perpendicular to the path (flight line) of the aircraft.

Littoral drift The material that is moved by waves and currents in coastal areas.

Littoral transport The movement of sediment along a coastline. It is comprised of two components: longshore transport (beach drift) and onshore-offshore transport.

Load *See* **Sediment load.**

Loess Silt deposits laid down by wind over extensive areas of the midlatitudes during glacial and postglacial times.

Longshore current A current that moves parallel to the shoreline. Velocities generally range between 0.25 and 1 m/sec.

Longshore transport The movement of sediment parallel to the coast.

Low impact development Land use development designed specifically to minimize environmental impact in terms of energy use, air pollution, stormwater runoff, and land consumption. Applies to architecture, landscape architecture, and landscape planning.

Lowland corridor The belt of lowland that lies between uplands; often alluvial plain in a drainage network.

Magnetic declination The deviation in degrees east or west between magnetic north and true north.

Magnitude and frequency concept The principle that large, landscape-changing events, such as major floods, occur at low frequencies, whereas small events with little capacity for landscape change occur at high frequencies.

Manning formula A formula used to estimate the velocity of streamflow based on the gradient, hydraulic radius, and roughness of the channel; an empirical formula widely used in engineering for sizing channels and pipes.

Marsh A wetland dominated by herbaceous plants, typically cattails, reeds, and rushes.

Mass balance The relative balance in a system, based on the input and output of material such as sediment or water; the state of equilibrium between the input and output of mass in a system.

Mass movement A type of hillslope process characterized by the downslope movement of rock debris under the force of gravity; includes soil creep, rock fall, landslides, and mudflows; also termed *mass wasting*.

Meander A bend or loop in a stream channel.

Meander belt The corridor formed by a stream's meander system; defined by a set of lines drawn along the outer edge of active meanders.

Meander belt axis A line drawn down the center of a stream's meander belt.

Metastability *See* **Conditional stability.**

Microclimate The climate of small spaces such as an inner city, residential area, or mountain valley.

Misfit streams Streams that are either too large or too small for their valleys, such as small urban streams overloaded with stormwater discharge.

Mitigation A measure used to lessen the impact of an action on the natural or human environment.

Mitigation banking In wetland mitigation planning, the practice of building surplus acreage of compensation credits through replacement, enhancement, restoration, and/or preservation of wetland.

Model Any device, including conceptual constructs, mathematical formulas, or hardware apparatus, used in problem solving and analysis.

Monoculture An ecosystem dominated by few species with large populations, such as a farmfield or plantation forest.

Montmorillonite A type of clay that is notable for its capacity to shrink and expand with wetting and drying.

Moraine The material deposited directly by a glacier; also, the material (load) carried in or on a glacier. As landforms, moraines usually have hilly or rolling topography.

Morphogenetic region The concept that global landscapes are the product of different climatic regimes.

Mosaic A term used in landscape ecology to describe the patchy character of habitat as a result of fragmentation by land use.

Mudflow A type of mass movement characterized by the downslope flow of a saturated mass of clayey material.

National Pollution Discharge Elimination System U.S. federal program requiring a permit for the release of pollutants into natural water, including contaminated stormwater from communities over 100,000 population.

Nearshore circulation cell The circulation pattern of water and sediment formed by the combined action of rip currents, waves, and longshore currents.

Net sediment transport The balance between the quantities of sediment moved in two (opposite) directions along a shoreline.

Niche A term used to define an organism's way of life within its habitat; what it does to survive and reproduce.

Nonpoint source Water pollution that emanates from a spatially diffuse source such as the atmosphere or agricultural land.

NPDES *See* **National Pollution Discharge Elimination System.**

Nutrients Various types of materials that become dissolved in water and induce plant growth. Phosphorus and nitrogen are two of the most effective nutrients in aquatic plants.

Ogallala Aquifer The largest aquifer in North America, stretching from Nebraska to Texas.

Open forest A forest structure with a strong upper one or two stories and limited undergrowth; a forest that is largely open at ground level.

Open space Term applied to underdeveloped land, usually land designated for parks, greenbelts, water features, nature preserves and the like.

Open system A system characterized by a throughflow of material and/or energy; a system to which energy or material is added and released over time.

Opportunities and constraints A type of study often carried out in planing projects to determine the principal advantages and drawbacks to a development program proposed for a particular site.

Outflooding Flooding caused by a stream or river overflowing its banks.

Outwash plain A fluvioglacial deposit comprised of sand and gravel with a flat or gently sloping surface; usually found in close association with moraines.

Overdraft A condition of groundwater withdrawal in which the safe aquifer yield has been exceeded and the aquifer is being depleted.

Overland flow Runoff from surfaces on which the intensity of precipitation or snow melt exceeds the infiltration capacity; also called Horton overland flow, for hydrologist Robert E. Horton.

Oxbow A crescent-shaped lake or pond in a river valley formed in an abandoned segment of channel.

Ozone One of the minor gases of the atmosphere; a pungent, irritating form of oxygen that performs the important function of absorbing ultraviolet radiation.

Palustrine wetland Wetlands associated with inland sites that are not dependent on stream, lake, or oceanic water.

Parallels The east-west running lines of the global coordinate system. The equator, the Arctic Circle. and the Antarctic Circle are parallels; all parallels run parallel to one another.

Parent material The particulate material in which a soil forms. The two types of parent material are *residual and transported.*

Partial area concept A stormwater runoff model based on the observation that only a fraction of the area of a drainage basin contributes overland flow to stream discharge.

Passive solar collector A solar collector that operates without the aid of powered machinery.

Peak annual flow The largest discharge produced by a stream or river in a given year.

Peak discharge The maximum flow of a stream or a river in response to an event such as a rainstorm or over a period of time such as a year.

Peak flow *See* **Peak discharge.**

Pedon The smallest geographic unit of soil defined by soil scientists of the U.S. Department of Agriculture.

Percolation rate The rate at which water moves into soil through the walls of a test pit; used to determine soil suitability for wastewater disposal.

Percolation test A soil-permeability test performed in the field to determine the suitability of a material for wastewater disposal; the test most commonly used by sanitarians and planners to size soil-absorption systems.

Perennial stream A stream that receives inflow of groundwater all year; a stream that has a permanent baseflow.

Performance concept The concept of setting standards on how an environment or land use is expected to perform; includes formulation of goals, standards, and controls.

Periglacial environment An area where frost-related processes are a major force in shaping the landscape.

Permafrost A ground-heat condition in which the soil or subsoil is permanently frozen; long-term frozen ground in periglacial environments.

Permeability The rate at which soil or rock transmits groundwater (or gravity water in the area above the watertable); measured in cubic feet (or meters) of water transmitted through a specified cross-sectional area when under a hydraulic gradient of 1 foot per 1 foot (or 1 m per 1 m).

Photopair A set of overlapping aerial photographs that are used in stereoscopic interpretation of aerial photographs.

Photosynthesis The process by which green plants synthesize water and carbon dioxide and, with the energy from absorbed light, convert it into plant materials in the form of sugar and carbohydrates.

Physiography A term from physical geography traditionally used to describe the composite character of the landscape over large regions; today used mainly in planning and landscape architecture to describe the physical, mainly natural, character of a site or planning region.

Piping The formation of horizontal tunnels in a soil due to sapping, that is, erosion by seepage water. Piping often occurs in areas where gullying is or was active and is limited to soils resistant to cave-in.

Plane coordinate system A grid coordinate system designed by the U.S. National Ocean Survey in which the basic unit is a square measuring 10,000 feet on a side.

Planned unit development (PUD) A planning strategy aimed at reducing urban sprawl and related impacts by clustering development into carefully planned units.

Plant production The rate of output of organic material by a plant; the total amount of organic matter added to the landscape over some period of time, usually measured in grams per square meter per day or year.

Plume The stream of exhaust (smoke) emanating from a stack or chimney.

Point source Water pollution that emanates from a single source such as a sewage plant outfall.

Pollutant loading The amount of pollution released to runoff from different land uses measured in pounds or kilograms per acre or square kilometer.

Pollution Contamination of the environment from foreign substances or from increased levels of natural substances.

Porosity The total volume of pore (void) space in a given volume of rock or soil; expressed as the percentage of void volume to the total volume of the soil or rock sample.

Primary productivity The total amount of organic matter added to the landscape by plant photosynthesis, usually measured in grams per square meter per day or year.

Principal point The center of an aerial photograph, located at the intersection of lines drawn from the fiducial marks on the photo margin.

Principle of limiting factors The biological principle that the maximum obtainable rate of photosynthesis is limited by whichever basic resource of plant growth is in least supply.

Principle of stream orders The relationship between stream order and the number of streams per order. The relationship for most drainage nets is an inverse one, characterized by many low-order streams and fewer and fewer streams with increasingly higher orders. *See also* **Stream order.**

Progradation The process of seaward growth of a shoreline.

Pruning In hydrology the cutting back of a drainage net by diverting or burying streams; usually associated with urbanization or agricultural development.

Quadrat sampling A field sampling technique in which small plots, called quadrats, are laid out in the landscape and from which the sample is drawn.

Radiation The process by which radiant (electromagnetic) energy is transmitted through free space; the term used to describe electromagnetic energy, as in infrared radiation or short-wave radiation.

Radiation beam The column of solar radiation flowing into or through the atmosphere.

Rainfall erosion index A set of values representing the computed erosive power of rainfall based on total rainfall and the maximum intensity of the 30-minute rainfall.

Rainfall intensity The rate of rainfall measured in inches or centimeters of water deposited on the surface per hour or minute.

Rainshadow The dry zone on the leeward side of a mountain range of orographic precipitation.

Rainsplash Soil erosion from the impact of raindrops.

Rainwash Soil erosion by overland flow; erosion by sheets of water running over a surface; usually occurs in association with rainsplash; also called *wash*.

Rating curve A graph that shows the relationship between the discharge and stage of various flow events on a river. Once this relationship is established, it may be used to approximate discharge using stage data alone.

Rational method A method of computing the discharge from a small drainage basin in response to a given rainstorm. Computation is based on the coefficient of runoff, rainfall intensity, and basin area.

Recharge The replenishment of groundwater with water from the surface.

Recharge zone An area where groundwater recharge is concentrated.

Recurrence The number of years on the average that separate events of a specific magnitude, for example, the average number of years separating river discharges of a given magnitude or greater.

Regime The characteristic pattern of a process or system over time as in the seasonal regime of streamflow or the wet/dry regime of precipitation in California.

Regulatory floodway A zone designated by the U.S. federal flood policy as the lowest part of the floodplain where the deepest and most frequent floodflows occur.

Relief The range of topographic elevation within a prescribed area.

Replacement mitigation The practice of building new wetland to offset wetland acreage lost or damaged because of land use development.

Residence time The time taken to exchange the water in an aquifer or a lake for new water. Also called exchange time.

Restoration planning An area of planning that addresses damaged environments, such as degraded wetland habitats and disturbed stream channels.

Retention A strategy used in stormwater management in which runoff is retained on site in basins, underground, or released into the soil.

Retrogradation Shoreline retreat due to erosion, water level rise, land subsidence, or some combination of the three.

Riffle A short segment of stream channel characterized by rapid and often rough flow.

Riparian A reference to the environment along the banks of a stream; often more broadly applied to the larger lowland corridor on the stream valley floor.

Riparian wetland Wetlands that form on the edge of a major water feature such as a lake or stream.

Rip current A relatively narrow jet of water that flows seaward through the breaking waves. It serves as a release for water that builds up near shore.

Riprap Rubble such as broken concrete and rock placed on a surface to stabilize it and reduce erosion.

Risk management An area of planning that involves preparation and response to hazards, such as floods, hurricanes, and toxic waste accidents.

Riverine wetland Wetlands associated with streams and stream channels.

Runoff In the broadcast sense the flow of water from the land as both surface and subsurface discharge; in the more restricted and common use, surface discharge in the form of overland flow and channel flow.

Safe well yield   The maximum pumping rate that can be sustained by a well by lowering the water level below the pump intake.

Sand bypassing   The transfer of sand around an obstacle, such as a harbor breakwater, by artificial means, for example, dredging and barging.

Sapping   An erosional process that usually accompanies gullying in which soil particles are eroded by water seeping from a bank.

SAS   *See* **Soil-absorption system.**

Scatter diagram   A graph characterized by a series of plotted points showing the relationship between two quantitative variables.

Scattering   The process by which minute particles suspended in the atmosphere diffuse incoming solar radiation.

Scouring   The principal process of channel erosion in streams characterized by heavy particles bumping and skidding against the streambed.

Scrub-scrub wetland   Wetland dominated by shrubs and short trees.

Secure landfill   A class of landfills designed and constructed expressly for the disposal of hazardous waste.

Sediment load   The material transported by streams: bed load, suspended load, and dissolved load.

Seepage   The process by which groundwater or interflow water seeps from the ground.

Sensible heat   Heat that raises the temperature of a substance and thus can be sensed with a thermometer. In contrast to latent heat, it is sometimes called the heat of dry air.

Sensitive environment   Special environments, such as wetlands or coastal lands, that require protection from development because of their aesthetic and ecological value.

Septic system   Specifically, a sewage system that relies on a septic tank to store and/or treat wastewater; generally, an on-site (small-scale) sewage disposal system that depends on the soil to dispose of wastewater.

Septic tank   A vat, usually placed underground, used to store wastewater.

Setback   A term used in site planning to indicate the critical distance that a structure or facility should be separated from an edge, such as a backshore slope or lake shore.

Shoreland   The discontinuous belt of land around a waterbody that is not drained via stream basins.

Side-looking airborne radar (SLAR)   The radar system used in remote sensing; so named because the energy pulse is beamed obliquely on the landscape from the side of the aircraft.

Sieve method   A technique used to separate the various sizes of coarse particles in a soil sample.

Siltation   A term generally applied to the deposition of sediment in natural waters due to soil erosion and stormwater runoff.

Sink   Places or features in both terrestrial and aquatic environments, such as wetlands, reservoirs, and coastal embayments, where debris both natural and human collects and is stored.

Sinuousity   A measure of the curviness of a stream channel; the ratio of channel length to the length of the meander belt axis.

Site adaptive design   Landscape sensitive design that gives special consideration to site conditions, processes, and systems in laying out a land use system.

Site adaptive planning   Site planning that gives careful consideration to landscape features and systems, both natural and man-made. A nonprescriptive approach to land use planning.

SLAR   *See* **Side-looking airborne radar.**

Slope failure   A slope that is unable to maintain itself and fails by mass movement, such as a landslide, slump, or similar movement.

Slope form   The configuration of a slope, for example, convex, concave, or straight.

Sluiceway   A large drainage channel or spillway for glacier meltwater.

Slump   A type of mass movement characterized by a back rotational motion along a rupture plane.

Small circle   Any circle drawn on the globe that represents less than the full circumference of the earth. Thus the plane of a small circle does not pass through the center of the earth. All parallels except the equator are small circles.

Soil-absorption system   Small-scale wastewater disposal system that relies directly on soil to absorb and disperse sewage water from a house or larger building.

Soil creep   A type of mass movement characterized by a very slow downslope displacement of soil, generally without fracturing of the soil mass. The mechanisms of soil creep include freeze-thaw activity and wetting and drying cycles.

Soil-forming factors   The major factors responsible for the formation of a soil: climate, parent material, vegetation, topography, and drainage.

Soil-heat flux   The rate of heat flow into, from, or through the soil.

Soil material   Any sediment rock, or organic debris in which soil formation takes place.

Soil profile   The sequence of horizons, or layers, of a soil.

Soil structure   The term given to the shape of the aggregates of particles that form in a soil. Four main structures are recognized: blocky, platy, granular, and prismatic.

Soil texture   The cumulative sizes of particles in a soil sample; defined as the percentage by weight of sand, silt and clay-sized particles in a soil.

Solar constant   The rate at which solar radiation is received on a surface (perpendicular to the radiation) at the edge of the atmosphere. Average strength is 1372 joules/m^2 • sec, which can also be stated as 1.97 cal/cm^2 • min.

Solar gain   A general term used to indicate the amount of solar radiation absorbed by a surface or setting in the landscape.

Solar heating   The process of generating heat from absorbed solar radiation; a widely used term in the solar energy literature.

Solifluction   A type of mass movement in periglacial environments, characterized by the slow flowage of soil material and the formation of lobeshaped features; prevalent in tundra and alpine landscapes.

Solstice   The dates when the declination of the sun is at 23.27˚N latitude (the Tropic of Cancer) and 23.27˚S latitude (the Tropic of Capricorn)—June 21-22 and December 21-22, respectively. These dates are known as the winter and summer solstices, but which is which depends on the hemisphere.

Source control   Managing stormwater at its place of origin using various on-site techniques.

State plane coordinate system   *See* **Plane coordinate system.**

Stereoscope   A viewing device used to gain a three-dimensional image from a photopair.

Stormflow   The portion of streamflow that reaches the stream relatively quickly after a rainstorm, adding a surcharge of water to baseflow.

Stormwater   Surface runoff in response to heavy rainfall and/or snowmelt that rushes over the land to stream channels. Also used to refer to surface runoff or overland flow from developed areas.

Stormwater garden   A green infrastructure facility designed to hold and remediate stormwater by passing it through a constructed complex of wetland plants and soils.

Stratified sampling   A sampling technique in which the population or study area is divided into sets or subareas before drawing the sample.

Stream order   The relative position, or rank, of a stream in a drainage network. Streams without tributaries, usually the small ones, are first-order; streams with two or more first-order tributaries are second-order, and so on.

Subarctic zone   The belt of latitude between 55˚ and the Arctic and Antarctic circles.

Subbasin   A small drainage basin within the watershed of a stream, lake, or impoundment.

Sun angle   The angle formed between the beam of incoming solar radiation and a plane at the earth's surface or a plane of the same altitude anywhere in the atmosphere.

Sun pocket   A small space designed especially to take advantage of solar radiation and heating.

Surficial wetland   A wetland originating from impaired local drainage, usually surface or near surface runoff.

Surge   A large and often destructive wave caused by intensive atmospheric pressure and strong winds.

Suspended load   The particles (sediment) carried aloft in a stream of wind by turbulent flow; usually clay- and silt-sized particles.

Sustainability planning   An area of planning in which the objective is to achieve long-term and productive balance between land use and the environment.

Swamp   A wetland dominated by trees and/or shrubs. There are many varieties including cypress, mangrove, and conifer.

Taxon   Any unit (category) of classification system, usually biological.

Technical planning Data collection, analysis, and related activities used in support of the decision-making process in planning.

Temperate forest A forest of the midlatitude regions that could be described as climatically temperate, for example, broadleaf deciduous forests of Europe and North America, comprised of beeches, maples, and oaks.

Temperature inversion An atmospheric condition in which cold air underlies warm air. Inversions are highly stable conditions and thus not conducive to atmospheric mixing.

Texture The term used to describe the composite sizes of particles making up a soil sample, such as loam, sandy loam, and clay loam, based on the percentage by weight of sand, silt, and clay particles. *Also see* **Landscape texture.**

Thermal gradient The change in temperature over distance in a substance; usually expressed in degrees Celsius per centimeter or meter.

Thermal infrared system Line scanner capable of recording thermal infrared energy at wavelengths of 3–5 and 8–14 micrometers.

Thermokarst Sinkholelike depressions that result from differential melting of permafrost.

Threatened species According to the U.S. Endangered Species Act, a species with a rapidly declining population that is likely to become endangered.

Threshold The level of magnitude of a process at which sudden or rapid change is initiated.

Tolerance The range of stress or disturbance a plant is able to withstand without damage or death.

Toposequence Changes in soil composition and related features with topographic gradients on landforms.

Topsoil The uppermost of the soil, characterized by a high organic content; the organic layer of the soil.

Township and range A system of land subdivision in the United States that uses a grid to classify land units. Standard subdivisions include townships and sections.

Transect sampling A field sampling technique in which the sample is drawn from strips or transects laid out across the study area.

Transmission The lateral flow of groundwater through an aquifer; measured in terms of cubic feet (or meters) transmitted through a given cross-sectional area per hour or day.

Transpiration The flow of water through the tissue of a plant and into the atmosphere via stomatal openings in the foliage.

Transported soil Soil formed in parent material comprised of deposits laid down by water, wind, or glaciers.

Tree line The upper limit of tree growth on a mountain where forest often gives way to alpine meadow.

Tundra Landscape of cold regions, characterized by a light cover of herbaceous plants and underlain by permafrost.

Turbidity A measure of the clearness or transparency of water as a function of suspended sediment.

Turbulent flow Flow characterized by mixing motion in which the primary source of flow resistance is the mixing action between slow-moving and faster-moving molecules in a fluid.

Universal Soil Loss Equation A formula for estimating soil erosion by runoff based on rainfall, plant cover, slope, and soil erodibility.

Urban boundary layer A general term referring to the layer of air over a city that is strongly influenced by urban activities and forms.

Urban canyon City street lined with tall buildings; an urban terrain feature that has a pronounced effect on airflow, radiation, and microclimate as a whole.

Urban climate The climate in and around urban areas; it is usually somewhat warmer, foggier, and less well lighted than the climate of the surrounding region.

Urban design An area of professional activity by architects, landscape architects, and urban planners dealing with the forms, materials, and activities of cities.

Urbanization The term used to describe the process of urban development, including suburban residential and commercial development.

Variable source concept   *See* **Partial area concept.**

Variance   An allowable exception or alternative interpretation to a land use or environmental ordinance.

Vascular plants   Plants in which cells are arranged into a pipelike system of conducting, or vascular, tissue. Xylem and phloem are the two main types of vascular tissue.

Watertable   The upper boundary of the zone of groundwater. In fine-textured materials it is usually a transition zone rather that a boundary line. The configuration of the watertable often approximates that of the overlying terrain.

Wave base depth   The depth offshore where wave motion first touches bottom; roughly equal to 1.0 to 2.0 times wave height.

Wavelength   The distance from the crest of one wave to the crest of the next wave.

Wave refraction   The bending of a wave, which results in an approach angle more perpendicular to the shoreline.

Wellhead protection   Land use planning and management to control contaminant sources in the area contributing recharge water to community wells.

Wetland   A term generally applied to an area where the ground is permanently wet or wet most of the year and is occupied by water-loving (or tolerant) vegetation, such as cattails, mangrove, or cypress.

Wetland disposal system   A means of sewage disposal and treatment in which the waste is dispersed into and filtered through natural or constructed wetlands.

Wetted perimeter   The distance from one side of a stream to the other, measured along the bottom.

Windshield survey   A rapid and general sampling method for vegetation and land use based on observations from a moving automobile.

Zenith   For any location on earth, the point that is directly overhead to an observer. The zenith position of the sun is the one directly overhead.

Zenith angle   The angle formed between a line perpendicular to the earth's surface (at any location) and the beam of incoming solar radiation (on any date).

CREDITS

Introduction   Figure 0.1: ©Taylor S. Kennedy/NG Image Collection. Figure 0.2*a*: Courtesy of USGS. Figure 0.2*b*: Getty Images, Inc. Figure 0.3: Doug Babcock/Curt Perdina/ North Andover Flight Academy. Figure 0.4: Alexander Svirsky. Figure 0.A: Getty Images, Inc.

Chapter 1   Figure 1.2: The New York Public Library. Figure 1.3: ©Bettmann/©Corbis. Figure 1.4: ©Gary Braasch/©Corbis. Figure 1.8: Courtesy U. S. Army Engineer Waterways Experiment Station. Figure 1.B: LaightHawk/J. Henry Fair.

Chapter 2   Figure 2.3: ©Charlie Ott/Photo Researchers. Figure 2.5*a*: ©Kenneth Murray/Photo Researchers. Figure 2.6: Vivian Stockman/www.ohvec.org. Figure 2.10*a*: ©Bruce Coleman, Inc. Figure 2.11: National Geographic/Getty Images, Inc.. Figure 2.12b: ©John Buitenkant/Photo Researchers. Figure 2.14: Dick Durrance II/NG Image Collection. Figure 2.15: W. M. Marsh. Figure 2.19: ©Jim Hughs/Visuals Unlimited.

Chapter 3   Figure 3.1 (left): Courtesy U. S. Air Force. Figure 3.1 (right): Georg Gerster/ Photo Researchers. Figure 3.3 (left): ©Douglas Faulkner/Photo Researchers. Figure 3.3 (right): W. M. Marsh. Figure 3.4 (right): ©Bohemian Nomad Picturemakers/Corbis. Figure 3.9*a*: Courtesy Harvard Archives. Figure 3.9*b*: Courtesy Harvard Archives. Figure 3.10 (right): ©Roth Stein/ Library of Congress. Figure 3.11 (left): Getty Images, Inc.. Figure 3.11 (right): Getty Images, Inc. Figure 3.A: Courtesy of Charles Schlinger. Figure 3.C: Courtesy of Charles Schlinger.

Chapter 4   Figure 4.1 (left): ©Tom McHugh/Photo Researchers. Figure 4.7: J. G. Marsh and W. M. Marsh. Figure 4.8*a*: M. D. Simmons. Figure 4.8*b*: ©AP/Wide World Photos. Figure 4.9: Courtesy Robert L. Schuster/USGS. Figure 4.11: Courtesy U. S. Natural Resources Conservation Service.

Chapter 5   Figure 5.9: W. M. Marsh. Figure 5.10*a*: Getty Images, Inc. Figure 5.10*b*: ©JRC, Inc. / Alamy.

Chapter 6   Figure 6.4: ©Jack Dermid/Photo Researchers. Figure 6.9: ©Robert P. Carr/Bruce Coleman, Inc. Figure 6.A: ©Creatas/SuperStock.

Chapter 7   Figure 7.7: ©Bruce Clarke/Index Stock. Figure 7.A: Martin Kaufman. Page 162: N. L. Marsh.

Chapter 8   Figure 8.9*a*: Courtesy Abrams Aerial Survey Corporation. Figure 8.9*b*: Courtesy Abrams Aerial Survey Corporation. Figure 8.13: Courtesy USDA Soil Conservation Service. Figure 8.B: Patrick Condon.

Chapter 9   Figure 9.11*a* and Figure 9.B: W. M. Marsh. Page 195: W. M. Marsh.

Chapter 10   Figure 10.3: ©Warren Winter/Zuma Press. Figure 10.A and Figure 10.B: Charles Schlinger.

Chapter 11 Figure 11.5*b*: W. M. Marsh. Figure 11.6: ©Brian Larson. Figure 11.10*a*: ©Shirley Richards/Photo Researchers. Figure 11.10*b*: ©Jeff Greenberg/Photo Researchers. Figure 11.A: U.S. Geological Survey.

Chapter 12 Figure 12.5: W. M. Marsh. Figure 12.11: W. M. Marsh. Figure 12.12: N. L. Marsh. Figure 12:14: W. M. Marsh. Figure 12.A (left): ©Robert Destefano/Alamy. Page 256 (top inset): © North Wind Picture Archives/Alamy. Page 256 (bottom inset): National Geographic/Getty Images, Inc.

Chapter 13 Figure 13.6: W. M. Marsh. Figure 13.B: Scott Murdoch.

Chapter 14 Figure 14.6: W. M. Marsh. Figure 14.8*b*: W. M. Marsh. Figure 14.8 (right): Jake Rajs//Getty Images, Inc. Figure 14.13: Courtesy N. O. A. A. Restoration Center. Figure 14.A: B. K. Ferguson, W. M. Marsh and K. King.

Chapter 15 Figure 15.1: Jean-Pierre Pieuchot//Getty Images, Inc. Figure 15.3*a*: W. M. Marsh. Figure 15.12*a*: N. L. Marsh. Figure 15.13: Courtesy North Carolina Division of Marine Fisheries. Figure 15.15: W. M. Marsh. Figure 15.18: Jeff R. Clow/Getty Images, Inc. Figure 15.A: W. M. Marsh. Figure 15.B: W. M. Marsh.

Chapter 16 Figure 16.7: Williams and Woo, Inc. Figure 16.8*a*: DEA/A. Attini/Getty Images, Inc. Figure 16.8*b*: © Spring Images/Alamy. Figure 16.A: Brian Larson.

Chapter 17 Figure 17.8: © Jon Hicks/©Corbis.

Chapter 18 Figure 18.1: Tony Weyiouanna Sr. Figure 18.9*b*, Figure 18.12, Figure 18.A, and Figure 18.B: Troy L. Péwé.

Chapter 19 Figure 19.2: MPI/Getty Images, Inc.. Figure 19.3: Long Island State Park and Recreation Commission. Figure 19.4: AFP/Getty Images, Inc. Figure 19.5: ©Judie Long/Alamy. Figure 19.6: ©Skyscan Photolibrary/Alamy. Figure 19.7 (insert): Bill Barton/Getty Images, Inc. Figure 19.A and Figure 19.B: Jon Rodiek.

Chapter 20 Figure 20.2 (left): Courtesy F. R. Thompson, North Central Forest Experiment Station, U. S. Forest Service. Figure 20.6: ©G. Carleton Ray/Photo Researchers. Figure 20.A and Figure 20.B: Patrick Mooney.

Chapter 21 Figure 21.3: Courtesy Frank C. Golet, University of Rhode Island. Figure 21.7: K. King and W. M. Marsh. Figure 21.A: W. M. Marsh and G. F. Marx.

Chapter 22 Figure 22.1 (left): ©Gabe Palmer/Alamy. Figure 22.1 (right): Macduff Everton/Getty Images, Inc. Figure 22.2: ©Alex S. MacLean/Landslides.

(Unless otherwise indicated, all line art by William M. Marsh.)

INDEX